Sustainable Agriculture: Theories, Methods, Practices and Policies

Sustainable Agriculture: Theories, Methods, Practices and Policies

Editors

Moucheng Liu
Xin Chen
Yuanmei Jiao

MDPI

Basel • Beijing • Wuhan • Barcelona • Belgrade • Novi Sad • Cluj • Manchester

Editors

Moucheng Liu
Chinese Academy of Sciences
Beijing
China

Xin Chen
Zhejiang University
Hangzhou
China

Yuanmei Jiao
Yunnan Normal University
Kunming
China

Editorial Office
MDPI
St. Alban-Anlage 66
4052 Basel, Switzerland

This is a reprint of articles from the Special Issue published online in the open access journal *Agriculture* (ISSN 2077-0472) (available at: https://www.mdpi.com/journal/agriculture/special_issues/Sustainable_Agriculture_Methods_Practices).

For citation purposes, cite each article independently as indicated on the article page online and as indicated below:

Lastname, A.A.; Lastname, B.B. Article Title. *Journal Name* **Year**, *Volume Number*, Page Range.

ISBN 978-3-7258-1261-5 (Hbk)
ISBN 978-3-7258-1262-2 (PDF)
doi.org/10.3390/books978-3-7258-1262-2

Cover image courtesy of Moucheng Liu

Contents

agriculture

Editorial

Sustainable Agriculture: Theories, Methods, Practices and Policies

Moucheng Liu [1,*], Xin Chen [2] and Yuanmei Jiao [3]

1 Institute of Geographic Sciences and Natural Resources Research, Chinese Academy of Sciences, Beijing 100101, China
2 College of Life Sciences, Zhejiang University, Hangzhou 310058, China; chen-tang@zju.edu.cn
3 Faculty of Geography, Yunnan Normal University, Kunming 650500, China; ymjiao@sina.com
* Correspondence: liumc@igsnrr.ac.cn; Tel.: +86-15901262968

Due to the extensive degree of the consumption of resources and energy by industrial agriculture, there is a growing awareness of sustainable agriculture development that should not only increase yield to meet people's demands for food security, but should also improve product quality and promote the multi-functionality of the agricultural ecosystem. Although research and practices of sustainable agriculture have achieved remarkable results over the past 40 years, the development of the social economy and the current challenges of the world agriculture necessitate further research to improve the future of agriculture. Therefore, summarizing our experiences and lessons, analyzing existing problems, and contemplating the future of development will lead to innovations in the realms of theories and practical methods, which will promote the practice and policy of sustainable agriculture.

This Special Issue includes 16 articles related to the theoretical methods, policies and systems of agricultural sustainability, as well as practical experience, focusing particularly on the former two, with the aim to contribute to the sustainable development of agriculture and food security.

1. Theories and Methods of Agricultural Sustainable Development

Food supply is one of the important purposes of sustainable agricultural development. Gilmar et al. [1] used an innovative in vitro technique to increase the cultivation of micro tubers, improve the yield and quality of seeds and crops, and ensure regional food supply.

With the objective to examine how to assess the sustainability of agricultural development, Wang et al. [2] constructed a comprehensive system of evaluation indexes for cropland–livestock systems from three aspects: arable land, animal husbandry, and the environment. They used a coupling coordination degree model to evaluate the coupling coordination relationship between cropland and livestock and its influencing factors in 31 provinces in China during 2000–2020. Their results clarified that reducing the decoupling of cultivated land and animal husbandry can reduce agricultural non-point source pollution, and that the combination of cultivated land and animal husbandry can promote agricultural sustainability. Yu et al. [3] proposed that an accurate measurement of agricultural total factor productivity (AGTFP) is crucial to measure the level of sustainable agricultural development, and that the inclusion of agricultural carbon sink in the AGTFP is more conducive to improving green total factor productivity, reducing carbon dioxide in agricultural production and improving the carbon sink capacity of farmland.

In addition, in terms of research methods, Wang et al. [4] used a novel hybrid model by integrating the Fuzzy set, Delphi, and the Grey theory, among others, to calculate the relationship between various factors in agricultural green supply chain management so as to grasp the relationship and key factors between each link, promote the reform of supply chain management, and promote agricultural sustainability. Based on the data gathered in 2016, from the China Family Panel Studies (CFPS) and from a variety of econometric

Citation: Liu, M.; Chen, X.; Jiao, Y. Sustainable Agriculture: Theories, Methods, Practices and Policies. *Agriculture* **2024**, *14*, 2459. https://doi.org/10.3390/agriculture14030473

Received: 1 March 2024
Revised: 3 March 2024
Accepted: 13 March 2024
Published: 15 March 2024

models, Wang et al. [5] clarify that digital technology can promote land circulation (land leasing behavior), integrate fragmented land, reduce land abandonment, and achieve sustainable livelihoods for farmers.

2. Policies of Agricultural Sustainable Development

Abandonment is a major problem which is faced by the development of modern agriculture. Based on the survey data of 12 rural provinces in China, Xue et al. [6] analyzed the relationship between agricultural machinery harvesting services and abandonment, and proposed that agricultural machinery harvesting services can reduce cropland abandonment, which provides suggestions for policymakers to reduce cultivated land abandonment and ensure food security. Xue et al. [7] examined the relationship between financial literacy and farmland abandonment in Ghana, and found that the financial literacy of rural residents (especially low-income farm households and female farmers) is low, and pointed out that the lower the financial literacy, the more serious the phenomenon of cultivated land abandonment. Hence, the authors proposed that agricultural sustainability can be promoted by increasing the financial literacy training of rural households.

In addition, the development of circular agriculture is also an important system to ensure the sustainable development of agriculture. By measuring the efficiency and changes in the agricultural economy in 31 provinces and cities in China from 2017 to 2020, Guo et al. [8] pointed out that the implementation of a rural revitalization strategy can improve the efficiency of agricultural circular economy, promote rural economic development, social progress and ecological protection, and realize rural modernization. Finally, this Special Issue also assesses two livestock development policies.

By examining the impact of the implementation of the Converting Food Crops to Forage Crops Policy (CFFP) in the pilot counties of Hebei Province from 2010 to 2020 on the development of the herbivorous livestock industry, Zhang et al. [9] clarified that the CFFP can produce high-quality feed, improve the productivity of animal husbandry, and promote the sustainable development of agriculture and animal husbandry. Moreover, Li et al. [10] systematically reviewed the relevant theoretical and practical research of grassland eco-compensation in China, summarized the five characteristics and shortcomings of grassland ecological compensation, and point out that future work should focus on the response mechanism of herdsmen's families and the improvement of compensation measures.

3. Practical Experience of Agricultural Sustainable Development

Different agricultural production technologies or models have been created around the world, effectively ensuring the sustainable development of regional agriculture. Marta M. Moreno et al. [11] analyzed the degradation of biodegradable (BD) plastic mulch with different compositions in different soil types, and proposed that different compositions of biodegradable (BD) plastic mulch should be selected for different soils. Taking Limpopo Province as an example, Tlou E. Mogale et al. [12] evaluated the productivity of different species of sorghum intercropping at different cowpea densities, and proposed that the combination of intercropping and no-tillage can improve crop yield and productivity, thereby promoting agricultural sustainability. Taking the rice–fish coculture system in Jingning, Qingtian, and Yongjia counties of Zhejiang Province as an example, Ye et al. [13] compared and evaluated the effects of various aquaculture and rice–fish coculture systems on the genetic diversity of aquatic animals, and proposed that the implementation of the rice–fish coculture system can improve the genetic diversity and food security of aquatic species, thereby promoting agricultural sustainability.

For business entities, it is necessary to pay attention to the capacity building of smallholder farmers on the one hand, and the synergy between small-scale and large-scale farming systems on the other. Bader Alhafi Alotaibi et al. [14] analyzed the views and opinions of 193 rice farmers in Pakistan with regard to the government's public extension services and improving rice production efficiency, and concluded that Pakistan's public extension services lack attention to small-scale rice farmers. The authors proposed that small-scale rice farmers are the main body of agricultural production and yet lack modern agricultural technology and knowledge; therefore, the government should pay more attention to and train small-scale farmers to ensure national food security and promote agricultural sustainability. From the perspective of system dynamics, Mary Scholes et al. [15] analyzed and evaluated the small-scale and large-scale farming systems of mangoes and nuts in the Vhembe district of Limpopo South Africa, and found that small-scale and large-scale farming systems can work together to achieve food security at all levels. They also proposed that large-scale and small-scale farming systems work collaboratively rather than independently.

Lastly, taking Pakistani farmers as an example, Rabia Mazhar et al. [16] investigated the effects of three contract-farming regimes—long-term, medium-term, and short-term contracts—on the land-improvement investment, productivity, and technical efficiency of contract farmers in Punjab, Pakistan. Additionally, the authors clearly proposed that the implementation of long-term contract-farming regime is important for sustainable land development and management.

In summary, this Special Issue provides a comprehensive overview of the theories, methods, policies, and practices of sustainable agricultural development, with the aim at providing new insights and contributions to sustainable agricultural development, modern agriculture, and rural revitalization. The papers in this Special Issue represent some of the latest and most promising research results in this field, and we are confident that this Special Issue will facilitate further research. Here, the invited editors would like to express their heartfelt gratitude to all the contributors, authors, and reviewers who have contributed to the high-level research presented in this Special Issue.

Conflicts of Interest: The authors declare no conflicts of interest.

References

1. Peña-Rojas, G.; Carhuaz-Condori, R.; Andía-Ayme, V. Improved Production of Mashua (*Tropaeolum tuberosum*) Microtubers MAC-3 Morphotype in Liquid Medium Using Temporary Immersion System (TIS-RITA®). *Agriculture* **2022**, *12*, 943. [CrossRef]
2. Chen, J.; Gao, X.; Zhang, Y. Analysis on Coupling Coordination Degree for Cropland and Livestock from 2000 to 2020 in China. *Agriculture* **2023**, *13*, 1304. [CrossRef]
3. Yu, Z.; Lin, Q.; Huang, C. Re-Measurement of Agriculture Green Total Factor Productivity in China from a Carbon Sink Perspective. *Agriculture* **2022**, *12*, 2025. [CrossRef]
4. Wang, M.; Zhang, K. Improving Agricultural Green Supply Chain Management by a Novel Integrated Fuzzy-Delphi and Grey-WINGS Model. *Agriculture* **2022**, *12*, 1512. [CrossRef]
5. Zuo, X.; Hong, Z. The Impact of Digital Technology on Land Rent-Out Behavior: Information Sharing or Exclusion? *Agriculture* **2022**, *12*, 1046. [CrossRef]
6. Xue, P.; Han, X.; Wang, Y. Can Agricultural Machinery Harvesting Services Reduce Cropland Abandonment? Evidence from Rural China. *Agriculture* **2022**, *12*, 901. [CrossRef]
7. Ankrah Twumasi, M.; Dogbe, B.S.; Ankrah, E.K. Assessing Financial Literacy and Farmland Abandonment Relationship in Ghana. *Agriculture* **2023**, *13*, 580. [CrossRef]
8. Guo, C.; Zhang, R.; Zou, Y. The Efficiency of China's Agricultural Circular Economy and Its Influencing Factors under the Rural Revitalization Strategy: A DEA–Malmquist–Tobit Approach. *Agriculture* **2023**, *13*, 1454. [CrossRef]
9. Zhang, H.; Wang, G.; Liu, J. The Influence of Converting Food Crops to Forage Crops Policy Implementation on Herbivorous Livestock Husbandry Development—Based on Policy Pilot Counties in Hebei, China. *Agriculture* **2022**, *12*, 1872. [CrossRef]
10. Li, Z.; Su, B.; Liu, M. Research Progress on the Theory and Practice of Grassland Eco-Compensation in China. *Agriculture* **2022**, *12*, 721. [CrossRef]
11. Villena, J.; Moreno, M.; González-Mora, S. Degradation Pattern of Five Biodegradable, Potentially Low-Environmental-Impact Mulches under Laboratory Conditions. *Agriculture* **2022**, *12*, 1910. [CrossRef]
12. Mogale, T.E.; Ayisi, K.K.; Munjonji, L. Yield Responses of Grain Sorghum and Cowpea in Binary and Sole Cultures under No-Tillage Conditions in Limpopo Province. *Agriculture* **2022**, *12*, 733. [CrossRef]
13. Ye, Y.; Ren, W.; Zhang, S. Genetic Diversity of Fish in Aquaculture and of Common Carp (*Cyprinus carpio*) in Traditional Rice–Fish Coculture. *Agriculture* **2022**, *12*, 997. [CrossRef]
14. Alzahrani, K.; Ali, M.; Azeem, M.I. Efficacy of Public Extension and Advisory Services for Sustainable Rice Production. *Agriculture* **2023**, *13*, 1062. [CrossRef]

15. Materechera, F.; Scholes, M. Scenarios for Sustainable Farming Systems for Macadamia Nuts and Mangos Using a Systems Dynamics Lens in the Vhembe District, Limpopo South Africa. *Agriculture* **2022**, *12*, 1724. [CrossRef]
16. Mazhar, R.; Azadi, H.; Van Passel, S. Does Contract Length Matter? The Impact of Various Contract-Farming Regimes on Land-Improvement Investment and the Efficiency of Contract Farmers in Pakistan. *Agriculture* **2023**, *13*, 1651. [CrossRef]

agriculture

MDPI

Article

Does Contract Length Matter? The Impact of Various Contract-Farming Regimes on Land-Improvement Investment and the Efficiency of Contract Farmers in Pakistan

Rabia Mazhar [1], Hossein Azadi [2], Steven Van Passel [3], Rando Värnik [4], Marcin Pietrzykowski [5], Rytis Skominas [6], Zou Wei [1,*] and Bi Xuehao [1]

1 College of Public Administration, Nanjing Agricultural University, Nanjing 210095, China; 2018209033@njau.edu.cn (R.M.); 2020209030@stu.njau.edu.cn (B.X.)
2 Department of Economic and Rural Development, Gembloux Agro-Bio Tech, University of Liège, 4000 Liege, Belgium; hossein.azadi@uliege.be
3 Department of Engineering Management, University of Antwerp, Prinsstraat 13, 2000 Antwerp, Belgium; steven.vanpassel@uantwerpen.be
4 Institute of Agricultural and Environmental Sciences, Rural Economics, Estonian University of Life Sciences, 51014 Tartu, Estonia; rando.varnik@emu.ee
5 Department of Ecological Engineering and Forest Hydrology, Faculty of Forestry, University of Agriculture in Krakow, al. 29 Listopada 46, 31-425 Krakow, Poland; m.pietrzykowski@urk.edu.pl
6 Institute of Hydraulic Engineering, Vytautas Magnus University Agriculture Academy, 53361 Kaunas, Lithuania; rytis.skominas@vdu.lt
* Correspondence: zw@njau.edu.cn

Abstract: Land-tenure security is integral to local communities' socioeconomic development. It has been a center of debate in academia and for legislators and advocates to implement reforms to enhance efficient and sustainable development in land management. Yet, knowledge gaps remain in how various contract-farming regimes contribute to land-improvement investment and technical efficiency. This study used a data set of 650 farm households collected through a two-stage stratified sampling to investigate the influence of three contract-farming regimes: long-term, medium-term, and short-term contracts, on the land-improvement investment, productivity, and technical efficiency of contract farmers in Punjab, Pakistan. The study used multivariate probit and ordinary least square regression models to examine the posit relationships. The findings highlight that farmers with long-term land contracts have higher per hectare yield, income and profit than those with medium-term and short-term contracts. The results confirm that farmers with medium- and long-term contracts tend to invest more in land-improvement measures, i.e., organic and green manure. Further, the study findings demonstrate that long-term land tenures are more effective when farmers make decisions regarding the on-farm infrastructure, like tube-well installation, tractor ownership, and holding farm logistics. Last, the study results confirm that long-term contracts are more robust regarding technical efficiency. Moreover, the findings support the Marshallian inefficiency hypothesis and extend the literature on contract farming, land-improvement investment, and land use policy, and offer coherent policy actions for stakeholders to improve farmers' productivity, technical efficiency, and income.

Keywords: contract-farming regimes; land-improvement investment; land-use policy; productivity; technical efficiency; organic farming

Citation: Mazhar, R.; Azadi, H.; Van Passel, S.; Värnik, R.; Pietrzykowski, M.; Skominas, R.; Wei, Z.; Xuehao, B. Does Contract Length Matter? The Impact of Various Contract-Farming Regimes on Land-Improvement Investment and the Efficiency of Contract Farmers in Pakistan. *Agriculture* **2023**, *13*, 1651. https://doi.org/10.3390/agriculture13091651

Academic Editors: Xin Chen, Moucheng Liu and Yuanmei Jiao

Received: 31 July 2023
Revised: 16 August 2023
Accepted: 18 August 2023
Published: 22 August 2023

1. Introduction

The vast prevalence of insecure land-use rights is a primary impediment to developing organic farming in Pakistan. Myriad farmers do not have proper land-use titles and are vulnerable to lease termination, eviction, and seizure at any time [1,2]. Unstable land-tenure arrangements protect farmers from investing in land-improvement measures and farm investment discourages them from switching to sustainable land-management practices [1]. Previous studies have rigorously explored the impact of contract farming on technical efficiency [3–9], the uptake of sustainable farm practices [10], and farmers' income [11–13]. Yet, no research examined the impact of various contract-farming regimes (e.g., one, three, and five-year contracts) on land-improvement investment and efficiency. The country has enormous potential for organic farming [14]; knowledge gaps remain in how various contract-farming regimes contribute to land-improvement investment and technical efficiency. By analyzing the impact of different contract-farming regimes, this study offers valuable insights into the role of contract length in determining farmers' land-improvement investment behavior and the performance of contract farming in the country.

Contract farming can economically contribute to Pakistan's economy [14]. Organic agriculture needs fewer external inputs, like pesticides, chemical fertilizers, and herbicides. Thus, on the input side, it can reduce costs and potentially contribute to farm profit [15]. Organic farming generates numerous employment opportunities; it can help local and remote communities thrive through income generation and employment. It can help instigate a series of organic-related processes that start from the production, processing, marketing, and distribution to the local and far-off markets [16,17], thereby shedding far-reaching impacts on the country's exports and foreign income earned. Progress toward organic farming encourages adopting sustainable land-management practices and improves land productivity. Hence, it translates into improved per hectare yield and higher income [18,19]. Likewise, evidence shows that organic farming promotes rural development strategies by providing the livelihood of rural communities and upscale farmers' livelihood strategies [14,20]. It encourages the development of farming entrepreneurs and clusters, bringing more significant economic benefits through value-addition and improving rural–urban vertical linkages [10,21]. Thus, organic farming has enormous potential to economically contribute to developing economies by increasing productivity, rural development, and supporting a sustainable development agenda [22].

Many researchers noted that land-use rights are crucial to enhancing contract farming [6,23,24]. Organic farming needs longer term contract security to ensure the investment payoff of investors since they need to improve soil health, which takes a couple of years. However, investors or farming entrepreneurs need strong landlord commitment for contract longevity due to the financial risks associated with their investments [8,25]. Given this, secure land arrangements encourage entrepreneurs to invest in farms and apply sustainable land-management techniques and measures to enhance soil health and productivity [26]. In Pakistan, about 38% of the land is owned by absentee landlords, usually considered rent seekers. These absentee landlords operate through third parties without connections to farming and agriculture [27]. Most of them lend their land to small farmers or local communities on short-term (one-year), medium-term (less than three years), and long-term (above five years) contracts. Most short-term contracts are highly insecure and informal, under which farmers cannot decide how to use the land in the short-term [28]. Organic farming demands secure land rights for employing measures like composting, organic manure, green manure, crop rotation, and agroforestry, leading to a sustainable farm ecosystem. Further, secure land-tenure arrangements encourage farmers to invest in farm infrastructure and participate in farm cooperatives to apply innovative technologies to improve productivity and income [29]. Likewise, studies find that land-tenure security is positively related to access to farm credit. Thus, it makes it easier for farmers to provide collateral on loans and offer them the financial ability to invest in organic-promoting practices [30].

Land-tenure security has an enormous role in the development of organic farming. Organic agriculture has environmental implications and ensures the socioeconomic well-being of farmers and local communities [29]. However, the absence of secure land rights disincentivizes farmers to switch to organic farming, which restricts its development [31] and limits opportunities for sustainable development [27]. Secure land tenure can promote sustainable development in agriculture, which contributes to food security, poverty reduction, and climate-change mitigation [31–33].

Given the above debate, the following research questions arise on the role of three operational contract-farming regimes: short term, medium term, and long term. Do characteristics of farmers vary across land-tenure regimes? Do farmers with long-term contracts have higher land-improvement investments? Do various land-tenure regimes have significant variations in farm yield? Do long-term contracts have higher technical efficiency? By answering these questions, the study contributes to the literature on contract farming, land-tenure security, land-improvement investment, and sustainable land-use practices in developing countries.

2. Landscape of Current Land-Use Rights in Pakistan

The following are current land-use arrangements being practiced in Pakistan (the data were obtained online from the Government of Punjab Land Record Website: https://landportal.org/library/resources/guide-land-and-property-rights-pakistan, accessed on 15 June 2023).

1. Ownership: Four ownership categories are recognized at the national level: public, private, common, and cooperative. A system of property laws and customary practices governs land ownership. Ownership, however, could also be impacted by regional or local traditions and practices;
2. Leasing: Short and long-term leases are frequently employed in Pakistani agriculture. Most leases are likely to be informal and based on verbal agreements; however, there are some situations when formal legal contracts are used;
3. Tenancy: Many farmers in Pakistan own a small plot and work as sharecroppers or under unofficial agreements with landowners; they are frequently tenants rather than landowners. Tenants' rights to the property and decision-making authority are mostly restricted and landowners often make crucial choices about the preparation of the land, the choice of crops, and the sale of the products made there;
4. Land redistribution: Instead of being proprietors, many farmers in Pakistan are tenants who frequently cultivate small plots as sharecroppers or under the terms of unofficial agreements with landowners. Tenants frequently have limited rights to control the land and decision-making authority, with landlords having the final say in crop selection, land preparation, and the sale of goods made on the property;
5. Land disputes: In Pakistan, unfortunately, disagreements about property rights, inheritance, and boundary lines sometimes result in land disputes. Farmers, especially those marginalized or without political clout, may struggle to secure land-use rights because of these disagreements.

Pakistan's land-use rights require further reforms to promote agricultural growth and long-term food security. Additionally, there are several reasons why Pakistan's current system of land-use rights is not conducive to the development of organic farming [15]. The leading cause is that most of the fertile land was given to large-scale, commercial farmers that employ cutting-edge technology to grow crops and increase earnings. To get high yields, these farmers frequently use chemical fertilizers, pesticides, and herbicides, which makes it challenging for organic farmers to secure suitable land [34].

In rural locations, landlords typically possess small parcels of land that they lease to tenants for short periods, usually a few years or less. These leases are only temporary, which prevents farmers from making the long-term expenditure required for organic farming [2]. Low levels of external inputs, such as chemical fertilizers, are needed in organic farming. However, it takes time for such procedures to show results, reducing the incentive for landlords to permit their renters to use the property this way. Likewise, poor farmers find it challenging to adopt sustainable agricultural techniques due to a lack of access to capital, training, and extension services [35–37]. Without sufficient technical understanding, farmers may not have the abilities and knowledge to utilize organic farming methods to their fullest potential. Farmers who want to transition to organic farming face another obstacle: a shortage of institutional financing [17]. Most banks and financial organizations are still reluctant to lend money to farmers without some form of collateral. Therefore, farmers with short-term contracts are more insecure are discouraged from embracing innovative agricultural practices that improve soil health, the environment, and local communities' incomes [36,38].

Research Gap

Many studies have explored different dimensions of contract farming. Barret et al. [39] explored the determinants of contract-farming participation and noted that contract participation improves household welfare. Likewise, Fialor et al. [3] studied the effect of contract farming on productivity and illustrated that contract participation improves crop productivity. Further, contract farming enhances the uptake of improved inputs that, in turn, boost productivity and income. Dubert et al. [4] examined the relationship between contract-farming participation and the uptake of sustainable farm practices. The findings indicate that contract farmers use more sustainable farm practices than conventional farmers.

Studies have also explored the interplay between contract farming, ecological change, and reciprocal social transformation [6]. The literature rigorously explored the connection between contract farming and productivity [12,40,41], farmers' income [12,13,42,43], sustainable production [4,5,44,45], loan repayment [36], market integration [7,11,46], and welfare [47,48]. Recent studies have examined the relationship between contract farming and production risk-management strategies [49,50]. No research has examined the impact of various contract-farming regimes (e.g., short, medium, and long-term contracts) on land-improvement investment and efficiency in developing countries. Hence, this study contributes to bridging the literature gap between contract-farming regimes, land-improvement investment, and the efficiency of contract farmers in a developing country context. Through rigorous empirical analysis and comprehensive data collection, this research aims to contribute to the existing literature on contract farming and provide evidence-based recommendations for policymakers and stakeholders in Pakistan's agricultural industry.

3. Materials and Methods

3.1. Sample and Data Collection

The study was conducted in Punjab, Pakistan. Six of nine rice-growing districts in the province (refer to Figure 1) were purposely selected for data collection since these districts account for 80 percent of the total basmati rice production. The data-collection stage took place between January and March 2022, with the target population being the farmers in the "Kalar track", a specialized geographically indicated area renowned for basmati rice production and export. Wheat and vegetable crops are produced in the region, yet rice is the dominant cash crop.

We selected the farmers using an equal-size stratified (two-stage) cluster design. The nine 'kallar track' districts embodied the first cluster, followed by the village (the second stage). We used probability proportion to size (PPS) to allocate villages across the selected districts based on the area under rice production. Thus, it ensures equal sampling proportion in each cluster. Next, systematic PPS was adopted to select the villages within each district using published information on the total number of households in each village. In total, 34 villages from six districts were selected from high-intensity rice districts in Punjab. Following the first stage of selected villages, we randomly selected rice farmers. Based on the prior research and surveys conducted in Punjab (Pakistan Integrated, Household Survey, 1991 (PIHS 1991) of the World Bank (Online available: https://microdata.worldbank.org/index.php/catalog/543, accessed on 15 June 2023)), we set the nonresponse rate at 33% for the second-stage selection. Thus, we adapted and prescribed 30 farmers from each village, of which 20 were finally selected for the final interview. Questionnaires with missing entries were discarded and the final data set of 650 households was obtained for further analysis. We used the following formula to calculate the sample size in this study (see Equation (1)).

$$n = \frac{N}{1 - N(e^2)} \quad (1)$$

where n is the sample size, N represents the population, and e denotes the expected error., There are approximately 100,000 rice growers in the sampled districts. Thus, we used this number as the total population to calculate the sample size by taking the value of the expected error to be 4 percent.

We adopted inclusion criteria for respondents based on three prescribed factors. The criteria were: being an export-oriented rice farmer, engaging in organic rice farming for at least 5 years, and knowing about contract-farming participation. Further, we tested the normality of the dependent variable (see Supplementary Materials S1).

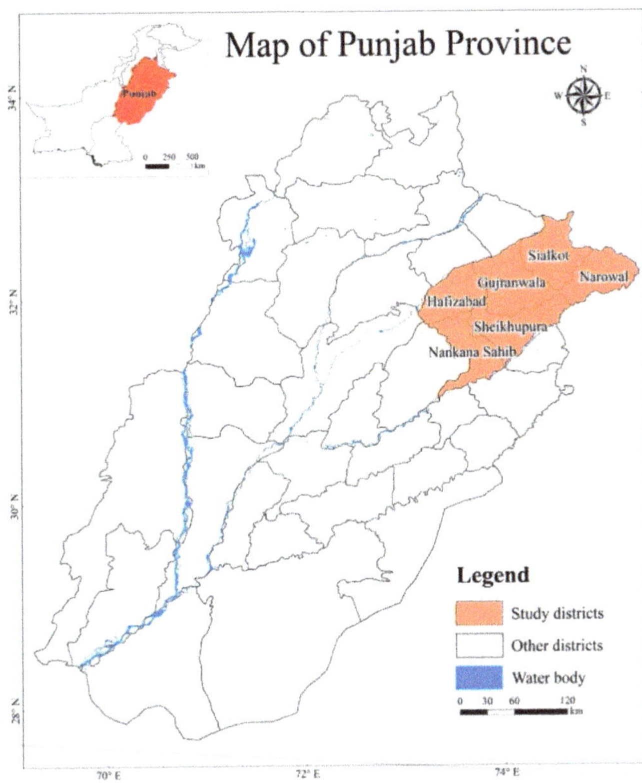

Figure 1. Location of the selected districts.

3.2. Conceptual Framework

This study examines the impact of land-rights arrangements on agricultural production efficiency and investment in land enhancements and yield improvements. The study adapts and builds on the model developed by Akram et al. [28]. The analysis includes a farm-level production function that accounts for fixed factors, such as given below:

$$y = f(x, t, n, ; z) \quad (2)$$

where labor (x), land (t), input(s) (n), and y represent the yield, which is dependent on given factors such as investments in land-enhancing activities. Various variable inputs include green and organic manure, usually from poultry and farm animals, whereas z denotes the farm household characteristics.

Profit maximization is the primary goal of farmers; herein, it is measured by output prices (p), unit labor costs (w), and land costs $r(\theta, \delta)$, which are given as:

$$\pi = \max_{x,t,n}[py(x,t,n;z) - wx - r(\theta,\delta) - cn] \tag{3}$$

This study presents three distinct regimes of land rights, including short-term, medium-term, and long-term contracts, and calculates the cost of land based on these factors, as given below:

$$r(\theta,\delta) = (1-\theta)\bar{r} + \theta\delta py \tag{4}$$

where the shared output ratio (δ), for short-term stands $\theta = 1$, while for the medium term it is $\theta = 0$. Likewise, short-term land cost is p_y, and long-term contracts is r.

Using the following function, profit maximization can be indicated as price function, household endowments, and the three forms of land-use rights given by θ and δ, as given below:

$$\pi = \pi(p,w,c,z,\theta,\delta) \tag{5}$$

By directly applying the profit function, as shown in Equation (3).

$$y = y(p,w,c,z,\theta,\delta) \tag{6}$$

Equation (6) illustrates farmer characteristics and prices that influence the demand for inputs.

3.3. Empirical Specifications

The empirical estimation used is the simple and formal specification form of Equation (6), representing inputs, outputs, and productivity. Initially, the paper compares the farmers' characteristics (e.g., land size, per acre yield, and profit) across the given land-tenure regimes. Next, the study provides empirical estimates on the impact of various contract-farming regimes on land-improvement investment m (green manure G, organic manure M) using a multivariate probit model. This estimation assesses the possible substitutability and complementarity in the investment as an instrument variable. Hence, it helps determine the distinctive effect of land-tenure regimes on per hectare yield, profit, and investment, including farm and farmer-specific characteristics.

Given the land-improvement investment decision, probit specifications were applied to cover the investment for various measures, as follows:

$$\begin{aligned} J_{im} &= B_{im}Q_{im} + \gamma_{im}Z_{im} + \mu_{im} \\ J_{im} &\begin{cases} J_{im}\ if\ J_{im} > 0 \\ 0\ \text{otherwise} \end{cases} \quad m = M, G \end{aligned} \tag{7}$$

Here J_{im} indicates the anticipated profit for farmer i that invests in land improvement m. The term J_{im} refers to measures of observed variables representing land-improvement investment; otherwise, it assumes a value equal to zero. Likewise, the term μ_{im} refers to errors that may have identical distribution. Vector Q_{im} denotes land-tenure regimes and terms θ and δ represent that land is operated under long-term, medium-term, or short-term contracts. Further, the vector Z_{im} represents household and family characteristics like age, education, and farm size.

4. Results and Discussion

4.1. Farm-Level Characteristics

This section compares the various characteristics of farmers based on the three contract-farming regimes, namely long-term (up to five years or more), medium-term (up to three years), and short-term contracts (one year). Table 1 compares the various characteristics of contract farmers, including yield, profit, income, farm size, and other variable inputs used among long-term and short-term contracts. The independent sample *t*-test is used to compute the statistical significance between the means of characteristics of farmers under these two contract regimes.

Table 1. Characteristics of farmers under long-term and short-term contracts.

Variables	Long-Term	Short-Term	T-Value
	Mean	Mean	
Yield (kg/ha)	2381.83	2147.26	4.16 ***
Income per ha	317,200.28	291,675.71	2.29 **
Profit per ha	109,542.71	89,617.29	6.14 ***
Farm size (ha)	7.73	3.81	4.63 ***
Public–private partnership	0.19	0.03	0.38 ***
Farming experience	8.09	4.39	1.75 *
Subsidy financial incentive	0.68	0.57	1.83 *
Organic manure application	0.76	0.53	3.48 ***
Green manure application	0.87	0.83	2.89 ***
Improved seed	5.93	4.27	1.87 *
Hired labor	20.16	15.73	3.57 ***
Family labor	6.39	5.94	1.12
Livestock holding	2.18	3.27	1.26
Household-head age	39.43	43.83	2.18 **
Household-head education	9.17	8.43	2.36 **
Formal credit received	0.53	0.39	1.81 *
Crop rotation	0.87	0.48	1.71 *
Tube-well ownership	0.78	0.48	0.56
Farm advisory	0.59	0.48	0.82

Note: * $p < 0.10$, ** $p < 0.05$, *** $p < 0.01$.

Among others, farmers with long-term contracts differ from short-term ones in terms of per hectare yield, income per hectare, profit, farm size, and farming experience. Comparatively, it indicates that farmers with long-term contracts have higher mean values. Further, farmers with long-term contracts also have higher green and organic manure values than short-term contracts. Interestingly, public–private partnerships and subsidies or financial incentives exist in long-term contracts. It indicates long-term contracts offer more flexibility and freedom in farming decision-making and entails economic incentives for farmers than short-term contracts. Thus, under long-term agreements, farmers tend to apply more land-improvement measures, like organic manure and green manure, because they can harvest the economic gain of such practices in the longer run.

Table 2 compares the various characteristics of farming under medium-term and short-term contracts. It compares varying factors, including yield, profit, income, farm size, and other variable inputs used in long-term and short-term contracts. *t*-test is used to compute the statistical significance between the means of characteristics of farmers under these two contract regimes.

The *t*-test results indicate that, among others, farmers with medium-term contracts are different from short-term contracts in terms of per hectare yield, income per hectare, and farm size. Regarding the farm size, it inculcates that farmers tried to operate on relatively large farms under medium-term contracts. One of the reasons might be farming experience and realizing the presence of economies of scale. Likewise, mean subsidy or economic incentives values are significantly higher for medium-term contracts. It highlights that farmers feel more secure in medium-term contracts than in short-term ones. Further, farmers with medium-term contracts also have higher values of organic manure than short-term contracts and adopt improved seeds and availed formal credit. Medium-term contracts offer more flexibility to realize economies of scale and expand production and profit.

Table 2. Characteristics of farmers under short-term and long-term land use contracts.

Variables	Medium-Term	Short-Term	T-Value
	Mean	Mean	
Yield (kg/ha)	2196.72	2147.26	1.83 *
Income per ha	299,524.27	291,675.71	1.57 *
Profit per ha	93,748.72	89,617.29	1.08
Farm size (ha)	5.72	3.81	2.35 **
Public–private partnership	0.07	0.03	1.07
Farming experience	4.08	4.39	0.37
Subsidy financial incentive	0.57	0.38	1.87 *
Organic manure application	0.42	0.53	2.27 **
Green manure application	0.87	0.83	1.04
Improved seed	5.31	4.27	2.98 ***
Hired labor	15.32	15.73	1.06
Family labor	5.81	5.94	1.03
Livestock holding	0.73	0.79	0.93
Household-head age	42.17	43.83	1.13
Household-head education	8.24	8.43	1.18
Formal credit received	0.61	0.39	1.78 *
Crop rotation	0.76	0.71	1.24
Tube-well ownership	0.61	0.48	1.87 *
Farm advisory	0.54	0.48	3.40 ***

Note: * $p < 0.10$, ** $p < 0.05$, *** $p < 0.01$.

Table 3 compares the various characteristics of farmers based on the two contract-farming regimes: long-term (up to five years or more) and medium-term (two years or more) contracts. Table 3 compares the various characteristics of contract farmers, including yield, profit, income, and farm and production-related variable inputs used among long-term and medium-term contracts. A *t*-test is used to compute the statistical significance between the means of characteristics of farmers under these two contract regimes.

Table 3. Characteristics of farmers under long-term and medium-term land-use contracts.

Variables	Long-Term	Medium-Term	T Value
	Mean	Mean	
Yield (kg/ha)	2381.83	2196.72	4.17 ***
Income per ha	317,200.28	299,524.27	1.69 *
Profit per ha	109,542.71	93,748.72	1.98 *
Farm size (ha)	7.73	5.72	1.78 *
Public–private partnership	0.19	0.07	3.94 ***
Farming experience	8.09	4.08	1.95 *
Subsidy financial incentive	0.68	0.57	2.54 **
Organic manure application	0.76	0.42	4.64 ***
Green manure application	0.87	0.83	3.17 ***
Improved seed	5.93	5.31	1.08
Hired labor	20.16	15.32	5.18 ***
Family labor	6.39	5.81	1.21
Livestock holding	2.18	0.73	1.02
Household-head age	39.43	42.17	1.45
Household-head education	9.17	8.24	1.24
Formal credit received	0.53	0.53	0.98
Crop rotation	0.78	0.76	1.24
Tube-well ownership	0.78	0.49	2.67 **
Farm advisory	0.59	0.54	1.87 *

Note: * $p < 0.10$, ** $p < 0.05$, *** $p < 0.01$.

The results in Table 3 reveal that farmers with long-term contracts differ from medium-term contracts in per hectare yield, income per hectare, profit, farm size, and farming experience. It indicates that farmers with long-term contracts have higher mean values for the above characteristics than medium-term contracts. Further, farmers with long-term contracts also have significantly higher green and organic manure values than short-term contracts. Remarkably, there also exist public–private partnerships and subsidies or financial incentives in long-term contracts, lacking under medium-term contracts. Likewise, farmers under long-term contracts have higher values for improved seed, formal credit, and participation in farm advisory services. Further, farmers with long-term contracts have installed on-farm tube wells and used hired labor. It indicates that under such contracts, farmers tend to invest more in farm infrastructure to seek longer run benefits arising from such investments. It reinforces that long-term contracts entail greater economic security, providing farmers with freedom in decision-making and improving on-farm investment, increasing farm productivity and income. Moreover, it enhances the investment in land-improvement measures, like organic and green manure.

4.2. *Econometric Estimations for Land Investment*

In this section, we used a multivariate probit model to estimate the effects of various land-tenure regimes on demand for various production-related variable inputs. For this purpose, we study the impact of two land-improvement measures: organic manure and green manure. Further, we used the output delivery function to capture the impact of various land-tenure regimes. Given this, the model assumes a volatile instrument approach and accounts for the contract-farming regimes endogenous to tenure agreements.

4.2.1. Land-Tenure Regimes and Land-Improvement Investment

The first regression phase for land-tenure regimes was based on Equation (4), while the second-phase results represent the instruments used for land-tenure regimes (Table 4). We eliminated one of the three contract-farming regimes—short-term contract—to employ a linear probability model for further estimation. The connection between household characteristics and various land-tenure regimes was calculated. Regarding the key instrument variable, farm location and market connection have a significant and positive relationship with the uptake of long-term land-use contracts. In contrast, the distance to the market has a negative connection with the uptake of both medium and long-term contracts. These instruments identify the collective impact of farm location, market information, and distance on the uptake of various contract regimes. These findings are aligned with prior studies [3,50,51]. Among other factors, farm size, public–private partnership, farm advisory service, agricultural subsidy/economic incentives, farm logistics, livestock holding, farmer-based organizations (FBO) membership, tractor ownership, and organic farming experience are positively associated with the uptake of long-term tenure. These results indicate that land plots near input–output markets are more likely to rent under medium- and long-term contracts. More informed farmers are likely to choose long-term agreements. These findings endorse the prior studies [5,52], advocating that contract farming is a more common phenomenon in areas near big cities and commercial zones.

Interestingly, public–private partnerships and subsidies positively influence the uptake of long-term contracts, which implies that these encourage farmers' prolonged stay in the agriculture business. Moreover, investment in agriculture-allied businesses, like livestock holding, tractor ownership, and farm logistics, tends to induce the prevalence of long-term contracts. These findings support existing evidence on the determinants of contract farming in developing countries [5,6,36,52].

Table 5 reports the consequences of the second phase of investments in land-improvement measures using Equation (5). Considering Marshall's theory of inefficiency [53], the third land-tenure regime, the 'short-term contract' was deleted. Further, we assessed the effect of farmers' characteristics, production-related inputs, land-tenure regimes, and organizational factors on land-improvement investment. According to the results, the correlation coefficient (p) is significant and uncorrelated with land-improvement investment, complementing the suitability of the probit model used herein. We extracted the insignificant residual variables from the first-stage regression—RESO and RESF—for long-term and medium-term contract participation. The results reported herein nullify the presence of inconsistent coefficients and concurrency in variation [54]. Further, Wald test statistics confirm the consistency and robustness of the estimates and model through a residual vector, which is given in Table 5.

Table 4. Estimates of land-use rights using the probit model: marginal effects.

Variables	Medium-Term	Long-Term
	Coefficient	**Coefficient**
Farm size	0.127 **	0.164 ***
Public–private partnership	0.106	0.217 *
Tube-well ownership	0.085 ***	0.138 ***
Farm advisory	0.083 **	0.156 ***
Subsidy/financial incentive	0.0128	0.148 ***
Distance to market	−0.067 **	−0.125 **
Farm location	0.206	0.178 ***
Market connection	0.173	0.149 ***
Farm logistic	0.097	0.115 ***
Household size	0.037	0.039
Household-head age	0.036	0.064
Household-head education	0.065	0.046
Livestock holding	0.116	0.201 ***
FBO membership	0.126 *	0.174 **
Tractor ownership	0.089 ***	0.075 ***
No. of tillage operations	0.043	0.031
Mechanical harvesting	0.078	0.056
Organic experience	0.043 **	0.106 ***
R^2	0.47.82	
Adjusted R^2	0.46.17	
Breush–Pegan Test (χ^2)	11.53	0.001
Goodness of fit (χ^2)	78.27	0.006

Note: * $p < 0.10$, ** $p < 0.05$, *** $p < 0.01$.

The results show that the residual of both contract-farming regimes—long term and medium term—the equation is equivalent to zero, thus validating individual t-test results. These results support the exogenous theory of land-contract regimes [55]. Table 5 presents the coefficients of land-improvement investment, including long-term and medium-term land contracts. In this estimation, we controlled farm-specific characteristics to improve the robustness of the estimates. The results indicate that long-term contracts enhance the investment in organic and green manure, while medium-term contracts are only related to investment in organic manure. It advocates that long-term contracts induce more significant investment in land-improvement measures. These results align with previous studies [56,57], advocating that improving longer-term contract security would foster land-improvement investment and promote sustainable land-use practices in developing countries.

Farming experience, public–private partnership, FBO membership, and provision of subsidies or financial incentives are related to organic and green manure investment. It inculcates that organizational factors hold significant potential for improving land management and sustainable development in agriculture. Given this, providing targeted subsidies to contract farmers and enhancing public–private partnerships can help foster land-improvement investment vis-à-vis smallholder contract farmers' land-use efficiency, income, and sustainability in developing countries. These findings support the empirical work of [15], complementing that institutional factors promote sustainable land-management practices in smallholder agriculture.

Among the farmer and farm wealth factors, livestock holding, tractor ownership, logistic ownership, and organic farming experience are related to investment in organic and green manure. These findings relate to previous studies [58–60]. This implies that more wealthy, resourceful, and experienced farmers tend to invest more in sustainable land-management measures, which reflect through investment in organic and green manure application.

Table 5. Farmers' investments in land-improving measures: probit model results (marginal effects).

Variables	Organic Manure	Green Manure
Long-term tenure	0.382 ***	0.235 ***
Medium-term tenure	0.162 *	0.025
Farm size	0.261	0.173
Farming experience	0.126 **	0.184 *
Tube-well ownership	0.028	0.073
Public–private partnership	0.195 ***	0.153 ***
Farm advisory	0.108 *	0.083
Market distance	0.075	−0.138
FBO membership	0.237 **	0.114 ***
Subsidy/financial incentive	0.217 **	0.107 *
Household size	0.093	0.117
Training participation	0.037 **	0.112 *
Household-head age	0.136	0.157
Market connection	0.035	0.075
Household-head education	0.183 ***	0.136 ***
Livestock holding	0.205 **	0.183 **
Tractor ownership	0.158 ***	0.276 ***
Farm logistic	0.145 *	0.096
No. of tillage operations	0.362	0.283
Mechanical harvesting	0.082	0.236
Organic experience	0.381 ***	0.425 **
RESF	0.213	0.157
RESO	0.194	0.237
R^2	47.59	
Cross equation correlation (pMG)	0.237 ***	
Joint statistics χ^2	132.13 (0.001)	
Breush–Pegan Test (χ^2)	28.72 (0.007)	
Goodness of fit (χ^2)	75.65 (0.016)	

Note: * $p < 0.10$, ** $p < 0.05$, *** $p < 0.01$.

4.2.2. Land-Tenure Regimes and Yield

Table 6 reports the determinants of farm yield. It illustrates the impact of contract-farming regimes, farm, farmer-related, and other variables on farm productivity. Using the probit model, we controlled farm and farmer-related variables and used instruments following the previous section. Thus, the instrument covers the medium-term contract as farm distance increases from the market. It implies that if the farm distance from the market is less, the chances of a medium- or short-term contract are less likely. In either situation, the owner is likely to operate the farm or chooses to engage in a longer-term, more stable contract. We used control variables exogenous to contract-farming regimes and inserted the predicted value of first-phase regression results to compute farm productivity. The findings support the previous studies [1,28,59]. The results indicate that long-term and medium-term contracts positively affect farm productivity. It implies that farmers have more per hectare yield under these contracts than short-term contracts. These results support the previous section's results, reinstating that longer-term land contracts are more efficient regarding farmers' efficiency and farm yield. Likewise, the results align with the Marshallian inefficiency hypothesis [53], complementing that short-term contracts are the least effective among the given land-contract regimes. The presence of a public–private partnership and subsidies or economic incentives primarily encourages long-term and medium-term farmers' engagements in agriculture and land-improvement investment. Farm size, farming experience, tube-well ownership, FBO membership, training participation, livestock holding, tractor ownership, and mechanical harvesting significantly and positively affect farm yield per hectare.

Table 6. Determinants of farm yield: OLS estimates.

Variables	Coefficient	T-Value
Long-term tenure	0.137 ***	4.81
Medium-term tenure	0.712 **	2.46
Farm size	0.621 ***	3.39
Farming experience	0.274 *	1.74
Tube-well ownership	0.136 **	2.18
Farm advisory	0.155	0.09
Market distance	0.028	0.78
FBO membership	0.125 ***	3.98
Household size	0.093	0.73
Training participation	0.083 **	2.17
Household-head age	0.092	0.37
Market connection	0.093	1.06
Household-head education	0.027	0.93
Livestock holding	0.671 *	1.78
Tractor ownership	0.127 **	2.45
Farm logistic	0.194	0.87
No. of tillage operations	0.383	0.14
Mechanical harvesting	0.138 ***	4.28
Organic experience	0.183	0.91
Public–private partnership	0.129 **	0.164 ***
FBO membership	0.148 *	1.83
Constant	0.26	4.62
R^2	0.561	
Adjusted R^2	0.546	
p-Value	0.000	
Breush–Pegan Test (χ^2)	14.27	0.035
Goodness of fit (χ^2)	81.93	0.048

Note: * $p < 0.10$, ** $p < 0.05$, *** $p < 0.01$.

Table 7 presents the technical efficiency scores and production performance levels of long-term and medium-term contract farmers in Punjab, Pakistan. The results indicate that among the unmatched samples, the efficiency score is 88.2% and 79.5% for long-term and medium-term contracts, respectively. In the matched sample, full sample measures show technical efficiency scores of 87.9% and 75.8% for long-term and medium-term contracts. These indicate that farmers under long-term contracts produce more output (12.1%) than medium-term contracts. Simultaneously, the results illustrate that 12.1% of the potential yield per hectare is lost due to technical inefficiency, which might be due to inefficient inputs. It confirms that improving the state of land-tenure regimes through contract security and long-term stability would help realize higher productivity and economic benefits to the farmers. Likewise, improving the efficiency of crop inputs would help realize the minimization of losses due to technical inefficiency. These results follow prior studies on contract security and technical efficiency [3,5,60], implying that longer-term contract security plays the foremost role in productivity and technical efficiency.

Table 7. Mean and standard deviation technical efficiency in PSM matching estimations.

	Long-Term	Medium-Term	Difference in Means	t-Test
	Mean	Mean		
	TE—Probit Model (n = 450)			
Unmatched	0.882	0.795	0.019	3.87 ***
ATT	0.879	0.758	0.027	3.68 ***

Note: *** $p < 0.01$.

Table 8 illustrates the technical efficiency scores and levels of production performance of long-term and short-term contract farmers. In the matched sample, full sample measures show technical efficiency scores of 86.2% and 74.6% for long-term and short-term contracts. These indicate that farmers under long-term contracts produce more output (11.8%) than short-term contracts. Simultaneously, the results illustrate that 13.8% of the potential yield per hectare is lost due to technical inefficiency. It confirms that improving the state of land-tenure regimes through contract security and long-term stability and improving the input use efficiency would help realize higher productivity and economic benefits to the farmers. Further, the two-sample t-test confirms that long-term, medium-term, and long-term and short-term contract regimes statistically differ regarding technical efficiency. It reinstates the findings that a longer term contract is more secure and improves productivity, technical efficiency, and investment in land-improvement measures.

Table 8. Mean and standard deviation of technical efficiency in PSM matching estimations.

	Long-Term	Short-Term	Difference in Means	t-Test
	Mean	Mean		
	TE—Probit Model (n = 400)			
Unmatched	0.873	0.755	0.031	4.13 ***
ATT	0.862	0.746	0.022	3.77 ***

Note: *** $p < 0.01$.

5. Conclusions, Policy Implications, and Way Forward

Land-tenure security plays an integral role in the socioeconomic development of local communities. It has been a center of debate in academia and for legislators and advocates to implement reforms to enhance efficiency and sustainable development in land management. Likewise, in the face of mounting challenges of climate change and productivity, it has been crucial to validate the policy reforms on the role of land-tenure length and current land-tenure regimes in developing countries. This study investigates the influence of three contract-farming regimes, long-term, medium-term, and short-term contracts, on the land-improvement investment, productivity, and technical efficiency of contract farmers in Punjab, Pakistan. The study used a data set of 650 farm households gathered through face-to-face interviews. The study provides interesting insights into the role given contract-farming regimes and offers practical policy suggestions for stakeholders.

The findings of the study are fourfold. First, the results suggest that farmers with long-term land contracts have higher per hectare yield, income, and profit than those with medium-term and short-term contracts. Likewise, findings demonstrate that farmers have higher PPPs and subsidies or financial incentives under longer-term contracts. Second, the results confirm that farmers with medium- and long-term contracts tend to invest more in land-improvement measures, i.e., organic and green manure. Further, under these contracts, farmers have more yield and higher demand for crop- and land-improvement measures, i.e., hired labor and improved seeds. The findings support the Marshallian inefficiency hypothesis and reinstate that short-term land tenure is more inefficient than a long-term contract. Third, the study findings demonstrate that long-term land tenures are more effective when farmers make decisions regarding the investment in land-improvement measures (e.g., organic and green manure application) and on-farm infrastructure, like installation of a tube well, tractor ownership, and holding a farm logistic. Last, the study results confirm that long-term contracts are more robust regarding technical efficiency. Hence, the empirical evidence supports the notion that farmers with long-term and secure land-use rights tend to invest more in land-improvement measures. Likewise, it reinstates that long-term contracts are more fruitful regarding yield, productivity, and economic efficiency. Further, the findings clarify that long-term lease agreements offer higher institutional incentives to farmers and encourage the adoption of the latest technology to boost productivity and farm income.

Based on the study findings, the following policy actions are suggested to improve land-use rights, land-improvement investments, and sustainable development in developing countries. First, there is a need to enhance land-use rights. Land-use reform could be integral to turning the current land-use regimes into robust lease agreements. This could be accomplished by clearly defining the land-use rights under various contract and lease agreements and protecting and enforcing such laws. This would help promote longer term land-improvement investment and sustainable development. Second, complex regulations need to be simplified and streamlined. For that purpose, there is a need to revisit current bureaucratic complexities, which are the foremost barriers to land-improvement investment. A robust and efficient set of land regulations would help navigate investors toward land-improvement investment by reducing the costs and related complexities. Third, there is a dire need to improve land-tenure security, particularly for medium and short-term contract farmers. Short contracts are extremely insecure and hinder land investment in developing countries. Since investment needs a more extended payoff period, the government should reform land lease arrangements that protect the rights of land investors for the broader interests of the local communities and society. Four, a public–private partnership (PPP) has enormous potential to harness a significant investment in land-improvement measures. The government can encourage investors by easing regulatory and legal frameworks and financial incentives, like tax relief, to promote sustainable land management and the broader interests of society. Moreover, promoting sustainable land-use practices by incentivizing through subsidizing green-promoting farm implements would help realize minimal risks to the environment and local communities. In sum, implementing these actions can help foster land-improvement investment, promote economic activities for local communities, and support efforts for sustainable development in developing countries.

Supplementary Materials: The following supporting information can be downloaded at: https://www.mdpi.com/article/10.3390/agriculture13091651/s1.

Author Contributions: Conceptualization, R.M. and H.A.; methodology, R.M., S.V.P., R.V., M.P. and R.S.; software, R.M.; validation, R.M., B.X. and Z.W.; formal analysis, R.M. and H.A.; investigation, R.M.; resources, Z.W. and H.A.; data curation, R.M. and H.A.; writing—original draft preparation, R.M.; writing—review and editing, R.M., S.V.P., R.V., M.P. and H.A.; visualization, R.M.; supervision, Z.W.; project administration, Z.W.; funding acquisition, Z.W. All authors have read and agreed to the published version of the manuscript.

Funding: This research was funded by the National Natural Science Foundation of P.R. China (grant number 42071221): under project named: The Impact of Environment Regulation of Hog reduction on Rural Land Use Change.

Institutional Review Board Statement: Not applicable.

Data Availability Statement: Not applicable.

Acknowledgments: We acknowledge and thank the three anonymous reviewers for their suggestion.

Conflicts of Interest: The authors declare no conflict of interest.

References

1. Akram, N.; Akram, M.W.; Wang, H.; Mehmood, A. Does Land Tenure Systems Affect Sustainable Agricultural Development? *Sustainability* **2019**, *11*, 3925. [CrossRef]
2. Ali, A.; Abdulai, A.; Goetz, R. Impacts of Tenancy Arrangements on Investment and Efficiency: Evidence from Pakistan. *Agric. Econ.* **2012**, *43*, 85–97. [CrossRef]
3. Bidzakin, J.K.; Fialor, S.C.; Awunyo-Vitor, D.; Yahaya, I. Contract Farming and Rice Production Efficiency in Ghana. *J. Agribus. Dev. Emerg. Econ.* **2020**, *10*, 269–284. [CrossRef]
4. Dubbert, C.; Abdulai, A.; Mohammed, S. Contract Farming and the Adoption of Sustainable Farm Practices: Empirical Evidence from Cashew Farmers in Ghana. *Appl. Econ. Perspect. Policy* **2021**, *45*, 487–509. [CrossRef]
5. Le Ngoc, H. Contract Farming Effects on Technical Efficiency of the Export-Oriented Rice Production Sector in Vietnam. In Proceedings of the International Association of Agricultural Economists (IAAE), 2018 Conference, Vancouver, BC, Canada, 28 July–2 August 2018; European Environment Agency: Copenhagen, Denmark, 2018.

6. Mitra, A.; Rao, N. Contract Farming, Ecological Change and the Transformations of Reciprocal Gendered Social Relations in Eastern India. *J. Peasant Stud.* **2021**, *48*, 436–457. [CrossRef]
7. Ba, H.A.; de Mey, Y.; Thoron, S.; Demont, M. Inclusiveness of Contract Farming along the Vertical Coordination Continuum: Evidence from the Vietnamese Rice Sector. *Land Use Policy* **2019**, *87*, 104050. [CrossRef]
8. Vicol, M. Is Contract Farming an Inclusive Alternative to Land Grabbing? The Case of Potato Contract Farming in Maharashtra, India. *Geoforum* **2017**, *85*, 157–166. [CrossRef]
9. Sokchea, A.; Culas, R.J. Impact of Contract Farming with Farmer Organizations on Farmers' Income: A Case Study of Reasmey Stung Sen Agricultural Development Cooperative in Cambodia. *Australas. Agribus. Rev.* **2015**, *23*, 1–11.
10. Mishra, A.K.; Kumar, A.; Joshi, P.K.; D'Souza, A.; Tripathi, G. How Can Organic Rice Be a Boon to Smallholders? Evidence from Contract Farming in India. *Food Policy* **2018**, *75*, 147–157. [CrossRef]
11. Sartorius, K.; Kirsten, J. A framework to facilitate institutional arrangements for smallholder supply in developing countries: An agribusiness perspective. *Food Policy* **2007**, *32*, 640–655. [CrossRef]
12. Khan, M.F.; Nakano, Y.; Kurosaki, T. Impact of Contract Farming on Land Productivity and Income of Maize and Potato Growers in Pakistan. *Food Policy* **2019**, *85*, 28–39. [CrossRef]
13. Nhân, T.Q.; Takeuchi, I.; Hoang, D.V. Rice Contract Farming-the Potential Key to Improve Rice Growers' Income: A Farm Level Study in An Giang Province. *J. Sci. Dev.* **2013**, *11*, 1062–1072.
14. Anjum, A.S.; Zada, R.; Tareen, W.H. Organic Farming: Hope for the Sustainable Livelihoods of Future Generations in Pakistan. *J. Pure Appl. Agric.* **2016**, *1*, 20–29.
15. Mazhar, R.; Ghafoor, A.; Xuehao, B.; Wei, Z. Fostering Sustainable Agriculture: Do Institutional Factors Impact the Adoption of Multiple Climate-Smart Agricultural Practices among New Entry Organic Farmers in Pakistan? *J. Clean. Prod.* **2020**, *283*, 124620. [CrossRef]
16. Hassan, S.Z.; Jajja, M.S.S.; Asif, M.; Foster, G. Bringing More Value to Small Farmers: A Study of Potato Farmers in Pakistan. *Manag. Decis.* **2020**, *59*, 829–857. [CrossRef]
17. Sher, A.; Mazhar, S.; Zulfiqar, F.; Wang, D.; Li, X. Green Entrepreneurial Farming: A Dream or Reality? *J. Clean. Prod.* **2019**, *220*, 1131–1142. [CrossRef]
18. Wani, S.A.; Wani, M.A.; Mehraj, S.; Padder, B.A.; Chand, S. Organic Farming: Present Status, Scope and Prospects in Northern India. *J. Appl. Nat. Sci.* **2017**, *9*, 2272–2279. [CrossRef]
19. Sujatha, R.V.; Eswara Prasad, Y.; Suhasini, K. Comparative Analysis of Efficiency of Organic Farming Vs Inorganic Farming-A Case Study in Karimnagar District of Andhra Pradesh. *Agric. Econ. Res. Rev* **2006**, *19*, 232.
20. Jouzi, Z.; Azadi, H.; Taheri, F.; Zarafshani, K.; Gebrehiwot, K.; Van Passel, S.; Lebailly, P. Organic Farming and Small-Scale Farmers: Main Opportunities and Challenges. *Ecol. Econ.* **2017**, *132*, 144–154. [CrossRef]
21. Akram, M.W.; Akram, N.; Hongshu, W.; Mehmood, A. An Assessment of Economic Viability of Organic Farming in Pakistan. *Custos E Agronegocio Online* **2019**, *15*, 141–169.
22. Eyhorn, F.; Muller, A.; Reganold, J.P.; Frison, E.; Herren, H.R.; Luttikholt, L.; Mueller, A.; Sanders, J.; Scialabba, N.E.-H.; Seufert, V. Sustainability in Global Agriculture Driven by Organic Farming. *Nat. Sustain.* **2019**, *2*, 253. [CrossRef]
23. Väth, S.J.; Gobien, S.; Kirk, M. Socio-Economic Well-Being, Contract Farming and Property Rights: Evidence from Ghana. *Land Use Policy* **2019**, *81*, 878–888. [CrossRef]
24. Bahati, I.; Martiniello, G.; Abebe, G.K. The Implications of Sugarcane Contract Farming on Land Rights, Labor, and Food Security in the Bunyoro Sub-Region, Uganda. *Land Use Policy* **2022**, *122*, 106326. [CrossRef]
25. Adams, T.; Gerber, J.-D.; Amacker, M.; Haller, T. Who Gains from Contract Farming? Dependencies, Power Relations, and Institutional Change. *J. Peasant Stud.* **2019**, *46*, 1435–1457. [CrossRef]
26. Benjamin, E.O. Smallholder Agricultural Investment and Productivity under Contract Farming and Customary Tenure System: A Malawian Perspective. *Land* **2020**, *9*, 277. [CrossRef]
27. Sher, A.; Mazhar, S.; Azadi, H.; Lin, G. Smallholder Commercialization and Urban-Rural Linkages: Effect of Interest-Free Agriculture Credit on Market Participation of Rice Growers in Pakistan. *Land* **2021**, *10*, 7. [CrossRef]
28. Akram, M.W.; Akram, N.; Hongshu, W.; Andleeb, S.; ur Rehman, K.; Kashif, U.; Mehmood, A. Impact of Land Use Rights on the Investment and Efficiency of Organic Farming. *Sustainability* **2019**, *11*, 7148. [CrossRef]
29. Lobley, M.; Butler, A.; Reed, M. The Contribution of Organic Farming to Rural Development: An Exploration of the Socio-Economic Linkages of Organic and Non-Organic Farms in England. *Land Use Policy* **2009**, *26*, 723–735. [CrossRef]
30. Saqib, S.E.; Kuwornu, J.K.M.; Panezia, S.; Ali, U. Factors Determining Subsistence Farmers' Access to Agricultural Credit in Flood-Prone Areas of Pakistan. *Kasetsart J. Soc. Sci.* **2018**, *39*, 262–268. [CrossRef]
31. Holden, S.T.; Ghebru, H. Land Tenure Reforms, Tenure Security and Food Security in Poor Agrarian Economies: Causal Linkages and Research Gaps. *Glob. Food Secur.* **2016**, *10*, 21–28. [CrossRef]
32. Sher, A.; Qiu, Y. Pakistan's Solar Mission: Do Solar Finance and Subsidy Remove the Barriers to Solar Installations? *Renew. Energy* **2022**, *190*, 993–1005. [CrossRef]
33. Sher, A.; Mazhar, S.; Qiu, Y. Toward sustainable agriculture: The impact of interest-free credit on marketing decisions and technological progress in Pakistan. *Sustain. Dev.* **2023**, *34*, 1–16. [CrossRef]
34. Friedrich, T.; Derpsch, R.; Kassam, A. Overview of the Global Spread of Conservation Agriculture. In *Sustainable Development of Organic Agriculture*; Apple Academic Press: Palm Bay, FL, USA, 2017; pp. 75–90.

35. Krippner, G.R. Democracy of Credit: Ownership and the Politics of Credit Access in Late Twentieth-Century America. *Am. J. Sociol.* **2017**, *123*, 1–47. [CrossRef]
36. Khan, M.F.; Kurosaki, T.; Sakurai, T. Contract Farming, Loan Repayment Ability and Access to Credit of Small Farmers in Pakistan. *Jpn. J. Agric. Econ.* **2020**, *22*, 123–128.
37. Sher, A.; Mazhar, S.; Abbas, A.; Iqbal, M.A.; Li, X. Linking Entrepreneurial Skills and Opportunity Recognition with Improved Food Distribution in the Context of the CPEC: A Case of Pakistan. *Sustainability* **2019**, *11*, 1838. [CrossRef]
38. Chandio, A.A.; Jiang, Y.; Wei, F.; Rehman, A.; Liu, D. Famers' Access to Credit: Does Collateral Matter or Cash Flow Matter?—Evidence from Sindh, Pakistan. *Cogent Econ. Financ.* **2017**, *5*, 1369383. [CrossRef]
39. Barrett, C.B.; Bachke, M.E.; Bellemare, M.F.; Michelson, H.C.; Narayanan, S.; Walker, T.F. Smallholder Participation in Contract Farming: Comparative Evidence from Five Countries. *World Dev.* **2012**, *40*, 715–730. [CrossRef]
40. Mishra, A.K.; Kumar, A.; Joshi, P.K.; D'souza, A. Production Risks, Risk Preference and Contract Farming: Impact on Food Security in India. *Appl. Econ. Perspect. Policy* **2018**, *40*, 353–378. [CrossRef]
41. Razzaq, A.; Liu, H.; Zhou, Y.; Xiao, M.; Qing, P. The Competitiveness, Bargaining Power, and Contract Choice in Agricultural Water Markets in Pakistan: Implications for Price Discrimination and Environmental Sustainability. *Front. Environ. Sci.* **2022**, *10*, 917984. [CrossRef]
42. Wainaina, P.W.; Okello, J.J.; Nzuma, J.M. *Impact of Contract Farming on Smallholder Poultry Farmers' Income in Kenya*; European Environment Agency: Copenhagen, Denmark, 2012.
43. Girma, J.; Gardebroek, C. The Impact of Contracts on Organic Honey Producers' Incomes in Southwestern Ethiopia. *For. Policy Econ.* **2015**, *50*, 259–268. [CrossRef]
44. Carney, J.A. Struggles over Crop Rights and Labour within Contract Farming Households in a Gambian Irrigated Rice Project. *J. Peasant Stud.* **1988**, *15*, 334–349. [CrossRef]
45. Ren, Y.; Peng, Y.; Campos, B.C.; Li, H. The Effect of Contract Farming on the Environmentally Sustainable Production of Rice in China. *Sustain. Prod. Consum.* **2021**, *28*, 1381–1395. [CrossRef]
46. Miyata, S.; Minot, N.; Hu, D. Impact of Contract Farming on Income: Linking Small Farmers, Packers, and Supermarkets in China. *World Dev.* **2009**, *37*, 1781–1790. [CrossRef]
47. Bellemare, M.F. As You Sow, so Shall You Reap: The Welfare Impacts of Contract Farming. *World Dev.* **2012**, *40*, 1418–1434. [CrossRef]
48. Bellemare, M.F.; Bloem, J.R. Does Contract Farming Improve Welfare? A Review. *World Dev.* **2018**, *112*, 259–271. [CrossRef]
49. Mishra, A.K.; Rezitis, A.N.; Tsionas, M.G. Estimating Technical Efficiency and Production Risk under Contract Farming: A Bayesian Estimation and Stochastic Dominance Methodology. *J. Agric. Econ.* **2019**, *70*, 353–371. [CrossRef]
50. Singh, S. Contracting out Solutions: Political Economy of Contract Farming in the Indian Punjab. *World Dev.* **2002**, *30*, 1621–1638. [CrossRef]
51. Rehber, E. *Contract Farming: Theory and Practice*; European Environment Agency: Copenhagen, Denmark, 2007.
52. Kumar, J.; Kumar, K.P. Contract Farming: Problems, Prospects and Its Effect on Income and Employment. *Agric. Econ. Researh Rev.* **2008**, *21*, 243–250.
53. Arcand, J.-L.; Ai, C.; Éthier, F. Moral Hazard and Marshallian Inefficiency: Evidence from Tunisia. *J. Dev. Econ.* **2007**, *83*, 411–445. [CrossRef]
54. Wooldridge, J.M.. *Econometric Analysis of Cross Section and Panel Data*; MIT Press: Cambridge, MA, USA, 2010; ISBN 0-262-29679-9.
55. Hansmann, H.; Kraakman, R. Property, Contract, and Verification: The Numerus Clausus Problem and the Divisibility of Rights. *J. Leg. Stud.* **2002**, *31*, S373–S420. [CrossRef]
56. Nguyen, T.T.; Bauer, S.; Grote, U. Does Land Tenure Security Promote Manure Use by Farm Households in Vietnam? *Sustainability* **2016**, *8*, 178. [CrossRef]
57. Abdulai, A.; Owusu, V.; Goetz, R. Land Tenure Differences and Investment in Land Improvement Measures: Theoretical and Empirical Analyses. *J. Dev. Econ.* **2011**, *96*, 66–78. [CrossRef]
58. Cai, J.; Ung, L.; Setboonsarng, S.; Leung, P. *Rice Contract Farming in Cambodia: Empowering Farmers to Move beyond the Contract toward Independence*; ADBI Discussion Paper; Asian Development Bank Institute (ADBI): Tokyo, Japan, 2008; Available online: https://www.adb.org/sites/default/files/publication/156748/adbi-dp109.pdf (accessed on 15 June 2023).
59. Wang, H.H.; Wang, Y.; Delgado, M.S. The Transition to Modern Agriculture: Contract Farming in Developing Economies. *Am. J. Agric. Econ.* **2014**, *96*, 1257–1271. [CrossRef]
60. Ton, G.; Desiere, S.; Vellema, W.; Weituschat, S.; D'Haese, M. The Effectiveness of Contract Farming in Improving Smallholder Income Andfood Security in Low-and Middle-Income Countries: A Mixed-Method Systematic Review. *3ie Syst. Rev.* **2017**, *38*, 114.

agriculture

MDPI

Article

The Efficiency of China's Agricultural Circular Economy and Its Influencing Factors under the Rural Revitalization Strategy: A DEA–Malmquist–Tobit Approach

Chenghan Guo [1], Rong Zhang [2] and Yuntao Zou [2,3,4,*]

1 School of Business, University of Queensland, Toowoomba, QLD 4350, Australia; chenghan.guo@uqconnect.edu.au
2 FutureFront Interdisciplinary Research Institute, Wuhan 430074, China; zhangr4444@gmail.com
3 School of Computer of Science and Technology, Huazhong University of Science and Technology, Wuhan 430074, China
4 School of Energy and Power Engineering, Huazhong University of Science and Technology, Wuhan 430074, China
* Correspondence: zouyuntao@hust.edu.cn

Abstract: In 2018, the Chinese government proposed the Rural Revitalization Strategy with the objective of bolstering economic development, social progress, and ecological protection in rural areas, thereby achieving rural modernization. This paper employs the Data Envelopment Analysis (DEA) method and the Malmquist index model to measure the efficiency and changes of the agricultural circular economy in 31 provinces and cities in China from 2017 to 2020. Using Tobit regression, we further examine the correlation analysis in the context of the rural revitalization policy. The study reveals that the efficiency of China's agricultural circular economy continued to grow between 2017 and 2020. The policy of the rural revitalization strategy significantly impacts the efficiency of the agricultural circular economy. Government financial support has a significant positive influence on the efficiency of the agricultural circular economy. Based on the research findings, we proposed several constructive suggestions.

Keywords: agriculture; circular economy; efficiency; rural revitalization; DEA; Malmquist; Tobit

Citation: Guo, C.; Zhang, R.; Zou, Y. The Efficiency of China's Agricultural Circular Economy and Its Influencing Factors under the Rural Revitalization Strategy: A DEA–Malmquist–Tobit Approach. *Agriculture* **2023**, *13*, 1454. https://doi.org/10.3390/agriculture13071454

Academic Editors: Xin Chen, Moucheng Liu and Yuanmei Jiao

Received: 19 June 2023
Revised: 17 July 2023
Accepted: 21 July 2023
Published: 23 July 2023

1. Introduction

1.1. China Agricultural Circular Economy

In the fifth plenary session of the 16th Central Committee of the Communist Party of China held in 2005, it was explicitly proposed that China should develop a circular agricultural economy. "An ecological agricultural model based on the circular economy contemplates the coordination of various production elements in rural areas such as soil, water, seeds, fertilizers, pesticides, electricity, oil, firewood, and grains, facilitating holistic planning, systemic conservation, and comprehensive development. This model encourages the recycling and extensive utilization of waste products from rural agricultural and livestock activities, as well as waste generated by urban industries and rural enterprises that use agricultural products as raw materials. This results in the transformation of waste into useful resources, generating significant economic, social, and environmental benefits. The model aims to continuously improve the productivity of various resources in agricultural production and the overall agricultural production capacity, leading to an increase in farmers' income. The circular economy-oriented ecological agriculture promotes the acceleration of agricultural technological progress, facilitates the adjustment of rural industrial structures, transforms agricultural growth modes, expands the scale of modern agriculture, extends the industrial chain, and broadens the employment space in urban and rural areas [1]". The development of this type of ecological agriculture not only produces safe and high-quality agricultural products, but is also beneficial for soil improvement and resource conservation, and promotes the sustainable development of agriculture.

In conclusion, the focus of the circular economy is the enhancement of production efficiency, that is, achieving the maximum output with the minimum input. The study by Wu et al. considers efficiency as a key factor in evaluating the level of the circular economy [2]. The inputs of the agricultural circular economy include the aforementioned production factors such as "soil, water, seeds, fertilizers, pesticides, electricity, oil, firewood, and grains". Improving the efficiency of the agricultural circular economy implies increasing output while effectively controlling emissions and environmental pollution during the production process, thus achieving sustainable development. In their research on the agricultural circular economy, Xin et al. [3] evaluated the level of development of the agricultural circular economy by constructing efficiency models. Their research findings indicate an upward trend in the level of agricultural resource recycling, economic benefits, and ecological benefits, but the effect of controlling the reduction of agricultural resource usage is less than ideal. Ul Haq et al. employed efficiency as a measure to evaluate the circular economy efficiency of tea gardens in Turkey [4]. Similarly, when assessing the level of development of the circular economy in China, Fan et al. also centered their evaluation around the concept of efficiency [5]. Shahbaz and colleagues also utilized efficiency as a metric to study the level of agricultural circular economy in Pakistan [6].

1.2. Rural Revitalization Strategy

The Rural Revitalization Strategy of China is a significant strategic initiative proposed by the Chinese government in 2018. It aims to stimulate economic development, social progress, and ecological protection in rural areas, realizing the harmonious and integrated development of urban and rural regions. "Rural revitalization represents a comprehensive revitalization encompassing the rejuvenation of industries, talents, culture, ecology, and organizations. The overarching goal of implementing the rural revitalization strategy is the modernization of agriculture and rural areas. The primary guideline is to prioritize the development of agriculture and rural areas. The overall requirements are the prosperity of industries, ecological livability, civilized ethos, effective governance, and affluent life. The institutional guarantee is the establishment of a sound urban-rural integration development system, mechanism, and policy framework [7]". The implementation of this strategy aims to resolve numerous issues faced by rural areas, such as population outflow, rural poverty, underdeveloped infrastructure, and the deterioration of the ecological environment.

To achieve this objective, the Chinese government has implemented a series of policies and measures. The first is the financial guarantee supported by policies, which includes direct fiscal expenditure, tax incentives, financial support, and land policies. These policies and measures aim to encourage participation in rural revitalization from all sectors and provide the necessary financial guarantees. From 2016 to 2019, the national general public budget allocated a cumulative expenditure of CNY 16.07 trillion related to agriculture and rural areas, with an average annual growth of 8.8%, higher than the average increase in the national general public budget expenditure [8]. In 2023, further increases were made to the scale of the central fiscal subsidies for promoting rural revitalization, with CNY 175 billion allocated, representing an increase of CNY 10 billion from the previous year [9]. These policies include the following:

Rural Infrastructure Construction: The government has increased its investment in rural infrastructure construction, including improvements in rural roads, water supply, electricity, and communication. This contributes to enhancing the accessibility of transportation and living conditions in rural areas, promoting industrial development, and facilitating employment and entrepreneurship among farmers [10,11].

Industrial Upgrading in Rural Areas: This pertains to facilitating the adjustment and transformation of the rural industrial structure and accelerating the modernization of agriculture on a large scale. In December 2018, the "Guidance of the State Council on Accelerating the Transformation and Upgrading of Agricultural Mechanization and

Agricultural Machinery Equipment Industry" [12] was promulgated, emphatically stipulating the steady implementation of agricultural machinery purchase subsidy policies to maximize policy benefits. Moreover, the government has been encouraging the development of new business models, such as rural characteristic industries, modern agriculture, and rural e-commerce. The industrial upgrading also manifests in vigorously promoting the corporatization and industrialization of agriculture. In 2022, the Chinese government officially issued the "Notice of the State Council on Printing and Distributing the '14th Five-Year Plan' for Promoting Modernization of Agriculture and Rural Areas", explicitly proposing to accelerate the modernization process of agriculture and rural areas with Chinese characteristics [13].

Development of Social Undertakings: This strategy entails bolstering support for rural education, healthcare, culture, and other social undertakings to enhance public service levels in rural areas. This includes constructing rural schools, healthcare institutions, and cultural facilities; improving rural educational and medical conditions; and raising the educational level and quality of life of farmers. The government has also begun to adopt a service procurement approach, purchasing services such as sanitation, public legal assistance, public cultural activities, public sports programs, medical and health services, educational services, disability assistance, elderly care, and youth services, to support the development of the rural revitalization strategy [14].

Protection of Rural Ecological Environment: The government has intensified efforts towards the protection of the rural ecological environment and the promotion of greener and sustainable agricultural production methods [7]. These efforts encompass the advocacy for organic agriculture, ecological agriculture, and circular agriculture; strengthening of farmland water conservancy construction; improving the quality of the rural environment; and protecting the integrity and stability of rural ecosystems.

2. Literature Review and Objectives of This Paper

2.1. Research Related to Agricultural Policy and Efficiency

The foundation of the agricultural circular economy is predicated on the modernization of agricultural production. The primary objective of the rural revitalization strategy is to establish aesthetically pleasing, economically prosperous, and habitable rural communities, ultimately achieving rural modernization. However, the relationship between agricultural modernization and rural modernization, particularly the overlapping process where agricultural modernization expands into rural modernization, is inherently complex. This complexity can be attributed to the law of diminishing returns on land, as proposed by Malthus [15].

Contrasting the transition process from industrialization to urbanization, Scott [16] demonstrated that industrial modernization catalyzes the agglomeration of industrial elements, subsequently fostering urbanization. Urbanization, through its induced agglomeration of industrial elements and deepening division of labor, further stimulates industrialization, culminating in a mutually beneficial and reciprocally enhancing relationship between industrialization and urbanization. This premise is also validated by the research conducted by Hubendick [17] and others on the interplay between industrialization and urbanization.

The progression from agricultural socialization to rural modernization presents a different narrative. Agricultural modernization, which is an extension of industrial modernization, results in the agglomeration of industrial elements [18]. This trajectory, however, does not entirely align with the goals of agricultural modernization. For instance, while agricultural modernization enhances production efficiency, in accordance with Malthus's law of diminishing returns, the dual action of diminishing returns and increasing efficiency inevitably results in a requisite reduction in the scale of inputs. This subsequently triggers a decrease in rural employment opportunities and an increase in unemployment rate [19]. Conversely, escalating the agriculture-related inputs may potentially lead to a decrease in production efficiency [20].

Although the law of diminishing returns was not traditionally accepted by mainstream economists in China, it has been increasingly recognized in recent years. Jiang and Wang [21] explored the relationships among industrialization, urbanization, and agricultural modernization in Jilin. They posited that the improvement in the level of agricultural modernization could facilitate the transfer of surplus labor, gradually enlarging the demand scale for agricultural means of production, and advancing the level of urbanization. However, the elevation of urbanization levels often signifies a decline in rural modernization. Additionally, Yao and Liu [22], in their research on China's grain production, suggested that even in developing China, the law of diminishing returns is in effect. They assert that long-term growth in grain yield must be achieved through efficiency improvements.

Countries often adopt increased government investment and subsidies in their efforts to support agriculture, and rural revitalization strategies likewise emphasize government financial backing. Nevertheless, due to the law of diminishing returns and the effect of diminishing marginal returns, financial support faces potential risks of reducing production efficiency. For instance, de Jorge et al. found a correlation between the subsidies received by R&D companies and low efficiency in their study of Spanish manufacturing. They advised caution when using subsidies to stimulate enterprise innovation efficiency [23]. In China, Yao and Leng et al. [24] found that even within strategic emerging industries receiving strong government support and subsidies, fiscal subsidies had a significant inhibitory effect. They recommended adjustments in the direction of fiscal subsidies to enhance their benefits. Gao et al. [25] discovered that since China intensified fiscal and financial support in 2004, the direct effect of fiscal and financial support on agriculture has improved, but the spatial spillover effect turned from positive to negative. Kumbhakar and Lien [26] studied unbalanced panel data of Norwegian grain farms from 1991 to 2006, finding that agricultural subsidies negatively impacted agricultural production efficiency. According to the research by Guan Zhengfei et al., fiscal subsidies have a significant negative impact on agricultural productivity growth in the Netherlands, while debt, on the contrary, promotes productivity growth [27].

2.2. Research Objectives of This Paper

In conclusion, the development of China's agricultural circular economy requires agricultural modernization, whereas the goal of the rural revitalization strategy is the modernization of rural areas. Theoretically, these two concepts are incompatible. Many scholars have turned their attention to this issue. However, the rural revitalization strategy, which began implementation in 2018 and was affected by the outbreak of COVID-19 in 2020, has had a relatively short duration of undisturbed execution. Consequently, studies analyzing its influence on the overall efficiency of China's agricultural circular economy are rather limited. Much of the research remains at the qualitative level, with some focusing only on particular regions or specific dimensions.

This paper aims to fill this gap. Through the use of spatial econometrics, quantitative research and empirical analysis of the efficiency of the agricultural circular economy under the rural revitalization strategy are undertaken. The goals are as follows:

1. To measure and assess the efficiency of China's agricultural circular economy under the rural revitalization strategy, and analyze its development trend.
2. To conduct empirical research on the correlation between the efficiency of the agricultural circular economy and related policies of rural revitalization.

3. Materials and Methods

3.1. Data and Sources

The Rural Revitalization Strategy in China was formally proposed in 2018. Therefore, in this study, panel data from 2017 to 2020 are chosen as the research basis to compare the changes in the efficiency of agricultural circular economy before and after the strategy. This study selects panel data from 31 provinces, municipalities, and autonomous regions out of all 34 provincial-level administrative units in China.

The three excluded provincial-level administrative units are Taiwan, Hong Kong, and Macau, for the following reasons:

1. There are significant differences in the formulation and implementation of agricultural policies.
2. The statistical calibers of relevant data vary significantly.
3. The agricultural economies of these three provinces and cities is relatively small.

Thus, including Taiwan, Hong Kong, and Macau in the research scope would interfere with the research results and violate the consistency assumption in the DEA method. We believe that the selected 31 provinces and cities can represent the overall picture of China's agricultural circular economy.

Data source: China Statistical Yearbook, China Rural Statistical Yearbook.

3.2. Research Methodology

3.2.1. Research Process

The research process of this paper is as follows:

Data Collection: Gather agricultural and rural data from 31 provinces.

Data Envelopment Analysis (DEA): Use this method to calculate the efficiency of the agricultural circular economy. The output of this stage is the dependent variable.

Identification of Independent Variables: Include variables related to the Rural Revitalization Strategy, such as Degree of Financial Support for Agriculture, Degree of Agribusiness Development, Percentage of Rural Population, Degree of Energy Support, Degree of Water Infrastructure Support, and Degree of Informatization.

TOBIT Regression Model: Use this model for the correlation analysis to verify which policies are correlated with the efficiency of the agricultural circular economy.

DEA-Malmquist Method: Use this method to calculate the index model of the agricultural circular economy in China's 31 provinces.

Analytical Evaluation: Analyze whether the efficiency of the agricultural circular economy is improving or declining under the influence of the Rural Revitalization Strategy.

The flowchart of the study is shown below, as shown in Figure 1:

Figure 1. Research process on the efficiency and influencing factors of agricultural circular economy.

3.2.2. Measuring the Efficiency of China's Agricultural Circular Economy Using DEA Method

The Data Envelopment Analysis (DEA) model is a method for input–output analysis based on relative efficiency, proposed by Charnes et al. in 1978 [28]. It does not require the assignment of a priori weights to inputs and outputs, and can measure the relative efficiency of decision-making units (DMUs) with multiple inputs and outputs; as a result, it is widely used in efficiency assessment. The DEA model comprises several DMUs, each of which has the same input and output indicators. The efficiency frontier surface is determined through computation, which is then used to evaluate the efficiency of each DMU. Fundamental DEA models include the CCR model (named after its authors A. Charnes, W.W. Cooper, and E. Rhodes) [28] and the BCC model (named after its authors R.D Banker, A Charnes, and W.W. Cooper) [29]. The CCR model assumes constant returns to scale, and the resulting overall technical efficiency can be decomposed into pure technical efficiency and scale efficiency. On the other hand, the BCC model assumes variable returns to scale. The differences between the two are minor.

Currently, the DEA method is widely adopted in economic research, particularly in studies on the circular economy. Ul Haq et al. utilized the DEA method to evaluate the green economy efficiency of tea gardens in the Rize province of Turkey, establishing an efficiency model and identifying areas for improvement [4]. Zhao et al. applied the DEA method to analyze panel data from 286 prefecture-level cities in China, conducting a comprehensive study on China's green economy and its driving factors [30]. Streimikis and others also specifically examined the use of the DEA method in the green economy and agricultural pollution scenarios, finding that the DEA method has a broad-ranging impact [31].

The development of the DEA method has led to various models. For instance, the super-efficiency model was proposed by Andersen and Petersen in 1993 to solve the issue of further comparisons when multiple DMUs are on the frontier (i.e., efficiency equals 1) in the DEA model [32]. In the super-efficiency model, the super-efficiency score of a DMU can exceed 1, making it especially suitable for comparative studies between different DMUs [33]. However, the super-efficiency model has some drawbacks: it can often result in infeasible solutions during computation [34]; the results may change due to alterations in the scale of input or output data, implying it does not have scale-invariance; and it violates the weak disposability assumption in DEA when calculating the super-efficiency score by excluding the DMU under assessment, which could affect the model's theoretical consistency [35].

This study primarily analyzes the influence of policies on China's agricultural circular economy efficiency at a macro level. Taking into account both the strengths and weaknesses, we chose not to adopt the super-efficiency model, but instead applied the classic basic CCR model to measure the agricultural circular economy efficiency of 31 provinces in China. The calculation formula for the input-oriented CCR model is as follows:

Minimize:

$$\theta - \varepsilon \times (\Sigma(i = 1\ to\ n)\ \hat{s}-\_i + \Sigma(r = 1\ to\ s)\ \hat{s}+\_r)$$

Subject to:

$$\Sigma(j = 1\ to\ m)\ \lambda\_j \times x\_ij - x\_ik + \hat{s}-\_i = 0\ for\ all\ i$$

$$\Sigma(j = 1\ to\ m)\ \lambda\_j \times y\_rj - y\_rk - \hat{s}+\_r = 0\ for\ all\ r$$

$$\Sigma(j = 1\ to\ m)\ \lambda\_j = 1$$

$$\lambda\_j \geq 0\ for\ all\ j$$

$$\hat{s}-\_i \geq 0\ for\ all\ i$$

$$\hat{s}+\_r \geq 0\ for\ all\ r$$

In the above formulation:

θ represents the efficiency score to be evaluated.

x_ij is the ith input of the jth DMU.

y_rj is the rth output of the jth DMU.

λ_j are the decision variables, representing the weights for constructing a virtual decision-making unit (VDMU).

$s\hat{}-\_i$ are the slack variables for inputs, representing the efficiency loss of the ith input.

$s\hat{}+\_r$ are the slack variables for outputs, representing the efficiency gain of the rth output.

ε is a non-Archimedean infinitesimal, employed to ensure the resolution of the multiple-objective linear programming problem.

The aim of this model is to minimize the efficiency score (θ) and the sum of all slack variables for inputs/outputs. The constraints ensure that the efficiency loss of all inputs and outputs for all DMUs does not exceed their actual values in the evaluation of DMU_k. In addition, all weights (λ) and slack variables should be greater than or equal to zero.

3.2.3. Assessment of Changes in the Efficiency of China's Agricultural Circular Economy from 2017–2020 Using the DEA–Malmquist Index Model

The CCR model can only evaluate the efficiency of multiple DMUs within a single period or the efficiency of a single DMU across multiple periods. Each instance of the CCR model is a relative measure; hence, CCR models from different periods cannot be directly compared. Swedish economist and statistician Sten Malmquist proposed the Malmquist index for analyzing consumer changes over time [36]. By 1982, Caves et al. first proposed the Malmquist Total Factor Productivity Index (referred to as the Malmquist TFP index) [37]. They defined the total factor productivity index using a Malmquist input or output function. In 1992, Färe et al. developed a nonparametric (linear programming) method for calculating the Malmquist productivity index to evaluate the growth of total factor productivity [38]. As the Malmquist index can better analyze panel data, it can reflect the dynamic changes in relative efficiency at different periods [39], measure dynamic continuously changing characteristics, and analyze efficiency changes more effectively. The DEA–Malmquist model has been widely applied in various fields, especially in the construction of efficiency evaluation systems [40–42].

This paper employs the Malmquist index model to evaluate changes in the efficiency of China's agricultural circular economy. The Malmquist index model can evaluate multiple DMUs across multiple periods, thereby deriving the change index for total factor productivity (TFPCH). TFPCH is used to measure the dynamic trend of total factor productivity (TFP) of a DMU from time t to time t + 1, using a non-parametric distance function, that is, the ratio of distance functions before and after the two periods.

$$\mathrm{M_t} = \frac{\mathrm{D^t}\left(\mathrm{X^{t+1}}, \mathrm{Y^{t+1}}\right)}{\mathrm{D^t}(\mathrm{X^t}, \mathrm{Y^t})}$$

$$\mathrm{M_{t+1}} = \frac{\mathrm{D^{t+1}}\left(\mathrm{X^{t+1}}, \mathrm{Y^{t+1}}\right)}{\mathrm{D^{t+1}}(\mathrm{X^t}, \mathrm{Y^t})}$$

The expression for TFPCH is derived from the square root of the product of $\mathrm{M_t}$ and $\mathrm{M_{t+1}}$, denoted as $\mathrm{M_{t,t+1}}$, and its expression form is as follows:

$$TFPCH = \mathrm{M_{t,t+1}} = \sqrt{\frac{\mathrm{D^t}\left(\mathrm{X^{t+1}}, \mathrm{Y^{t+1}}\right)}{\mathrm{D^t}(\mathrm{X^t}, \mathrm{Y^t})} \times \frac{\mathrm{D^{t+1}}\left(\mathrm{X^{t+1}}, \mathrm{Y^{t+1}}\right)}{\mathrm{D^{t+1}}(\mathrm{X^t}, \mathrm{Y^t})}}$$

If TFPCH > 1, this implies an increase in the level of total factor productivity from period t to t + 1; if TFPCH = 1, it signifies no change in the level of total factor productivity from period t to t + 1; if TFPCH < 1, this indicates a decline in the level of total factor productivity from period t to t + 1.

The total factor productivity index (TFPCH) can further be decomposed into the product of the index of technical efficiency change (EFFCH) and the index of technological progress (TECHCH):

$$TFPCH = Effch \times Techch = \frac{D^{t+1}\left(X^{t+1}, Y^{t+1}\right)}{D^{t}\left(X^{t}, Y^{t}\right)} \times \sqrt{\frac{D^{t}\left(X^{t+1}, Y^{t+1}\right)}{D^{t+1}\left(X^{t+1}, Y^{t+1}\right)} \times \frac{D^{t}\left(X^{t}, Y^{t}\right)}{D^{t+1}\left(X^{t}, Y^{t}\right)}}$$

3.2.4. Study of the Factors Influencing the Efficiency of China's Agricultural Circular Economy Using Tobit Regression Model

Correlation research aims to determine whether there is a mutual connection between two or more sets of data, and to carry out a quantitative analysis of any potential links. The most common method is regression analysis. There are many methods of regression analysis, and this study primarily analyzes the correlation between the efficiency of the agricultural circular economy and the policy of the rural revitalization strategy. The dependent variable chosen, i.e., the variable to be explained, is the comprehensive efficiency of the agricultural circular economy calculated by the DEA method, whose value is between 0 and 1 [43]. Therefore, this study will employ the Tobit regression model.

The Tobit regression model was originally proposed by economist James Tobin in 1958, from which it derived its name [44]. The Tobit regression model is a type of linear regression model characterized by the truncation phenomenon in its dependent variable. Truncation refers to the inability to observe certain values, i.e., these values are restricted within a certain range. The Tobit regression model can transform such truncated data into a probability model, thereby statistically analyzing truncated data [45]. The mathematical formula for the Tobit model is as follows:

Firstly, we define a latent variable y*, representing the true but unobserved value of the observed variable y. We assume that y* follows a linear regression model:

$$y^* = X\beta + \varepsilon$$

In this, y* is a continuous latent variable, X is a matrix containing independent variables, β represents regression coefficients, and ε is the error term. Next, we define the observed variable y as follows:

$$y = max(0,\ y*)$$

This equation implies that if y* is less than or equal to 0, the observed y value is 0; otherwise, it equals y*. Next, to take truncation into account, a truncation variable c is introduced. If y* is less than the truncation point c, the observed y value is c; otherwise, it equals y*. This can be represented as:

$$y = max(c,\ y*)$$

The mathematical formula for the Tobit model encompasses both a linear regression model and the treatment of the observed value truncation. Through methods such as maximum likelihood estimation, parameters can be estimated and inferences made in the Tobit model.

The Tobit model offers the following advantages: it takes into account the impact of truncated data and can effectively handle issues with such data; it uses the maximum likelihood estimation method to estimate parameters, which provides high estimation accuracy and credibility. There are also some drawbacks to the Tobit regression model: the model assumes that the error term follows a normal distribution, so it may not be applicable for data with a skewed distribution.

The DEA–Tobit combination method is extensively employed in research within operations research, econometrics, and management science. Aldieri et al. utilized the DEA-Tobit method to study the energy economic policies of 136 countries, providing beneficial recommendations for energy policy modeling [46]. Shuai et al. applied the DEA–Tobit method to simulate the role of environmental regulations in China's green economy [47]. Dalei et al. examined the efficiency of refining in India using the DEA–Tobit method [48]. The logical reasoning behind the DEA–Tobit method is quite clear: it initially uses the DEA method to calculate the "outcome", i.e., the level of efficiency, and then applies the Tobit model to test associated factors or "causes". This closed-loop research process has led to its widespread application.

4. Results

4.1. Results of the Study on the Efficiency of Agricultural Circular Economy in 31 Provinces and Cities in China

1. Input and output indicators

In constructing the DEA-CCR model for agricultural circular economic efficiency, we select the number of rural personnel in each province and city to represent human capital input. The quantity of fertilizer applied, the amount of pesticide used, and the volume of diesel consumed represent the physical inputs. The area of crops sown serves as a representation of land input. On the output side, the total output value of agriculture, forestry, animal husbandry, and fishery is selected as an indicator of total agricultural income, while per capita disposable income in rural areas represents individual rural income. The inputs and outputs are summarized in Table 1.

Table 1. List of inputs and outputs.

Indicator Categories	Indicators
Input indicators	Rural Population
	Consumption of Chemical Fertilizers
	Consumption of Pesticides
	Consumption of Diesel Fuel
	Sown area of crops
output indicators	Gross Output Value of Agriculture, Forestry, Animal Husbandry and Fishery and Related Indices
	Per Capita Disposable Income of Rural Households by Region

It is worth noting that a more reasonable model for agricultural circular economy efficiency should include certain undesired output indicators, such as the amount of wastewater discharged and air pollution. However, the data collection poses certain challenges. The pollutant emission data for each province or city cannot be readily distinguished from data for industrial or agricultural emissions, necessitating further analysis. Secondly, agricultural economic efficiency itself implies achieving more output with less pesticide, diesel, and fertilizer use. Reducing these inputs often correlates with less pollutant emissions. Therefore, we did not choose undesired output indicators such as pollutant emissions when selecting input and output indicators.

This study employs DEARUN software to compute the CCR model of agricultural circular economic efficiency for 31 provinces and cities across China over four periods from 2017 to 2020. In the CCR model results, "crste" represents overall efficiency, "vrste" signifies pure technical efficiency, and "scale" denotes scale efficiency. A value of 1 for these three elements indicates DEA efficiency, suggesting a relatively ideal state. "Return of scale" represents scale returns, where a value of "CRS" signifies constant returns to scale for the corresponding province or city, "IRS" represents increasing returns to scale, and "DRS" denotes decreasing returns to scale.

The pure technical efficiency is presented in Table 2, the scale efficiency is shown in Table 3, the scale returns are displayed in Table 4, and the overall efficiency is summarized in Table 5.

Table 2. China's 31 provinces and cities agricultural circular economy pure technical efficiency statistics.

vrste	2017	2018	2019	2020
Number of "1"	18	19	19	21
Mean values	0.917676056	0.927100049	0.934294486	0.942051641
Beijing	1	1	1	1
Tianjin	1	1	1	1
Hebei	0.717395847	0.800793331	0.790225422	0.786663162
Shanxi	0.625652622	0.623417207	0.620294113	0.634481269
Inner Mongolia	0.981032738	0.993619342	0.991089659	1
Liaoning	0.979851655	1	1	1
Jilin	0.679533142	0.674452166	0.671892708	0.738799541
Heilongjiang	1	1	1	1
Shanghai	1	1	1	1
Jiangsu	1	1	1	1
Zhejiang	1	1	1	1
Anhui	0.709850702	0.704738368	0.717505651	0.712786742
Fujian	1	1	1	1
Jiangxi	0.823196531	0.816229789	0.839371298	0.826546699
Shandong	1	1	1	1
Henan	0.800646469	0.801151749	0.837054209	1
Hubei	1	1	1	1
Hunan	0.913105503	0.906257436	0.980111828	0.999582155
Guangdong	1	1	1	1
Guangxi	0.901249912	0.933033675	0.937775498	0.891603519
Hainan	1	1	1	1
Chongqing	0.896524944	0.889127827	0.913959851	0.943742014
Sichuan	1	1	1	1
Guizhou	1	1	1	1
Yunnan	0.685418699	0.874419103	0.943392358	0.958876921
Tibet	1	1	1	1
Shaanxi	1	1	1	1
Gansu	0.734498958	0.722861528	0.720456486	0.71051886
Qinghai	1	1	1	1
Ningxia	1	1	1	1
Xinjiang	1	1	1	1

4.2. Empirical Study of the Factors Influencing the Efficiency of Agricultural Circular Economy

This paper carries out an empirical study of agricultural circular economic efficiency using the Tobit model. The dependent variable is the overall technical efficiency value of the agricultural circular economy for the 31 provinces and cities computed earlier. The independent variables are selected considering the key policies of the rural revitalization strategy and the ease of data accessibility, with the following variables chosen. As the dependent variable is a dimensionless efficiency value, the selected independent variables are also processed for dimension lessness:

Degree of financial support for agriculture: This represents the direct financial support from the government, calculated as the ratio of expenditure on agriculture, forestry, and water to the general public budget expenditure for each province and city (Supplementary Table S1).

Degree of agribusiness: This represents the scale and industrial transformation of agricultural production, calculated as the ratio of the number of agricultural legal entities to the total number of legal entities in each province and city (Supplementary Table S2). The data on the number of agricultural legal entities for each province in 2018 are missing and are supplemented using linear interpolation.

Percentage of rural population: This represents the direction of the flow of human resources, calculated as the ratio of the rural population to the total population in each province and city (Supplementary Table S3).

Table 3. 2017–2020 China's 31 provinces and cities agricultural circular economy scale efficiency statistics.

Scale	2017	2018	2019	2020
Number of "1"	14	15	14	14
Mean values	0.960846732	0.965154953	0.961535153	0.959233546
Beijing	1	1	1	1
Tianjin	1	1	1	1
Hebei	0.99031636	0.956360219	0.952135066	0.973404655
Shanxi	0.917604772	0.919805225	0.920142271	0.926550722
Inner Mongolia	0.974153006	0.983459883	0.983719805	1
Liaoning	0.999195714	1	0.993975762	0.97716915
Jilin	0.971682712	0.99707285	0.998150439	0.995625405
Heilongjiang	1	1	1	1
Shanghai	1	1	1	1
Jiangsu	1	1	1	0.98745086
Zhejiang	1	1	1	1
Anhui	0.998621612	0.997803217	0.981784367	0.978347235
Fujian	1	1	1	1
Jiangxi	0.995480757	0.995906898	0.988500945	0.998238031
Shandong	0.885800967	0.896802698	0.85146753	0.834198369
Henan	0.909562634	0.913604138	0.884266745	0.776282684
Hubei	1	1	1	1
Hunan	0.989602481	0.99569094	0.984981999	0.992456398
Guangdong	1	1	1	1
Guangxi	0.985197402	0.999295353	0.978909885	0.980587728
Hainan	1	1	1	1
Chongqing	0.944172535	0.980221365	0.994293388	0.999763172
Sichuan	1	1	1	1
Guizhou	1	1	1	1
Yunnan	0.98250859	0.999916476	0.997496304	0.999888668
Tibet	0.734372097	0.745466154	0.739718529	0.729462273
Shaanxi	1	1	1	1
Gansu	0.829970586	0.845979091	0.866269487	0.89128308
Qinghai	0.801390551	0.822906814	0.839865076	0.830865163
Ningxia	0.876615927	0.869512234	0.851912157	0.864666337
Xinjiang	1	1	1	1

Table 4. 2017–2020 China's 31 provinces and cities agricultural circular economy return of scale statistics.

Return of Scale	2017	2018	2019	2020
Number of "CRS"	14	16	14	14
Number of "IRS"	9	8	8	7
Number of "DRS"	8	7	9	10

Table 5. 2017–2020 China's 31 provinces and cities agricultural circular economy overall technical efficiency statistics.

crste	2017	2018	2019	2020
Number of "1"	14	15	14	14
Mean values	0.882431746	0.89561	0.89927	0.90367
Beijing	1	1	1	1
Tianjin	1	1	1	1
Hebei	0.710448844	0.765847	0.752401	0.765742
Shanxi	0.574101832	0.573422	0.570759	0.587879
Inner Mongolia	0.955675991	0.977185	0.974955	1
Liaoning	0.979063574	1	0.993976	0.977169
Jilin	0.660290607	0.672478	0.67065	0.735568
Heilongjiang	1	1	1	1
Shanghai	1	1	1	1
Jiangsu	1	1	1	0.987451
Zhejiang	1	1	1	1
Anhui	0.708872252	0.70319	0.704436	0.697353
Fujian	1	1	1	1
Jiangxi	0.819476306	0.812889	0.829719	0.82509
Shandong	0.885800967	0.896803	0.851468	0.834198
Henan	0.728238112	0.731936	0.740179	0.776283
Hubei	1	1	1	1
Hunan	0.903611472	0.902352	0.965393	0.992042
Guangdong	1	1	1	1
Guangxi	0.887909072	0.932376	0.917998	0.874295
Hainan	1	1	1	1
Chongqing	0.846474228	0.871542	0.908744	0.943519
Sichuan	1	1	1	1
Guizhou	1	1	1	1
Yunnan	0.67342976	0.874346	0.94103	0.95877
Tibet	0.734372097	0.745466	0.739719	0.729462
Shaanxi	1	1	1	1
Gansu	0.609612531	0.611526	0.624109	0.633273
Qinghai	0.801390551	0.822907	0.839865	0.830865
Ningxia	0.876615927	0.869512	0.851912	0.864666
Xinjiang	1	1	1	1

Degree of energy support: This represents the policies in the aspect of energy, which is computed as the ratio of electricity usage in rural areas to the total electricity usage in each province and city (Supplementary Table S4).

Degree of water support: This represents the supportive capacity of water infrastructure to agricultural production and to some extent reflects the effort in building agricultural water facilities. It is calculated as the ratio of the area of irrigated arable land to the total area of arable land in each province and city in a given year (Supplementary Table S5).

Degree of informatization: This represents the level of informatization in rural areas. It is calculated as the ratio of the number of Internet access point in the rural areas of each province and city to the total number of Internet access point in that province and city in a given year (Supplementary Table S6).

Using the degree of financial support for agriculture, the degree of energy support, the degree of water support, the degree of informatization, the degree of agribusiness, and the percentage of rural population, a total of six variables as independent variables, and the overall technical efficiency as the dependent variable for Tobit regression analysis, it can be seen from the table above that the model formula is:

Comprehensive Efficiency = 1.179 + 0.993 × Degree of Financial Support for Agriculture—0.043 × Degree of Energy Support—0.157 × Degree of Water Infrastructure Construction + 0.111 × Degree of Informatization—1.044 × Degree of Agribusiness—0.665 × Percentage of Rural Population.

In this paper, SPSSAU software was used to construct the Tobit model, and the results of the likelihood ratio test are as follows.

As seen in Table 6, the likelihood ratio test result of this model is $p < 0.05$, indicating that the null hypothesis is rejected, meaning that the selected independent variables in this model are valid and the construction of the model is meaningful. The Akaike Information Criterion (AIC) is a standard proposed by the Japanese statistician Hirotugu Akaike in 1974 to measure the goodness of fit of statistical models [49]. The Bayesian Information Criterion (BIC) was proposed by Schwarz in 1978, similar to AIC, and is used to prevent overfitting caused by excessive model complexity during model selection [50]. The relatively small AIC and BIC values in the likelihood ratio test of this model indicate a good relative representativeness of the model.

Table 6. Results of the Tobit model likelihood ratio test for factors influencing the efficiency of China's agricultural circular economy.

Model	−2 Times the Log-Likelihood Value	Cardinality	df	*p*	AIC	BIC
Intercept distance	−152.395					
Final model	−194.017	41.622	6	0	−180.017	−160.275

The final results of the Tobit model are presented in Table 7.

Table 7. Summary of Tobit model analysis results.

	Regression Coefficient
Intercept distance	1.179 ** (16.174)
Degree of financial support for agriculture	0.993 * (2.109)
Degree of energy support	−0.043 (−0.427)
Degree of water support	−0.157 * (−2.162)
Degree of informatization	0.111 (0.952)
Degree of agribusiness	−1.044 ** (−2.860)
Percentage of rural population	−0.665 ** (−4.608)
log(Sigma)	−2.201 ** (−34.666)
Sample size	124
McFadden R^2	−0.273
Dependent variable: crste	

* $p < 0.05$, ** $p < 0.01$, z-values in parentheses.

4.3. Study on the Change Trend of Efficiency of Agricultural Circular Economy

The CCR model of the circular economy in agriculture across China's 31 provinces and cities, as previously calculated, is applicable only for efficiency comparison among these provinces and cities within the same period. CCR models across different periods are not directly comparable; for instance, efficiency values from 2017 cannot be compared directly to those from 2018. To study the changing trends in the efficiency of the circular economy in agriculture over different periods, the Malmquist index model must be employed. This paper continues to use the indicators and data applied in the construction of the CCR model for the circular economy in agriculture across China's 31 provinces and cities. The DEARUN software was utilized to construct CCR–Malmquist adjacent reference models for three periods—2017–2018, 2018–2019, and 2019–2020—aiming to investigate the changing trends in the efficiency of the circular economy in agriculture across China's 31 provinces and cities under the rural revitalization strategy.

The elements in the Malmquist index model include: "Effch", which represents the change in technical efficiency; "Techch", the change in technological progress; "Pech", the change in pure technical efficiency; "Sech", the change in scale efficiency; and "Tfpch", the change in total factor productivity. A value greater than 1 in any of these indicators implies an improvement compared to the previous period. As can be seen from Table 8, all "Tfpch" values in the CCR–Malmquist index model across three periods from 2017 to 2020 exceed 1, indicating that the total factor productivity of the circular economy in agriculture across China's 31 provinces and cities continuously improved during this period.

Table 8. Summary of CCR–Malmquist adjacent reference index model of agricultural circular economy efficiency in 31 provinces and cities of China, 2017–2020.

Period	DMU	Effch	Techch	Pech	Sech	Tfpch
2017–2018	Beijing	1	1.081556	1	1	1.081556
2017–2018	Tianjin	1	1.075355	1	1	1.075355
2017–2018	Hebei	1.081744	1.074394	1.117542	0.967967	1.162219
2017–2018	Shanxi	0.992801	1.051735	0.997933	0.994857	1.044163
2017–2018	Inner Mongolia	1.018594	1.08879	1.012749	1.005772	1.109036
2017–2018	Liaoning	1.023702	1.04966	1.020373	1.003262	1.074539
2017–2018	Jilin	1.016423	1.062699	0.996948	1.019534	1.080151
2017–2018	Heilongjiang	1	1.059278	1	1	1.059278
2017–2018	Shanghai	1	1.089818	1	1	1.089818
2017–2018	Jiangsu	1	1.031345	1	1	1.031345
2017–2018	Zhejiang	1	1.065401	1	1	1.065401
2017–2018	Anhui	0.997167	1.029834	0.993751	1.003438	1.026917
2017–2018	Fujian	1	1.073229	1	1	1.073229
2017–2018	Jiangxi	0.995182	1.063731	0.99266	1.002541	1.058606
2017–2018	Shandong	1.013658	1.04469	1	1.013658	1.058959
2017–2018	Henan	1.012738	1.037978	1.001169	1.011555	1.051199
2017–2018	Hubei	1	1.028943	1	1	1.028943
2017–2018	Hunan	1.00012	1.039655	0.99289	1.007282	1.039779
2017–2018	Guangdong	1	1.017633	1	1	1.017633
2017–2018	Guangxi	1.052359	1.008734	1.035134	1.01664	1.06155
2017–2018	Hainan	1	1.033159	1	1	1.033159
2017–2018	Chongqing	1.023135	1.056493	0.993506	1.029823	1.080935
2017–2018	Sichuan	1	1.039342	1	1	1.039342
2017–2018	Guizhou	1	1.102977	1	1	1.102977
2017–2018	Yunnan	1.292435	1.053826	1.273906	1.014545	1.362002
2017–2018	Tibet	0.992139	1.048887	1	0.992139	1.040642
2017–2018	Shaanxi	1	1.063093	1	1	1.063093
2017–2018	Gansu	0.999462	1.073364	0.987078	1.012546	1.072787
2017–2018	Qinghai	1.002575	1.052801	1	1.002575	1.055512
2017–2018	Ningxia	0.978936	1.087319	1	0.978936	1.064416
2017–2018	Xinjiang	1	1.122112	1	1	1.122112
2018–2019	Beijing	1	1.082752	1	1	1.082752
2018–2019	Tianjin	1	1.055464	1	1	1.055464
2018–2019	Hebei	0.989706	1.103949	0.988895	1.00082	1.092585
2018–2019	Shanxi	0.998536	1.098215	1.008736	0.989889	1.096608
2018–2019	Inner Mongolia	0.996003	1.097631	0.997685	0.998314	1.093244
2018–2019	Liaoning	0.999686	1.09312	1	0.999686	1.092777
2018–2019	Jilin	1.004196	1.108883	1.003234	1.000959	1.113536
2018–2019	Heilongjiang	1	1.109718	1	1	1.109718
2018–2019	Shanghai	1	1.092227	1	1	1.092227
2018–2019	Jiangsu	1	1.069138	1	1	1.069138
2018–2019	Zhejiang	1	1.102197	1	1	1.102197
2018–2019	Anhui	1.012681	1.097253	1.022323	0.990568	1.111167
2018–2019	Fujian	1	1.100032	1	1	1.100032
2018–2019	Jiangxi	1.025257	1.094478	1.030817	0.994606	1.122122

Table 8. *Cont.*

Period	DMU	Effch	Techch	Pech	Sech	Tfpch
2018–2019	Shandong	0.967644	1.096541	1	0.967644	1.061061
2018–2019	Henan	1.027851	1.090147	1.045754	0.982881	1.120509
2018–2019	Hubei	1	1.090092	1	1	1.090092
2018–2019	Hunan	1.078462	1.0899	1.080877	0.997765	1.175415
2018–2019	Guangdong	1	1.12946	1	1	1.12946
2018–2019	Guangxi	0.990902	1.110025	1.00564	0.985345	1.099926
2018–2019	Hainan	1	1.103107	1	1	1.103107
2018–2019	Chongqing	1.039285	1.083151	1.029009	1.009986	1.125702
2018–2019	Sichuan	1	1.090826	1	1	1.090826
2018–2019	Guizhou	1	1.104277	1	1	1.104277
2018–2019	Yunnan	1.078133	1.07993	1.07888	0.999308	1.164309
2018–2019	Tibet	0.972491	1.070905	1	0.972491	1.041445
2018–2019	Shaanxi	1	1.084852	1	1	1.084852
2018–2019	Gansu	1.015698	1.092255	1.002747	1.012915	1.1094
2018–2019	Qinghai	0.993838	1.075644	1	0.993838	1.069016
2018–2019	Ningxia	0.9649	1.067222	1	0.9649	1.029763
2018–2019	Xinjiang	1	1.064995	1	1	1.064995
2019–2020	Beijing	1	1.03478	1	1	1.03478
2019–2020	Tianjin	1	1.052888	1	1	1.052888
2019–2020	Hebei	1.023046	1.106028	0.996327	1.026818	1.131518
2019–2020	Shanxi	1.028521	1.103685	1.02954	0.99901	1.135163
2019–2020	Inner Mongolia	1.024548	1.114029	1.008589	1.015822	1.141375
2019–2020	Liaoning	0.990719	1.078225	1	0.990719	1.068218
2019–2020	Jilin	1.098236	1.077455	1.097502	1.000669	1.183301
2019–2020	Heilongjiang	1	1.105457	1	1	1.105457
2019–2020	Shanghai	1	1.044847	1	1	1.044847
2019–2020	Jiangsu	1	1.077465	1	1	1.077465
2019–2020	Zhejiang	1	1.065462	1	1	1.065462
2019–2020	Anhui	1.001982	1.102422	0.995283	1.006731	1.104607
2019–2020	Fujian	1	1.061432	1	1	1.061432
2019–2020	Jiangxi	0.994623	1.115323	0.986843	1.007884	1.109325
2019–2020	Shandong	0.99169	1.097124	1	0.99169	1.088007
2019–2020	Henan	1.057024	1.115072	1.191705	0.886985	1.178658
2019–2020	Hubei	1	1.11242	1	1	1.11242
2019–2020	Hunan	1.027548	1.127912	1.019346	1.008046	1.158983
2019–2020	Guangdong	1	1.087595	1	1	1.087595
2019–2020	Guangxi	0.957942	1.112637	0.952254	1.005974	1.065842
2019–2020	Hainan	1	1.066561	1	1	1.066561
2019–2020	Chongqing	1.038458	1.102373	1.031424	1.00682	1.144768
2019–2020	Sichuan	1	1.15826	1	1	1.15826
2019–2020	Guizhou	1	1.124415	1	1	1.124415
2019–2020	Yunnan	1.014632	1.133587	1.016396	0.998265	1.150174
2019–2020	Tibet	0.975249	1.076658	1	0.975249	1.05001
2019–2020	Shaanxi	1	1.127938	1	1	1.127938
2019–2020	Gansu	1.006549	1.098318	0.990637	1.016062	1.105511
2019–2020	Qinghai	0.971136	1.088518	1	0.971136	1.057099
2019–2020	Ningxia	1.004343	1.089843	1	1.004343	1.094576
2019–2020	Xinjiang	1	1.119021	1	1	1.119021

5. Discussion

5.1. Agri-Circular Economy Efficiency Is Significantly Affected by China's Rural Revitalization Strategy

The Tobit model of Table 7 is plotted as a forest diagram in Figure 2.

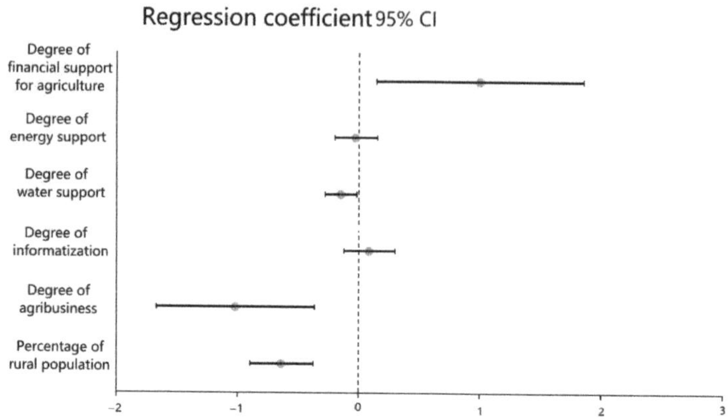

Figure 2. Tobit model regression coefficient 95% CI forest plot.

The regression coefficient of degree of financial support for agriculture is 0.993, showing significance at the 0.05 level (z = 2.109, p = 0.035 < 0.05), indicating that degree of financial support for agriculture has a significant positive effect on the efficiency of the circular economy in agriculture. Jiao and Liu [51] confirmed a significant positive impact of fiscal expenditure on agricultural production efficiency in northeastern China through analysis of panel data from 1971 to 2007. Chen et al. [52] also showed that fiscal expenditure significantly positively affected the efficiency of Henan's agricultural circular economy, following research on the province's panel data from 2013 to 2019. Zhou et al. [53], in their study using the DEA method, suggested that the government should not only strengthen fund management but also expand the scale of fiscal support for agriculture. Wei et al. analyzed panel data from 30 provinces and cities from 2003 to 2011, and similarly concluded that fiscal expenditure supporting agricultural production and assisting agriculture had a significant positive effect on agricultural modernization [54].

The regression coefficient for the degree of energy support is −0.043, but it does not show significance (z = −0.427, p = 0.669 > 0.05), indicating that the degree of energy support does not impact the technical efficiency.

The regression coefficient for the degree of water support is −0.157, showing significance at the 0.05 level (z = −2.162, p = 0.031 < 0.05), indicating that the degree of water support has a significant negative effect on the technical efficiency. The rural revitalization strategy's policy on water support is beneficial to the development of the agricultural circular economy in the long run. However, in the short term, a large amount of investment in water construction can directly crowd out some input resources. Therefore, the construction of agricultural water infrastructure is necessary, but its impact on agricultural economic efficiency is not necessarily positive. Yan et al. [55] showed that rural water resources in China face problems such as weak rural water infrastructure, uneven spatial and temporal distribution of rural water resources, and low investment efficiency, with the investment efficiency of China's rural water supply decreasing by an average of 1.2% from 2011 to 2015. When Wang et al. [56] evaluated China's agricultural water projects, they found that the benefits of the water construction investment scale in the eastern provinces were decreasing, while those in the western provinces were increasing. Lei et al. found that the supply efficiency of the national agricultural water facilities showed an overall declining trend, following analysis of panel data from 27 provinces and cities in China from 2009 to 2018 [57].

The regression coefficient for the degree of informatization is 0.111, but it does not show significance (z = 0.952, p = 0.341 > 0.05), indicating that the degree of informatization does not impact the technical efficiency.

The regression coefficient for the degree of agribusiness is -1.044, presenting significance at the 0.01 level (z = −2.860, p = 0.004 < 0.01), suggesting that the degree of agribusiness has a significant negative impact on technical efficiency. The variable of corporatization degree is used to examine whether the mode of agricultural production is shifting towards a more efficient, large-scale corporate model. It is generally believed that fewer and larger agricultural enterprises can improve efficiency. The independent variable reflecting the degree of agricultural corporatization in this Tobit model, which is the ratio of agricultural legal persons to the total number of legal persons, reasonably has a negative impact on the overall technical efficiency of the agricultural economy. Meena et al. found that the cost of transition from family-based to corporatized agriculture in India was higher [58]. Motes et al. argued that modern agriculture has shown a reverse Malthusian phenomenon of the land margin, with a continuous increase in food output, but this was due to the low production efficiency in these areas in the past [59]. Studies by Bojnec et al. on the overall technical efficiency of agriculture in Central and Eastern Europe also found that scaling up improved efficiency [60]. Wang et al. empirically demonstrated that the larger scale of production was key to enhancing productivity in China's scaled agriculture [61]. In fact, expanding the scale of production on limited agricultural resources, such as arable land, often implies a reduction in the number of agricultural enterprises. Da-You et al. posited that the presence of leading enterprises was of significant importance to the process of agricultural industrialization in a region [62]. The Chinese government has repeatedly expressed its intention to support leading agricultural enterprises and encouraged small and medium-sized enterprises to merge into larger ones to enhance production efficiency.

The regression coefficient for the percentage of rural population is −0.665, showing significance at the 0.01 level ($z = -4.608$, $p = 0.000 < 0.01$), indicating that the proportion of the rural population has a significant negative impact on technical efficiency. An increase in the population would increase agricultural production input. To improve the efficiency of the agricultural circular economy, it is essential to enhance the quality of talents and release more human resources to society. As early as 1798, the Malthusian model proposed the negative relationship between population size and agricultural resources [15]. Kögel and Prskawetz argued that improving agricultural productivity can escape the Malthusian trap, but it requires institutional guidance to reduce fertility rates [63]. Bilsborrow believed that one of the key factors to improving agricultural productivity is the decline in the population growth rate [64].

In summary, the degree of financial support for agriculture has a significant positive impact on technical efficiency, while the degree of water support, the degree of agribusiness, and the percentage of rural population have a significant negative impact. However, the degree of energy support and the degree of informatization do not impact technical efficiency.

5.2. Technological Advances Promote the Efficiency of China's Agricultural Circular Economy Year by Year

Both the DEA model and Tobit model indicate that the policies related to the rural revitalization strategy significantly affect the efficiency of the agricultural circular economy. What is the trend in the efficiency of the agricultural circular economy under the influence of these policies? The dependent variable in the Tobit regression model comes from the CCR model of agricultural circular economy efficiency, reflecting the relative situation of the agricultural circular economy efficiency of 31 provinces in the current year, and cannot be directly compared between different years. The CCR–Malmquist index model of China's agricultural circular economy measured in this paper from 2017 to 2020 can directly reflect the change in efficiency. After analyzing statistics on the data in Table 8 to form Table 9, the total factor productivity change rates (Tfpch) for all three periods of 2017–2018, 2018–2019, and 2019–2020 for 31 provinces and cities were found to all be greater than 1, indicating that the efficiency of the agricultural circular economy in these 31 provinces and cities has improved during this period.

Table 9. 2017–2020 China's 31 provinces and cities agricultural circular economy CCR–Malmquist index model statistics.

	2017–2018	2018–2019	2019–2020
Number of effch < 1	6	9	6
Number of techch < 1	0	0	0
Number of pech < 1	7	2	5
Number of sech < 1	4	13	7
Number of Tfpch > 1	31	31	31

From Table 9, it is clear that the growth of the total factor productivity index (Tfpch) in the agricultural circular economy primarily results from the technological progress change index (Techch) for all three periods in all 31 provinces and cities being greater than 1. This suggests that the main driving force of growth stems from technological progress. However, some provinces and cities still have room for improvement in terms of the technical efficiency change index (Effch).

The same evidence can be found in the analysis based on the CCR model. Table 10 presents a comprehensive statistical breakdown of the efficiency and its decomposition of the agricultural circular economy in 31 provinces and cities from 2017 to 2020. Notably, the number of provinces and cities achieving a technological efficiency of 1 significantly surpasses those achieving a scale efficiency of 1, with a steady upward trend year by year. This conclusively demonstrates that the advancement in the efficiency of the agricultural circular economy over these years can be attributed to technological upgrades and optimization.

Table 10. 2017–2020 China's 31 provinces and cities agricultural circular economy efficiency and decomposition of the results of statistics.

	2017	2018	2019	2020
Number of crste's value of 1	14	15	14	14
Mean value of crste	0.882432	0.895606	0.899268	0.903665
Number of vrste's value of 1	18	19	19	21
Mean value of vrste	0.917676	0.9271	0.934294	0.942052
Number of scale's value of 1	14	15	14	14
Mean value of scale	0.960847	0.965155	0.961535	0.959234
Number of CRS	14	16	14	14
Number of IRS	9	8	8	7
Number of DRS	8	7	9	10

5.3. Reasonable Policies Support the Efficiency of Agricultural Circular Economy

In conjunction with further analysis using the Tobit regression model, we believe that a significant factor contributing to the enhancement of the agricultural circular economy's efficiency is appropriate government fiscal support. The Tobit model indicates that the positive impact of the degree of financial support for agriculture on the overall technical efficiency of the agricultural circular economy is at the 5% level. The expenditure on agriculture, forestry, and water in all 31 provinces and cities has been increasing year by year, and its proportion in the general public budget expenditure at the provincial and municipal levels has also been steadily rising, as shown in Figures 3 and 4.

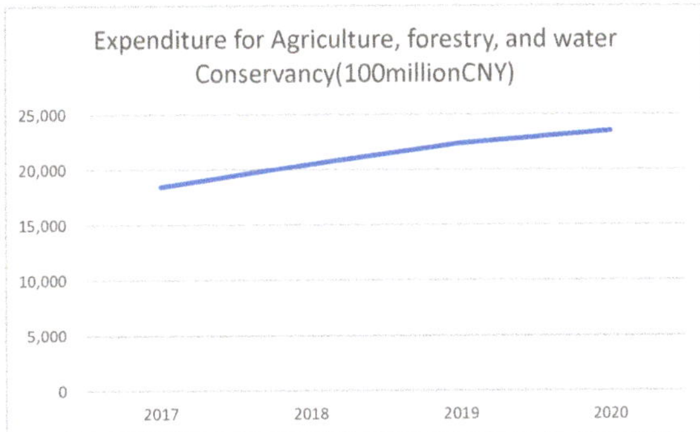

Figure 3. Trends of expenditure for agriculture, forestry, and water conservancy in 31 provinces and cities in China, 2017–2020.

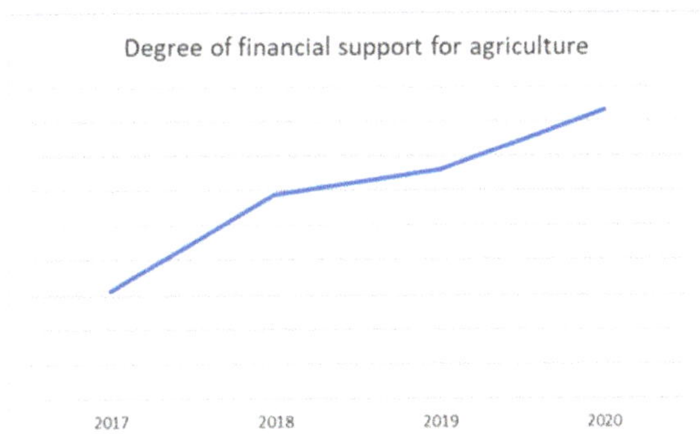

Figure 4. Trend of the degree of financial support for agriculture in 31 provinces and cities of China from 2017 to 2022.

The degree of water support from 2017 to 2020 is shown in Figure 5. During this period, the degree of water support was steady with a slight increase, not blindly pursuing scale, and did not excessively crowd out resources, affecting the efficiency of the agricultural circular economy.

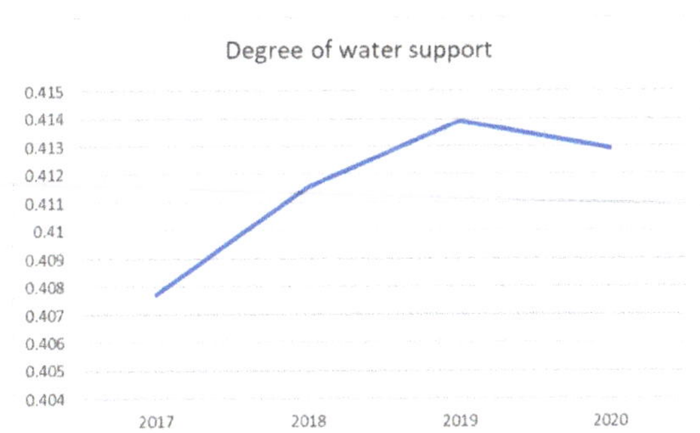

Figure 5. Development trend of the degree of water support in 31 provinces and cities of China from 2017 to 2020.

The degree of agribusiness has a significant negative impact on the efficiency of the agricultural circular economy, and the development trend of socialization degree from 2017 to 2020, as shown in Figure 6. Compared with the number of corporate legal persons in all industries in China, the proportion of agricultural enterprises is decreasing year by year. On the one hand, the Chinese government has made it clear that it wishes to promote agricultural modernization, and on the other hand, the increase in the number of agricultural enterprises is limited. These two are not contradictory. Instead, they indicate that the government's policy is more committed to the scaling up and technological upgrading of agricultural enterprises, rather than simply pursuing an increase in quantity. This has promoted the growth of the efficiency of the agricultural circular economy. In 2021, the Ministry of Agriculture and Rural Affairs specifically issued the "Opinions of the Ministry of Agriculture and Rural Affairs on Promoting the Growth and Strengthening of Leading Enterprises in Agricultural Industrialization", which also confirms our research results.

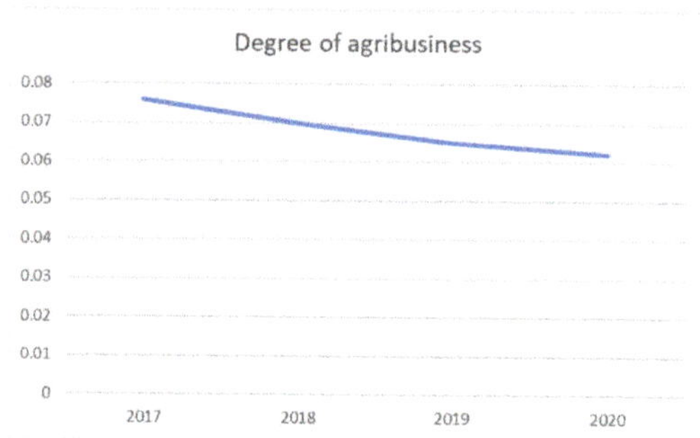

Figure 6. Development trend of the degree of agribusiness in 31 provinces and cities in China from 2017 to 2020.

The percentage of rural population also has a significantly negative impact on the efficiency of the agricultural circular economy. The changing trend of the percentage of rural population in China from 2017 to 2020 is as shown in Figure 7. The decreasing trend from 2017 to 2020 supports the improvement of the efficiency of the agricultural circular economy, indicating that the related policies of the rural revitalization strategy are more focused on improving the quality of agricultural talents to release more labor and improve efficiency.

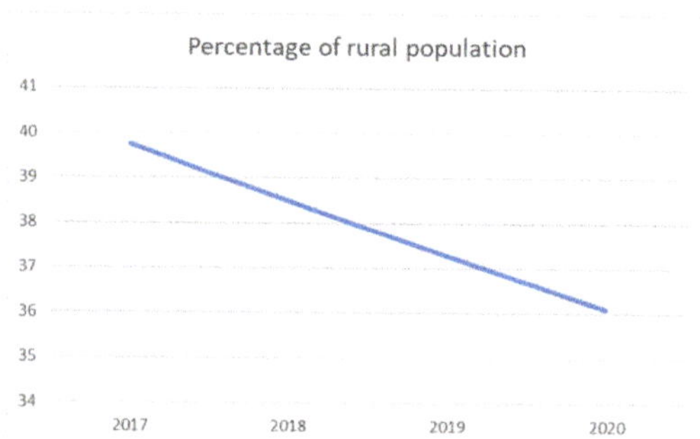

Figure 7. Development trend of percentage of rural population in 31 provinces and cities in China from 2017 to 2020.

In summary, the formulation and implementation of various policies under the rural revitalization strategy, considering the incompatibility of agricultural modernization and rural modernization, are quite rational. Combined with the analysis results of the DEA–Malmquist model, it can be seen that the annual increase in financial support does not blindly pursue the expansion of investment scale, but is mainly used for upgrades in agricultural technology, management level, rationality of asset structure, etc. Other variables with a negative impact are stable or declining, creating a favorable foundation for the improvement of the efficiency of the agricultural circular economy.

5.4. There Are Significant Differences between the Efficiency of Agricultural Circular Economy in 31 Provinces and Cities in China

As shown in Figure 8, the distribution of the overall technical efficiency of the agricultural circular economy in the 31 provinces and cities has been very stable over the past four years, with the number of provinces having an overall technical efficiency of 1 (i.e., DEA efficient) ranging between 14 and 15. Among these, 13 provinces and cities, including Beijing, Tianjin, Heilongjiang, Shanghai, Zhejiang, Fujian, Hubei, Guangdong, Hainan, Sichuan, Guizhou, Shaanxi, and Xinjiang, have maintained an overall technical efficiency of 1 (i.e., DEA efficient) for four consecutive years. This suggests that these provinces and cities have significantly higher levels of agricultural production technology, management level, resource utilization rate, etc., compared to other provinces and cities.

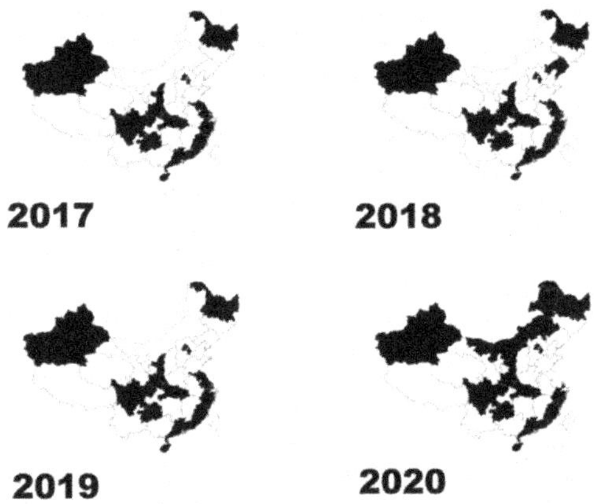

Figure 8. 2017–2020 Distribution of provinces and cities in 31 Chinese provinces and municipalities where the overall technical efficiency of agricultural circular economy reaches DEA effectiveness.

Additionally, due to geographical and climatic influences, agricultural production varies substantially across different regions. However, the 13 provinces and cities that have achieved DEA effectiveness, distributed across seven regions including South China, Central China, North China, East China, Northeast, Southwest, and Northwest China, somewhat indicate that the heterogeneity-induced errors among various DMUs in the DEA model are not significant. The model is thus deemed highly reliable, showing minimal influence from regional and climatic differences. Simultaneously, in the economically advanced eastern and southeastern coastal regions, the efficiency of the agricultural circular economy is generally higher.

5.5. There Is Room to Improve the Scale Efficiency of Agricultural Circular Economy

As shown in Figure 9, the distribution of scale efficiency in the agricultural circular economy across the 31 provinces and cities from 2017 to 2020 closely aligns with the overall efficiency distribution. The 13 provinces and cities of Beijing, Tianjin, Heilongjiang, Shanghai, Zhejiang, Fujian, Hubei, Guangdong, Hainan, Sichuan, Guizhou, Shaanxi, and Xinjiang have consistently achieved a scale efficiency of 1, indicating DEA effectiveness, for four consecutive years.

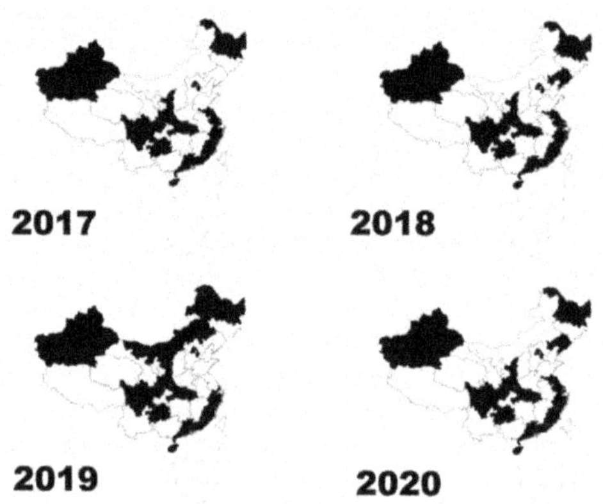

Figure 9. 2017–2020 Distribution of provinces in 31 Chinese provinces where the scale efficiency of agricultural circular economy reaches DEA effectiveness.

From 2017 to 2020, technical efficiency reached 1 in 18–21 of the 31 provinces and cities. The distribution, as illustrated in Figure 10, covers various regions in China, similar to the distribution of overall efficiency. This indicates that the primary driving force for the improvement in overall technical efficiency in China's agricultural circular economy comes from the enhancement of technical efficiency.

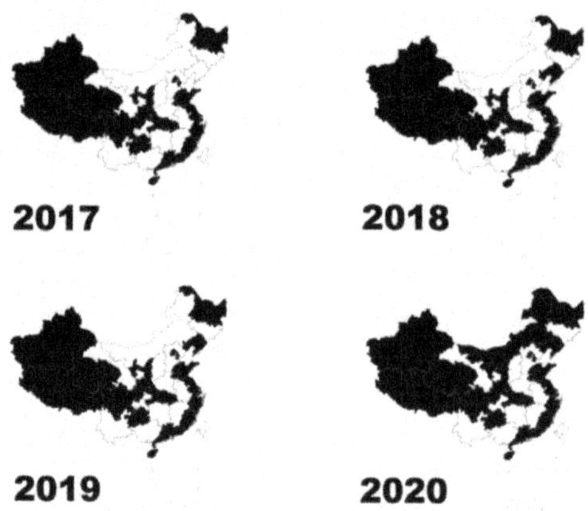

Figure 10. 2017–2020 Distribution of provinces in 31 Chinese provinces where the pure technical efficiency of agricultural circular economy reaches DEA effectiveness.

In conclusion, there exists substantial room for improvement in the scale efficiency of the agricultural circular economy. Enhancing scale efficiency should be a key focus of future policy considerations.

6. Conclusions, Recommendations, and Shortcomings

6.1. Conclusions and Recommendations

1. From the results presented in Section 4.3, it can be observed that the overall trend of China's agricultural circular economy efficiency has been increasing year by year around the implementation of the rural revitalization strategy in 2018. This indicates that the relevant policies are rational and can ensure the simultaneous realization of agricultural modernization and rural modernization. The future focus should be on implementing

various policies for rural revitalization and actively researching how to transform policy investments into productivity.

2. Sections 4.2 and 5.1 explicitly indicate that agricultural fiscal support has a significant positive impact on the efficiency of China's agricultural circular economy. Future policies should aim to maintain the growth of agricultural fiscal expenditure.

3. Drawing on Sections 4.1 and 5.5, the primary factor hindering the efficiency of the agricultural circular economy is inadequate scale efficiency. Provinces lagging in scale efficiency should adjust their input scales according to their specific circumstances, improve organizational management levels, resource utilization, etc., which can achieve an overall efficiency improvement at a relatively small cost.

4. From Sections 4.1 and 5.5, it can also be inferred that the level of agricultural technology and management are robust safeguards for the enhancement of agricultural circular economy efficiency, and are important links in the rural revitalization strategy. Therefore, investment related to agricultural technology should be further strengthened.

5. Section 5.5's graphical representation shows that provinces with higher comprehensive efficiency in the agricultural circular economy are highly stable. This suggests that these provinces have significant advantages in areas such as agricultural technological advancement and upgrade, as well as management level, offering lessons for other provinces.

6.2. Innovation Point

This paper presents the following innovative contributions:

1. The research exploring the correlation between the rural revitalization strategy and the efficiency of agricultural circular economy is a novel perspective.
2. The approach of extracting independent variables related to policy from the rural revitalization strategy represents an innovative method.
3. While most previous studies on the level of agricultural economy have focused on specific regions, investigating economic differences between these regions, the novelty of this paper lies in its national scope. It explores development trends and influential factors at the national level.

6.3. Shortcomings

1. There is considerable heterogeneity in agricultural production across different provinces in China. Some provinces have one harvest per year (such as in the Northeast), while others have three (such as Hainan). Some are predominantly involved in animal husbandry, while others focus mainly on crop farming. Climate and water resource variations also exist. In future research, there is a plan to eliminate the impact of this heterogeneity, with a preliminary idea of establishing an intermediate model to mitigate these differences. However, practical research has shown that even with the existence of heterogeneous factors, the final results are still relatively evenly distributed, which indicates that the validity of the model is assured, reflecting a macroscopic view of the agricultural circular economy.

2. Due to limitations in data acquisition, the model of agricultural circular economy efficiency still has some shortcomings. For instance, variables related to the environment, such as the emissions of waste, have not yet been introduced into the model.

3. There is a need for further refinement in the research. Both the Rural Revitalization Strategy and the agricultural circular economy are complex systems. When performing coupled analysis, it is necessary to further improve the granularity of the research. For example, a more in-depth analysis of the differentiated causes for the 31 provinces' agriculture could be obtained through super-efficiency DEA models and slack analysis. The main objective of this study was to conduct a macroscopic analysis of the overall efficiency of China's agricultural circular economy, and this part was not included.

4. Some influencing factors have not yet been included in the correlation analysis due to incomplete data acquisition. For instance, the impact of the Rural Revitalization Strategy's efforts on healthcare, education, transportation, etc., on the agricultural circular economy has not been addressed in this research.

Supplementary Materials: The following supporting information can be downloaded at: https://www.mdpi.com/article/10.3390/agriculture13071454/s1, Table S1: 2017–2020 China's 31 provinces and cities degree of financial support for agriculture Calculation result statistics; Table S2: 2017–2020 China's 31 provinces and cities degree of agribusiness Calculation result statistics; Table S3: 2017–2020 China's 31 provinces and cities percentage of rural population calculation result statistics; Table S4: 2017–2020 China's 31 provinces and cities degree of energy support calculation result statistics; Table S5: 2017–2020 China's 31 provinces and cities degree of water support calculation result statistics; Table S6: 2017–2020 China's 31 provinces and cities degree of informatization calculation result statistics.

Author Contributions: Conceptualization, C.G.; methodology, R.Z.; software, R.Z.; validation, C.G.; formal analysis, C.G. and R.Z.; investigation, C.G., Y.Z. and R.Z.; resources, Y.Z.; data curation, C.G., Y.Z. and R.Z.; writing—original draft preparation, C.G.; writing—review and editing, Y.Z. and R.Z. All authors have read and agreed to the published version of the manuscript.

Funding: This research received no external funding.

Institutional Review Board Statement: Not applicable.

Data Availability Statement: The data used in this paper are from China National Bureau of Statistics, China Statistical Yearbook and China Ministry of Agriculture and Rural Affairs, China Rural Statistical Yearbook.

Conflicts of Interest: The authors declare no conflict of interest.

References

1. The State Council of the People's Republic of China. The Eleventh Five-Year Plan for National Economic and Social Development of the People's Republic of China. 2006. Available online: https://www.gov.cn/gongbao/content/2006/content_268766.htm (accessed on 15 July 2023).
2. Wu, H.Q.; Shi, Y.; Xia, Q.; Zhu, W.D. Recycling. Effectiveness of the policy of circular economy in China: A DEA-based analysis for the period of 11th five-year-plan. *Resour. Conserv. Recycl.* **2014**, *83*, 163–175. [CrossRef]
3. Xin, Y.; Zhou, X.J. Development benefit evaluation of agricultural circular economy in China. *J. South. Agric.* **2013**, *44*, 1220–1224.
4. Ul Haq, S.; Boz, I.; Shahbaz, P.; Yıldırım, Ç.J.E.S.; Research, P. Evaluating eco-efficiency and optimal levels of fertilizer use based on the social cost and social benefits in tea production. *Environ. Sci. Pollut. Res.* **2020**, *27*, 33008–33019. [CrossRef]
5. Fan, Y.; Fang, C. Circular economy development in China-current situation, evaluation and policy implications. *Environ. Impact Assess. Rev.* **2020**, *84*, 106441. [CrossRef]
6. Shahbaz, P.; Haq, S.u.; Boz, I. Linking climate change adaptation practices with farm technical efficiency and fertilizer use: A study of wheat–maize mix cropping zone of Punjab province, Pakistan. *Environ. Sci. Pollut. Res.* **2022**, *29*, 16925–16938. [CrossRef]
7. The State Council of the People's Republic of China. The Central Committee of the Communist Party of China and the State Council Opinions on Implementing the Rural Revitalization Strategy. 2018. Available online: https://www.gov.cn/gongbao/content/2018/content_5266232.htm (accessed on 15 July 2023).
8. The National People's Congress of the People's Republic of China. The Budget Working Committee of the Standing Committee of the National People's Congress, the National Committee on Finance and Economy of the National People's Congress, the National People's Congress Committee on Agriculture and Rural Affairs on the Fiscal Agricultural and Rural Funds Research Report on the Distribution and Utilization of Financial Agricultural and Rural Funds. 2021. Available online: http://www.npc.gov.cn/npc/c5871/202206/b157d2e15e0a4000ae33874d009d947f.shtml (accessed on 15 July 2023).
9. People's Daily Online. Subsidies of 175 Billion Yuan of Financial Strength to Help Promote Rural Revitalization across the Board. 2023. Available online: http://finance.people.com.cn/n1/2023/0408/c1004-32659855.html (accessed on 15 July 2023).
10. Ministry of Transport of the People's Republic of China. 2023. Available online: https://www.gov.cn/zhengce/2023-04/23/content_5752770.htm (accessed on 15 July 2023).
11. The State Council of the People's Republic of China. China to Improve Rural Infrastructure Amid Rural Revitalization Drive. 2023. Available online: http://english.www.gov.cn/premier/news/2018/03/05/content_281476067447206.htm (accessed on 15 July 2023).
12. The State Council of the People's Republic of China. Guiding Opinions of the State Council on Accelerating the Transformation and Upgrading of Mechanised Agriculture and Agricultural Machinery Equipment Industry. 2019. Available online: https://www.gov.cn/gongbao/content/2019/content_5355467.htm (accessed on 7 July 2023).
13. The State Council of the People's Republic of China. The State Council's Decision on Issuing the "14th Five Year Plan" Promotion Notice on Agricultural and Rural Modernization Planning. Available online: https://www.gov.cn/gongbao/content/2022/content_5675948.htm (accessed on 15 July 2023).
14. People's Daily Online. Expand the Scope of Government Purchase Services and Improve the Level of Management Science Standardization. 2023. Available online: https://www.gov.cn/zhengce/202306/content_6885168.htm (accessed on 15 July 2023).

15. Malthus, T. An Essay on the Principle of Population. 1798. Available online: https://la.utexas.edu/users/hcleaver/368/368MalthusPopCh10table.pdf (accessed on 15 July 2023).
16. Scott, A.J. Industrialization and urbanization: A geographical agenda. *Ann. Assoc. Am. Geogr.* **1986**, *76*, 25–37. [CrossRef]
17. Hubendick, B. Industrialization and Urbanization. In *The Global Environment: Science, Technology and Management*; Wiley-VCH: Hoboken, NJ, USA, 1997.
18. Yang, D.T.; Zhu, X. Modernization of agriculture and long-term growth. *J. Monet. Econ.* **2013**, *60*, 367–382. [CrossRef]
19. Hardeman, E.; Jochemsen, H. Are there ideological aspects to the modernization of agriculture? *J. Agric. Environ. Ethic* **2012**, *25*, 657–674. [CrossRef]
20. Knickel, K.; Ashkenazy, A.; Chebach, T.C.; Parrot, N. Agricultural modernization and sustainable agriculture: Contradictions and complementarities. *Int. J. Agric. Sustain.* **2017**, *15*, 575–592. [CrossRef]
21. Jiang, H.M.; Wang, Z.H. Empirical Analysis on the Relationship Among Industrialization, Urbanization and Agricultural Modernization in Jilin Province. *Sci. Geogr. Sin.* **2012**, 32, 591–595.
22. Yao, S.; Liu, Z. Determinants of Grain Production and Technical Efficiency in China. *J. Agric. Econ.* **2010**, *49*, 171–184. [CrossRef]
23. de Jorge, J.; Suárez, C. Influence of R&D subsidies on efficiency: The case of Spanish manufacturing firms. *Cuad. Econ. Dir. Empresa* **2011**, *14*, 185–193.
24. Yao, L.X.; Leng, N.M. An Analysis on the Incentive Effects of Fiscal Subsidies and Tax Incentives on the Innovation Efficiency of Strategic Emerging Industries. *East China Econ. Manag.* **2018**, 32, 7. [CrossRef]
25. Gao, Y.; Wen, T.; Wen, Y.; Wang, X. A spatial econometric study on effects of fiscal and financial supports for agriculture in China. *Agric. Econ.* **2013**, *59*, 315–332.
26. Kumbhakar, S.C.; Lien, G. Impact of subsidies on farm productivity and efficiency. In *The Economic Impact of Public Support to Agriculture. Studies in Productivity and Efficiency*; Springer: New York, NY, USA, 2010; pp. 109–124.
27. Guan, Z.; Lansink, A.O. The source of productivity growth in Dutch agriculture: A perspective from finance. *Am. J. Agric. Econ.* **2006**, *88*, 644–656.
28. Charnes, A.; Cooper, W.W.; Rhodes, E. Measuring the efficiency of decision making units. *Eur. J. Oper. Res.* **1978**, 2, 429–444. [CrossRef]
29. Banker, R.D.; Charnes, A.; Cooper, W.W. Some models for estimating technical and scale inefficiencies in data envelopment analysis. *Manag. Sci.* **1984**, *30*, 1078–1092. [CrossRef]
30. Zhao, X.; Ma, X.; Shang, Y.; Yang, Z.; Shahzad, U. Green economic growth and its inherent driving factors in Chinese cities: Based on the Metafrontier-global-SBM super-efficiency DEA model. *Gondwana Res.* **2022**, *106*, 315–328. [CrossRef]
31. Streimikis, J.; Saraji, M.K. Green productivity and undesirable outputs in agriculture: A systematic review of DEA approach and policy recommendations. *Econ. Res.-Ekon. Istraživanja* **2022**, 35, 819–853. [CrossRef]
32. Andersen, P.; Petersen, N.C. A procedure for ranking efficient units in data envelopment analysis. *Manag. Sci.* **1993**, *39*, 1261–1264. [CrossRef]
33. Zhao, X.; Mahendru, M.; Ma, X.; Rao, A.; Shang, Y. Impacts of environmental regulations on green economic growth in China: New guidelines regarding renewable energy and energy efficiency. *Renew. Energy* **2022**, *187*, 728–742. [CrossRef]
34. Li, S.; Jahanshahloo, G.R.; Khodabakhshi, M. A super-efficiency model for ranking efficient units in data envelopment analysis. *Appl. Math. Comput.* **2007**, *184*, 638–648. [CrossRef]
35. Zhang, Q.; Chen, W.; Ling, W. Tech-eco efficiency evaluation of hydrogen production industry under carbon dioxide emissions regulation in China. *Int. J. Hydrogen Energy* **2022**, *47*, 41183–41194. [CrossRef]
36. Malmquist, S. Index numbers and indifference surfaces. *Trab. Estad.* **1953**, *4*, 209–242. [CrossRef]
37. Caves, D.W.; Christensen, L.R.; Diewert, W.E. The economic theory of index numbers and the measurement of input, output, and productivity. *Econometrica* **1982**, *50*, 1393–1414. [CrossRef]
38. Färe, R.; Grosskopf, S.; Lindgren, B.; Roos, P. Productivity changes in Swedish pharamacies 1980–1989: A non-parametric Malmquist approach. *J. Prod. Anal.* **1992**, 3, 85–101. [CrossRef]
39. Grifell-Tatjé, E.; Lovell, C. A note on the Malmquist productivity index. *Econ. Lett.* **1995**, *47*, 169–175. [CrossRef]
40. Pastor, J.T.; Lovell, C.A.K. A global Malmquist productivity index. *Econ. Lett.* **2005**, *88*, 266–271. [CrossRef]
41. Coelli, T.J.; Rao, D.S.P. Total factor productivity growth in agriculture: A Malmquist index analysis of 93 countries, 1980–2000. *Agric. Econ.* **2005**, *32*, 115–134. [CrossRef]
42. Kortelainen, M. Dynamic environmental performance analysis: A Malmquist index approach. *Ecol. Econ.* **2008**, *64*, 701–715. [CrossRef]
43. Baba, V. Methodological issues in modeling absence: A comparison of least squares and Tobit analyses. *J. Appl. Psychol.* **1990**, *75*, 428. [CrossRef]
44. Tobin, J. Estimation of relationships for limited dependent variables. *Econom. J. Econom. Soc.* **1958**, *26*, 24–36. [CrossRef]
45. Amore, M.D.; Murtinu, S. Tobit models in strategy research: Critical issues and applications. *Glob. Strat. J.* **2021**, *11*, 331–355. [CrossRef]
46. Aldieri, L.; Gatto, A.; Vinci, C.P. Is there any room for renewable energy innovation in developing and transition economies? Data envelopment analysis of energy behaviour and resilience data. *Resour. Conserv. Recycl.* **2022**, *186*, 106587. [CrossRef]
47. Shuai, S.; Fan, Z. Modeling the role of environmental regulations in regional green economy efficiency of China: Empirical evidence from super efficiency DEA-Tobit model. *J. Environ. Manag.* **2020**, *261*, 110227. [CrossRef]

48. Dalei, N.N.; Joshi, J.M. Estimating technical efficiency of petroleum refineries using DEA and tobit model: An India perspective. *Comput. Chem. Eng.* **2020**, *142*, 107047. [CrossRef]
49. Akaike, H. A new look at the statistical identification model. *IEEE Trans. Autom. Control.* **1974**, *19*, 716–723. [CrossRef]
50. Schwarz, G. Estimating the dimension of a model. *Ann. Stat.* **1978**, *6*, 461–464. [CrossRef]
51. Jiao, F.Y.; Liu, L.N. Contribution of fiscal expenditure on supporting agriculture to promoting agricultural economic growth—Empirical study based on Northeast panel data. In Proceedings of the 2009 International Conference on Management Science and Engineering, Moscow, Russia, 14–16 September 2009.
52. Chen, S.; Yang, J.; Kang, X. Effect of Fiscal Expenditure for Supporting Agriculture on Agricultural Economic Efficiency in Central China—A Case Study of Henan Province. *Agriculture* **2023**, *13*, 822. [CrossRef]
53. Zhou, H.M.; Li, M.X. Efficiency evaluation of fiscal expenditure on supporting agriculture in Hunan Province based on DEA model. *Res. Agric. Mod.* **2016**, *37*, 284–289.
54. Li, W.; Ma, Y. Research on the Effects of Fiscal Expenditure for Agriculture on the Agricultural Modernization: Empirical Analysis Based on Dynamic Panel Data Model. *J. Xi'an Univ. Financ. Econ.* **2014**, *3*, 5–9. [CrossRef]
55. Yan, J. Spatiotemporal analysis for investment efficiency of China's rural water conservancy based on DEA model and Malmquist productivity index model. *Sustain. Comput. Inform. Syst.* **2019**, *21*, 56–71. [CrossRef]
56. Wang, B.; Zhang, H.; Luo, W. Evaluation of operational efficiency of agricultural water conservancy projects in China based on inter-provincial panel data. *J. Econ. Water Resour.* **2019**, *37*, 59–66.
57. Yu, L.; Yang, G. Supply Efficiency and Factors of Farmland Water Conservancy Facilities Based on Sbm-Malmquist-Tobit Model. *Resour. Ind.* **2022**, *24*, 77–89.
58. Meena, T. Corporatization of Agriculture and Its Effect. 2016. Available online: https://www.readcube.com/articles/10.2139/ssrn.2823387 (accessed on 15 July 2023).
59. Motes, W. Modern Agriculture and Its Benefits–Trends, Implications and Outlook. 2010. Available online: https://www.researchgate.net/publication/268178065_Modern_Agriculture_and_Its_Benefits_-_Trends_Implications_and_Outlook (accessed on 15 July 2023).
60. Bojnec, Š.; Fertő, I.; Jámbor, A.; Tóth, J. Determinants of technical efficiency in agriculture in new EU member states from Central and Eastern Europe. *Acta Oeconomica* **2014**, *64*, 197–217. [CrossRef]
61. Wang, Y.; Dong, F.; Xu, J. Production Efficiency of Scaled-up Agricultural Operations in China: An Empirical Analysis. In Proceedings of the 2018 Annual Meeting, Washington, DC, USA, 5–7 August 2018.
62. Xu, D.; Sun, H. Leading Companies in Industrialization of Agriculture in Guizhou:Effect on Increasing Farmers' Income. 2010. Available online: http://www.cqvip.com/qk/96372x/201003/33646595.html (accessed on 15 July 2023).
63. Kögel, T.; Prskawetz, A. Agricultural productivity growth and escape from the Malthusian trap. *J. Econ. Growth* **2001**, *6*, 337–357. [CrossRef]
64. Bilsborrow, R.E. Population pressures and agricultural development in developing countries: A conceptual framework and recent evidence. *World Dev.* **1987**, *15*, 183–203. [CrossRef]

agriculture

Article

Analysis on Coupling Coordination Degree for Cropland and Livestock from 2000 to 2020 in China

Jianxing Chen [1], Xuesong Gao [1,2,*], Yanyan Zhang [1,2], Petri Penttinen [1], Qi Wang [1], Jing Ling [1,2], Ting Lan [1,2], Dinghua Ou [1,2] and Yang Li [1,2]

1 College of Resources, Sichuan Agricultural University, Chengdu 611130, China; cjx98@stu.sicau.edu.cn (J.C.); yanyan.zhang@sicau.edu.cn (Y.Z.); petri.penttinen@helsinki.fi (P.P.); wq94@stu.sicau.edu.cn (Q.W.); 11963@sicau.edu.cn (J.L.); tlan@sicau.edu.cn (T.L.); 14340@sicau.edu.cn (D.O.); 14835@sicau.edu.cn (Y.L.)

2 Key Laboratory of Investigation and Monitoring, Protection and Utilization for Cultivated Land Resources, Ministry of Natural Resources, Chengdu 611130, China

* Correspondence: xuesonggao@sicau.edu.cn

Abstract: The decoupling of cropland and livestock due to the industrialization of livestock production is a difficult problem for sustainable agricultural development in many global locations, including China. As population and urbanization increase, this decoupling is likely to become more serious. To date, the relationship between cropland and livestock has been mainly studied from a single perspective, and mostly at the regional and the local scales. Thus, the objective of our study is to systematically assess the coupling relationship between cropland and livestock from multiple aspects on a large scale. Here, we used a complex system covering cropland, livestock and environment subsystems to comprehensively analyze the spatio-temporal variation of the coupling coordination between cropland and livestock and its influencing factors in China over the past two decades. Elaborating on the data, we constructed a comprehensive system of evaluation indexes for cropland–livestock systems. We used a coupling coordination degree model to evaluate the coupling coordination relationship between cropland and livestock in 31 provinces of China during 2000–2020. The results show that the range of cropland–livestock and cropland–livestock–environment coupling coordination degree was 0.4–0.9. In most of the provinces, there was no risk of cropland and livestock decoupling; however, the coupling coordination degree needed to be increased. More attention should be paid to the coordinated development of cropland and livestock coupling in urbanized areas such as Beijing and Tianjin, where cropland and livestock decoupling was more likely to occur. Among the assessed 29 factors, 15 and 16 had an impact on the cropland–livestock and the cropland–livestock–environment coupling coordination degrees, respectively. Our study provides science-based evidence to support estimating the coupling relationship between cropland and livestock in the future.

Keywords: cropland–livestock systems; index system; coupling coordination degree; influence factor

Citation: Chen, J.; Gao, X.; Zhang, Y.; Penttinen, P.; Wang, Q.; Ling, J.; Lan, T.; Ou, D.; Li, Y. Analysis on Coupling Coordination Degree for Cropland and Livestock from 2000 to 2020 in China. *Agriculture* **2023**, *13*, 1304. https://doi.org/10.3390/agriculture13071304

Academic Editors: Xin Chen, Moucheng Liu and Yuanmei Jiao

Received: 19 May 2023
Revised: 13 June 2023
Accepted: 15 June 2023
Published: 26 June 2023

1. Introduction

Crops and livestock play a synergistic role in global food production and the livelihoods of farmers [1]. For centuries, crops and livestock have formed a coupled system of planting and breeding with a circular flow of material and nutrient elements. In the coupled system, livestock manure has been used as a nutrient source for crops [2], livestock can be used as draft animals in crop cultivation, and the cropland provides feed for the livestock. In traditional settings, the coupled livestock–cropland system has a high material circulation rate [3,4]. However, due to rapid urbanization and the sharp increase in population, the demand for livestock products has increased, and the response to the demand has led to drastic changes. One of the biggest changes is the emergence of large-scale, intensive and specialized industrial livestock production systems. Industrial livestock production has brought about changes in spatial allocation and land use, resulting in the spatial separation and lack of functional interaction between livestock and cropland and a decrease in the nutrient cycling rate between the two systems [5–7]. Furthermore, intensive and spatially separated livestock have increased the costs of applying livestock manure as a source of nutrients to croplands. Concomitantly, industrial fertilizers with low unit nutrient cost have replaced manure on croplands. This future hinders the effective recycling of nutrients in livestock manure, aggravating the decoupling of livestock and cropland [8–10].

The concentration of livestock in areas with little or no cropland has a great impact on the environment. The environmental impact is mainly related to the poor management of livestock manure, which can lead to the contamination of surface and ground waters with nutrients, organic matter and heavy metals [1,5,7]. The aggregation of livestock production systems and the accompanying large amounts of fertilizer and improperly managed manure may increase the adverse impacts on water quality, especially the enrichment of phosphorus, nitrogen and other nutrients in the water, leading to the eutrophication of water bodies [4,11]. In China, the high intensity of livestock production and its increasing proximity to urban areas has resulted in more than 1 billion people being exposed to intense nitrogen pollution in the air and water [8]. To alleviate the negative effects, circular agriculture with combined cropland and livestock has been proposed as a key strategy to promote sustainable agricultural intensification [9,12,13]. Since manure contributes to soil health and versatility by providing nutrients and improving soil properties, the partial replacement of industrial fertilizers with manure can improve crop productivity, enhance interactions within and among soil microbial communities, increase carbon sequestration on the surface, increase soil organic matter content, and reduce agricultural greenhouse gas emissions and nitrogen pollution [14–18].

Hence, in order to analyze the connection between cropland and livestock and alleviate environmental pollution, various aspects of coupled cropland–livestock systems have been studied using a variety of methods. At present, research on the relationship between cropland and livestock mainly focuses on the current situation and the recoupling of the cooperative relationship between cropland and livestock. Single indicators, such as livestock density and the nitrogen or phosphorus load of livestock into farmland, have been applied to establish the relationship between regional cropland and livestock [7,19,20]. The coupling relationship between cropland and livestock has been measured using the nutrient balance method or models based on nutrient balance theory [21–23]. Zhang et al. (2019) used the nutrient balance method to propose a cropland-based livestock production system from the perspective of agricultural production and human consumption to rebuild the linkage between livestock and cropland in China [24]. Kamilaris et al. (2020) and Li et al. (2021) used an objective optimization method to find the optimal flow mode of regional livestock manure to reconstruct the coupling relationship between intensive large-scale livestock and agricultural production [25,26]. Scenario analysis has been applied to explore the synergistic relationship between cropland and livestock in the future [27–29]. Index analysis, nutrient balance methods, evaluation models and scenario simulation are the most commonly used methods to research the coupling cooperative relationship between cropland and livestock.

Similar to the United States and many other developed countries, in China, the world's largest livestock breeding country, livestock production agglomeration and decoupling between livestock and cropland are increasing [8,24]. The Chinese coupling relationship between cropland and livestock has been analyzed using surveys and statistics to analyze the main obstacles affecting the interaction between cropland and livestock from the perspectives of material flow and environmental factors [30–32]. The changes in the coupling relationship between cropland and livestock in China in the past decades have been analyzed using the nutrient element flow balance method [20,27]. Zhao et al. (2015) employed the coupling coordination degree model to investigate the spatiotemporal variation characteristics of the coupling of farming and animal husbandry in agricultural areas located in the Tarim River Basin of China. Based on the findings, some suggestions were proposed [33]. The coupling relationship between cropland and livestock has been mainly explored from the perspectives of element flow, nutrient management, index analysis, and environmental and economic benefits. However, there is a lack of comprehensive analysis of the cropland–livestock coupling in which environmental factors are considered.

The coupling coordination degree model, developed based on the coupling theory, can reflect the degree of interaction and coupling between systems accurately. This model has been found suitable for evaluating the level of coupling and coordination development between systems in research on the regional coupling relationship between cropland and livestock [33,34]. Therefore, our assumption is that the use of the coupling coordination degree model can reflect the phenomenon of decoupling between cropland and livestock in various regions of China. This phenomenon has shown a trend of gradual expansion.

Our aims were to deepen the understanding of cropland and livestock coupling coordination in China over the past 20 years and to explore its spatiotemporal changes and main influencing factors. Based on previous studies, the innovations in our study: (1) we considered the environmental factors as an independent subsystem in the coupling relationship between cropland and livestock, constructed a comprehensive evaluation indicator system of cropland and livestock system; (2) we investigated the spatiotemporal changes of the coupled and coordinated relationship between cropland and livestock in China under the multiple influence factors.

2. Materials and Methods

2.1. Study Area

According to the main grain-producing areas in China, 31 provinces were divided into six regions [20]: Northeast, North China, Middle-lower Reaches of the Yangtze River, Northwest, Southwest and Southeast (Figure 1). Data for Hong Kong, Macao and Taiwan are not included in this paper.

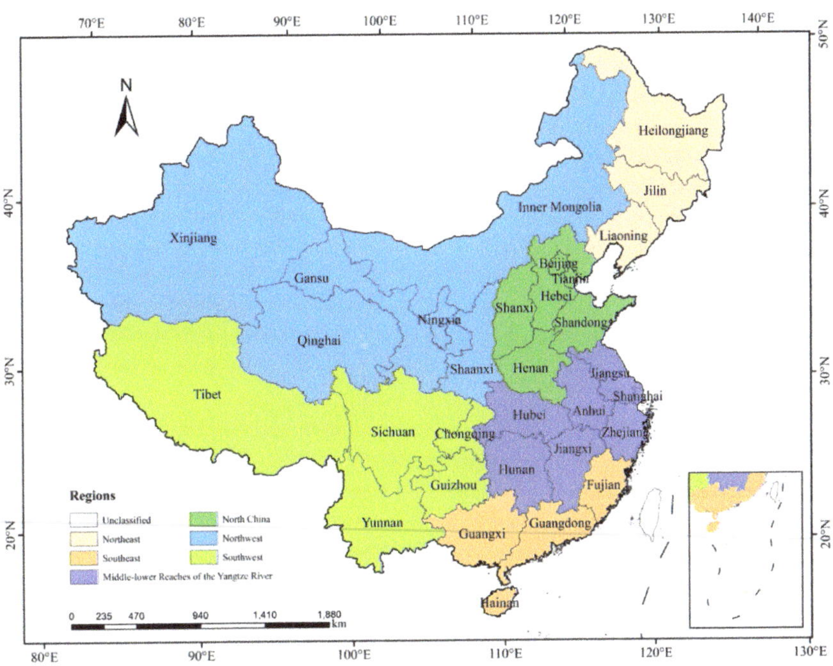

Figure 1. Division of the provinces of China into six study regions.

2.2. Research Framework

To explore the coupling relationship between cropland and livestock in China in the past two decades, we constructed a research framework consisting of four processes (Figure 2). The comprehensive evaluation of cropland–livestock systems and cropland–livestock–environment systems included the construction of an index system, weight calculation and comprehensive evaluation. In the coupling coordination degree model, the coupling degree (C), comprehensive reconciliation index (T) and coupling coordination degree (D) of cropland–livestock and cropland–livestock–environment systems were calculated. The spatial autocorrelation analysis of coupling coordination degree included calculating global Moran's I and local spatial autocorrelation analysis. The main influencing factors and the degree of coupling coordination degree were explored using the Geographical detector model.

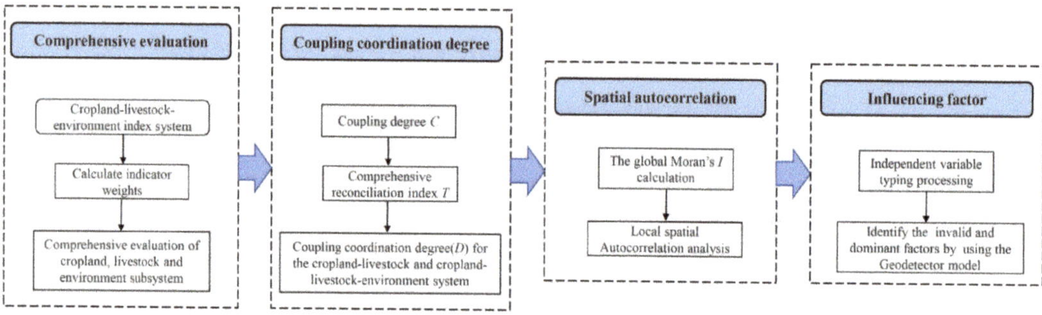

Figure 2. Research framework.

2.3. *Indicator System Construction*

2.3.1. CLE Indicator System

Following the principles of consistent objectives, comprehensiveness, validity, independence and measurability, and referring to previous studies, the comprehensive index system of cropland–livestock–environment was constructed (Table S5). The index system was finally constructed by using the multicollinearity method to screen 29 indexes in cropland, livestock and environment subsystems (Table 1). The cropland subsystem contains eight indicators, among which the output of farm crops is the sum of grain, cotton, oilseed, flax, sugar crop, tobacco, vegetable, and fruit yields. The grain crop straw yield index was calculated as shown in formula (1). The livestock subsystem contains eight indicators, among which the number of captive livestock, livestock density, the ratio of large-scale livestock farms, and livestock urine and manure production were defined and calculated in detail in Section S.1 of Supplementary Materials S1 [35–41]. In addition, this study's focus is on livestock species, including pigs, cattle, sheep (both wool sheep and goats), and poultry (layers, broilers, ducks, and rabbits). Considering the comprehensiveness and operability principle of index selection, the environment subsystem included 13 evaluation indexes from both natural and social aspects, among which annual total precipitation, annual total sunshine hours and annual average temperature were obtained from the meteorological data of major cities.

For the index of grain crop straw yield: in this study, the grain crop straw yield is the sum of rice, wheat and maize straw yields. The straw yield of grain crop was calculated based on the grain yield and the straw-to-grain ratio of three grain crops in different main grain-producing areas (Table S1) [35]. The formula is as follows:

$$TS = \sum_{i=1}^{n} x_i \times R_i \quad (1)$$

where TS is the theoretical resource amount of straw (air-dried base), i is the ith grain crop, x is the economic yield of grain crops, and R is the straw-to-grain ratio of grain crops.

Table 1. Cropland–livestock–environment comprehensive evaluation index system.

Subsystem	Criteria	Indicator	Unit	Explaining
Cropland (C)	Input	Cultivated area	10^3 ha	
		Consumption of chemical fertilizers	10^4 tons	The quantity of chemical fertilizers applied in agriculture per year (volume of effective component)
		Irrigated area of cultivated land	10^3 ha	
		Total sown area of farm crops	10^3 ha	
	Output	Gross output value of farming	billion yuan	
		Output of farm crops	10^4 tons	
		Output of major grain per hectare	kg/ha	Output of Grain/Sown area of grain crops
		Grain crop straw yield	10^4 tons	Only include the straw of rice, wheat and corn
Livestock (L)	Input	Number of captive livestock	10^4 pig equivalents	Only confined livestock is included
		Livestock density	pig equivalent/km^2	Number of captive livestock/regional land area
		Number of pigs raised	10^4 heads	
		Total output of feed	10^4 tons	
		Ratio of large-scale livestock farms	%	Number of large-scale livestock farms/total number of livestock farms × 100
	Output	Gross output value of animal husbandry	billion yuan	
		Livestock urine production	10^4 tons	
		Livestock dung production	10^4 tons	
Environment (E)	Natural	Annual total precipitation	mm	
		Annual total sunshine hours	hours	
		Annual average temperature	°C	
		Per capita water resources	cu. m/person	
		Area of afforested land	10^4 ha	
	Social	Population	10^4 persons	
		Length of highways	km	
		GDP	billion yuan	
		Per capita GDP	yuan	GDP/population
		R&D investments	%	Expenditure on R&D/GDP × 100
		General public budget revenue	billion yuan	
		Environmental protection investment	%	investment in the treatment of environmental pollution/GDP × 100
		Rural family Engel's coefficient	%	Food, tobacco and liquor expenditure/living expenditure × 100

2.3.2. Data Sources

The data sources of this study include mainly statistical yearbooks and parameters or coefficients collected in technical guidelines and previous studies. (1) Statistical yearbooks: China Statistical Yearbook, China Rural Statistical Yearbook, China Animal Husbandry and Veterinary Yearbook, China Feed Industry Yearbook, China Environmental Statistics Yearbook and China Agricultural Yearbook, including data on cultivated area, number of captive livestock and total output of feed. (2) Collected parameters and coefficients: the data in this study, such as the straw-to-grain ratio of crops, conversion coefficient of pig equivalent and livestock feeding period were obtained from national technical guidelines and previous studies. Furthermore, due to the lack of data, per capita water resources and area of afforested land in 2000 were obtained from the China Statistical Yearbook in 2004. The data on environmental protection investment in 2000 and 2020 was based on the 2014 and 2018 China Environmental Statistics Yearbook, respectively. Cultivated area data for 2000 and 2015 came from statistical yearbooks of the provinces. The original data of each indicator from 2000 to 2020 are shown in Supplementary Material S2.

2.4. Comprehensive Evaluation

To avoid the influence of subjective factors on the results, the weight of each index was calculated using the entropy weight method, and the comprehensive evaluation values of cropland, livestock and environment subsystems were calculated using the weighted summation method [42]. The evaluation steps were as follows:

First, annual total precipitation, annual total sunshine hours and annual average temperature were transformed using the reciprocal distance method as follows:

$$T = \frac{1}{\sqrt[2]{(m_i - m_0)^2}} \tag{2}$$

where T is the transformed index, mi is the original index, and m_0 is the average of m_i.

Second, to eliminate the influence of dimension, magnitude and positive and negative orientation, the data were standardized using the following formulas:

$$y_{\theta ij} = \frac{x_{\theta ij} - min(x_{\theta ij})}{max(x_{\theta ij}) - min(x_{\theta ij})} \tag{3}$$

$$y_{\theta ij} = \frac{max(x_{\theta ij}) - x_{\theta ij}}{max(x_{\theta ij}) - min(x_{\theta ij})} \tag{4}$$

where $y_{\theta ij}$ is the standardized data, $x_{\theta ij}$ refers to the value of indicator j of city i in year θ, and min and max are the minimum and maximum values, respectively.

Third, weight calculation was conducted through the use of the entropy weight method. W_j is the weight of each index in the subsystems, calculated in three steps using the entropy weight method, as described earlier [43]. The calculation results are shown in Table S6.

Last, the comprehensive evaluation values of cropland, livestock and environment subsystems were calculated by using Equation (5):

$$Z_{\theta i} = \sum_{j=1}^{n} (y_{\theta ij} W_j) \tag{5}$$

where $Z_{\theta i}$ represents the comprehensive evaluation value of province i in year θ. $Z_{\theta i}(C)$, $Z_{\theta i}(l)$ and $Z_{\theta i}(E)$ were used to represent the comprehensive evaluation value of cropland, livestock and environment subsystems, respectively. The calculation results are shown in Table S7.

2.5. Coupling Coordination Degree Model

The coupling coordination degree model can reflect the interaction between systems or among subsystems within a system to estimate the development of coupling coordination of the system [42]. After continuous development, the current coupling coordination degree model includes eight model types with minor differences; the original model is the most commonly used coupling coordination degree model [43]. Hence, on the basis of the original model, we constructed the coupling coordination degree models of cropland–livestock and cropland–livestock–environment systems.

The coupling coordination degree model of cropland–livestock systems:

$$C_2 = Z_{(C)} Z_{(L)} / (\frac{Z_{(C)} + Z_{(L)}}{2})^2 \tag{6}$$

$$T_2 = \alpha Z_{(C)} + \beta Z_{(L)} \quad (7)$$

$$D_2 = \sqrt{C_2 \times T_2} \quad (8)$$

where C_2 is the coupling degree, reflecting the degree of mutual influence between systems, and $C_2 \in [0, 1]$. When C_2 is larger, the degree of coupling between the systems is greater. T_2 is the comprehensive reconciliation index. D_2 is the coupling coordination degree, the value of which is positively correlated with the degree of coupling coordination between systems. $Z_{(C)}$ and $Z_{(L)}$ represent the comprehensive evaluation value of cropland subsystems and livestock subsystems, respectively. α and β show the weight of the importance of the two subsystems, where $\alpha = \beta = 1/2$.

The coupling coordination degree model of cropland–livestock–environment systems:

$$C_3 = \left\{ Z_{(C)} Z_{(L)} Z_{(E)} / \left(\frac{Z_{(C)} + Z_{(L)} + Z_{(E)}}{3} \right)^3 \right\}^{1/3} \quad (9)$$

$$T_3 = \alpha Z_{(C)} + \beta Z_{(L)} + \gamma Z_{(E)} \quad (10)$$

$$D_3 = \sqrt{C_3 \times T_3} \quad (11)$$

where D_3 is the coupling coordination degree for the cropland–livestock–environment system, $Z_{(C)}$, $Z_{(L)}$ and $Z_{(E)}$ represent the comprehensive evaluation value of cropland, livestock and environment subsystems, respectively, where $\alpha = \beta = \gamma = 1/3$. The coupling degree C and comprehensive reconciliation index T are shown in Tables S8 and S9).

In previous studies, different methods were used to classify the coupling coordination degree [44,45]. In this study, the coupling coordination degree was divided into 10 types via the use of a continuous uniform distribution function (Table 2).

Table 2. The types and criteria of coupling coordination degree.

Category	D_n Value	Coupling Coordination Type
Uncoordinated development	$0 \leq D_n \leq 0.1$	Extreme decoupled maladjustment
	$0.1 < D_n \leq 0.2$	Severe decoupled maladjustment
	$0.2 < D_n \leq 0.3$	Moderate decoupled maladjustment
	$0.3 < D_n \leq 0.4$	Mild decoupled maladjustment
Transformation development	$0.4 < D_n \leq 0.5$	On the verge of decoupled maladjustment
	$0.5 < D_n \leq 0.6$	Barely coupled coordination
Coordinated development	$0.6 < D_n \leq 0.7$	Basic coupled coordination
	$0.7 < D_n \leq 0.8$	Intermediate coupled coordination
	$0.8 < D_n \leq 0.9$	Good coupled coordination
	$0.9 < D_n \leq 1$	Excellent coupled coordination

2.6. Spatial Autocorrelation

Spatial autocorrelation reflects the correlation of a phenomenon or feature with neighboring regions, while global correlation is used to describe the spatial clustering or differentiation characteristics of a phenomenon or attribute in the whole domain [46]. Moran's I is an index commonly used for spatial autocorrelation analysis. We used global and local Moran's I to explore the spatial correlation of the coupling coordination degree for cropland–livestock and cropland–livestock–environment systems [34,47]. The formula of global Moran's I is as follows:

$$\text{Moran's } I = \frac{\sum_{f=1}^{n}\sum_{t=1}^{n} W_{ft}\left(D_f - \overline{D}\right)\left(D_t - \overline{D}\right)}{S^2 \sum_{f=1}^{n}\sum_{t=1}^{n} W_{ft}} \quad (12)$$

where D_f and D_t represent the coupling coordination degree of region f and t, n represents the total number of regions, and W_{ft} represents the spatial weight matrix; $S^2 = \frac{1}{n}\sum_{f=1}^{n}(Df - \overline{D})$, and $\overline{D} = \frac{1}{n}\sum_{f=1}^{n} Df$. The value range of global Moran's I is $[-1,1]$, and the larger the absolute value, the greater the global spatial correlation. Some spatial phenomena or features not only have global correlation characteristics, but also have local regional spatial aggregation or heterogeneity. Thereby, the local Moran's I index was introduced to analyze the local spatial autocorrelation of the coupling coordination degree for cropland–livestock and cropland–livestock–environment systems. The formula of local Moran's I is as follows:

$$\text{Local Moran's } I = \frac{\left(D_f - \overline{D}\right)}{S^2}\sum_{f=1}^{n} W_{ft}\left(D_f - \overline{D}\right) \quad (13)$$

Hainan was not included in the spatial analysis because it has no border with other provinces.

2.7. Geographical Detector

Geographical detector is a new analytical method for exploring the factors behind spatial differentiation. Geographical detector has been applied in many fields of natural and social sciences. We used geographic detector to detect the influencing factors and effects of the coupling coordination degree for cropland–livestock and cropland–livestock–environment systems [48–50]. The specific calculation methods are shown in Section S.2 of the supplementary materials.

3. Results

3.1. Analyzing the Development of Chinese Cropland and Livestock in 2000–2020

The total sown area of farm crops in China increased gradually, accompanied by increases in grain and grain crop straw yields from 2000 to 2020. The consumption of chemical fertilizers increased till 2015 to the maximum application amount of 602.25 million tons and then decreased (Figure S1). Across the country, the biggest increases in farm crops sown area were in Heilongjiang, Jilin, Inner Mongolia and Xinjiang, with Heilongjiang adding 5.58 million hectares. Spatially, the largest changes in terms of the output of grain and grain crop straw yield were in North China, Northeast and the Middle-lower Reaches of the Yangtze River; the largest increases were in Heilongjiang, Henan and Shandong. In Henan, Shandong and Heilongjiang, which have the largest farm crops sown area, the amounts of applied chemical fertilizers were largest in Henan and Shandong (Figure S2).

During the past 20 years, the number of captive livestock in China fluctuated; the maximum was 1550.14 million heads (pig equivalent) in 2005, corresponding to a livestock manure production of 2.79 billion tons (Figure S3). The number of captive livestock in most provinces varied less than the livestock manure production. In North China, the number of captive livestock and the livestock manure production of Hebei, Henan and Shandong provinces have changed greatly (Figure S4).

3.2. Coupling Coordination Degree Analysis

The coupling coordination degree for the cropland–livestock system ranged from 0.4 to 0.9, i.e., from being on the verge of decoupled maladjustment to good coupled coordination. Most provinces were in basic and intermediate coupled coordination. In the Middle-lower Reaches of the Yangtze River and Northeast, the coupling coordination degree values of cropland–livestock systems were all greater than 0.5, indicating that there was no risk of decoupled maladjustment in the regions. In some years, the coupling coordination degree values of cropland–livestock system exceeded 0.8, i.e., the level of good coupled coordination, in Heilongjiang, Hebei, Shandong and Jiangsu. From 2000 to 2020, Beijing, Qinghai and Tibet were on the verge of decoupled maladjustment state, as were Tianjin, Ningxia and Hainan in some years (Figure 3a). The within-region differences in the coupling coordination degree for the cropland and livestock system varied between regions. The within-region differences were smallest in Northeast and largest in North China and Northwest (Figure 3b).

Region	Province	2000	2005	2010	2015	2020
Northeast	Heilongjiang	0.7912	0.819	0.8511	0.8832	0.8689
	Jilin	0.6828	0.7297	0.7236	0.7270	0.7048
	Liaoning	0.7021	0.6952	0.6849	0.6946	0.6762
North China	Beijing	0.4909	0.4762	0.4685	0.4797	0.4715
	Tianjin	0.4702	0.4954	0.4844	0.4754	0.5060
	Hebei	0.8034	0.7713	0.8065	0.7976	0.7649
	Henan	0.7837	0.7429	0.7415	0.7154	0.7384
	Shandong	0.7981	0.7718	0.7960	0.801	0.7520
	Shanxi	0.6433	0.6224	0.6125	0.6360	0.6271
Middle-lower Reaches of the Yangtze River	Shanghai	0.5480	0.5296	0.5203	0.5163	0.5295
	Jiangsu	0.8126	0.7815	0.7711	0.7937	0.7848
	Zhejiang	0.6852	0.6690	0.6554	0.6477	0.6320
	Anhui	0.7614	0.7521	0.7530	0.7644	0.7522
	Hubei	0.7837	0.7429	0.7415	0.7154	0.7384
	Hunan	0.7405	0.7273	0.7154	0.7359	0.7456
	Jiangxi	0.6856	0.6727	0.6547	0.6587	0.6552

Region	Province	2000	2005	2010	2015	2020
Northwest	Inner Mongolia	0.7566	0.7668	0.7673	0.7919	0.7951
	Shaanxi	0.6561	0.6318	0.6446	0.6619	0.6657
	Ningxia	0.4887	0.4987	0.5232	0.5349	0.5313
	Gansu	0.6225	0.6124	0.6364	0.6596	0.6513
	Qinghai	0.4139	0.4239	0.4093	0.4134	0.4169
	Xinjiang	0.7459	0.7527	0.7796	0.8075	0.8044
Southwest	Chongqing	0.5994	0.6072	0.6101	0.6147	0.6208
	Sichuan	0.6663	0.6680	0.6499	0.6619	0.6697
	Guizhou	0.6156	0.6030	0.6412	0.6607	0.6698
	Yunnan	0.6868	0.6880	0.6600	0.6796	0.6678
	Tibet	0.4550	0.4640	0.4625	0.4628	0.4680
Southeast	Fujian	0.6440	0.6280	0.6103	0.6238	0.6052
	Guangdong	0.7179	0.6948	0.6635	0.6750	0.6831
	Guangxi	0.7156	0.7126	0.7065	0.7137	0.7241
	Hainan	0.4921	0.4597	0.4772	0.5096	0.5144

0–0.1	0.1–0.2	0.2–0.3	0.3–0.4	0.4–0.5	0.5–0.6	0.6–0.7	0.7–0.8	0.8–0.9	0.9–1
Extreme decoupled maladjudtment	Severe decoupled maladjudtment	Moderate decoupled maladjudtment	Mild decoupled maladjudtment	On the verge of decoupled maladjudtment	Grudging coupled coordination	Basic coupled coordination	Intermediate coupled coordination	Good coupled coordination	Excellent coupled coordination

(a)

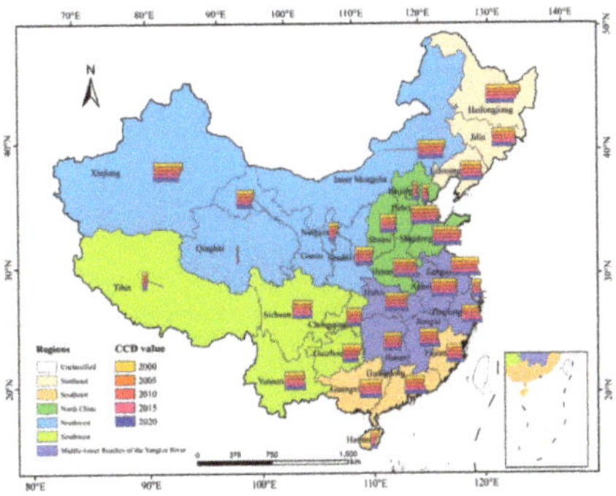

(b)

Figure 3. The coupling coordination degree for the cropland–livestock system in 31 provinces of China from 2000 to 2020. (**a**): the coupling coordination degree values for the cropland–livestock system in 31 provinces; (**b**): the spatial distribution of the coupling coordination degree.

At the provincial scale, the coupling coordination degree value for the cropland–livestock–environment system ranged from 0.4 to 0.8, i.e., from being on the verge of decoupled maladjustment to the intermediate coupled coordination level. Most of the provinces were in a barely coupled coordination state. The provinces in Southwest, Northeast, the Middle-lower Reaches of the Yangtze River and North China coupling coordination degree values for the cropland–livestock–environment system were all greater than 0.5, i.e., there was no risk of decoupled maladjustment. In some years, the coupling coordination degree values in Ningxia, Qinghai and Hainan were between 0.4 and 0.5, indicating that these three regions were on the verge of decoupled maladjustment state (Figure 4a). There were significant spatiotemporal differences in the coupling coordination degree for the cropland–livestock–environment system. In terms of time, the coupling coordination degree of cropland–livestock–environment systems in Qinghai, Hebei, Shandong and other provinces showed great differences over the years. Spatially, the differences in the coupling coordination degree values for the cropland–livestock–environment system among provinces in Southwest, Northeast, Middle- lower Reaches of the Yangtze River and North China were small. While the differences among provinces were great in the Northwest (Figure 4b).

When the environment subsystem was added to the cropland–livestock system, the regional coupling coordination degree changed significantly (Figure 5). The coupling coordination degree values of most provinces decreased by two levels (Figures 3 and 4). Among them, the good coupled coordination state of Heilongjiang, Hebei, Shandong and Jiangsu decreased to the basic or barely coupled coordination state. However, the coupling coordination degree values of Beijing, Tianjin, Shanghai, Tibet and Qinghai increased, and the state changed from on the verge of decoupled maladjustment to barely coupled coordination in Beijing, Tianjin and Tibet. In addition, adding the environment subsystem changed the spatiotemporal difference of coupling coordination degree in most provinces and increased the degree of variation on the time scale, especially in the Middle-lower Reaches of the Yangtze River and North China.

Region	Province	2000	2005	2010	2015	2020
Northeast	Heilongjiang	0.6103	0.6633	0.6552	0.6535	0.6468
	Jilin	0.5330	0.5996	0.5870	0.5616	0.5906
	Liaoning	0.5709	0.6279	0.5954	0.5657	0.5741
North China	Beijing	0.5444	0.5311	0.5389	0.5495	0.5757
	Tianjin	0.5053	0.5160	0.5225	0.5241	0.5545
	Hebei	0.7052	0.6050	0.7266	0.7614	0.7422
	Henan	0.6317	0.5974	0.5973	0.5861	0.6267
	Shandong	0.7079	0.6464	0.7463	0.6683	0.6381
	Shanxi	0.5305	0.5284	0.5387	0.5362	0.5352
Middle-lower Reaches of the Yangtze River	Shanghai	0.5716	0.5703	0.5458	0.5489	0.6022
	Jiangsu	0.6411	0.6966	0.6574	0.6586	0.6771
	Zhejiang	0.5817	0.6092	0.5802	0.5703	0.5869
	Anhui	0.6070	0.6027	0.5945	0.6092	0.6400
	Hubei	0.6317	0.5974	0.5973	0.5861	0.6267
	Hunan	0.5693	0.5996	0.5703	0.5930	0.6193
	Jiangxi	0.5298	0.5756	0.5355	0.5251	0.5454

Region	Province	2000	2005	2010	2015	2020
Northwest	Inner Mongolia	0.6030	0.6378	0.6222	0.6622	0.6570
	Shaanxi	0.6274	0.6210	0.6157	0.5752	0.6029
	Ningxia	0.4632	0.4635	0.4717	0.4872	0.5197
	Gansu	0.5137	0.5182	0.5154	0.5252	0.5889
	Qinghai	0.4253	0.5497	0.4454	0.4495	0.5273
	Xinjiang	0.5600	0.6244	0.5913	0.6186	0.6349
Southwest	Chongqing	0.5026	0.5293	0.5362	0.5600	0.5415
	Sichuan	0.5855	0.5826	0.5753	0.5580	0.5763
	Guizhou	0.6107	0.6195	0.5854	0.5447	0.5788
	Yunnan	0.6224	0.6506	0.5640	0.5775	0.5867
	Tibet	0.5496	0.5526	0.5520	0.5541	0.5643
Southeast	Fujian	0.5503	0.5738	0.5245	0.5373	0.5654
	Guangdong	0.6103	0.6097	0.5883	0.5821	0.6068
	Guangxi	0.5794	0.5837	0.5464	0.5777	0.6057
	Hainan	0.4315	0.5220	0.4740	0.4646	0.5146

0-0.1	0.1-0.2	0.2-0.3	0.3-0.4	0.4-0.5	0.5-0.6	0.6-0.7	0.7-0.8	0.8-0.9	0.9-1
Extreme decoupled maladjustment	Severe decoupled maladjustment	Moderate decoupled maladjustment	Mild decoupled maladjustment	On the verge of decoupled maladjustment	Grudging coupled coordination	Basic coupled coordination	Intermediate coupled coordination	Good coupled coordination	Excellent coupled coordination

(a)

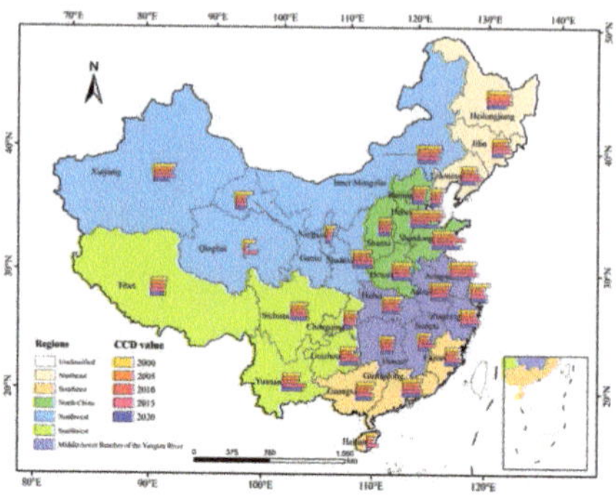

(b)

Figure 4. The coupling coordination degree for the cropland–livestock–environment system in 31 provinces of China from 2000 to 2020. (**a**): the coupling coordination degree values for the cropland–livestock–environment system in 31 provinces; (**b**): the spatial distribution of the coupling coordination degree.

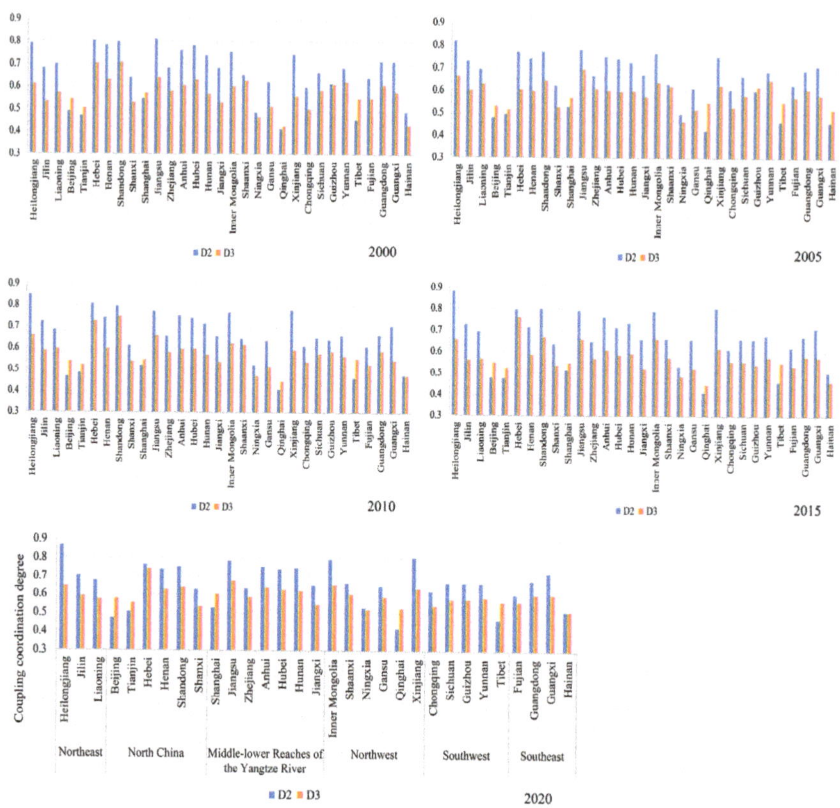

Figure 5. Coupling coordination degree comparison of cropland and livestock system in 31 provinces of China from 2000 to 2020.

3.3. Spatial Correlation Analysis

Global Moran's I was used to determine the global spatial autocorrelation of the coupling coordination degree for the cropland–livestock and cropland–livestock–environment systems. The global Moran's I p values of the coupling coordination degree for cropland–livestock and cropland–livestock–environment systems in 2000 were below 0.1, and the Z values were over 1.65, and in 2010, the p value of the coupling coordination degree for the cropland–livestock–environment system was below 0.05, and the Z value was over 1.96, indicating a significant spatial positive correlation (Table 3). Nevertheless, in other years the p values were all over 0.1, and the Z values below 1.65, indicating that there was no spatial correlation.

Table 3. Global spatial autocorrelation results.

Year (CLS CDD)	Global Moran's I	p Value	Z Value	Year (CLS CDD)	Global Moran's I	p Value	Z Value
2000	0.1755	0.0567	1.9057	2000	0.1451	0.0986	1.6517
2005	0.1459	0.1026	1.6326	2005	0.0300	0.5609	0.5816
2010	0.1378	0.1185	1.5621	2010	0.1864	0.0389	2.0654
2015	0.1112	0.1858	1.3232	2015	0.1098	0.1751	1.3560
2020	0.0914	0.2534	1.1442	2020	0.0529	0.4186	0.8089

CLS, cropland–livestock system; CLES, cropland–livestock–environment system.

There were four types of local spatial autocorrelation (Figure 6). At the national level, high–high clusters (H–H) were mainly distributed in Northeast, North China and the Middle-lower Reaches of the Yangtze River, including Jilin, Liaoning, Shandong, Henan, Anhui and Heilongjiang, indicating that the coupling coordination relationship between cropland and livestock in these cluster areas was better than in the surrounding provinces. Due to the consistent high–low outlier (H–L) characteristics, the coupling coordination degree values for the cropland–livestock system in Xinjiang were always higher than in the surrounding areas. The spatial relationship of the coupling coordination degree for the cropland–livestock system between Sichuan and neighboring provinces was complicated; the autocorrelation type changed from H–L to low–low cluster (L–L) and then back to H–L, indicating that the coupling coordination degree for the cropland–livestock system in Southwest and Northwest, centered on Sichuan province, had changed considerably.

The local spatial correlation differences of the coupling coordination degree for cropland–livestock–environment systems over time were large (Figure 7). The local spatial correlation was not significant in the Middle-lower Reaches of the Yangtze River and Southeast. Shandong, Hunan and Anhui formed a large H–H cluster in 2000 and 2010 and Shandong and Hunan in 2015 and 2020, indicating that the coupling coordination relationships between cropland, livestock and environment in the cluster areas were better than that in the surrounding provinces. From 2000 to 2020, the spatial correlation of the coupling coordination degree for the cropland–livestock–environment system in Tibet, Shaanxi, Sichuan, Liaoning, Shanxi, Tianjin and Shanghai changed significantly; the spatial correlation of Shaanxi, Tianjin and Shanghai changed only in one year.

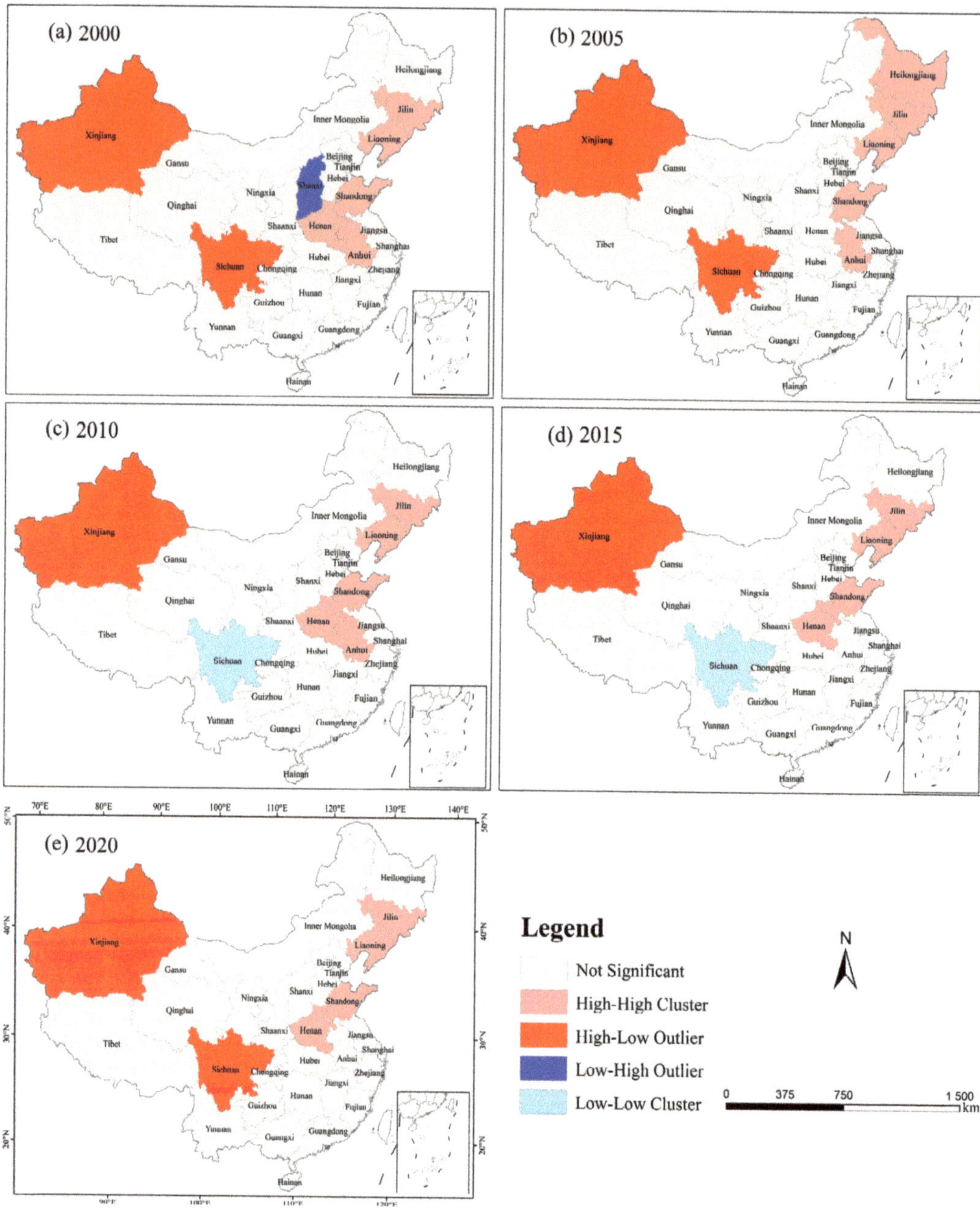

Figure 6. Local spatial association cluster maps of cropland–livestock coupling degree from 2000 to 2020.

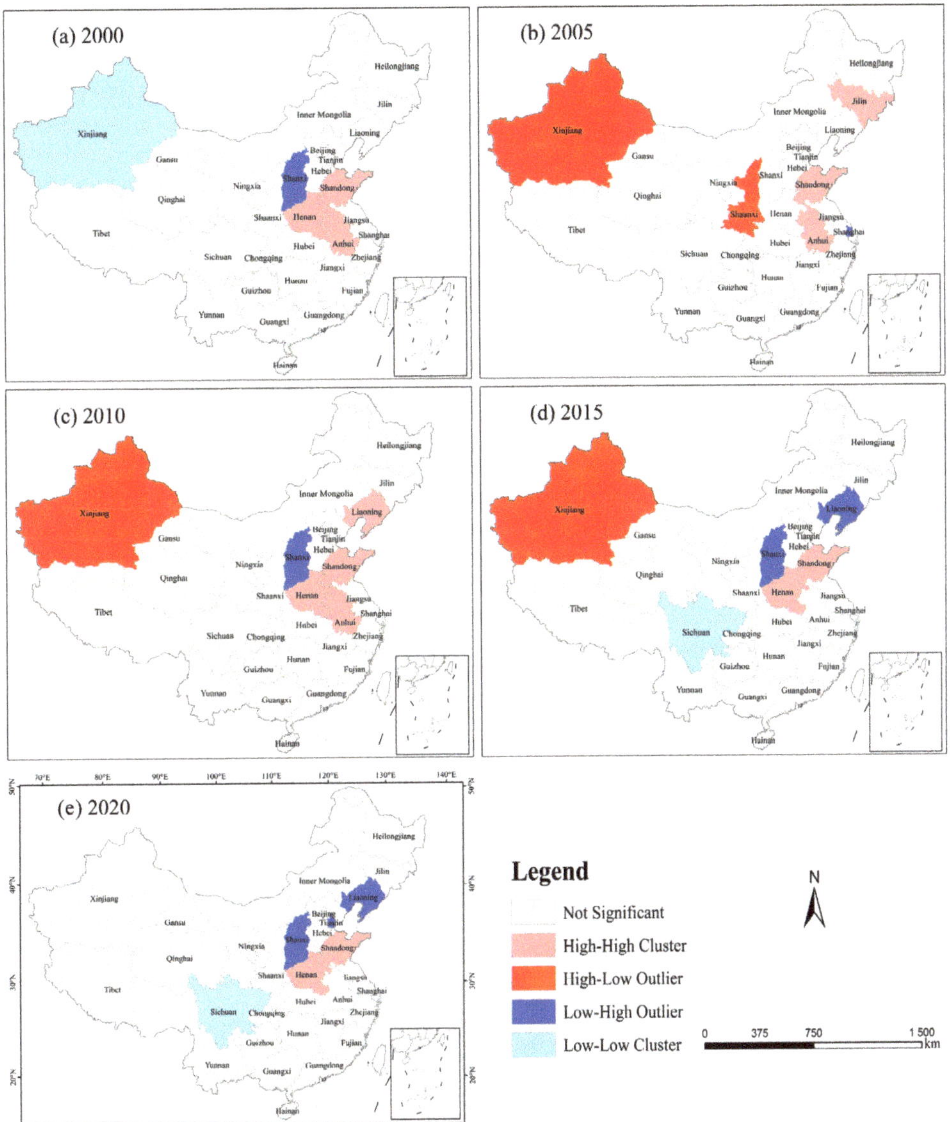

Figure 7. Local spatial association cluster maps of cropland–livestock–environment coupling degree from 2000 to 2020.

3.4. *Influencing Factors*

The q value and statistical significance reflect the influence degree of the factors on the coupling coordination degree. Except for the livestock density, all the factors had significant effects on the coupling coordination degree for the cropland–livestock system (Table 4). The q value of the irrigated area of cultivated land was 0.8643, indicating that it had the greatest influence on the coupling coordination degree. The livestock density had a negligible influence on the coupling coordination degree within the cropland–livestock system. The six primary factors with a substantial influence on the coupling coordination degree for the cropland–livestock system belong to the cropland subsystem, indicating its dominant role in affecting the degree of coupling coordination.

Table 4. Factors influencing the coupling coordination degree of the cropland–livestock system.

Factor	q Value	Factor	q Value
Irrigated area of cultivated land	0.8643 ***	Livestock urine production	0.6877 ***
Total sown area of farm crops	0.8097 ***	Cultivated land area	0.6751 ***
Output of farm crops	0.7995 ***	Livestock bung production	0.6590 ***
Consumption of chemical fertilizers	0.7887 ***	Number of pigs raised	0.5476 ***
Gross output value of farming	0.7669 ***	Total output of feed	0.3496 ***
Grain crop straw yield	0.7602 ***	Ratio of large-scale livestock farms	0.2838 ***
Gross output value of animal husbandry	0.7345 ***	Output of major grain per hectare	0.1506 **
Number of captive livestock	0.6970 ***	Livestock density	0.0217

*** and ** indicate statistical significance at q-value levels 0.01 and 0.05 respectively.

Sixteen factors influenced the coupling coordination degree of the cropland–livestock–environment system ($q < 0.01$) (Table 5). The factors with greatest influence were of the cropland subsystem; among all the factors, the irrigated area of cultivated land with q value 0.5414 had the biggest influence on the coupling coordination degree. Among the six factors of the livestock subsystem, the gross output value of animal husbandry has the biggest effect; and among the four factors of the environment subsystem, population had the greatest impact. The five most significant factors were all in the cropland subsystem, whereas general public budget revenue has the least significant impact. The factors not mentioned among the aforementioned 16 had negligible influence on the coupling coordination degree of the cropland–livestock–environment system.

Table 5. Factors influencing the coupling coordination degree of the cropland–livestock–environment system.

Factor	q Value	Factor	q Value
Irrigated area of cultivated land	0.5414 ***	General public budget revenue	0.2697 ***
Total sown area of farm crops	0.4370 ***	Total output of feed	0.2402
Output of farm crops	0.4357 ***	Annual average temperature	0.2269
Grain crop straw yield	0.4292 ***	Annual total sunshine hours	0.1885
Consumption of chemical fertilizers	0.4250 ***	Area of afforested land	0.1117
Gross output value of animal husbandry	0.4074 ***	Ratio of large-scale livestock farms	0.1107
Gross output value of farming	0.4052 ***	Output of major grain per hectare	0.1045
Population	0.3837 ***	Annual total precipitation	0.0958
GDP	0.3409 ***	Rural family Engel's coefficient	0.0831
Length of highways	0.3249 ***	Per capita GDP	0.0803
Livestock bung production	0.3266 ***	Per capita water resources	0.0411
Number of pigs raised	0.3251 ***	Livestock density	0.0397
Livestock urine production	0.3055 ***	R&D investments	0.0217
Number of captive livestock	0.3036 ***	Environmental protection investment	0.0209
Cultivated land area	0.3015 ***		

*** indicate statistical significance at q-value levels 0.01 respectively.

4. Discussion

Prior to the 1990s, China primarily focused on agricultural production, employing traditional coupled livestock and cropland. This approach yielded low agricultural productivity, although there existed a significant degree of integration between cropland and livestock. Following the 1990s, due to a continuous expansion in the inexpensive fertilizer market, there has been a significant decrease in the use of livestock manure as organic fertilizer in cropland, resulting in cropland decoupling from livestock and gradually worsening non-point source pollution in agriculture since 2000, with the significant development in agricultural mechanization and intensification, as well as increased public concern for environmental pollution. Consequently, new models for integrated crop-livestock farming have emerged, leading to the recoupling of livestock and cropland [8,24,51]. How has the development of rebuilding the linkage between livestock and cropland been since 2000?

Therefore, it is necessary to conduct a comprehensive analysis of the coupling coordination relationship between cropland and livestock and explore its driving factors [4,10]. We built a comprehensive evaluation indicator system of cropland and livestock system from three aspects: cropland, livestock and environment. Furthermore, we explored the spatiotemporal variation of the coupling coordination relationship for cropland–livestock systems in China during 2000–2020 based on a coupling coordination model and determined its key driving factors.

4.1. Enhancing the Coupling Coordination Relationship between Cropland and Livestock

Similar to a previous study [20], the results showed that most of the provinces in China were not at risk of the decoupling between cropland and livestock. From 2005 to 2020, the best coupling relationship between cropland and livestock was in Heilongjiang, possibly because the cultivated area in Heilongjiang is the biggest in China, chemical fertilizers are applied less and fewer livestock is raised than in the other provinces (Figure 3, Figures S2 and S4). Six provinces, including Beijing and Tianjin, are facing the risk of cropland and livestock decoupling, partly because there is less cultivated land to absorb livestock manure, large-scale farms account for a relatively high proportion, and the difference in the comprehensive evaluation values between cropland and livestock subsystems was significant in these provinces (Figure 3 and Table S7). The cropland and livestock system in Shanghai was also at risk of being decoupled, indicating that the risk of excess manure production in more urbanized areas is higher, making the cropland and livestock decoupling more likely [20,52]. However, the cropland–livestock coupling coordination degree was higher in Shanghai than that in Beijing and Tianjin, possibly because the Shanghai government has adopted and implemented stricter management measures for livestock manure [53]. Since 2000, the Central Government of China has paid more and more attention to the environmental pollution caused by livestock farming and issued a series of policies and regulations to manage and restrict livestock farming. However, due to that, farmers lack environmental protection awareness and fail to realize the importance of nutrient management and there is a lack of attention by some local governments; therefore, the coupling coordination degree of cropland and livestock in most provinces of China has not improved significantly from 2000 to 2020 (Figure 3) [27,53,54]. As individuals increasingly prioritize healthy dietary habits by reducing their consumption of animal-derived products, and with the implementation of more scientifically informed spatial planning for livestock production, there is an opportunity to align the development of cropland and livestock towards greater coupling coordination. However, the continuing trend of urbanization and rising population density has resulted in the gradual relocation of livestock farms from rural areas to suburban fringes, distancing them from cropland, which could potentially exacerbate the decoupling of cropland and livestock in certain regions [5,8,9,55]. Therefore, it is necessary to further enhance the degree of coupling coordination between cropland and livestock, especially in relatively developed areas with less cultivated land. Additionally, there is a need to strengthen the executive force of regulations and laws related to livestock production in the future.

4.2. The Role of Environmental Factors in Cropland–Livestock Coupling

When considering the environmental factors as an independent subsystem in the coupling relationship between cropland and livestock, the regional degree of cropland and livestock coupling had obviously changed. The coupling coordination degree value of most of the provinces decreased and was generally low, with a large decoupling risk. However, the coupling coordination degree value of Beijing, Tianjin and Shanghai increased (Figure 5). This was partly because of the rapid economic development of areas such as Beijing, with relatively high per capita GDP, convenient transportation and abundant resources for environmental pollution control [20,52]. Environmental factors such as population, GDP, traffic conditions and terrain were the main reasons for the decoupling of regional cropland and livestock system [8,56]. Furthermore, taking environmental, social and economic factors into full consideration is the key to the spatial planning of livestock production and the reconstruction of the spatial connection between cropland and livestock [56] Therefore, environmental factors must be considered when analyzing the regional coupling relationship between cropland and livestock.

4.3. Key Drivers for the Coupling of Cropland and Livestock

The factors with the highest influence on cropland–livestock and cropland–livestock–environment coupling coordination were all from the cropland subsystem, indicating that the cropland subsystem plays a leading function in the coupling and coordination of cropland–livestock and cropland–livestock–environment systems. Interestingly, the irrigated area of cultivated land was the most influential factor in the coupling coordination relationship of both cropland–livestock and cropland–livestock–environment systems (Tables 4 and 5). Possibly the irrigated area of cultivated land is related to the consumption of freshwater resources in the region, which not only has a great impact on crop yield and gross output value of farming but also directly affects the supply of freshwater resources for livestock in the region [3,57,58]. Population, GDP, the length of highways and the general public budget in the environment subsystem had an impact on the coupling coordination relationship for cropland–livestock–environment systems (Table 5). Similar to earlier conclusions [8,11,56], population, GDP and the length of highways that determines the transportation distance of livestock were the influencing factors for cropland–livestock systems of the coupling relationship (Tables 4 and 5). Contrary to earlier results, livestock density was not an influencing factor for cropland–livestock or cropland–livestock–environment systems. The difference to earlier results may be due to a different definition of livestock density or to our comprehensive approach where more indicators were considered [5,7,19].

4.4. Limitations and Outlook of the Study

Different from previous studies where the synergistic effect of regional cropland and livestock were measured from limited perspectives, e.g., breeding quantity, animal density, crop-livestock nutrient balance and land carrying capacity [7,20,53], we used the coupling coordination degree model to comprehensively establish the coupling coordination relationship between cropland, livestock and environment subsystems. Although we constructed a relatively comprehensive evaluation index system from the three aspects and carried out a correlation analysis of indicators, a lack of evaluation of the suitability and risk of indicators with uncertainties remains. In addition, due to the limited access to data at the provincial level, the survey data, such as the area of livestock farms and livestock manure treatment methods, were not considered in the indicator system. Consequently, to make the quantitative results of the coupling relationship between cropland and livestock more accurate in future research, a comprehensive evaluation of the indicators related to the coupling relationship between cropland and livestock should be carried out, and a comprehensive evaluation indicator system for cropland–livestock could be constructed at different scales (e.g., county, city or district) and dimensions [59]. The data used in this study come from various statistical yearbooks in China, which are the most credible data sources in China. Calculated coefficients, e.g., straw-to-grain ratio, pig equivalent conversion coefficient and livestock feeding period, from previous studies and Chinese technical guidelines similar to the national statistics, are also reliable. However, the differences in the values of coefficients between our calculations and previous studies and national technical guidelines have resulted in uncertainty in the research [24].

5. Conclusions

There was no risk of decoupling between cropland and livestock in most of the provinces in China during 2000–2020, but the degree of coupling coordination was low down. Beijing, Tianjin and other more urbanized areas were more likely to undergo the decoupling of cropland and livestock, but it was also easier to reestablish the contact of cropland and livestock and increase the degree of coupling coordination in those areas. The spatial autocorrelation of cropland and livestock coupling coordination among provinces in China was not significant. In Hebei, Henan, Shandong and Sichuan, which are major agricultural and livestock breeding provinces, cropland and livestock system were not at the risk of decoupling. This indicates that the coupling coordination degree between cropland and livestock system is higher in areas with comprehensive development of planting and breeding industries. Our results showed that the cropland subsystem had the greatest influence on the coupling coordination between cropland and livestock systems. The irrigated area of cultivated land was the most influential factor in the coupling coordination relationship of both cropland–livestock and cropland–livestock–environment systems. Clearly, our research examined the national scale and did not involve analyzing the coordinated relationship between cropland and livestock under different policy contexts. Such an analysis would be more suitable for studying regional scales under uniform contexts. To fully understand the critical influencing factors of the coupling relationship between cropland and livestock, we suggest that in future research, an indicator system is constructed from multiple scales and multiple dimensions to compare the spatiotemporal characteristics of the coupling between cropland and livestock at different scales and to explore the key influencing factors and major barriers at multiple scales.

Supplementary Materials: The following supporting information can be downloaded at: https://www.mdpi.com/article/10.3390/agriculture13071304/s1. The Supplementary Material S1 includes indicator definition and calculation, Geographical detector model introduction, figures and tables. Table S1: Straw-to-grain ratio of major grain crop in different regions; Table S2: Confined ratio of pastoral provinces from 2000 to 2020; Table S3: Different scale livestock farms division standard; Table S4: Daily bung/urine excretion and nitrogen/phosphorus content(fresh based) by various livestock and feeding period; Table S5: Cropland–livestock–environment original comprehensive evaluation index system; Table S6: Weight of each indicator; Table S7: Comprehensive evaluation value of cropland, livestock and environment subsystem from 2000 to 2020; Table S8: Coupling degree values for the cropland–livestock and cropland–livestock–environment system from 2000 to 2020; Table S9: The comprehensive reconciliation index values for the cropland–livestock and cropland–livestock–environment system from 2000 to 2020. Figure S1: The development of Chinese crop farming from 2000 to 2020; Figure S2: The crop farming development of 31 provinces in China from 2000 to 2020; Figure S3: The development of Chinese livestock farming from 2000 to 2020; Figure S4: The livestock farming development of 31 provinces in China from 2000 to 2020. The Supplementary Material S2 is the original data.

Author Contributions: Conceptualization, J.C. and X.G.; methodology, J.C. and Y.Z.; software, Q.W. and D.O.; validation, J.C., Y.L. and X.G.; formal analysis, J.C. and X.G.; investigation, J.C.; resources, J.C., J.L., T.L. and X.G.; writing—original draft preparation, J.C.; writing—review and editing, X.G., Y.Z., P.P., Q.W., J.L., T.L. and Y.L.; visualization, J.C. and X.G.; supervision, X.G. and Y.Z.; project administration, X.G. and Y.Z.; funding acquisition, X.G., J.L. and T.L. All authors have read and agreed to the published version of the manuscript.

Funding: This research was funded by the Ministry of Science and Technology (grant number 2022YFD1901400).

Institutional Review Board Statement: Not applicable.

Data Availability Statement: The data presented in this study are available upon request from the corresponding author.

Acknowledgments: The authors acknowledge the anonymous referees for the helpful comments that improved this manuscript and the editor of this journal.

Conflicts of Interest: The authors declare no conflict of interest.

References

1. Garrett, R.D.; Niles, M.T.; Gil, J.D.B.; Gaudin, A.; Chaplin-Kramer, R.; Assmann, A.; Assmann, T.S.; Brewer, K.; de Faccio Carvalho, P.C.; Cortner, O.; et al. Social and Ecological Analysis of Commercial Integrated Crop Livestock Systems: Current Knowledge and Remaining Uncertainty. *Agric. Syst.* **2017**, *155*, 136–146. [CrossRef]
2. Karmakar, S.; Laguë, C.; Agnew, J.; Landry, H. Integrated Decision Support System (DSS) for Manure Management: A Review and Perspective. *Comput. Electron. Agric.* **2007**, *57*, 190–201. [CrossRef]
3. Herrero, M.; Thornton, P.K.; Notenbaert, A.M.; Wood, S.; Msangi, S.; Freeman, H.A.; Bossio, D.; Dixon, J.; Peters, M.; van de Steeg, J.; et al. Smart Investments in Sustainable Food Production: Revisiting Mixed Crop-Livestock Systems. *Science* **2010**, *327*, 822–825. [CrossRef] [PubMed]
4. Jin, S.; Zhang, B.; Wu, B.; Han, D.; Hu, Y.; Ren, C.; Zhang, C.; Wei, X.; Wu, Y.; Mol, A.P.J.; et al. Decoupling Livestock and Crop Production at the Household Level in China. *Nat. Sustain.* **2021**, *4*, 48–55. [CrossRef]
5. Gerber, P.; Chilonda, P.; Franceschini, G.; Menzi, H. Geographical Determinants and Environmental Implications of Livestock Production Intensification in Asia. *Bioresour. Technol.* **2005**, *96*, 263–276. [CrossRef]
6. Herrero, M.; Thornton, P.K. Livestock and Global Change: Emerging Issues for Sustainable Food Systems. *Proc. Natl. Acad. Sci. USA* **2013**, *110*, 20878–20881. [CrossRef]
7. Nesme, T.; Senthilkumar, K.; Mollier, A.; Pellerin, S. Effects of Crop and Livestock Segregation on Phosphorus Resource Use: A Systematic, Regional Analysis. *Eur. J. Agron.* **2015**, *71*, 88–95. [CrossRef]
8. Bai, Z.; Fan, X.; Jin, X.; Zhao, Z.; Wu, Y.; Oenema, O.; Velthof, G.; Hu, C.; Ma, L. Relocate 10 Billion Livestock to Reduce Harmful Nitrogen Pollution Exposure for 90% of China's Population. *Nat. Food* **2022**, *3*, 152–160. [CrossRef]
9. Lemaire, G.; Franzluebbers, A.; de Faccio Carvalho, P.C.; Dedieu, B. Integrated Crop–Livestock Systems: Strategies to Achieve Synergy between Agricultural Production and Environmental Quality. *Agric. Ecosyst. Environ.* **2014**, *190*, 4–8. [CrossRef]
10. Naylor, R.; Steinfeld, H.; Falcon, W.; Galloway, J.; Smil, V.; Bradford, E.; Alder, J.; Mooney, H. Losing the Links Between Livestock and Land. *Science* **2005**, *310*, 1621–1622. [CrossRef]
11. Sharara, M.; Sampat, A.; Good, L.W.; Smith, A.S.; Porter, P.; Zavala, V.M.; Larson, R.; Runge, T. Spatially Explicit Methodology for Coordinated Manure Management in Shared Watersheds. *J. Environ. Manag.* **2017**, *192*, 48–56. [CrossRef]
12. Ryschawy, J.; Choisis, N.; Choisis, J.P.; Joannon, A.; Gibon, A. Mixed Crop-Livestock Systems: An Economic and Environmental-Friendly Way of Farming? *Animal* **2012**, *6*, 1722–1730. [CrossRef] [PubMed]
13. Vilela, L.; Marchao, R.L.; Guimaraes Junior, R.; Pulrolnik, K. Evolution of Integrated Crop-Livestock and Crop-Livestock-Forestry Systems in Brazil. In Proceedings of the World Congress on Integrated Crop-Livestock-Forestry Systems, Online, 3–6 May 2021; pp. 922–929.
14. Gross, C.D.; Bork, E.W.; Carlyle, C.N.; Chang, S.X. Biochar and Its Manure-Based Feedstock Have Divergent Effects on Soil Organic Carbon and Greenhouse Gas Emissions in Croplands. *Sci. Total Environ.* **2022**, *806*, 151337. [CrossRef] [PubMed]
15. Peyraud, J.-L.; Taboada, M.; Delaby, L. Integrated Crop and Livestock Systems in Western Europe and South America: A Review. *Eur. J. Agron.* **2014**, *57*, 31–42. [CrossRef]
16. Tang, Q.; Cotton, A.; Wei, Z.; Xia, Y.; Daniell, T.; Yan, X. How Does Partial Substitution of Chemical Fertiliser with Organic Forms Increase Sustainability of Agricultural Production? *Sci. Total Environ.* **2022**, *803*, 149933. [CrossRef] [PubMed]
17. Xia, L.; Lam, S.K.; Yan, X.; Chen, D. How Does Recycling of Livestock Manure in Agroecosystems Affect Crop Productivity, Reactive Nitrogen Losses, and Soil Carbon Balance? *Environ. Sci. Technol.* **2017**, *51*, 7450–7457. [CrossRef]
18. Leal, V.N.; Santos, D.d.C.; Paim, T.d.P.; Santos, L.P.d.; Alves, E.M.; Claudio, F.L.; Calgaro Junior, G.; Fernandes, P.B.; Salviano, P.A.P. Economic Results of Forage Species Choice in Crop–Livestock Integrated Systems. *Agriculture* **2023**, *13*, 637. [CrossRef]
19. Saam, H.; Mark Powell, J.; Jackson-Smith, D.B.; Bland, W.L.; Posner, J.L. Use of Animal Density to Estimate Manure Nutrient Recycling Ability of Wisconsin Dairy Farms. *Agric. Syst.* **2005**, *84*, 343–357. [CrossRef]
20. Zheng, L.; Zhang, Q.; Zhang, A.; Hussain, H.A.; Liu, X.; Yang, Z. Spatiotemporal Characteristics of the Bearing Capacity of Cropland Based on Manure Nitrogen and Phosphorus Load in Mainland China. *J. Clean. Prod.* **2019**, *233*, 601–610. [CrossRef]
21. Gao, M.; Qiu, J.; Li, C.; Wang, L.; Li, H.; Gao, C. Modeling Nitrogen Loading from a Watershed Consisting of Cropland and Livestock Farms in China Using Manure-DNDC. *Agric. Ecosyst. Environ.* **2014**, *185*, 88–98. [CrossRef]
22. MacDonald, G.K.; Bennett, E.M.; Potter, P.A.; Ramankutty, N. Agronomic Phosphorus Imbalances across the World's Croplands. *Proc. Natl. Acad. Sci. USA* **2011**, *108*, 3086–3091. [CrossRef]
23. Vizzari, M.; Santucci, A.; Casagrande, L.; Pauselli, M.; Benincasa, P.; Farneselli, M.; Antognelli, S.; Morbidini, L.; Borghi, P.; Bodo, G. Potential Nitrogen Load from Crop-Livestock Systems: An Agri-Environmental Spatial Database for a Multi-Scale Assessment. In *Computational Science and Its Applications—ICCSA 2015, Proceedings of the 15th International Conference, Banff, AB, Canada, 22–25 June 2015*; Gervasi, O., Murgante, B., Misra, S., Gavrilova, M.L., Rocha, A.M.A.C., Torre, C., Taniar, D., Apduhan, B.O., Eds.; Springer International Publishing: Cham, Switzerland, 2015; pp. 45–59.
24. Zhang, C.; Liu, S.; Wu, S.; Jin, S.; Reis, S.; Liu, H.; Gu, B. Rebuilding the Linkage between Livestock and Cropland to Mitigate Agricultural Pollution in China. *Resour. Conserv. Recycl.* **2019**, *144*, 65–73. [CrossRef]
25. Kamilaris, A.; Engelbrecht, A.; Pitsillides, A.; Prenafeta-Boldú, F.X. Transfer of Manure as Fertilizer from Livestock Farms to Crop Fields: The Case of Catalonia. *Comput. Electron. Agric.* **2020**, *175*, 105550. [CrossRef]
26. Li, J.; Akdeniz, N.; Kim, H.H.M.; Gates, R.S.; Wang, X.; Wang, K. Optimal Manure Utilization Chain for Distributed Animal Farms: Model Development and a Case Study from Hangzhou, China. *Agric. Syst.* **2021**, *187*, 102996. [CrossRef]

27. Bai, Z.; Ma, W.; Ma, L.; Velthof, G.L.; Wei, Z.; Havlík, P.; Oenema, O.; Lee, M.R.F.; Zhang, F. China's Livestock Transition: Driving Forces, Impacts, and Consequences. *Sci. Adv.* **2018**, *4*, eaar8534. [CrossRef] [PubMed]
28. Bai, Z.; Ma, L.; Jin, S.; Ma, W.; Velthof, G.L.; Oenema, O.; Liu, L.; Chadwick, D.; Zhang, F. Nitrogen, Phosphorus, and Potassium Flows through the Manure Management Chain in China. *Environ. Sci. Technol.* **2016**, *50*, 13409–13418. [CrossRef]
29. Zhang, T.; Hou, Y.; Meng, T.; Ma, Y.; Tan, M.; Zhang, F.; Oenema, O. Replacing Synthetic Fertilizer by Manure Requires Adjusted Technology and Incentives: A Farm Survey across China. *Resour. Conserv. Recycl.* **2021**, *168*, 105301. [CrossRef]
30. Billen, G.; Lassaletta, L.; Garnier, J. A Biogeochemical View of the Global Agro-Food System: Nitrogen Flows Associated with Protein Production, Consumption and Trade. *Glob. Food Secur.* **2014**, *3*, 209–219. [CrossRef]
31. Chadwick, D.; Wei, J.; Yan'an, T.; Guanghui, Y.; Qirong, S.; Qing, C. Improving Manure Nutrient Management towards Sustainable Agricultural Intensification in China. *Agric. Ecosyst. Environ.* **2015**, *209*, 34–46. [CrossRef]
32. Norse, D.; Ju, X. Environmental Costs of China's Food Security. *Agric. Ecosyst. Environ.* **2015**, *209*, 5–14. [CrossRef]
33. Zhao, J.; Liu, X.-P.; Liu, X.H.; Tian, T. Analysis on Coupling Coordination Degree Between Agriculture and Animal Husbandry Systems in Tarim River Basin. *Arid Land Geogr.* **2015**, *38*, 1077–1084.
34. Weng, Q.; Lian, H.; Qin, Q. Spatial Disparities of the Coupling Coordinated Development among the Economy, Environment and Society across China's Regions. *Ecol. Indic.* **2022**, *143*, 109364. [CrossRef]
35. Liu, X.; Li, S. Temporal and spatial distribution characteristcs of crop straw nutrient resources and returning to farmland in China. *Trans. Chin. Soc. Agric. Eng.* **2017**, *33*, 1–19.
36. LIU, X.; WANG, X.; LI, S. Phosphorus loading rates from livestock and poultry faeces, and environmental evaluation in China. *J. Agro-Environ. Sci.* **2019**, *38*, 2594–2608.
37. Liu, X.; Li, S. Temporal and spatial distribution of nutrient resource from livestock and poultry feces and its returning to cropland. *Trans. Chin. Soc. Agric. Eng.* **2018**, *34*, 1–14.
38. Acutis, M.; Alfieri, L.; Giussani, A.; Provolo, G.; Guardo, A.D.; Colombini, S.; Bertoncini, G.; Castelnuovo, M.; Sali, G.; Moschini, M.; et al. ValorE: An Integrated and GIS-Based Decision Support System for Livestock Manure Management in the Lombardy Region (Northern Italy). *Land Use Policy* **2014**, *41*, 149–162. [CrossRef]
39. Qian, Y.; Song, K.; Hu, T.; Ying, T. Environmental Status of Livestock and Poultry Sectors in China under Current Transformation Stage. *Sci. Total Environ.* **2018**, *622–623*, 702–709. [CrossRef]
40. Bao, W.; Liu, J.; An, J.; Xie, G. Discussion on value-taking of relative parameters for assessment of livestock and poultry excrement resource in China. *Trans. Chin. Soc. Agric. Eng.* **2018**, *34*, 314–322.
41. Song, D.; Hou, S.; Wang, X.; Liang, G.; Zhou, W. Nutrient resource quantity of animal manure and its utilization potential in China. *J. Plant Nutr. Fertil.* **2018**, *24*, 1131–1148.
42. Li, J.; Yuan, W.; Qin, X.; Qi, X.; Meng, L. Coupling Coordination Degree for Urban Green Growth between Public Demand and Government Supply in Urban Agglomeration: A Case Study from China. *J. Environ. Manag.* **2022**, *304*, 114209. [CrossRef]
43. Bian, D.; Yang, X.; Xiang, W.; Sun, B.; Chen, Y.; Babuna, P.; Li, M.; Yuan, Z. A New Model to Evaluate Water Resource Spatial Equilibrium Based on the Game Theory Coupling Weight Method and the Coupling Coordination Degree. *J. Clean. Prod.* **2022**, *366*, 132907. [CrossRef]
44. Dong, F.; Li, W. Research on the Coupling Coordination Degree of "Upstream-Midstream-Downstream" of China's Wind Power Industry Chain. *J. Clean. Prod.* **2021**, *283*, 124633. [CrossRef]
45. Gan, L.; Shi, H.; Hu, Y.; Lev, B.; Lan, H. Coupling Coordination Degree for Urbanization City-Industry Integration Level: Sichuan Case. *Sustain. Cities Soc.* **2020**, *58*, 102136. [CrossRef]
46. Zhang, Y.; Fu, Y.; Kong, X.; Zhang, F. Prefecture-Level City Shrinkage on the Regional Dimension in China: Spatiotemporal Change and Internal Relations. *Sustain. Cities Soc.* **2019**, *47*, 101490. [CrossRef]
47. Zhang, Z.; Li, Y. Coupling Coordination and Spatiotemporal Dynamic Evolution between Urbanization and Geological Hazards—A Case Study from China. *Sci. Total Environ.* **2020**, *728*, 138825. [CrossRef]
48. Wang, J.-F.; Zhang, T.-L.; Fu, B.-J. A Measure of Spatial Stratified Heterogeneity. *Ecol. Indic.* **2016**, *67*, 250–256. [CrossRef]
49. Wang, J.-F.; Li, X.-H.; Christakos, G.; Zhang, T.; Gu, X.; Zheng, X.-Y. Geographical Detectors-Based Health Risk Assessment and Its Application in the Neural Tube Defects Study of the Heshun Region, China. *Int. J. Geogr. Inf. Sci.* **2010**, *24*, 107–127. [CrossRef]
50. WANG, J.; XU, C. Geodetector: Principle and prospective. *Acta Geogr. Sin.* **2017**, *72*, 116–134.
51. Han, D.; Wiesmeier, M.; Conant, R.T.; Kühnel, A.; Sun, Z.; Kögel-Knabner, I.; Hou, R.; Cong, P.; Liang, R.; Ouyang, Z. Large Soil Organic Carbon Increase Due to Improved Agronomic Management in the North China Plain from 1980s to 2010s. *Glob. Chang. Biol.* **2018**, *24*, 987–1000. [CrossRef] [PubMed]
52. Li, Y.; Sun, Z.; Accatino, F. Satisfying Meat Demand While Avoiding Excess Manure: Studying the Trade-off in Eastern Regions of China with a Nitrogen Approach. *Sci. Total Environ.* **2022**, *816*, 151568. [CrossRef]
53. Li, J.; Yang, W.; Liu, L.; Liu, X.; Qiu, F.; Ma, X. Development and Environmental Impacts of China's Livestock and Poultry Breeding. *J. Clean. Prod.* **2022**, *371*, 133586. [CrossRef]
54. Zhang, W.; Cao, G.; Li, X.; Zhang, H.; Wang, C.; Liu, Q.; Chen, X.; Cui, Z.; Shen, J.; Jiang, R.; et al. Closing Yield Gaps in China by Empowering Smallholder Farmers. *Nature* **2016**, *537*, 671–674. [CrossRef] [PubMed]
55. Neset, T.-S.; Cordell, D.; Mohr, S.; VanRiper, F.; White, S. Visualizing Alternative Phosphorus Scenarios for Future Food Security. *Front. Nutr.* **2016**, *3*, 47. [CrossRef]

56. Jin, X.; Bai, Z.; Oenema, O.; Winiwarter, W.; Velthof, G.; Chen, X.; Ma, L. Spatial Planning Needed to Drastically Reduce Nitrogen and Phosphorus Surpluses in China's Agriculture. *Environ. Sci. Technol.* **2020**, *54*, 11894–11904. [CrossRef] [PubMed]
57. Fan, Y.; He, L.; Liu, Y.; Wang, S. Optimal Cropping Patterns Can Be Conducive to Sustainable Irrigation: Evidence from the Drylands of Northwest China. *Agric. Water Manag.* **2022**, *274*, 107977. [CrossRef]
58. Yin, L.; Tao, F.; Chen, Y.; Wang, Y. Reducing Agriculture Irrigation Water Consumption through Reshaping Cropping Systems across China. *Agric. For. Meteorol.* **2022**, *312*, 108707. [CrossRef]
59. Andrade, E.P.; Bonmati, A.; Esteller, L.J.; Vallejo, A.A. Assessment of Social Aspects across Europe Resulting from the Insertion of Technologies for Nutrient Recovery and Recycling in Agriculture. *Sustain. Prod. Consum.* **2022**, *31*, 52–66. [CrossRef]

agriculture

Article

Efficacy of Public Extension and Advisory Services for Sustainable Rice Production

Khodran Alzahrani, Mubashar Ali, Muhammad Imran Azeem and Bader Alhafi Alotaibi *

Department of Agricultural Extension and Rural Society, College of Food and Agriculture Sciences, King Saud University, Riyadh 11451, Saudi Arabia
* Correspondence: balhafi@ksu.edu.sa; Tel.: +966-504240201

Abstract: Agriculture is an integral constituent of Pakistan's economy and the primary source of livelihood for nearly 65% of the population living in rural areas. Rice is the second major staple food after wheat and a significant source of foreign exchange earnings through Basmati exports. Pakistan has established an extensive network of agricultural extension to educate the farming community about modern agricultural practices for enhancing the agricultural productivity of major food crops grown in the country. The present study was undertaken to evaluate rice farmers' views about public extension services and to identify their perspective regarding various ways of enhancing rice production in Pakistan. A multi-stage simple random sampling technique was employed, and data were collected from 193 rice farmers with the help of structured interviews using a pre-tested questionnaire. The findings revealed that a vast majority of the rice farmers were poorly satisfied with the public extension services. The results of the Spearman Rank-Order Correlation showed that landholding size had a significant effect on deciding extension contact; public extension agents are more likely to visit and serve those rice farmers who possess large landholders and therefore have the tendency to intentionally neglect small-scale rice farmers. For enhancing rice production in Pakistan, farmers believed that the provision of subsidized agricultural inputs and a minimum support price for rice is indispensable. Based on our findings, we suggest that to make public extension services more effective, public extension agents should particularly focus on the capacity building of small-scale farmers rather than large-scale farmers. Moreover, there is a need to broaden the scope of public extension services from simple crop protection measures to a set of comprehensive sustainable agricultural practices for increasing agricultural productivity, resource-use efficiency, as well as resilience toward adverse impacts of climate change.

Keywords: public extension; capacity building; smallholder farmers; Basmati rice

Citation: Alzahrani, K.; Ali, M.; Azeem, M.I.; Alotaibi, B.A. Efficacy of Public Extension and Advisory Services for Sustainable Rice Production. *Agriculture* **2023**, *13*, 1062. https://doi.org/10.3390/agriculture13051062

Academic Editors: Moucheng Liu, Xin Chen and Yuanmei Jiao

Received: 14 April 2023
Revised: 12 May 2023
Accepted: 13 May 2023
Published: 16 May 2023

1. Introduction

Pakistan is the sixth most populous country of the world, with a population over 190 million people [1,2]. In terms of Purchasing Power Parity, it is the 24th largest economy of the world, while it ranks 44th in terms of nominal Gross Domestic Product (GDP) [3–5]. Agriculture has been the mainstay of Pakistan's economy. Although the share of the agriculture sector in the national economy has consistently declined over the last few decades, it is still an integral part of the economy. It contributes about 22% to the national GDP. Moreover, it is a major source of employment for the country's workforce; about 37% of the labor force is employed in this sector. Over 65% of the people living in rural and remote areas rely on to sustain their livelihoods. It is also the main source of foreign exchange earnings; nearly three-fourths of the exports are agro-based products. Various domestic manufacturing industries are dependent on agriculture for the provision of raw materials [6–8]. The total arable land in the country is around 30.9 million hectares (Mha), out of which 24.1 Mha is under different crops [9]. Important crops of the country include wheat, rice, cotton, sugarcane, and maize. Besides its importance for the economy, it is vital to the national food security and economic stability of both rural and urban populations amid rapid population growth [10].

Rice (*Oryza sativa* L.) is an important cash crop in Pakistan. It is the second major staple food after wheat in the country [11,12]. The total area under rice cultivation in different parts of the country is around 3.53 million hectares; much of this area is located in the Punjab province that is also the main province in terms of rice production [6]. Within Punjab, there are certain areas that are well known for Basmati rice cultivation. These areas are collectively known as the "Collar" tract (locally known as Kalar tract) of rice, and they include the following districts of Punjab: Gujranwala, Hafizabad, Sialkot, Narowal, Sheikhupura, Nankana, Gujrat, and Mandi Bahauddin. Both the soils and climatic conditions of these areas particularly suit Basmati rice cultivation compared with other areas in the country [13–16]. Pakistan's total rice production stands at nearly 9.32 million tons [6]. Besides being a staple food in the country, rice exports are also a considerable source of foreign exchange earnings. Pakistan is among one of the world's largest exporters of rice. According to the Trade Development Authority of Pakistan, the total worth of rice exports was estimated to be around $2.04 billion during the 2021 fiscal year [17].

Although the term "sustainable development" has multiple definitions and interpretations, the most popular and widely used definition is: "Sustainable development is development that meets the needs of the present without compromising the ability of future generations to meet their own needs" [18]. Since the publication of the Brundtland Report by WECD, the concept of sustainable development has significantly evolved to add more focus toward resource-use optimizations, conservation of natural resources, environmental protection, and social equality and inclusion [19]. Some scholars [20] even argue for revisiting the prevailing notion of sustainable development as an analytical framework to guide international development endeavors in the context of recent unprecedented health and economic crises. In order to achieve sustainable development, since 2015, all UN member states have adopted a set of 17 Sustainable Development Goals (also known as the Sustainable Development Agenda 2030). Being universal in nature, these global goals aim to alleviate poverty, protect the environment, and ensure sustainable, resilient, and prosperous societies across the globe [21–23]. In the context of food and agriculture, the concept of sustainable development refers to all those sustainable agricultural and food practices that aim to ensure food security for all the people on planet Earth without their overexploitation and to reduce the global carbon footprint of agriculture, contributing toward climate change adaptation and mitigation [24,25]. Agriculture, forestry, and other land-use practices collectively contribute about 24–30% of the total global greenhouse gas (GHG) emissions [26,27]. Therefore, the adoption of sustainable agricultural practices that aim to reduce GHG emissions can play a crucial role in climate change mitigation. Therefore, the institutional function of agricultural extension in this context is to promote the adoption of sustainable and climate-smart agricultural practices among the farming community by raising their level of awareness, knowledge, and skills using all possible means, especially in the developing and under-developed countries, where a large proportion of the farmers are relatively less educated and still use traditional agricultural practices that are not only resource-intensive but are also becoming less profitable and non-competitive.

Pakistan has established an extensive network of public agricultural extension across the country in order to disseminate agricultural information, educate and train the farming community about modern sustainable agricultural practices for enhancing agricultural productivity and economic growth, and alleviate poverty in the rural areas [28–30]. Public agricultural extension refers to the extension and advisory services that are provided by the government's Agriculture Department without any service fee. Each province has its own independent public Agriculture Department with its affiliated agricultural extension wing that works under the aegis of respective provincial ministries of agriculture. The Agriculture Department has its offices in every tehsil (an administrative subdivision of a district) for the provision of extension and advisory services. Moreover, the Agriculture Department is systematically linked with a network of "Adaptive Research Stations" that are tasked with the creation and testing of innovative agricultural practices applicable under local agro-climatic conditions and which cater to farmers' needs to enable them to be competitive in continuously evolving agricultural markets. At the Markaz level (a collection of a specific number of villages), the Agriculture Department deploys an extension officer (a university graduate) along with two subordinate agricultural field assistants who are mainly involved in the provision of extension and advisory services to the farming community. Figure 1 provides an overview of the functional organization of public agricultural extension in the province of Punjab, Pakistan. The private sector is also actively engaged in the delivery of extension services to the farmers [31–34]. Major private companies that participate in the provision of extension services to the farmers are mainly the input suppliers of seeds, herbicides, and pesticides. Unlike public extension, these profit-oriented companies and enterprises particularly focus on educating and training farmers regarding the use of their products for maximizing their product sales. However, despite such a large network of both public and private extension, agricultural productivity of the major crops grown in the country is relatively low compared with other neighboring nations and under similar farming systems [8,35–38].

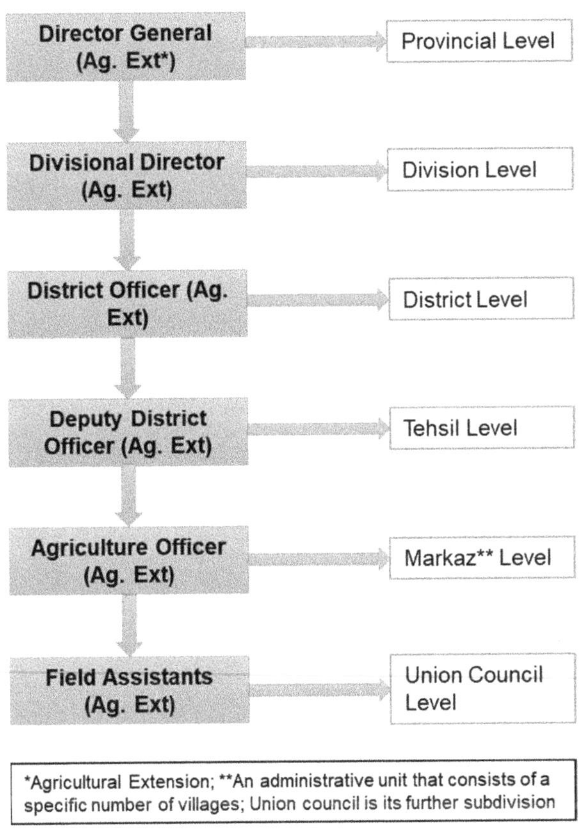

Figure 1. Organizational hierarchy of agricultural extension in Punjab, Pakistan.

Low agricultural productivity can be attributed to several distinct factors; however, a significant driver of this is farmers' low adoption of modern production practices owing to their lack of or poor technical knowledge and farm management skills [39–44]. Amid the severe financial crisis in the country, it is difficult to justify huge public investment for maintaining a large public extension network and institutions without having any significant impact on national agricultural development and food security. One way to assess the impact is to explore the performance of the public Agriculture Department in terms of the extension and advisory services it provides to the farming community that are aimed at their education and capacity development regarding modern agricultural techniques and practices. In this context, the present study was designed to assess rice farmers' views about public extension services and to analyze their own perspective about different ways of enhancing rice production, both for meeting domestic needs and for exporting to other countries.

2. Materials and Methods

2.1. Description of the Study Area

The present study was conducted in the Gujranwala district of Punjab, Pakistan. Gujranwala is further subdivided into different areas known as towns and tehsils for administrative purposes. These include Qila Didar Singh, Aroop, Khiali Shahpur, Nandipur, Wazirabad, Kamoke, and Nowshera Virkan. Each tehsil comprises several villages, and it is the smallest administrative unit [45]. The total area of Gujranwala is around 3622 sq.km [45]. As per the 2017 census, the total population of the district is about 5.01 million [45,46]. The district experiences a semi-arid climate with fluctuations throughout the year. Temperatures during the summer season may reach up to 42 °C. During the winter season, the temperature may drop to 7 °C. The highest amount of precipitation occurs during the monsoon season (July–Aug). During other periods of the year, the average precipitation is about 25 mm. Most of the rural people of the area are engaged in farming. The total cultivated area of the district is about 0.778 million acres [45,47]. Pakistan's best-quality Basmati Rice, which is known for its peculiar aroma, is transplanted on vast tracts of land in this area. Overall, the main crops grown in the area include wheat, rice, maize, millet, and oilseed crops, such as sunflower and canola [48]. Rice is the major Kharif season crop, covering about 93% of the cultivated area. In the Rabi season, 80% of the cultivated area is under a wheat crop. Perennial canals as well as abstraction of groundwater through tube wells are the main major prime sources of irrigation. According to official sources, about 38% of the farmers are classified as small landholders, 54% as medium, and around 8% as large-scale commercial farmers [47]. Figure 2 shows the map of the study area.

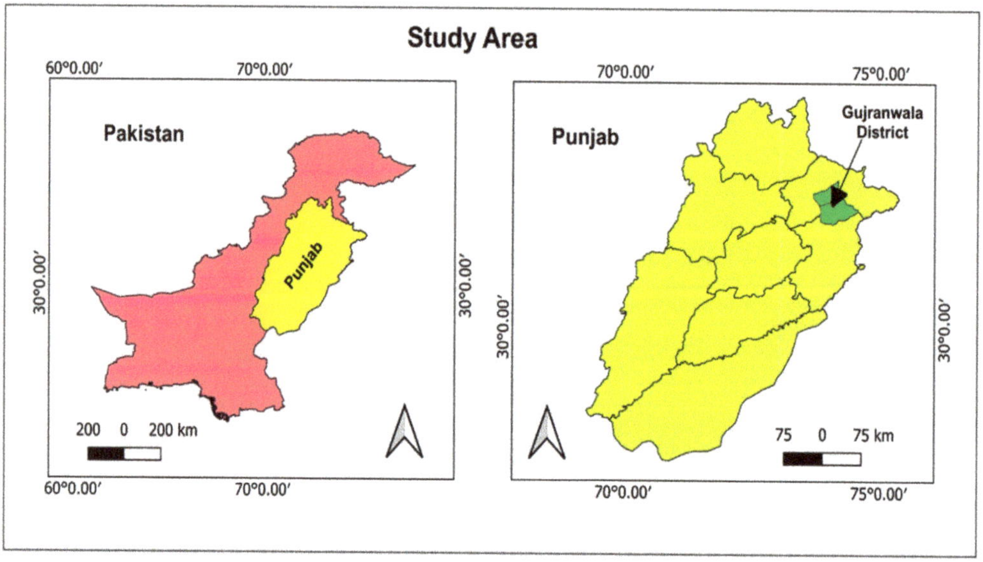

Figure 2. Map of the study area.

2.2. Research Design

A cross-sectional survey was employed as a research design to implement the study. In terms of sampling, we adopted a multi-stage random sampling approach. In the first stage, two tehsils (Kamoke and Nowshera Virkan) of the Gujranwala district were randomly selected. In the second stage, 20 villages from both the tehsils (10 from each one) were randomly selected. In the third stage, 200 farmers (10 farmers from each of the 20 randomly selected villages) were selected for final data collection. The research questionnaire was developed by a group of researchers in the Department of Agricultural Extension and Rural Society at the King Saud University, Riyadh, Saudi Arabia. The approval of the Research Ethics Committee of Deanship of Scientific Research at King Saud University was also obtained before initiating the process of data collection. Moreover, the informed consent of the farmers was taken verbally before collecting the data. We clearly explained to them that their participation was not mandatory, and the collected data would only be used for academic purposes. Data were collected using structured interviews; questions using the same wording and order were asked of all the farmers to ensure a standardized pattern. Out of the 200 selected farmers, 7 were not available for the interview. Each interview lasted for around 35–40 min. Before final data collection, a pilot study involving 30 farmers was conducted to test the questionnaire and to measure the internal consistency of the Likert Scale designed for determining farmers' views about public extension services. The Cronbach alpha run for reliability analysis yielded a score of 0.83. Several studies report that an alpha coefficient value above 0.70 indicates a high level of internal consistency on the Likert Scale [49–53].

2.3. Research Instrument

The research questionnaire was divided into three different sections. In the first section, questions related to demographic and socio-economic characteristics were included. It contained the following questions: age, level of formal education, farming experience, landholding size, type of land ownership, and sources of income of the rice farmers. The second section included questions related to rice farmers' views about the extension and advisory services delivered by the public Agriculture Department (AD). The rice farmers were asked about different extension services provided by the AD that are aimed at educating them and improving their knowledge and skills regarding rice cultivation. A five-point Likert scale (1 = Strongly Disagree; 2 = Disagree; 3 = Undecided; 4 = Agree; 5 = Strongly Agree) was used to determine their views about public extension services. In the last section, questions related to public extension officers' visit to the farmers, farmers' various sources of agricultural information, and their perspective about ways of enhancing rice production in Punjab were included.

2.4. Data Analysis

Both descriptive and inferential statistics were used to analyze the data. Demographic and socio-economic characteristics of the farmers were summarized using frequencies and percentages. Farmers' views about public extension services were also tabulated using percentages. "Strongly disagree" and "disagree" categories were merged into one category of "disagree", while "strongly agree" and "agree" categories on the Likert scale were merged into one category of "agree". Based on their scores on the Likert scale, a new ordinal variable was computed to classify farmers into three different groups (1 = Poorly satisfied; 2 = Moderately satisfied; 3 = Highly satisfied). In order to find a correlation between ordinal demographic and socio-economic variables (age, education level, farming experience, landholding size, extension agents' visit to farmers) and farmers' views about public extension services, the Spearman Rank-Order correlation was employed. For nominal variables (land ownership and income sources), Mann–Whitney and Kruskal–Wallis tests were used. All the analyses were run using Statistical Package for Social Sciences (IBM SPSS v27.0). Figure 3 presents an overview of the methodological framework.

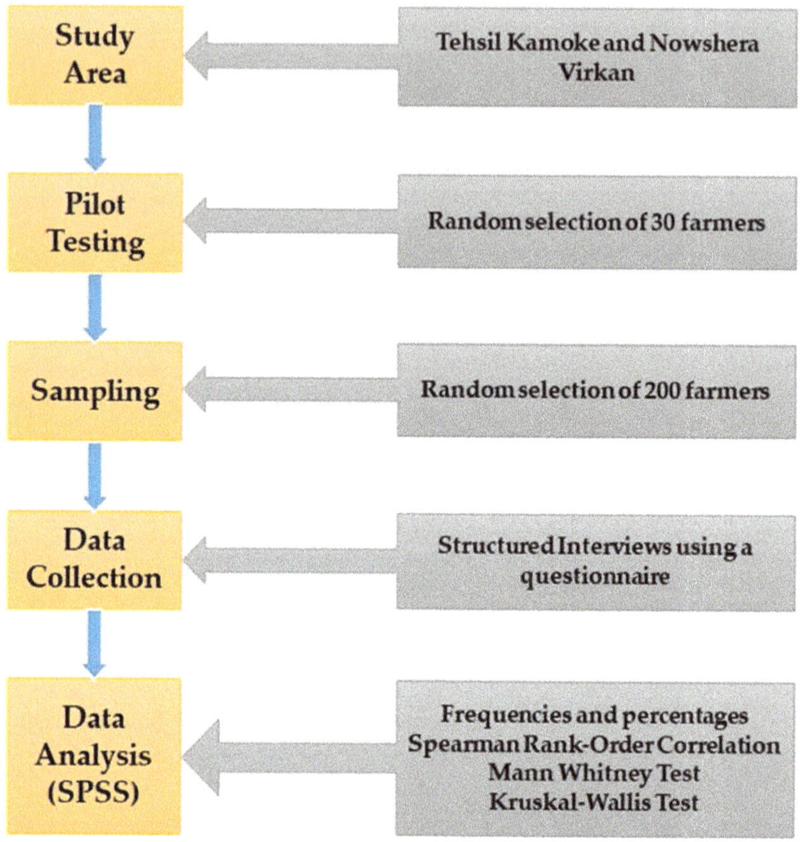

Figure 3. Methodological framework.

3. Results

3.1. *Socio-Economic Characteristics of the Rice Farmers*

Table 1 shows the demographic and socio-economic characteristics of the rice farmers in the study area. About 24% of the farmers were below 40 years of age. Nearly two-fifths of them (39%) were between 41–50 years of age. Around 37% of the farmers were above 50 years of age. About 18% of the farmers had no formal education. Around 26% of them said that they had attended formal schooling, but only up to primary level. Nearly 24% of the sampled farmers had educational qualifications up to middle standard. A low percentage of the farmers reported an education level of Matric or beyond; only 15% of the rice farmers attained an education level of at least Matric, whereas about 17% of them reported that they had obtained higher educational qualifications. About 43% of the respondents had farming experience between 11–20 years; around one-fourth of them had farming experience of more than 20 years. Collectively, about three-fourths of the farmers had less than 20 years of farming experience. The majority of the farmers (60%) in the study area possessed agricultural lands below 20 acres (8.09 ha); out of these, about 31% reported landholding sizes below 10 acres (4.04 ha). About 22% of the farmers had land of more than 30 acres (12.14 ha). A vast majority of the rice farmers (89%) were owners of their lands; only around 11% of them reported that they had rented agricultural lands for farming. The sources of income of around 43% of the rice farmers were both agriculture and other small side businesses. Around 38% of the farmers indicated that agriculture was their sole source of income. Apart from this, about one-fifth of them reported that they had other businesses as a main source of income, not agriculture.

Table 1. Rice farmers' socio-economic characteristics.

Variable	Frequency (n = 193)	Percent	Variable	Frequency (n = 193)	Percent
Age			Education Level		
Below 30 years	12	6.2	Illiterate	34	17.6
31–40 years	35	18.1	Primary (Grade 5)	51	26.4
41–50 years	76	39.4	Middle (Grade 8)	47	24.4
51–60 years	53	27.5	Matric (Grade 10)	29	15.0
Above 60 years	17	8.8	Above Matric	32	16.6
Farming Experience			Income Sources		
Below 10 years	61	31.6	Purely agriculture	73	37.8
11–20 years	83	43.0	Other businesses	37	19.2
Above 20 years	49	25.4	Both	83	43.0
Landholding Size			Land Ownership		
Below 10 acres	60	31.1	Owner	171	88.6
11–20 acres	55	28.5	Tenant	22	11.4
21–30 acres	35	18.1			
Above 30 acres	43	22.3			

3.2. *Rice Farmer's Views about Public Extension Services*

Table 2 depicts the results of rice farmers' views about public extension services. Around 61% of the rice farmers disagreed with the statement that the public Agriculture Department (AD) conducted field demonstrations about new crop varieties and modern production practices during the rice season. They (61%) also disagreed regarding the random selection of farmer's lands for field demonstration; only one-fifth of the farmers agreed that field demonstrations were performed on randomly selected farmer fields. Over half (57%) of the farmers believed that the AD did not provide information about land preparation for nursery sowing and rice transplanting. About 64% disagreed with the statement that the AD provided information regarding irrigation and fertilizer application methods for rice crops. Over half (54%) of the farmers expressed disagreement regarding the provision of information about crop protection measures; around 42% believed that the AD provided information to control insects, pests, and diseases. Around 68% disagreed with the statement that the AD provided information regarding good post-harvest practices for rice crop management. The majority of the farmers (79%) disagreed that the AD educated the rice farmers to acquire better marketing skills. They (79%) disagreed that the AD helped rice farmers to sell their rice crop after harvesting at a profitable price. A vast majority (81%) agreed that the AD did not conduct extension programs for educating rice farmers before the start of the rice season. Around 73% of the farmers were also convinced that the AD did not use both electronic and print media effectively for the dissemination of information about modern production practices related to rice crops. About 79% of the farmers indicated that the AD did not provide timely information. The majority of them (81%) also believed that information provided by the AD was not relevant to rice farmers' needs. A vast majority (83%) was convinced that the AD did not play a role in the timely provision of subsidized agricultural inputs. About 71% believed that extension staff were not available when they visited the AD. About three-fourths of the rice farmers were convinced that the overall performance of the AD was not satisfactory. Based on the farmers' views about public extension services, they were classified into three distinct categories. About 76% of the farmers were poorly satisfied with the extension services provided by the AD, 10% were moderately satisfied, while 14% of them were highly satisfied (Figure 4).

Table 2. Rice farmers' views about public extension services.

Statements	Disagree (%)	Undecided (%)	Agree (%)	Mean (n = 193)	Standard Deviation
Agriculture Department (AD) conducts field demonstrations about new crop varieties and modern production practices during the rice season.	60.6	6.7	32.6	1.72	0.927
Field demonstrations are conducted on randomly selected farmer fields without any bias.	61.1	18.1	20.7	1.60	0.812
AD provides information about land preparation for nursery sowing and rice transplanting.	56.5	11.4	32.1	1.76	0.912
AD provides information about irrigation and fertilizer application methods.	64.2	6.7	29.0	1.65	0.902
AD provides information about rice crop protection measures to control insects, pests, and diseases.	53.9	4.7	41.5	1.88	0.971
AD provides information about post-harvest practices for rice crop management.	67.9	5.7	26.4	1.59	0.880
AD educates farmers to acquire marketing skills to earn more profit.	78.8	9.8	11.4	1.33	0.671
AD helps rice farmers to sell their produce at a profitable price.	79.3	4.1	16.6	1.37	0.754
AD conducts extensions programs for educating rice farmers before the start of the rice season.	80.8	2.6	16.6	1.36	0.751
AD effectively uses both print and electronic media to disseminate information about modern rice production practices.	72.5	15.5	11.9	1.39	0.692
AD provides extension services in a timely manner.	79.3	5.7	15.0	1.36	0.730
Extension services provided by AD are relevant to rice farmers' needs.	80.8	4.7	14.5	1.34	0.718
AD plays a role in the timely provision of agricultural inputs at subsidized rates.	82.9	4.1	13.0	1.30	0.687
When you visit AD, extension staff is available.	71.0	8.3	20.7	1.50	0.817
The overall performance of AD is satisfactory.	75.1	9.8	15.0	1.40	0.737

Statements were measured using a 5-point Likert scale (1 = Strongly Disagree to 5 = Strongly Agree). However, in the final analysis, these categories were combined into three categories (1 = Disagree; 2 = Undecided; 3 = Agree).

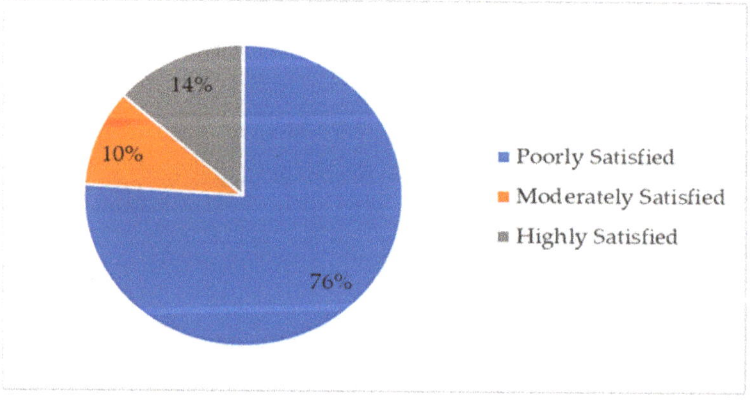

Figure 4. Rice farmers' classification according to their satisfaction regarding public extension services.

3.3. Extension Agents' Visit to Farmers, Farmers' Sources of Information, and Their Perspective about Ways of Enhancing Rice Production

Table 3 describes the results of extension agents' visit to rice farmers, their sources of agricultural information, and their perspective about different ways of enhancing rice production. About 60% of the farmers indicated that extension agents never visited their farms. Around 17% of them revealed that they were visited by the public extension agents only once in 6 months. Only 13% of the farmers indicated that extension agents visited their farms on a monthly basis.

Regarding sources of information, nearly half (47%) of the farmers disclosed that neighboring farmers were their main source of agricultural information. About 16% of the farmers indicated extension agents as their source of agricultural information. The internet was also used as a source of information by around 15% of the rice farmers. Radio as a source of agricultural information was only used by a small percentage (3.6%) of the farmers.

Rice farmers were also asked about their perspective on enhancing rice production. About 39% of the farmers believed that the provision of subsidized agricultural inputs to the farmers can play a significant role in enhancing rice production in Pakistan. About 35% were convinced that a suitable support price set by the government for the rice crop can also enhance rice production. Farmers' capacity building was seen as less important; only 14% of the farmers believed that building the capacity of farmers can be helpful in raising rice production. Around 12% of them believed that the provision of credit services to the farmers is important for enhancing rice production.

Table 3. Extension agents' visit to farmers, farmers' sources of information, and their perspective about enhancing rice production.

Variable	Frequency (n = 193)	Percent
Extension Agents' Visit to Farmers		
Monthly	25	13.0
Once in 3 months	21	10.9
Once in 6 months	32	16.6
Never	115	59.6
Farmers' Sources of Agricultural Information		
Extension Agents	31	16.1
Neighboring Farmers	91	47.2
Print Media	19	9.8
TV	17	8.8
Radio	7	3.6
Internet	28	14.5
Ways of Enhancing Rice Production		
Farmers' capacity building	26	13.5
Provision of subsidized agricultural inputs	75	38.9
Provision of credit services	24	12.4
Ensuring a support price by government for rice crops	68	35.2

3.4. Relationship of Socio-Economic Characteristics with Farmers' Satisfaction Regarding Public Extension Services

Table 4 shows the results of the Spearman Rank-Order Correlation. The non-parametric correlation was run to find relationships between ordinal socio-economic variables (age, education, farming experience, landholding size, and extension agents' visit to farmers) and rice farmers' views about public extension services (ordinal variable computed from 15 Likert Scale items). The analysis revealed that age and farming experience were not statistically significantly correlated with the farmers' views about public extension services. There was a significant positive correlation between education level and farmers' views about public extension services (r_s = 0.180; p = 0.012). However, the value of correlation coefficient suggests that it is a weak correlation.

Table 4. Spearman Rank-Order Correlation.

Independent Variables [a]	Correlation Coefficient (r_s)	*p*-Value
Age	0.044	0.545
Education Level	0.180 *	0.012
Farming Experience	0.088	0.226
Landholding Size	0.744 **	<0.001
Extension Agents' Visits to Farmers	−0.638 **	<0.001
Landholding size [b]	−0.774 **	<0.001

[a] Dependent variable is farmers' satisfaction regarding public extension services (1 = poorly satisfied; 2 = moderately satisfied; 3 = highly satisfied). * Correlation is significant at the 0.05 level (2-tailed). ** Correlation is significant at the 0.01 level (2-tailed). [b] Dependent variable is extension agents' visits to farmers (1 = monthly; 2 = once in 3 months; 3 = once in 6 months; 4 = never).

The analysis indicated that there was a significant positive correlation between landholding size and farmers' views about public extension services ($r_s = 0.744$; $p \leq 0.001$). The value of the correlation coefficient reveals that it is a strong correlation. In other words, those rice farmers who have more land area are more likely to be highly satisfied with the public extension services. Extension agents' visits to farmers was also significantly negatively correlated with farmers' views about public extension services ($r_s = -0.638$; $p \leq 0.001$). The analysis of the correlation coefficient suggests that it is a moderate correlation. The more frequently extension agents visit the farmers, the more likely the farmers are to be highly satisfied with the public extension services.

A Spearman correlation test was also run in order to identify the relationship between landholding size and extension agents' visits to the farmers. The analysis revealed that there was a significant negative correlation between landholding size and extension agents' visits to the farmers. The value of the correlation coefficient suggests that it is a strong correlation ($r_s = -0.774$; $p = \leq 0.001$). The farmers who possess a large land area are more likely to be frequently visited by the extension agents.

Table 5 describes the results of non-parametric tests. The Mann–Whitney test was run in order to determine differences in farmers' views about public extension services as per their land ownership. The test revealed that there were no significant differences in the views of farmers who are owners of the land compared with those who are tenants ($U = 1838.50$; $p = 0.817$). To find differences in farmers' views about public extension services between those having different sources of income, the Kruskal–Wallis test was conducted. The results indicated that the views about public extension services by farmers having different income sources are not significantly different ($\chi^2 = 1.027$; $p = 0.598$).

Table 5. Results of Mann–Whitney and Kruskal–Wallis tests.

Independent Variable	Farmers' Satisfaction Regarding Public Extension Services [a] Mean Rank	Mann–Whitney *U*	*p*-Value
Land Ownership			
Owner (n = 171)	97.25	1838.50	0.817
Tenant (n = 22)	95.07		
Variable	Kruskal—Wallis Test Mean Rank	Chi Square	*p*-Value
Income Sources			
Purely agriculture (n = 73)	95.77	1.027	0.598
Other businesses (n = 37)	92.28		
Both (n = 83)	100.19		

[a] Dependent variable is farmers' satisfaction regarding public extension services (1 = poorly satisfied; 2 = moderately satisfied; 3 = highly satisfied).

4. Discussion and Implications

In this study, we attempted to assess rice farmers' views about public extension and advisory services as well as their perspective regarding different ways of enhancing rice production in Punjab, Pakistan. The analysis of the farmers' demographic and socio-economic profile reveals that the majority of the farmers are below 50 years of age and possess farming experience of 20 years or less. Age might have a relation with agricultural productivity as relatively young farmers are considered more innovative based on their ability to achieve higher overall agricultural productivity and profitability [54,55]. Farming experience can also enhance the farmers' technical capacity as well as their orientation for adopting improved agricultural practices [56–58]. About half of them have attained an educational level of 8th grade or below, whereas one-fifth have no formal education. In Pakistan, most of the people who are engaged in agriculture generally have low educational background [59,60]. This is also one of the major reasons for the low resource-use efficiency and overall productivity in the country. Several studies reported a significant relationship between education of the farmers and their technical efficiency [56,57].

Although a vast majority of the farmers are owners of their lands rather than being a tenant, the majority own relatively small lands (less than 8 hectares). According to the agricultural census of 2016–2017, around 90% of the farms in Punjab are less than 10 hectares and occupy about 69% of the land area in Punjab [61]. Farmers with large landholdings are generally more productive and have high economic potential [62–64]. Land ownership is also known to have an impact on the long-term sustainability of land; farmers who own land are not only more innovative, but they are also more concerned about the physical condition of their lands [65,66]. Nearly two-fifths of the farmers have agriculture as their sole source of income; however, a slightly greater proportion of them use both agriculture and other business activities to earn income. Fluctuations in the market and low profitability force farmers to explore other sources of livelihood generation in addition to agriculture to supplement their income.

Regarding rice farmers' views about public extension and advisory services, the findings reveal that a vast majority of the farmers are poorly satisfied with these services. They believe that Agriculture Department is not actively involved in the provision of various extension services, such as field demonstrations, land preparation for nursery sowing and rice transplanting, fertilizer and irrigation application methods, and post-harvest practices. They also believe that the Agriculture Department does not provide timely information and that the services provided are not relevant to rice farmers' evolving needs. The focus of the department is generally on providing information about crop protection measures, including insects, pests, and disease control and prevention. In Punjab, the Agriculture Department of the Government of Punjab is primarily responsible for the provision of agricultural information as well as educating and training farmers about modern production practices. Ideally, the department has the mandate to provide a diverse range of extension services covering all the aspects of crop production for the various crops grown in the province. However, our findings suggest that the Agriculture Department has failed in its efforts to provide advisory services to the rice farming community. One apparent reason for the rice farmers' dissatisfaction with public extension services is the lack of extension agents' contact with the rice farmers in the study area. Extension agents rarely conduct field visits and are not a major source of agricultural information for the farmers. Rather, rice farmers rely on neighboring farmers and other sources for agricultural information and advice. Farmers' access to extension and advisory services is known to positively influence their adoption of modern agricultural practices and can also increase their agricultural productivity as well as economic efficiency [58,60,65,67,68].

Public agricultural extension officials attribute the small extension workforce as the main cause of extension agents' lack of visits to the farmers. According to the Director General of Agriculture Punjab (Extension and Adaptive Research), it is not possible to frequently visit most of the farming community owing to the current small extension workforce in the province [33]. However, our findings suggest that landholding size is a significant factor in deciding access to extension. Extension agents prefer to frequently visit farmers with large landholdings, and they intentionally neglect small farmers. It also explains why the rice farmers with large landholdings expressed a high level of satisfaction regarding public extension services. Several studies identified this undesirable extension practice [60,67,69,70]; however, it seems that this biased practice still continues to prevail, negatively affecting the agricultural productivity and farm income of small-scale farmers.

An assessment of the rice farmers' perspective about various ways of enhancing rice production in Pakistan provides insights into the problems faced by the farming community. They think that the provision of subsidized agricultural inputs at affordable prices is indispensable for enhancing rice production under the prevailing circumstances. Since the start of 2022, the Pakistani currency (Pakistani Rupee) has depreciated by almost 55% in value against the US dollar and continues to depreciate at a rapid pace, leading the country toward economic collapse [71]. The depreciation of the Rupee has substantially increased the prices of basic agricultural inputs in the domestic market. High fuel and a corresponding rise in electricity prices has increased irrigation costs for rice. Rice is a water-intensive crop as it is transplanted on puddled soils instead of direct seeding in Pakistan. The rice seed market is also monopolized by the private sector. According to official data, out of 44,148 metric tons of total paddy seed requirement, about 40,037 metric tons are provided by private seed companies, whereas 4145 metric tons is obtained through import from other countries. The public institutions have a negligible share, procuring only 965 metric tons of seed [6]. Moreover, fertilizer prices are at an all-time high, especially phosphate and potash fertilizers, due to high energy costs. As the bulk of the agrochemicals are imported, their prices are also on a continuous rise. Use of agricultural inputs in recommended doses ensures increased agricultural productivity and farm income [72]. However, higher input prices considerably increase the costs of production, and it becomes increasingly difficult for resource-poor farmers to sustain their farming business due to a decline in yields and farm income. Decease in production levels of rice and high prices in the domestic market would affect national food security as rice is the second major staple food after wheat. Additionally, it would reduce foreign exchange earnings through a reduction in rice exports.

Another important strategy for enhancing rice production in Pakistan is to ensure a respectable Minimum Support Price (MSP) for rice growers. In the context of significantly rising input costs and a poor marketing system, this seems to be a valid demand by the farmers. The marketing system in Pakistan is not conducive for small-scale farmers as the supply chain consists of multiple intermediaries that not only add additional costs and inefficiencies but tend to exploit the smallholders by offering a low price for their produce that is not profitable for them [8]. Small farmers are not in a bargaining position because they have to immediately sell their crop in order to support their families and purchase inputs for the cultivation of the next crop. The announcement of a MSP by the government acts as a protection for smallholders as retailers are forced to buy from farmers at that price. In the case of wheat, the government announces the MSP each year to ensure that this main staple food crop is sown in enough areas by the farmers to meet the domestic food requirements; however, in case of rice, this is rare. Therefore, rice growers are making serious efforts to convince the government to use all viable channels of farmer-based organizations to offer price support. Besides cost subsidies, in many developing countries, governments employ MSP as an alternative subsidy scheme for both safeguarding smallholder farmers against market exploitation and price volatility and to encourage more production [73–76]. Extension departments can play a role in the implementation of MSP, especially in rural areas, as they are part of price control and regulatory committees formulated at the tehsil level. However, several studies reported that input subsidies can be a more effective policy intervention rather than the implementation of MSP because the latter may result in the loss of competitiveness in the international market [77–79].

Provision of credit services and farmers' capacity building are envisioned as relatively less important factors than subsidized agricultural inputs and minimum support price by the farmers for enhancing rice production. To understand the farmers' perspective, we need to contextualize the prevailing economic conditions; the country is on the brink of economic collapse, with low prospects of recovery, and inflation has risen to almost

27%, inflating prices of the agricultural inputs to record-high levels. Provision of credit facilities may not be able to provide any tangible relief with the provision of cost subsidies. The analysis of credit provision services also reveals that the current agricultural credit policy is biased toward large-scale and commercial farmers who already possess abundant resources. In Pakistan, more than fifty financial institutions provide agricultural loans to the farmers across the country. During the 2022 fiscal year, about 304 billion PKR were disbursed to large farmers (farmers above 12.5 hectares). On the other hand, only 170 billion rupees were given to small-scale and subsistence farmers, who constitute around 90% of the farmers [6]. This credit policy suggests that large farmers, who represent less than 10% of the country's farming community, have abundant agricultural credit, whereas small farmers have restricted access to credit services. Restricted access to credit may lower farmers' adoption capacity as well as the welfare of the farming families [60,80–85]. In terms of farmers' capacity building, the Agriculture Department should particularly focus on developing rice farmers' knowledge and skills to employ modern agricultural practices for improving agricultural productivity and resource-use efficiency. One of the main reasons for the low per-hectare agricultural yields of the major crops in Pakistan is the farmers' consistent use of traditional farming practices due to poor technical knowledge and management skills [42,83,86,87].

Agricultural development policymakers and sustainability practitioners can benefit from useful insights provided by the mindsponge theory for the creation, dissemination, and management of sustainable agricultural innovations and solutions. The mindsponge framework is a novel approach that elucidates how the human mind processes information received from different sources and forms eventual decisions using subjective cost-benefit judgements by applying various filtering mechanisms to align inflowing information with an existing core set of values. It also explains how the human mind influences thought processes and guides behaviors, and how it can be reinforced or modified using information as a resource. This framework helps us understand the innovation adoption process and enhances our understanding about why certain agricultural innovations, despite having potential usefulness and application, might be rejected or not adopted on a wide scale by the farming community if they are in contradiction with their existing core values [88,89]. Moreover, the mindsponge mechanism also forms the basis of the serendipity–mindsponge–3D knowledge management framework that can be effectively employed to understand and explain the processes of innovation creation, dissemination, and management in a rapidly changing era of technological advancements and information flooding. The 3D creativity management framework could be applied in different contexts on an individual, institutional, and national level for fostering innovations in an efficient and effective manner [90].

Another framework that effectively utilizes the mechanism of mindsponge theory and further extends the serendipity–mindsponge–3D knowledge management framework is mindspongeconomics or mindspongecon. Being a new framework of applied economics, it advocates the incorporation of environmental values into the planning and policy formulation processes for estimating the true worth of goods and services to affect potential decisions and corresponding behaviors of both producers and consumers of such services [91]. The agricultural development and environment policymakers of Pakistan can allocate substantial financial resources for cultivating environmental values in both the producers (farmers and growers) and consumers. Once they recognize the ongoing threat of climate change and understand that agriculture and the land-use sector are significant contributors of anthropogenic greenhouse gas emissions, it may affect their thinking to undertake individual actions to combat this challenge. Moreover, once environmental stewardship is assimilated into their core values, it would be easier to convince them to adopt sustainable agricultural production and consumption practices through relevant extension education and training programs. Additionally, the government and other relevant institutions that are involved in the generation of new agricultural innovations should work in close collaboration with each other to produce affordable and practical solutions for their large-scale adoption.

5. Conclusions

The current study was designed to identify rice farmers' views about public extension services and their perspective regarding various ways of enhancing rice production in Pakistan. A vast majority of the rice farmers were found to be poorly satisfied with the extension and advisory services provided by the Agriculture Department of the provincial government. Farmers that have large landholdings are highly satisfied with the public extension services because extension agents serve such farmers on a preferential basis. The public extension agents' biased orientation toward large-scale farmers poses serious implications for rural development and national food security. Small farmers not only constitute the bulk of the farming community, but they also generally lack advanced agricultural knowledge and skills and management skills. Extension agents' intentional neglect to serve these farmers would affect their agricultural productivity as well as farm income and profitability. Keeping in view the small extension work force as claimed by the extension officials, it is suggested that the Agriculture Department should recruit more agricultural graduates for the dissemination of agricultural innovations in rural areas. One viable option could be building public–private partnerships with the existing input supply companies that supply seeds and pesticides in the area. The government can grant them certain concessions in terms of tax and tariff reductions in return for their services to train the resource-poor farmers. Similarly, non-government organizations (NGOs) that are working on rural development could also be involved in this process by sharing their vision and recognizing their efforts.

Moreover, farmers' lack of ability to meet the food demands of a rapidly growing country's population due to low agricultural production compromises national food security. A decrease in rice production may also lower Basmati rice exports, which is a considerable source of export earnings. In order to streamline and make public extension services more effective and demand-driven for rice farmers in particular, and all farmers in general, the government should particularly focus on small-scale subsistence farmers to enhance their technical skills and managerial capacity. Equitable agricultural transformation and sustainable growth in this sector is difficult to achieve without small-scale farmers' capacity development. Besides, there is a need to broaden the scope of public extension services from simple crop protection measures to a more comprehensive set of sustainable agricultural and climate-smart practices that cover all the aspects of crop production and to ensure resilience of farming businesses in the wake of climate change. In addition to streamlining public extension services, governmental support for the provision of subsidized agricultural inputs at affordable prices and a minimum support price for rice, similar to that provided for wheat crops, is necessary to enhance rice production in Pakistan. The prevailing economic circumstances in the country suggest that if the government fails to provide any tangible relief in the form of cost subsidies, it will seriously affect the rural economy, and that would consequently have an impact on the urban consumers who rely on rural farmers for their food needs.

Author Contributions: Conceptualization, K.A. and M.A.; methodology, M.I.A., B.A.A. and M.A.; pilot testing and data collection, M.A. and M.I.A.; software and data analysis, M.I.A. and B.A.A.; resources, K.A.; writing—original draft. M.I.A. and B.A.A.; writing—review and editing, M.I.A., K.A., M.A. and B.A.A. All authors have read and agreed to the published version of the manuscript.

Funding: This research was funded by the Researchers Supporting Project (RSP2023R443), King Saud University, Riyadh, Saudi Arabia.

Institutional Review Board Statement: The study was approved by the Research Ethics Committee of Deanship of Scientific Research at King Saud University.

Data Availability Statement: The data is not publicly available. However, interested researchers may be given access to the data upon request to the Deanship of Scientific Research at King Saud University, Riyadh, Saudi Arabia.

Acknowledgments: The authors extend their appreciation to The Researchers Supporting Project number (RSP2023R443) King Saud University, Riyadh, Saudi Arabia.

Conflicts of Interest: The authors declare no conflict of interest.

References

1. Saleem, Z.; Hassali, M.A.; Versporten, A.; Godman, B.; Hashmi, F.K.; Goossens, H.; Saleem, F. A multicenter point prevalence survey of antibiotic use in Punjab, Pakistan: Findings and implications. *Expert Rev. Anti-Infect. Ther.* **2019**, *17*, 285–293. [CrossRef] [PubMed]
2. Dimitrova, R.; Fernandes, D.; Malik, S.; Suryani, A.; Musso, P.; Wiium, N. The 7Cs and developmental assets models of positive youth development in India, Indonesia and Pakistan. In *Handbook of Positive Youth Development*; Springer: Berlin/Heidelberg, Germany, 2021; pp. 17–33.
3. Khan, M.Z. Pakistan emerges as 24th largest economy in 75-year journey. *DAWN*, 14 August 2022.
4. Shahbaz, M. 75 Years Economic Journey of Pakistan: A Positive Perspective. *Daily Times*, 9 September 2022.
5. Muratalievna, O.D. Our distant relatives. *Acad. Globe Inderscience Res.* **2022**, *3*, 230–232.
6. MoF. *Pakistan Economic Survey 2021-22*; Ministry of Finance, Government of Pakistan: Islamabad, Pakistan, 2022.
7. World Bank. Pakistan Agriculture Food System: Knowledge Products. Available online: https://www.worldbank.org/en/country/pakistan/brief/pakistan-agriculture-food-systems (accessed on 5 January 2023).
8. Horst, A.; Watkins, S. *Enhancing Smallholder Incomes by Linking to High Value Markets in Pakistan's Punjab and Sindh Provinces*; World Bank: Washington, DC, USA, 2022.
9. PBS. *Pakistan Statistical Year Book 2020*; Pakistan Bureau of Statistics, Ministry of Planning Development and Special Initiatives, Government of Pakistan: Islamabad, Pakistan, 2021.
10. Spielman, D.J.; Malik, S.J.; Dorosh, P.; Ahmad, N. *Agriculture and the Rural Economy in Pakistan: Issues, Outlooks, and Policy Priorities*; University of Pennsylvania Press: Philadelphia, PA, USA, 2016.
11. Memon, N.A. Rice: Important cash crop of Pakistan. *Pak. Food J.* **2013**, *26*, 21–23.
12. Chandio, A.A.; Magsi, H.; Ozturk, I. Examining the effects of climate change on rice production: Case study of Pakistan. *Environ. Sci. Pollut. Res.* **2020**, *27*, 7812–7822. [CrossRef]
13. ADB. *Islamic Republic of Pakistan: Punjab Basmati Rice Value Chain*; Asian Development Bank: Mandaluyong, Philippines, 2018.
14. Bakhsh, K.; Ahmad, B.; Gill, Z.A.; Hassan, S. Estimating indicators of higher yield in radish cultivation. *Int. J. Agric. Biol.* **2006**, *8*, 783–787.
15. Bashir, K.; Khan, N.M.; Rasheed, S.; Salim, M. Indica rice varietal development in Pakistan: An overview. *Paddy Water Environ.* **2007**, *5*, 73–81. [CrossRef]
16. Akhter, M.; Haider, Z. Basmati rice production and research in Pakistan. In *Sustainable Agriculture Reviews 39*; Springer: Cham, Switzerland, 2020; pp. 119–136.
17. TDAP. Exports' Outlook for the Year 2020-21. Available online: https://tdap.gov.pk/agro-food-division/ (accessed on 20 December 2022).
18. WCED. *Report of the World Commission on Environment and Development: Our Common Future*; World Commission on Environment and Development: Cape Town, South Africa, 1987.
19. Tomislav, K. The concept of sustainable development: From its beginning to the contemporary issues. *Zagreb Int. Rev. Econ. Bus.* **2018**, *21*, 67–94.
20. Manioudis, M.; Meramveliotakis, G. Broad strokes towards a grand theory in the analysis of sustainable development: A return to the classical political economy. *N. Political Econ.* **2022**, *27*, 866–878. [CrossRef]
21. United Nations. Sustainable Development Goals. Available online: https://sdgs.un.org/goals (accessed on 26 April 2023).
22. IISD. SDG Knowledge Hub. Available online: http://sdg.iisd.org/ (accessed on 26 April 2023).
23. Fleming, A.; Wise, R.M.; Hansen, H.; Sams, L. The sustainable development goals: A case study. *Mar. Policy* **2017**, *86*, 94–103. [CrossRef]
24. Gil, J.D.B.; Reidsma, P.; Giller, K.; Todman, L.; Whitmore, A.; van Ittersum, M. Sustainable development goal 2: Improved targets and indicators for agriculture and food security. *Ambio* **2019**, *48*, 685–698. [CrossRef] [PubMed]
25. Mondejar, M.E.; Avtar, R.; Diaz, H.L.B.; Dubey, R.K.; Esteban, J.; Gómez-Morales, A.; Hallam, B.; Mbungu, N.T.; Okolo, C.C.; Prasad, K.A. Digitalization to achieve sustainable development goals: Steps towards a Smart Green Planet. *Sci. Total Environ.* **2021**, *794*, 148539. [CrossRef] [PubMed]
26. Mishra, A.; Bruno, E.; Zilberman, D. Compound natural and human disasters: Managing drought and COVID-19 to sustain global agriculture and food sectors. *Sci. Total Environ.* **2021**, *754*, 142210. [CrossRef] [PubMed]
27. Ortiz-Bobea, A.; Ault, T.R.; Carrillo, C.M.; Chambers, R.G.; Lobell, D.B. Anthropogenic climate change has slowed global agricultural productivity growth. *Nat. Clim. Change* **2021**, *11*, 306–312. [CrossRef]
28. Qamar, M.K. *Modernizing National Agricultural Extension Systems: A Practical Guide for Policy-Makers of Developing Countries*; FAO: Roma, Italy, 2005.
29. Shah, M.; Israr, M.; Khan, N.; Ahmad, N.; Shafi, M.; Raza, S. Agriculture extension curriculum: An analysis of agriculture extension students views in the agricultural Universities of Pakistan. *Sarhad J. Agric.* **2010**, *26*, 435–442.
30. Swanson, B.E.; Rajalahti, R. *Strengthening Agricultural Extension and Advisory Systems: Procedures for Assessing, Transforming, and Evaluating Extension Systems*; The World Bank: Washington, DC, USA, 2010.

31. Riaz, K.; Jansen, H.G. Spatial patterns of revealed comparative advantage of Pakistan's agricultural exports. *Pak. Econ. Soc. Rev.* **2012**, *50*, 97–120.
32. Baloch, M.A.; Thapa, G.B. Review of the agricultural extension modes and services with the focus to Balochistan, Pakistan. *J. Saudi Soc. Agric. Sci.* **2019**, *18*, 188–194. [CrossRef]
33. AESA. Agriculture Extension in Pakistan: Challenges and Ways Forward. Available online: https://www.aesanetwork.org/agriculture-extension-in-pakistan-challenges-and-ways-forward/ (accessed on 10 January 2023).
34. Davidson, A.P.; Ahmad, M. Effectiveness of public and private sector agricultural extension: Implications for privatisation in Pakistan. *J. Agric. Educ. Ext.* **2002**, *8*, 117–126. [CrossRef]
35. Abro, A.A.; Awan, N.W. Comparative analysis of profitability of major and minor crops in Pakistan. *J. Saudi Soc. Agric. Sci.* **2020**, *19*, 476–481. [CrossRef]
36. Ashraf, M.N.; Khan, T.Z.A. The Impact of Agricultural Mechanization Development Project on the Yield of Wheat Crop: A Case Study of Punjab, Pakistan. *J. Appl. Res. Multidiscip. Stud.* **2020**, *1*, 17–29.
37. Liu, J.; Wang, M.; Yang, L.; Rahman, S.; Sriboonchitta, S. Agricultural productivity growth and its determinants in south and southeast asian countries. *Sustainability* **2020**, *12*, 4981. [CrossRef]
38. Hanif, U. Pakistan's agriculture productivity among the lowest in the world. *The Express Tribune*, 24 January 2018.
39. Ullah, A.; Khan, A. Effect of extension-farmers contact on farmers' knowledge of different pest management practices in the rain-fed districts of Khyber Pakhtunkhwa, Pakistan. *Sarhad J. Agric.* **2019**, *35*, 602–609. [CrossRef]
40. Farooq, A.; Khan, M.Z. Knowledge Gap of Improved Management Practices of Sugarcane Growers in Khyber Pakhtunkhwa, Pakistan. *Sarhad J. Agric.* **2019**, *35*, 320–662. [CrossRef]
41. Khuhro, S.N.; Junejo, I.A.; Hullio, M.H.; Hassan, M.F.; Maitlo, S.A.; Sheikh, M. Knowledge attitude practice regarding pesticide application among vegetable growers of Dadu canal irrigated areas of northern Sindh Pakistan. *Pak. J. Agric. Res.* **2020**, *33*, 331. [CrossRef]
42. Usman, M.; Ch, K.M.; Ashraf, I.; Tanveer, A. Factors impeding the adoption of weed management practices in four cropping systems of the Punjab, Pakistan. *Int. J. Agric. Ext.* **2021**, *9*, 119–127. [CrossRef]
43. Khan, Y.S.; Hussain, S.; Aslam, M.N. E-Learning Needs Assessment in Agriculture Sector of Pakistan. *Asian J. Plant Sci. Res.* **2021**, *11*, 37–48.
44. Hassan, Z.; Shahbaz, B.; Ali, S.; Nazam, M. Impact assessment of plant clinics on farm income of farmers in the Punjab, Pakistan. *Int. J. Agric. Ext.* **2022**, *10*, 205–217.
45. GoP. District at Galance. Available online: https://gujranwala.punjab.gov.pk/district_glance (accessed on 10 January 2023).
46. PBS. Gujranwala District. Available online: https://www.pbs.gov.pk/census-2017-district-wise/results/047 (accessed on 10 January 2023).
47. Mahmood, A. The importance of Gujranwala. *DAWN*, 12 March 2021.
48. CRS. *Kharif crop estimates 2020-21*; Crop Reporting Service, Agriculture Department, Government of Punjab, Pakistan: Lahore, Pakistan, 2020.
49. Nunnally, J.; Bernstein, L. *Psychometric Theory*; McGraw-Hill Higher: New York, NY, USA, 1994.
50. Bland, J.M.; Altman, D.G. Statistics notes: Cronbach's alpha. *BMJ* **1997**, *314*, 572. [CrossRef] [PubMed]
51. DeVellis, R.F.; Thorpe, C.T. *Scale Development: Theory and Applications*; Sage Publications: Thousand Oaks, CA, USA, 2021.
52. Tavakol, M.; Dennick, R. Making sense of Cronbach's alpha. *Int. J. Med. Educ.* **2011**, *2*, 53. [CrossRef] [PubMed]
53. Taber, K.S. The use of Cronbach's alpha when developing and reporting research instruments in science education. *Res. Sci. Educ.* **2018**, *48*, 1273–1296. [CrossRef]
54. Hamilton, W.; Bosworth, G.; Ruto, E. Entrepreneurial younger farmers and the "young farmer problem" in England. *Agric. For.* **2015**, *61*, 61–69. [CrossRef]
55. Omobolanle, O.L. Analysis of extension activities on farmers' productivity in Southwest, Nigeria. *Afr. J. Agric. Res.* **2008**, *3*, 469–476.
56. Asogwa, B.; Umeh, J.C.; Ater, P. Technical efficiency analysis of Nigerian cassava farmers: A guide for food security policy. In Proceedings of the Poster Paper Prepared for Presentation at the International Association of Agricultural Economists Conference, Gold Coast, Australia, 12–18 August 2006; pp. 1–14.
57. Adeyemo, R.; Oke, J.; Akinola, A. Economic efficiency of small scale farmers in Ogun State, Nigeria. *Tropicultura* **2010**, *28*, 84–88.
58. Akpan, S.B.; Okon, U.E.; Jeiyol, E.N.; Nkeme, K.K.; John, D.E. Economic efficiency of Cassava based farmers in Southern Wetland Region of Cross River State, Nigeria: A translog model approach. *Int. J. Humanit. Soc. Sci.* **2013**, *3*, 173–181.
59. Aslam, M. Agricultural productivity current scenario, constraints and future prospects in Pakistan. *Sarhad J. Agric.* **2016**, *32*, 289–303. [CrossRef]
60. Elahi, E.; Abid, M.; Zhang, L.; Ul Haq, S.; Sahito, J.G.M. Agricultural advisory and financial services; farm level access, outreach and impact in a mixed cropping district of Punjab, Pakistan. *Land Use Policy* **2018**, *71*, 249–260. [CrossRef]
61. Agriculture Department. Overview of Pakistan Agriculture. Available online: https://www.agripunjab.gov.pk/overview (accessed on 15 December 2022).
62. Sadiq, G.; Ali, F.; Mahmood, K.; Shah, M.; Khan, I. Technical efficiency of maize farmers in various ecological zones of AJK. *Sarhad J. Agric.* **2009**, *25*, 607–610.

63. Nandi, J.A.; Gunn, P.; Yurkushi, E.N. Economic analysis of cassava production in Obubra local government area of cross river state, Nigeria. *Asian J. Agric. Sci.* **2011**, *3*, 205–209.
64. Chandio, A.A.; Jiang, Y.; Gessesse, A.T.; Dunya, R. The nexus of agricultural credit, farm size and technical efficiency in Sindh, Pakistan: A stochastic production frontier approach. *J. Saudi Soc. Agric. Sci.* **2019**, *18*, 348–354. [CrossRef]
65. Sheikh, A.; Rehman, T.; Yates, C. Logit models for identifying the factors that influence the uptake of new 'no-tillage' technologies by farmers in the rice–wheat and the cotton–wheat farming systems of Pakistan's Punjab. *Agric. Syst.* **2003**, *75*, 79–95. [CrossRef]
66. Ahmed, T.; Ahmad, B.; Ahmad, W. Why do farmers burn rice residue? Examining farmers' choices in Punjab, Pakistan. *Land Use Policy* **2015**, *47*, 448–458. [CrossRef]
67. Abid, M.; Scheffran, J.; Schneider, U.A.; Elahi, E. Farmer perceptions of climate change, observed trends and adaptation of agriculture in Pakistan. *Environ. Manag.* **2019**, *63*, 110–123. [CrossRef] [PubMed]
68. Ashraf, M.Q.; Khan, S.A.; Khan, R.; Iqbal, M.W. Determinants of adaptation strategies to climate change by farmers in district Sargodha, Pakistan. *Int. J. Econ. Environ. Geol.* **2019**, *9*, 16–20.
69. Jan, I.; Khan, H.; Jalaluddin, M. Analysis of agricultural extension system: A discrepancy between providers and recipients of the extension services empirical evidence from North-West Pakistan. *Sarhad J. Agric* **2008**, *24*, 349–354.
70. Baloch, A.M.; Thapa, B.G. Agricultural extension in Balochistan, Pakistan: Date palm farmers' access and satisfaction. *J. Mt. Sci.* **2014**, *11*, 1035–1048. [CrossRef]
71. Zubairi, T. PKR downslide continues with Rs7 loss in interbank. *DAWN*, 30 January 2023.
72. Chandio, A.A.; Jiang, Y. Determinants of credit constraints: Evidence from Sindh, Pakistan. *Emerg. Mark. Financ. Trade* **2018**, *54*, 3401–3410. [CrossRef]
73. Chintapalli, P.; Tang, C.S. Crop minimum support price versus cost subsidy: Farmer and consumer welfare. *Prod. Oper. Manag.* **2022**, *31*, 1753–1769. [CrossRef]
74. Kamboj, P. Trend analysis of area, production and productivity of basmati rice in India and Haryana. *Pharma Innov. J.* **2021**, *10*, 488–493.
75. Abro, A.A.; Panhwar, I.A. Impact of Various Factors on Crop Diversification Towards High Value Crops in Pakistan: An Empirical Analysis by using THI. *Sarhad J. Agric.* **2020**, *36*, 1010–1324. [CrossRef]
76. Zehra, N.; Sohail, F. *Dynamics of Food Prices in Major Cities of Pakistan*; Pakistan Institute of Development Economics: Islamabad, Pakistan, 2022.
77. Shahzad, M.A.; Razzaq, A.; Qing, P. On the wheat price support policy in Pakistan. *J. Econ. Impact* **2019**, *1*, 80–86. [CrossRef]
78. Nandy, S. Indian Agriculture in the Perspective of the Provisions of Domestic Subsidies in the Agreement on Agriculture under WTO. In *Indian Economy: Reforms and Development: Essays in Honour of Manoj Kumar Sanyal*; Springer: Berlin/Heidelberg, Germany, 2019; pp. 47–67.
79. Kumar, M.; Sharma, M.; Kumar, K. Minimum support price for agricultural commodities in India: A review. *Pharma Innov. J.* **2021**, *10*, 12–16. [CrossRef]
80. Gentle, P.; Maraseni, T.N. Climate change, poverty and livelihoods: Adaptation practices by rural mountain communities in Nepal. *Environ. Sci. Policy* **2012**, *21*, 24–34. [CrossRef]
81. Tippe, D.E.; Rodenburg, J.; Schut, M.; van Ast, A.; Kayeke, J.; Bastiaans, L. Farmers' knowledge, use and preferences of parasitic weed management strategies in rain-fed rice production systems. *Crop Prot.* **2017**, *99*, 93–107. [CrossRef]
82. Mahmood, N.; Arshad, M.; Kächele, H.; Ullah, A.; Müller, K. Economic efficiency of rainfed wheat farmers under changing climate: Evidence from Pakistan. *Environ. Sci. Pollut. Res.* **2020**, *27*, 34453–34467. [CrossRef]
83. Ullah, A.; Arshad, M.; Kächele, H.; Khan, A.; Mahmood, N.; Müller, K. Information asymmetry, input markets, adoption of innovations and agricultural land use in Khyber Pakhtunkhwa, Pakistan. *Land Use Policy* **2020**, *90*, 104261. [CrossRef]
84. Lakhan, G.R.; Channa, S.A.; Magsi, H.; Koondher, M.A.; Wang, J.; Channa, N.A. Credit constraints and rural farmers' welfare in an agrarian economy. *Heliyon* **2020**, *6*, e05252.
85. IFAD. *Investing in Rural People in Pakistan*; International Fund for Agricultural Development: Roma, Italy, 2019.
86. Raza, M.H.; Shahbaz, B.; Bell, M.A. Review based analysis of adoption gap and training needs of farmers in Pakistan. *Int. J. Agric. Ext.* **2017**, *4*, 185–193.
87. Ahmad, S.; Zhang, C.; Ekanayake, E. Smallholder Farmers' Perception on Ecological Impacts of Agroforestry: Evidence from Northern Irrigated Plain, Pakistan. *Pol. J. Environ. Stud.* **2021**, *30*, 2969–2979. [CrossRef]
88. Vuong, Q.-H. *Mindsponge Theory*; De Gruyter: Berlin, Germany, 2023.
89. Vuong, Q.H.; Napier, N.K. Acculturation and global mindsponge: An emerging market perspective. *Int. J. Intercult. Relat.* **2015**, *49*, 354–367. [CrossRef]
90. Vuong, Q.-H.; Le, T.-T.; La, V.-P.; Nguyen, H.T.T.; Ho, M.-T.; Van Khuc, Q.; Nguyen, M.-H. Covid-19 vaccines production and societal immunization under the serendipity-mindsponge-3D knowledge management theory and conceptual framework. *Humanit. Soc. Sci. Commun.* **2022**, *9*, 22. [CrossRef]
91. Khuc, Q.V. Mindspongeconomics. 2022. Available online: https://doi.org/10.31219/osf.io/hnucr (accessed on 10 January 2023).

agriculture

Article

Assessing Financial Literacy and Farmland Abandonment Relationship in Ghana

Martinson Ankrah Twumasi, Bright Senyo Dogbe, Ernest Kwarko Ankrah, Zhao Ding and Yuansheng Jiang *

College of Economics, Sichuan Agricultural University, Chengdu 611130, China
* Correspondence: yjiang@sicau.edu.cn

Abstract: Farmland abandonment has been a major concern for policymakers in most developing nations since it is associated with food security and poverty alleviation. In view of this, assessing its potential determinants is essential and timely. This study examines the relationship between financial literacy and farmland abandonment in Ghana using survey data (N = 572). The study employs endogenous switching regression (ESR) for its estimation. Our findings show that financial literacy is low among rural dwellers. Also, the findings depict that financial literacy is positively related to farmland abandonment reduction. Moreover, different household groups depict a heterogeneous relationship between financial literacy and farmland abandonment. Thus, the association between financial literacy and farmland abandonment reduction is more pronounced for low-income farm households and female farmers. We recommended that financial literacy programs can be organized or shown on national radios and television to provide financial education to the country's residents. Our findings could offer some implications for stimulating agricultural intensification while ensuring rural advancements.

Keywords: financial literacy; farmland abandonment; endogenous switching regression model; agricultural intensification; Ghana

Citation: Ankrah Twumasi, M.; Dogbe, B.S.; Ankrah, E.K.; Ding, Z.; Jiang, Y. Assessing Financial Literacy and Farmland Abandonment Relationship in Ghana. *Agriculture* **2023**, *13*, 580. https://doi.org/10.3390/agriculture13030580

Academic Editor: Yasuo Ohe

Received: 18 January 2023
Revised: 22 February 2023
Accepted: 24 February 2023
Published: 27 February 2023

1. Introduction

Abandonment of farmland is a multidimensional, complex process with interrelated environmental and socioeconomic drivers [1]. Although farmland abandonment comes with positive effects, including the provision of ecological services, such as soil recovery [2] and water retention [3], its diverse effect on humankind and economic development is outrageous. For example, the abandonment of farmland serves as a threat to food security [4], widens the urban-rural income gap [5,6], and causes agricultural landscapes' biodiversity loss and agroecosystem degradation [7]. As a result, the patterns and extent of farmland abandonment we currently face are the subjects of open debate in many parts of the world. Thus, farmland abandonment has attracted the attention of researchers and policymakers in many countries around the world.

Abandonment of agricultural land is usually associated with many factors leading to low farm productivity and profitability or causing high production costs [6–8]. These factors may include undesirable physical and climatic features such as poor soil quality, steep slopes, high altitude, and limited rainfall [6,9]; unfavorable socioeconomic conditions (e.g., low farm/household income) [8] and demographic change (e.g., an aging population) [10]; reduction in the land use net income emanating from the rise in the production cost of agricultural products and services [11,12]; and urbanization due to high economic industrialization [13,14]. However, these factors, characterized by environmental, economic, and social constraints, are likely to be curtailed if farmers are financially secure and literate. Therefore, we can ascertain that an improvement in farmers' financial literacy and economic performance can lead to a reduction in farmland abandonment, all other things remaining constant.

Accessing financial services and products in the financial market is vital for farmers' economic and financial well-being [15,16]. Thus, financial services accessibility or financial market participation has been hyped as an important avenue to intensify agriculture activities and ensure agricultural sustainability. For example, studies have revealed that financial services accessibility (e.g., savings, loans, and insurance) places farmers in a position to enhance their farm productivity/income because it can help them purchase needed inputs or use new and improved farming technologies such as climate-smart agriculture techniques, modern improved crop varieties and many more [15,17,18]. Also, farmers can manage farm risk [19,20] and adopt precision agricultural practices [21,22] to intensify their farming activities when they patronize financial services. While financial inclusion has been a great crusader to household welfare and agricultural development, about two billion adults residing mostly in developing nations have no bank accounts or are participants in the financial market. Challenges traceable to both the demand and supply sides of the financial markets, coupled with a host of other factors, are partially responsible for the low financial market patronage; nevertheless, a major impediment from the demand side is low financial literacy [23–25].

Financial literacy refers to how people create or conceive financial and economic understanding and make well-informed decisions to promote financial investment and ensure good use and management of financial products and services to create wealth while reducing debt [26]. This explanation indicates that acquiring high financial skills and knowledge, i.e., being financially literate may positively affect people's financial decisions and behaviors. Studies from Ankrah Twumasi [27] and Klapper and Lusardi [26] revealed that financially literate individuals are most likely to obtain beneficial financial information to promote their wealth accumulation strategies. One needs to be financially literate to properly diagnose questions leading to sound financial decision-making, especially when participating in the financial market. Overall, we can argue that financial inclusion, a promoter of household poverty alleviation and agricultural development or intensification, is achievable through financial literacy. Thus, improving farmers' financial and economic performance, which can cause a reduction in farmland abandonment, is possible through a financial inclusion enhancer from the demand side referred to as financial literacy.

Also, financial literacy may have a direct and indirect association with agricultural land use. On one side, financial literacy may directly improve farmland use by helping farmers make informed decisions about using their land in a way of preventing or alleviating excess costs [28–30]. A financially literate farmer will stick to farmland projects that are highly profitable due to their ability to assess the cost and benefit of that project/investment. On the other hand, financial literacy has an indirect effect on farmland use through financial market participation. Akoto [29] found out that financially literate farmers are more likely to patronize the credit market to secure loans to curb challenges pertaining to their farm production. The purchase of farm insurance to curb the risk to adopt risky but profitable projects on farmland is positively associated with financial literacy [30,31]. Therefore, better use of financial resources and risk management knowledge and skills may grow as farmers' financial literacy improves, hence, enabling farmers to utilize their farmland effectively and efficiently. The literature reviewed suggests that a potential connection exists between financial literacy and farmland abandonment reduction. However, all the literature addressing factors associated with farmland abandonment e.g., [9,13,14,18,29,32] indicates no presence of data addressing whether farm households' farmland abandonment reduction can be enhanced should citizens in developing countries such as Ghana improve their financial literacy. This vacuum in literature is filled using data from Ghana.

This study has two objectives to fulfill. First, we quantitatively assess the relationship between farmland abandonment and financial literacy. We proposed a hypothesis that farmland abandonment can reduce as financial literacy improves. As established from the literature that inadequate agricultural financial incentives are significant determinants of farmland abandonment, assessing the association between an income enhancer (financial literacy) and farmland abandonment is essential. We argue that financially literate farmers can improve their wealth accumulation and purchasing power through significant financial and investment decisions, empowering them to intensify agricultural production (e.g., adopting farm technologies and reducing farmland abandonment). Second, we examine the heterogeneous effect of how financial literacy impacts farmland abandonment based on household income and gender statutes of the farmers. We add to the existing literature in diverse ways. First, this study attempts to assess the quantitative nexus between financial literacy and farmland abandonment in Sub-Sahara Africa (SSA). Second, distinguish from prior farmland abandonment studies that prioritized agricultural credit [8,18] and NGOs grants and government subsidies [33,34] as an avenue for addressing financial barriers to farmland abandonment reduction, we reveal the essence of financial literacy and its possible effects in promoting farmland abandonment reduction in developing countries. Third, we used a suitable econometric approach to correct the potential endogeneity problem related to the treatment variable (financial literacy). Adequately dealing with potential endogeneity could bring consistency to our findings; thus, preventing unbiased estimation.

The remaining parts of the study take this form of arrangement. We presented the study's theoretical framework in Section 2. Sections 3 and 4 took the study's methodology, results, and discussions, while the conclusion and policy implication was presented in Section 5.

2. Theoretical Analysis

Financial literacy and how it influences farm household livelihood, a determinant of farmland abandonment, can theoretically be modified following the farm household model theory suggested by Huffman [35]. The theoretical model suggests that regarding a budget constraint, farm householders' utility can be characterized as a function of agricultural practices anytime the farmer maximizes utility. In the model, the household is assumed to maximize a unitary household utility function, and this can be presented as shown below:

$$Max\ U = U\,(G,\ A) \tag{1}$$

where $U,\ G\ and\ A$ are the utility, normal goods, and agricultural practices function for a household, respectively. We assume that the consumption of normal goods and intensification of agricultural practices (e.g., adopting farm technologies and reducing farmland abandonment) is subject to budget constraint, which is a function of income (I) and financial literacy (FL). Let us note that since the units of income (measured in financial units) and FL (measured in qualitative scales, such as low and high) are different, we cannot add the two together. Therefore, for the purpose of the study, we assume that income is expressed as high and low to meet the requirement of unit measurement. The presence of income and financial literacy improves the ability of the household to purchase goods (G) associated with the price (P_g) and agricultural practices required inputs associated with price (P_A). The scenario from the above led to a new model expressed as:

$$P_g G + P_A A \leq I + FL \tag{2}$$

Based on the study's objective, the farm household farmland abandonment decision depends on:

$$Farmland\ abandonment = f(FL,\ I,\ P_g, P_A) \tag{3}$$

Theories and literature depicting the direct link between financial literacy and household livelihood/business growth align with this model. According to Ankrah Twumasi [36] and Xu [14], an individual needs to be financially literate to make sound financial decisions. Thus, a financially literate person may easily acquire solutions to questions relating to investment and wealth accumulation, which can improve households' standard of living (e.g., smooth consumption, improved purchasing power, and business establishment). Also, financially literate individuals yearn to secure appropriate financial information; therefore, they are willing to participate in the financial market to maximize their wealth or incomes, which tends to empower them to acquire their needs [26,28]. For example, a financially literate farmer to whom financial services are made accessible (e.g., secure credit or farm/equipment insurance policy) may be able to obtain farm inputs [37] and adopt risky yet profitable agricultural technologies [20], thereby willing to intensify his/her agricultural participation, which can cause a reduction in farmland abandonment. Achieving a higher financial literacy status is likely to lead to an effect on one's income, enabling households to enjoy improved disposable income; hence, equipping them to obtain a higher indifference curve. All other things remaining constant, securing higher financial skills and knowledge (being financially literate) has a potential association with efficient and effective consumption of normal goods and intensification of agricultural practices [36]. In addition, the role of income cannot be overlooked when it comes to farmland abandonment. Studies have shown that household income enables farmers to acquire the necessary tools to improve and expand farmland utilization [38,39]. Also, other normal goods consumption (e.g., food, healthcare facility use, education, etc.) has an indirect relationship with farmland use since the share of household income to a booster of farmland use intensification may be used for other normal goods consumption [8].

As shown in Equation (3), the direct connection between financial literacy (promoter of financial services accessibility) and farm household agricultural practices is constrained by market failure in the financial markets, primarily because of high transaction costs [40]. Following Han [41], we categorized these transaction costs from the demand side into different financial, in-kind, and psychic divisions. The costs emanating from the financial side include transportation costs to attend financial literacy lectures and fees charged by financial experts when acquiring financial education. The opportunity cost of time spent searching for a financial expert and the booking or waiting time in the expertise office is attributed to in-kind costs. The psychic cost is the psychological stress of putting the acquired financial knowledge and skills into practice. Based on the above literature, individuals who have links with financially literate people are more likely to be financially literate themselves than their counterparts without such an advantage [36]. This reflects that financial literacy is an endogenous variable due to the presence of transaction costs; therefore, estimating Equation (3) by applying ordinary least squares (OLS) is likely to produce unreliable estimates. It is, therefore, tedious to account for the transaction costs in the model because of its nature of divisions. Thus, an endogeneity issue resulting from an omitted variable problem is present. Although we may account for the financial transaction costs, the other two costs (in-kind and psychic) are hard to be captured.

Prior research works examining the association between financial literacy and welfare enhancement [15], gambling behavior [42], and financial inclusion [43] have used the instrumental variable (IV) estimation approach. Consistent with these researchers, we also employed an IV estimation approach, using financial education (i.e., whether the farmer has a relative/friend with an economics or financial education background) as our instrument. Ankrah Twumasi [44] and Watanapongvanich [42] have used this variable as an instrument in their analysis. Details of the IV approach are explained in Section 4.2. We test the validity of the theoretical claim that acquiring high financial literacy improves the ability of farm households to intensify agricultural activities through a reduction in farmland abandonment and, if so, to what extent?

3. Why Ghana?

Ghana presents an interesting and relevant case study for assessing the association between financial literacy and farmland abandonment. In Ghana, the rate of financial literacy is currently at 32% [42], which is deemed relatively low. A recent global study on the financial literacy rate ranking of 144 countries placed Ghana in the 90th position [45]. The country, in recent years, has considered financial literacy policy a priority since it contributes to national development. Thus, several interventions and policies have been introduced by the national governments. For example, the Ministry of Finance and Economic Planning has launched the National Financial Literacy Week to raise awareness and enhance the public's understanding of the range of financial goods and services financial institutions offer.

Again, together with other NGOs (e.g., Danish International Development Agency (Danida) and the German Agency of International Cooperation (GIZ)), successive governments have introduced several financial education programs aimed at enhancing Ghanaians' understanding of financial services (e.g., loan acquisition, investment, and insurance cover). Despite the tremendous efforts on the part of stakeholders (successive governments, financial institutions, and charitable organizations) to witness significant improvement in the level of financial literacy of Ghanaian citizens, especially rural peasants, through training and educational programs, proof of how positively these activities are impacting their general economic welfare have been very little/minimal. A study in Ghana showed that farmers find themselves in debt after post-harvest sales because of low financial skills and education [46]. The researchers indicated that this menace partly explains why farmers are replacing their farming activities with off-farm jobs and youths are abandoning farming in Ghana. In addition to improving savings, recent studies on financial literacy in Ghana by Koomson [47] and Chowa [48] showed that improvements in the rate of financial literacy make households financially resilient. Regarding how instrumental the improvement of financial literacy is to agricultural intensification (e.g., land abandonment reduction), not much has been done in the case of Ghana and countries in SSA. We believe Ghana provides the right setting to undertake this study, given the details in the above background.

4. Methodology

4.1. Source of Data and Key Variables Definitions

The origin of the study is Ghana, and the data was collected from January 2018 to May 2018. Farmers engaged in crop cultivation were the targeted population. The collection of the data was done by employing questionnaires and face-to-face interview schedules. Every interview took about 15 to 20 min with a farmer. Engaging the respondents in in-depth interviews was for the purpose of gaining all the relevant data necessary for the study. A pre-test of the questionnaire was necessary to avoid any mistakes that would create misunderstanding for the respondents; therefore, we took a pre-test with 20 farmers in one of the selected regions. Some of the information we solicited for study include the farmers' socioeconomic and demographical characteristics (e.g., education level, age, credit accessibility, and health status), rate of financial literacy (see Table A1 in the Appendix A for the questions), farm information (e.g., abandoned farmland area, and farm size) and other variables that are important to attain the objective of the study.

The multi-stage sampling procedure was employed to reach an appropriate sample for the study. First, we choose four regions, i.e., Northern, Brong Ahafo (BA), Central, and Eastern. The purposive selection of the 4 regions led us to randomly select one district in each region at the next stage. These districts are East Gonja district, Atebubu Amantin district, Ekumfi district, and the Kwahu Afram Plains district in the Northern, Brong Ahafo, Central, and Eastern regions of Ghana, respectively. Let us note that these regions' record of having most rural dwellers engaged in agricultural activities led to their purposive selection [49]. 7 After getting the districts, we randomly chose three (3) communities from each selected district in the proceeding stage. Finally, with the help of a well-trained research team, we randomly chose 15–30 rural households comprising 600 farmers as our sample size. However, a total sample size of 572 was used for the analysis because some submitted questionnaires were not completed. A detailed sample procedure can be seen in the Appendix A in a framework form (Figure A1).

This study's aim means that we need to develop a measurement for the key variables (financial literacy and farmland abandonment). Concerning the financial literacy measurement, a set of 7 questions was selected after following existing literature e.g., [24,45,46] (see Table A1 in Appendix A). The 7 questions were used to obtain a score for the farmers. A farmer who answered all(none of) the questions rightly received a score of 7(0). These scores were converted into binary; i.e., using the median score (3) as a breakeven point, a farmer is assigned the value one (1) if his/her score is above 3 (the median score of the total financial literacy score), and zero (0) for a score equal to or below 3. This financial literacy measurement method has been employed in prior studies, including Ankrah Twumasi [27,44] and Andoh [50]. In terms of the farmland abandonment variable measurement, the total area of farmland abandoned in the past 12 months in acres was used. Here, farmland is considered abandoned if its abandonment is not based on natural restoration of vegetation or degradation of farmland facilities reasons but due to financial issues.

Also, taking existing studies about financial literacy and farmland abandonment into consideration e.g., [9,18,24,47,48,51] and our available data, other rich control variables such as gender, age, education years, self-reported health status, smartphone use, and many others of the household/respondent were included. As stated earlier, these variables may affect both the financial literacy and farmland abandonment of the farmers. We expect age, gender, and education to positively affect the two outcome variables. Age and education are elements of human capital; thus, the skills and knowledge gained through education and aging provide financial knowledge [26,27] and also help individuals to use their lands efficiently and effectively [9]. Male household heads tend to be more financially literate than their counterparts [47]; therefore, we expect the same result in this study. We also expect healthy individuals to intensify their agricultural activities; hence, likely to reduce farmland abandonment [52]. Also, people with smartphones access the online for farming ideas and financial information [25]; hence, we expect smartphone users to have a positive relationship with financial literacy and farmland abandonment. Farmers with their land registered, members of cooperative unions, and farm machinery users are expected to reduce farmland abandonment. Cooperative members have access to market and farming techniques, which tend to motivate them to intensify their farming activities [53,54]. Also, the use of machines for cultivation promotes productivity; serving as an encouragement to reduce farmland abandonment [55]. We expect credit-constrained farmers to increase farmland abandonment and reduce their financial literacy level. Xu [24] and Ankrah Twumasi [25] showed that financial illiterates are less likely to access financial services. Also, farmers without financial services access due to being financially illiterate tend to abandon farmlands [8,18]. Table 1 exhibits all the study variables, including their definitions, means, and standard deviations. The analyses pertaining to the study's aim were accomplished by employing STATA 15 and IBM SPSS version 26 statistical packages.

Table 1. Demographic and socioeconomic characteristics of respondents.

Variables	Description	Mean	Std. Dev
Farmland abandonment	Area of cropland abandonment in acres in 2017	0.96	2.04
Financial literacy	Farmer is financially literate (1 = Yes; 0 = No)	0.31	0.44
Gender	Farmer is a male (1 = Yes; 0 = No)	0.69	0.46
Age	Farmer's age	41.66	12.20
Education	Farmers' number of years of education	5.28	4.24
Self-reported health	Farmer's health status is good (1 = Yes; 0 = No)	0.43	0.51
Household Dependency ratio	Number of older adults (60 years and above) and children below 12 years in the farmer's family	3.29	1.17
Family size	Number of household size	6.60	3.20
Smartphone use	Farmer uses smartphone (1 = Yes; 0 = No)	0.29	0.33
Mechanization	Farmer used any farming machine (1 = Yes; 0 = No)	0.35	0.42
FBOs membership	Farmer is FBO member (1 = Yes; 0 = No)	0.41	0.49
Credit constraint	Farmer was credit constrained 2017 (1 = Yes; 0 = No)	0.34	0.47
Land size	Total farmland size of the farmer (acres)	3.85	1.74
Land registration	Farmer's household land is officially registered (1 = Yes; 0 = No)	0.36	19.82
Financial education (IV)	Farmer has a relative/friend with an economics or financial education background (1 = Yes, 0 = No)	0.27	0.35
Northern	Farmer resident is in Northern region (1 = Yes; 0 = No)	0.18	0.37
BA	Farmer resident is in BA region (1 = Yes; 0 = No)	0.26	0.43
Eastern	Whether the farmer resident is in the Eastern region (1 = Yes; 0 = No)	0.27	0.44
Central	Farmer resident is in Central region (1 = Yes; 0 = No)	0.29	0.45

4.2. Empirical Model

This study aims to investigate how farmers' financial literacy influences farmland abandonment in Ghana. However, since farmers' financial knowledge and skills acquisition to be financially literate is voluntary, the problem of selection bias becomes an issue to address. Also, the characteristics of the farmer/farm household that affects the financial literacy status may have an equal effect on the outcome variable (farmland abandonment). On this note, financial literacy becomes a potential endogenous variable and addressing this problem is essential to prevent estimation bias. To address this endogenous problem, the endogenous switching regression (ESR) model is adopted for estimation. The selection of the ESR model over other methods such as the Heckman Selection Model, Regression Adjustment (RA), and Propensity Score Matching (PSM) is its ability to take into consideration the observed and unobserved (e.g., inner motivation and risk traits) factors of the farmers when the estimation is done [56,57]. To ensure consistency in our estimation, dealing with the unobserved factors becomes essential [58,59]. Thus, selecting the ESR model over the others is the best in this study's analysis.

In the ESR method, three main equations are derived. Thus, one treatment selection equation and two separate outcome equations. The two separated outcome equations are (1) financially literate farmers and (2) financially illiterate farmers. The linear equation format is used for the outcome variable (area of farmland abandoned) estimation, while the treatment equation, which estimates the factors influencing farmers' financial literacy status, is achieved using the Probit model.

The assumption here is that a farmer has an expected utility (U^*_i) and he/she will seek financial knowledge and skills to improve their financial literacy if the expected utility for being financially literate (U^*_{i1}) is greater than the expected utility of being financially illiterate (U^*_{i2}). Thus, $U^*_{i1} - U^*_{i2} > 0 = FL^*_i$. The probability of a farmer seeking financial knowledge and skills to improve their financial literacy is FL^*_i. The linear equation for the outcome variable, which is predicted by the farmer/farm household characteristics and other factors (e.g., institutional factors like cooperative membership), is also expressed below (see Equation (4)). The utility difference, which is impossible to observe, requires a latent variable equation for its expression (see Equation (5)).

$$A^*_i = \gamma Z_i + \alpha Y_i + \varepsilon_i \tag{4}$$

$$FL^*_i = \beta X_i + \mu_i \qquad FL_i = \begin{cases} 1 \text{ if, } FL^*_i > 0 \\ 0 \text{ if, otherwise} \end{cases} \tag{5}$$

where A^*_i is the farmland abandoned area (outcome variable). Z_i and Y_i are the exogenous (e.g., gender, age, education level, family size, etc.) and endogenous (financial literacy, i.e., 1 = financially literate and 0 =otherwise) variables, respectively. γ, α, and β are the vector of parameters to be estimated. μ_i and ε_i denote the random disturbance terms. The variables in X_i and Z_i are equal; however, X_i contains the IV introduced in the theoretical analysis section, but this variable should not be included in the Z_i variables. Also, this IV should not directly correlate with the area of farmland abandoned but vis-à-vis the treatment (financial literacy) variable. Based on this reason and following previous literature (e.g., [27]), the variable financial education (i.e., whether the farmer has a relative/friend with an economics or financial education background) was chosen as this study's IV. We tested the validity of our selected IV using the Pearson correlation method (see Table A3 in the Appendix A). In Table A3, a respectively significant and insignificant correlation coefficient for financial literacy and farmland abandonment variables was observed, meaning that our IV is suitable.

As indicated above that the ESR outcome has two outcome equations, we express these two equations as follows. The expressions are divided into regimes [60]

$$\begin{array}{ll} \text{Regime 1 (financially literate)} & A_{1i} = Z_{1i}\gamma_1 + \varepsilon_{1i}, \text{ if } FL_i = 1 \\ \text{Regime 2 (financially illiterate)} & A_{2i} = Z_{2i}\gamma_2 + \varepsilon_{2i}, \text{ if } FL_i = 0 \end{array} \tag{6}$$

where the farmland abandonment status for a financially literate farmer is represented by A_{1i} and A_{2i} for a financially illiterate farmer. Also, (Z_{1i} and Z_{2i}) = explanatory variables, (γ_1 and γ_2) = vector of parameters to be calculated and (ε_{1i} and ε_{2i}) = error terms.

These indicators, μ_i, ε_{1i} and ε_{2i}, are assumed to have a tri-variate normal distribution with mean vector zero and covariance matrix:

$$cov(\mu_i, \varepsilon_1, \varepsilon_2) = \begin{bmatrix} \sigma_1^2 & \sigma_{12} & \sigma_{1\mu} \\ \sigma_{12} & \sigma_2^2 & \sigma_{2\mu} \\ \sigma_{1\mu} & \sigma_{2\mu} & \sigma_\mu^2 \end{bmatrix} \tag{7}$$

where the disturbance term's variance (ε_{1i} and ε_{2i} in Equation (6)) is represented by σ_1^2 and σ_2^2, while σ_μ^2 is for the variance of μ_i, the error term of Equation (4). Also, σ_{12}, $\sigma_{1\mu}$, and $\sigma_{2\mu}$ are the covariance of ε_{1i} and ε_{2i}, ε_{1i} and μ_i, and ε_{2i} and μ_i, respectively. The model assumes that σ_μ^2 = 1 because β can be estimated only up to a scale factor [61–63]. We proceed to calculate an inverse mill ratio (IMR) (λ_1 and λ_2) and the covariance term ($\sigma_{1\mu}$ and $\sigma_{2\mu}$) are calculated to provide a remedy for the selection bias issue in the ESR model. These estimated IMR and covariance terms are introduced in Equation (6). Thus, Equation (6) takes a new expression (Equation (8)).

$$E(I_{1i}|Y_i = 1) = Z_{1i}\gamma_1 + \sigma_{1\mu}\lambda_1 E(I_{2i}|Y_i = 1) = Z_{2i}\gamma_2 + \sigma_{2\mu}\lambda_2 \tag{8}$$

An appropriate method to ensure consistent standard error in this current model is by simultaneously estimating both the selection and outcome equations using a full information maximum likelihood (FIML) method [60,61]. Through the application of the FIML approach, the $\rho_1 = corr(\mu_i, \varepsilon_{1i})$ and $\rho_2 = corr(\mu_i, \varepsilon_{2i})$ are also determined. A non-zero ρ_1 and ρ_2 indicates that selection bias resulting from unobservable factors is present. As this study is concerned, the treatment effect of how financial literacy impacts farmland abandonment status is of interest. Thus, we need to estimate the average treatment effects on the treated (ATT) and average treatment effects on the untreated (ATU). Therefore, the following steps are considered.

$$\text{Financially literate had they been literate}: E(A_{1i}|FL_i = 1) = Z_{1i}\gamma_1 + \sigma_{1\mu}\lambda_1 \tag{9}$$

$$\text{Financially literate had they been illiterate}: E(A_{2i}|FL_i = 1) = Z_{2i}\gamma_2 + \sigma_{2\mu}\lambda_1 \tag{10}$$

$$\text{Financially illiterate had they been literate}: E(A_{1i}|FL_i = 0) = Z_{1i}\gamma_1 + \sigma_{1\mu}\lambda_2 \tag{11}$$

$$\text{Financially illiterate had they been illiterate}: E(A_{2i}|FL_i = 0) = Z_{2i}\gamma_2 + \sigma_{2\mu}\lambda_2 \tag{12}$$

The above expressions (the expected outcomes) can be utilized for consistent treatment effects, ATT, and ATU, derivation while considering unobserved and observed heterogeneity [64].

$$ATT = E(A_{1i}|FL_i = 1) - E(A_{2i}|FL_i = 1) = Z(\gamma_1 - \gamma_2) + \lambda_1(\sigma_{1\mu} - \sigma_{2\mu}) \tag{13}$$

$$ATU = E(A_{1i}|FL_i = 0) - E(A_{2i}|FL_i = 0) = Z(\gamma_1 - \gamma_2) + \lambda_2(\sigma_{1\mu} - \sigma_{2\mu}) \tag{14}$$

5. Results and Discussions

5.1. Descriptive Statistics

The analysis's variables, summary statistics, and definitions can be observed in Table 1. The reported mean farmland abandoned is 0.96 acres, and 31% of the farmers are financially literate. While a respective mean of approximately 5 and 42 years is reported for the farmer's years of education and age, 43% of the farmers believe they are in good health. The respective average dependency ratio and family size are approximately 3 and 7 people. The report from Table 1 displayed that 29% of the farmers use smartphones, and 35% of them have used farming machines on their farms. About 41% of the farmers are members of farm-based organizations (FBOs), and 34% reported being credit constrained. While the mean total farmland size of the farmer is 3.86 acres, only 36% of the farmers have their lands officially registered. The sampled group reveals that 27% of the household heads have relatives or friends with economics/financial education backgrounds. Finally, about 18, 26, 27, and 29% have their residence in the Northern, BA, Eastern, and Central regions.

Some key variables mean differences between financially illiterate and literate farmers are displayed (Table 2). The area of abandoned farmland for financially illiterate farmers is larger than their financially literate counterparts, according to Table 2. This difference supports Figure 1, which establishes that the abandoned farmland associated with financially literate farmers is lower compared to farmers who are financially illiterate irrespective of their household income level. Thus, farmers from high-income and low-income households with higher financial literacy rates have fewer abandoned farmlands. It can also be observed that financially literate farmers are educated, users of smartphones, less likely to be credit constrained, and had their land officially registered. The result further reveals that farmers with financially literate relatives or friends tend to be financially literate. While Table 2 results give a fair understanding of the study, it only displays a simple average difference that ignores the farmers' observed and unobserved factors. In that matter, our quantitative analysis of the connection between farmland abandonment and financial literacy requires a suitable econometric method such as the ESR model, which can capture the farmers' observed and unobserved factors to prevent biased estimation.

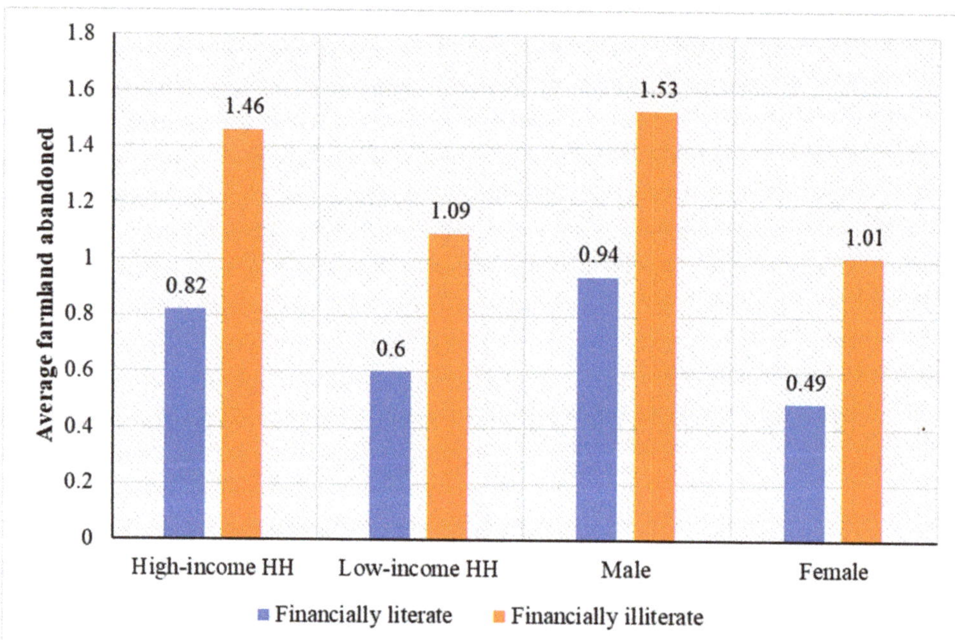

Figure 1. Distribution of average farmland abandoned by household income level and gender status. HH = Household.

Table 2. Main variables mean differences between financial literates and illiterates.

Variable	Literate	Illiterate	Differences (Normalized)
Farmland abandonment	0.71	1.27	−0.07 **
Gender	0.77	0.63	0.11 *
Age	43.75	40.52	0.09
Education	7.03	3.62	0.16 **
Self-reported health	0.44	0.43	0.02
Household Dependency ratio	3.98	2.66	0.30

Table 2. *Cont.*

Variable	Literate	Illiterate	Differences (Normalized)
Family size	5.34	7.95	−0.08
Smartphone use	0.35	0.27	0.13 *
Mechanization	0.37	0.34	0.06
FBOs membership	0.38	0.45	−0.11
Credit constraint	0.25	0.43	−0.23 ***
Land size	3.12	4.63	0.29
Land registration	0.43	0.31	0.20 *
Financial education (IV)	0.36	0.22	0.22 **
Observations	177	395	Total = 572

Source: survey results, 2018. Note: ***, **, and * respectively depict significant levels at 1, 5, and 10%.

5.2. *Empirical Analysis*

5.2.1. Determinants of Financial Literacy

Table 3, which was gathered from the selection equation of the ESR model (Table A2), displays the determining factors of financial literacy among the sample group. From the table, the variable, gender, is statistically significant, implying that male farmers in the study area are more financially literate than the female farmers in that area. Studies by Ankrah Twumasi [44] and Bucher-Koenen [65] support this finding. According to these researchers, females' engagement in STEM programs is generally low compared to males, thus making males more quantitatively efficient. The result also shows a positive and significant coefficient for the variable, education. Thus, a farmer's years of schooling increase their financial literacy level. Education equips individuals with vital fundamental financial skills and knowledge, which may affect their financial literacy level. Xu [24] and Lusardi and Mitchell [66] studies confirm this positive finding. For example, ref. [24] showed that educated people gain much understanding of financial services technicalities and terminologies and tend to have a higher probability of being financially literate.

Table 3. Financial literacy determinants.

Variables	Coefficients	Robust Standard Errors
Gender	0.024	0.010 *
Age	0.172	0.263
Education	0.291	0.075 ***
Self-reported health	0.016	0.086
Household Dependency ratio	−0.039	0.055
Family size	0.066	0.049
Smartphone use	0.078	0.027 **
Mechanization	0.044	0.091
FBOs membership	0.051	0.080
Credit constraint	0.096	0.047 *
Land size	−0.029	0.115
Land registration	0.088	0.030
Financial education (IV)	0.183	0.017 ***
Residual (smartphone use)	0.155	0.429
Constant	1.272	0.630 *
Regional dummies	Yes	Yes
Observations	572	

Source: survey results, 2018. Note: ***, **, and * respectively depict significant levels at 1, 5, and 10%. Northern = Reference region.

The results further revealed smartphone usage positively correlates with financial literacy. Smartphone use enables farmers to access innovations and essential financial information through the internet or text messages, which increases their financial literacy compared to non-smartphone users. This finding is in line with the conclusions of the studies of Khanal and Mishra [67] and Ma et al. [68]. They showed that internet-based information enlightens users' knowledge and skills about new things, such as financial services; hence, improving their financial literacy. Credit constraint is also seen to influence financial literacy negatively, indicating that credit-constrained farmers are likely to be financially illiterate. People are financially constrained because they lack the primary financial knowledge and skill needed in the financial market; hence, their negative tendencies toward engaging in the financial market or patronizing financial services [24,69]. Thus, farmers with credit access tend to have more knowledge about the financial markets and are exposed to the details of these services, which improves their knowledge [25].

Finally, financial education, used as an instrumental variable, had a positive and significant coefficient. This result indicates that the likelihood of farmers with a relative/friend with an economics or financial education background being financially literate is higher compared to farmers without financial education. This finding is consistent with [27,44], whose finding explained that the flow of financial knowledge and skills provided to individuals through friends and relatives enables them to make efficient financial decisions compared to those without financial education.

5.2.2. Financial Literacy and Farmland Abandonment Association Estimate

Table A2, shown in the Appendix A, reports the estimates of the ESR models; thus, the results for the treatment and outcome equations. It can be observed from the lower part of Table A2 that the sign of ρ_1 is statistically significant, implying the existence of selection bias; hence, the application of the ESR model to compute the analysis is suitable. Moreover, the Wald test for joint independence of the equation is significantly different from zero, portraying the rejection of the null hypothesis stating that the Equation (2) error term (μ_i) and the error terms of Equation (3) (ε_{1i} and ε_{2i}) does not correlate. We did not discuss the determinants (control variables in Table A2) of the outcome variable (farmland abandonment) because those results do not provide a detailed understanding of how farmland abandonment is affected by financial literacy. The ATT and ATU are regarded as significant results that reflect the nexus between financial literacy and farmland abandonment [70]. Therefore, the interpretation of how financial literacy affects farmland abandonment is based on the treatment effect results (Table 4).

Table 4. The impact of financial literacy on the abandonment of farmland.

Mean Area of Farmland Abandoned (ESR)			**Treatment Effect**	**t-Value**
	Financially literate	Financially illiterate		
Financially literate	0.682	1.142	ATT = −0.460	−4.19 ***
Financially illiterate	1.105	1.328	ATU = −0.223	−10.17 ***
Heterogeneity effects	−0.423	−0.186	−0.237	ATE = −0.646
Mean area of farmland abandoned PSM [a]				
Financially literate	0.707	1.086	ATT = −0.379	−2.97 ***

Source: survey results, 2018. Note: *** depict significant level at 1%. Northern = Reference region. [a] Nearest neighbor matching technique is used.

Table 4 presents the average treatment effects on the treated (ATT) and the untreated (ATU). In the context of this study, the ATT represents the average effect of being financially literate on the farmers who are financially literate in terms of farmland abandonment, while ATU represents the potential gains a financially illiterate farmer could have secured had they been financially literate. The estimates show that higher farmland abandonment reduction is associated with being financially literate. Financially literate farmers are observed to have 0.682 acres as their abandoned farmland, compared with 1.142 acres had they been financially illiterate, suggesting that being financially literate resulted in reducing farmland by about 40.3% (Table 4). In the same manner, financially illiterate farmers are observed to have 1.328 acres as their abandoned farmland, compared with 1.105 acres had they been financially literate, suggesting that being financially literate resulted in reducing farmland by about 17%. The heterogeneous effect result implies that the abandoned farmland effect on financially literate farmers is more profound than on their financially illiterate counterparts. Financially literate farmers may have the financial knowledge and skills to enjoy financial services (e.g., access to credit, insurance, and savings); hence, empowering the farmers' farm inputs purchasing power to boost productivity. When this happens farmers may be more likely to expand their production. Thus, farmland abandonment would be reduced. These findings echo Du [18] and Ankrah Twumasi's [25] results, which indicated that peasant households accessing financial services are less likely to practice farmland abandonment. It also confirms the theory underpinning this study, which states that financial literacy is a function of farmland abandonment [71].

The study conducted additional estimations to assess financial literacy's effect on farmland abandonment using the PSM method for robustness check purposes. As revealed in the lower section of Table 4, the PSM estimated ATT of financial literacy effect on abandoned farmland is −0.379, suggesting that an average farmer who is financially literate is more likely to reduce abandoned farmland by 0.379 acres than their financially illiterate counterparts. Both methods (ESR and PSM) show that financial literacy reduces farmland abandonment. Thus, results from the PSM and ESR are consistent.

5.2.3. Additional Estimates

Further estimates to heterogeneously assess farmland abandonments' impact on financial literacy are provided in Table 5. Here, the sampled group was categorized into divisions such as high and low-income households and the farmers' gender composition. In this study, households' median income was used as a breakeven point to differentiate between high- and low-income households. This implies that households whose income goes beyond (below) the breakeven point are classified as high (low) income households. Ankrah Twumasi [27] applied this method in their study.

Table 5. Disaggregated effect of financial literacy on farmland abandonment by household income and gender divisions.

Variables		Average Farmland Abandonment		ATT_{ESR}	t-Value	Change
		Financially Literate	Financially Illiterate			
Household income level	High	0.634	0.735	−0.101	−3.84 ***	13.74%
	Low	0.574	0.712	−0.138	−6.89 ***	19.38%
Gender	Male	1.005	1.184	−0.179	−2.31 *	15.12%
	Female	0.733	0.918	−0.185	−4.70 ***	20.15%

Source: Survey results, 2018. * and *** represent statistical significance at 10% and 1% alpha levels, respectively. All numbers in parentheses are robust standard errors.

The findings depict that financial literacy inversely affects farmland abandonment even after categorizing the farmers' attributes into different divisions. Particularly, the computed results reveal that the abandoned farmland effect on financially literate farmers is possible for farmers from both high- and low-income households. Nevertheless, the magnitude of percentage change is more prominent among low-income households than high-income household counterparts. The reason for this finding may be that, compared to farmers from high-income households, farmers from low-income households may see farm income as their main source of income, hence, more likely to reduce abandonment of farmland if their financial skills could help them patronize financial services (e.g., secure insurance policies and credit) to boost their production. The result agrees with Li [72], who showed that rural-urban migration reduces among farm households enjoying agricultural credit to improve their productivity.

Concerning the gender division, it can be observed that male and female farmers who are financially literate reduce farmland abandonment. To be precise, we observed that the female farmers' farmland reduction percentage is greater than their male counterparts. An explanation for this finding is the huge males' responsibility as family heads in most developing countries like Ghana. These responsibilities may push them to patronize agricultural credit fungibility (i.e., utilize a portion of the farm loans for household expenditures); therefore, causing farmland abandonment due to insufficient funds to cultivate the land [73]. Moreover, seeking off-farm work, an unfavorable determinant of farmland abandonment [14], is high among male farmers due to their high financial responsibility as family heads.

6. Conclusions, Policy Implications, and Limitations

Farmland abandonment has been a major concern for policymakers in most developing nations since it is associated with food security and poverty alleviation. Thus, assessing factors influencing its reduction is of good essence and timely. We assess how financial literacy affects farmland abandonment in this study. The report results show that 177 out of the 572 sampled groups were financially literate. After employing the ESR model for our estimation, the following emerged from our findings. The selection equation from the ESR model (determinants of financial literacy) displayed those variables, including gender, education, smartphone use, credit constrained, and financial education as financial literacy influencing factors. The finding again depicted that financially literate farmers' probability of reducing farmland abandonment was higher than their illiterate counterparts. Moreover, different household groups depicted a heterogeneous farmland abandonment effect of financial literacy.

Based on the study's results, we highlighted some policy implications that might benefit national governments and policymakers. In the first place, the negative association between financial literacy and abandonment of farmland establishes that financial literacy is an integral determinant of farmland abandonment reduction. Therefore, improving individuals' financial literacy is essential, especially for farmers. We recommend that financial literacy programs can be organized or shown on national radios and television to provide financial education to the country's residents. Also, community leaders can be supported by the government to organize conferences aimed at empowering the financial literacy level of the rural dweller, especially when access to information through radios and televisions is hard to find. Finally, the findings revealed that it is vital to encourage females' agricultural participation because their farmland abandonment reduction was profound relative to males. This study gives evidence of the essence of financial literacy in reducing farmland abandonment; thus, intensification of agriculture engagement can be promoted if farmers are financially literate.

The following limitations are important to be noted by future researchers. First, we restricted our research study area to only four regions in Ghana because of limited funds; hence, affecting our sample size. Future researchers with adequate funds should target the entire country. Secondly, other socio-political, socioeconomics, and environmental characteristics may play a major role in farmland abandonment; thus, forthcoming research works can examine the linkage between those attributes and farmland abandonment to provide more alternative policies aimed at promoting agricultural growth. Third, the focus group for this work was farmers in crop cultivation; however, other categories of farmers, including livestock and fishery (aquaculture), who are practicing farmland abandonment exist. In coming studies can assess the financial literacy's effect on these categories of farmers. Finally, using the median as a yardstick to measure financial literacy in a dummy variable format may be associated with some shortfalls, so readers must take caution in the study's interpretation. We edge future studies to improve on this measurement when the need arises.

Author Contributions: Conceptualization, E.K.A. and M.A.T.; data curation, M.A.T. and Z.D.; formal analysis, B.S.D. and M.A.T.; investigation, E.K.A.; methodology, Z.D., M.A.T. and B.S.D.; supervision, Y.J.; writing—review and editing, Y.J. and Z.D.; funding, Y.J. All authors have read and agreed to the published version of the manuscript.

Funding: This research received funding from the National Social Science Fund of China (Grant No: 20AJY011). Zhao Ding also acknowledges funding support from the Sichuan Province Science and Technology Support Program (Grant No. 2021JDR0305).

Institutional Review Board Statement: Not applicable.

Informed Consent Statement: Not applicable.

Data Availability Statement: The data used for the study is private but would be made available upon reasonable request.

Conflicts of Interest: The authors declare no conflict of interest.

Appendix A

Table A1. Questions about financial literacy and answers.

	Question	Answers
1.	Suppose you had GH¢100 in your savings account with a 2% annual interest. After 5 years, how much will you have in this account if you leave your money to gain interest? (Interest rate)	(a) more than GH¢102 (b) exactly GH¢102 (c) less than GH¢102 (d) I do not know
2.	When you save an amount of money, X, at a rate of 1% per annum and that savings suffer an increase in inflation after a year, will the value of that amount in savings be the same as it was the year of saving? (Inflation)	(a) Definitely (b) Not at all (c) I have no idea
3.	Is the following statement true or false? "Buying a single company stock usually provides a safer return than a stock mutual fund." (Diversification of risks)	(a) True (b) False (c) I do not know
4.	Which of these options is better: Borrowing GH¢500.00 and paying GH¢600 back in a month to a lender (N1) or borrowing the same GH¢500.00 from another lender (N2) and paying back the GH¢500.00 with a 15% interest in a month? (Borrowing)	(a) Borrowing from N1 (b) Borrowing from N2 (c) I have no idea
5.	If a man dies and bequeaths to his first son GH¢10,000 today and asked that another GH¢10,000 be given to other siblings 3 years from now, who becomes richer from the monies inherited (Time value of money)	(a) His first son (b) the sibling (c) Both beneficiaries (d) I have no idea
6.	Assume one's income doubles in a particular year, say 2010, and all commodity prices also double that same year. will one be able to buy more or less with that income today (Money illusion)	(a) More than today (b) The same (c) Less than today (d) I do not know
7.	A brand new farm machinery is less costly to insure than second-hand farm machinery? (insurance)	(a) Absolutely true (b) Totally untrue (c) I have no idea

Source: Ankrah Twumasi et al. [44], Lusardi et al [66], and Andoh et al. [50].

Table A2. Determinants of financial literacy and farmland abandonment.

Variables	First Stage Selection Equation	Second Stage Farmland Abandonment Equation	
	Financially Literate	Financially Literate	Financially Illiterate
Gender	0.024 (0.010) *	0.036 (0.018) *	0.061 (0.076)
Age	0.172 (0.263)	−0.097 (0.046) *	−0.043 (0.067)
Education	0.291 (0.075) ***	0.003 (0.000) **	0.011 (0.018)
Self-reported health	0.016 (0.086)	−0.055 (0.027) *	−0.113 (0.107)
Household Dependency ratio	−0.039 (0.055)	−0.086 (0.220)	0.063 (0.031) *
Family size	0.066 (0.049)	−0.006 (0.009)	−0.056 (0.024) *
Smartphone use	0.078 (0.027) **	0.048 (0.022) *	0.145 (0.177)
Mechanization	0.044 (0.091)	−0.086 (0.030) **	−0.064 (0.014) ***

Table A2. *Cont.*

Variables	First Stage Selection Equation	Second Stage Farmland Abandonment Equation	
	Financially Literate	Financially Literate	Financially Illiterate
FBOs membership	0.051 (0.080)	−0.079 (0.041)	−0.060 (0.013) ***
Credit constraint	0.096 (0.047) *	0.033 (0.112)	0.095 (0.011) ***
Land size	−0.029 (0.115)	0.042 (0.017) **	0.011 (0.007)
Land registration	0.088 (0.030)	0.021 (0.059)	0.061 (0.045) *
Financial education (IV)	0.183 (0.017) ***		
Residual (smartphone use)	0.155 (0.429)	0.063 (0.056)	0.018 (0.069)
Constant	1.272 (0.630) *	3.261 (1.282) **	1.774 (0.708) **
Regional dummies	Yes	Yes	Yes
σ_1	0.170 (0.147)		
σ_2	0.613 (0.451)		
ρ_1	0.072 (0.019) ***		
ρ_2	−0.032 (0.113)		
LR test of indep. eqns.: 4.26 **; Log likelihood = −887.839; Observations = 572			

Source: survey results, 2018. Note: Note: ***, **, and * respectively depicts significant level at 1%, 5%, and 10%. Northern = Reference region. All numbers in parentheses are robust standard errors.

Table A3. Pearson correlation analysis of the selected IV.

Variables	Correlation Coefficient	*p*-Value
Financial literacy	0.049 **	0.016
RE adoption	0.186	0.112

Source: Survey results, 2018. Note: ** $p < 0.5$.

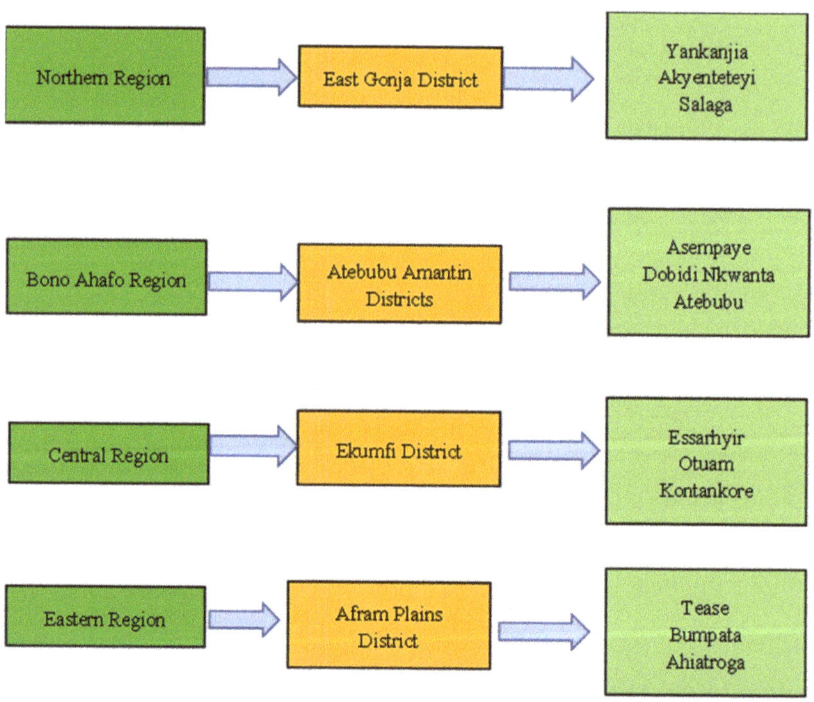

Figure A1. Diagram of household sample selection procedure.

References

1. Lambin, E.F.; Meyfroidt, P. Global land use change, economic globalization, and the looming land scarcity. *Proc. Natl. Acad. Sci. USA* **2011**, *108*, 3465–3472. [CrossRef]
2. Benayas, J.M.R.; Martins, A.; Nicolau, J.M.; Schulz, J.J. Abandonment of agricultural land: An overview of drivers and consequences. *CAB Rev. Perspect. Agric. Vet. Sci. Nutr. Nat. Resour.* **2007**, *2*, 57. [CrossRef]
3. Sileika, A.S.; Stålnacke, P.; Kutra, S.; Gaigalis, K.; Berankiene, L. Temporal and spatial variation of nutrient levels in the Nemunas River (Lithuania and Belarus). *Environ. Monit. Assess.* **2006**, *122*, 335–354. [CrossRef]
4. Viana, C.M.; Freire, D.; Abrantes, P.; Rocha, J.; Pereira, P. Agricultural land systems importance for supporting food security and sustainable development goals: A systematic review. *Sci. Total Environ.* **2022**, *806*, 150718. [CrossRef]
5. Wang, X.; Shao, S.; Li, L. Agricultural inputs, urbanization, and urban-rural income disparity: Evidence from China. *China Econ. Rev.* **2019**, *55*, 67–84. [CrossRef]
6. Zhou, T.; Koomen, E.; Ke, X. Determinants of Farmland Abandonment on the Urban–Rural Fringe. *Environ. Manag.* **2020**, *65*, 369–384. [CrossRef]
7. Blair, D.; Shackleton, C.M.; Mograbi, P.J. Cropland abandonment in South African smallholder communal lands: Land cover change (1950–2010) and farmer perceptions of contributing factors. *Land* **2018**, *7*, 121. [CrossRef]
8. Ankrah Twumasi, M.; Jiang, Y.; Ntiamoah, E.B.; Akaba, S.; Darfor, K.N.; Boateng, L.K. Access to credit and farmland abandonment nexus: The case of rural Ghana. *Nat. Resour. Forum* **2021**, *46*, 3–20. [CrossRef]
9. Gellrich, M.; Baur, P.; Koch, B.; Zimmermann, N.E. Agricultural land abandonment and natural forest re-growth in the Swiss mountains: A spatially explicit economic analysis. *Agric. Ecosyst. Environ.* **2007**, *118*, 93–108. [CrossRef]
10. Prishchepov, A.A.; Müller, D.; Dubinin, M.; Baumann, M.; Radeloff, V.C. Determinants of agricultural land abandonment in post-Soviet European Russia. *Land Use Policy* **2013**, *30*, 873–884. [CrossRef]
11. Li, H.; Song, W. Cropland abandonment and influencing factors in Chongqing, china. *Land* **2021**, *10*, 1206. [CrossRef]
12. Díaz, G.I.; Nahuelhual, L.; Echeverría, C.; Marín, S. Drivers of land abandonment in Southern Chile and implications for landscape planning. *Landsc. Urban Plan.* **2011**, *99*, 207–217. [CrossRef]
13. Deng, X.; Xu, D.; Qi, Y.; Zeng, M. Labor Off-Farm Employment and Cropland Abandonment in Rural China: Spatial Distribution and Empirical Analysis. *Int. J. Environ. Res. Public Health* **2018**, *15*, 1808. [CrossRef]
14. Xu, D.; Deng, X.; Guo, S.; Liu, S. Labor migration and farmland abandonment in rural China: Empirical results and policy implications. *J. Environ. Manag.* **2019**, *232*, 738–750. [CrossRef]
15. Ankrah Twumasi, M.; Zheng, H.; Asiedu-Ayeh, L.O.; Siaw, A.; Jiang, Y. Access to Financial Services and Its Impact on Household Income: Evidence from Rural Ghana. *Eur. J. Dev. Res.* **2022**, 1–22. [CrossRef]
16. Lahiani, A.; Mefteh-Wali, S.; Shahbaz, M.; Vo, X.V. Does financial development influence renewable energy consumption to achieve carbon neutrality in the USA? *Energy Policy* **2021**, *158*, 112524. [CrossRef]
17. Ankrah Twumasi, M.; Asante, D.; Asante, I.O.; Addai, B.; Jiang, Y. Assessing fish farm economic performance and access to financial services nexus: Empirical evidence from Ghana. *Aquac. Econ. Manag.* **2022**, 1–18. [CrossRef]
18. Du, J.; Zeng, M.; Xie, Z.; Wang, S. Power of Agricultural Credit in Farmland Abandonment: Evidence from Rural China. *Land* **2019**, *8*, 184. [CrossRef]
19. Lefebvre, M.; Nikolov, D.; Gomez-y-Paloma, S.; Chopeva, M. Determinants of insurance adoption among Bulgarian farmers. *Agric. Financ. Rev.* **2014**, *74*, 326–347. [CrossRef]
20. Gao, Y.; Shu, Y.; Cao, H.; Zhou, S.; Shi, S. Fiscal policy dilemma in resolving agricultural risks: Evidence from China's agricultural insurance subsidy pilot. *Int. J. Environ. Res. Public Health* **2021**, *18*, 7577. [CrossRef]
21. Pan, J. Financial support and the development of agricultural mechanization in China. *J. Adv. Oxid. Technol.* **2018**, *21*, 1–15.
22. Zheng, H.; Ma, W.; Zhou, X. Renting-in cropland, machinery use intensity, and land productivity in rural China. *Appl. Econ.* **2021**, *53*, 5503–5517. [CrossRef]
23. Lusardi, A.; Tufano, P. Debt literacy, financial experiences, and overindebtedness. *J. Pension Econ. Financ.* **2015**, *14*, 332–368. [CrossRef]
24. Xu, N.; Shi, J.; Rong, Z.; Yuan, Y. Financial literacy and formal credit accessibility: Evidence from informal businesses in China. *Financ. Res. Lett.* **2019**, *36*, 101327. [CrossRef]
25. Ankrah Twumasi, M.; Jiang, Y.; Wang, P.; Ding, Z.; Frempong, L.N.; Acheampong, M.O. Does financial literacy inevitably lead to access to finance services? Evidence from rural Ghana. *Ciência Rural* **2022**, *52*, 1–16. [CrossRef]
26. Klapper, L.; Lusardi, A. Financial literacy and financial resilience: Evidence from around the world. *Financ. Manag.* **2020**, *49*, 589–614. [CrossRef]
27. Ankrah Twumasi, M.; Asante, D.; Fosu, P.; Essilfie, G.; Jiang, Y. Residential renewable energy adoption. Does financial literacy matter? *J. Clean. Prod.* **2022**, *361*, 132210. [CrossRef]
28. Askar, M.W.; Quattara, B. Financial Literacy and Poverty Reduction: The Case of Indonesia. *Ikonomika* **2020**, *2*, 1–16.
29. Akoto, G.O.; Appiah, K.O.; Turkson, J.K. Financial literacy of cocoa farmers in Ghana. *Int. J. Account. Financ.* **2017**, *7*, 11. [CrossRef]
30. Ellis, E. Willingness to Pay for Index Based Crop Insurance in Ghana. *Asian Econ. Financ. Rev.* **2017**, *7*, 700–721. [CrossRef]
31. Wang, M.; Ye, T.; Shi, P. Factors Affecting Farmers' Crop Insurance Participation in China. *Can. J. Agric. Econ.* **2016**, *64*, 479–492. [CrossRef]

32. Su, G.; Okahashi, H.; Chen, L. Spatial pattern of farmland abandonment in Japan: Identification and determinants. *Sustainability* **2018**, *10*, 3676. [CrossRef]
33. Li, S.; Li, X. Global understanding of farmland abandonment: A review and prospects. *J. Geogr. Sci.* **2017**, *27*, 1123–1150. [CrossRef]
34. Renwick, A.; Jansson, T.; Verburg, P.H.; Revoredo-Giha, C.; Britz, W.; Gocht, A.; McCracken, D. Policy reform and agricultural land abandonment in the EU. *Land Use Policy* **2013**, *30*, 446–457. [CrossRef]
35. Huffman, W.E. Agricultural household models: Survey and critique. In *Multiple Job-Holding among Farm Families*; Iowa State University Press: Ames, IA, USA, 1991.
36. Ankrah Twumasi, M.; Jiang, Y.; Ding, Z.; Wang, P.; Abgenyo, W. The Mediating Role of Access to Financial Services in the Effect of Financial Literacy on Household Income: The Case of Rural Ghana. *SAGE Open* **2022**, *12*, 1–13. [CrossRef]
37. Martey, E.; Wiredu, A.N.; Etwire, P.M.; Kuwornu, J.K.M. The impact of credit on the technical efficiency of maize-producing households in Northern Ghana. *Agric. Financ. Rev.* **2019**, *79*, 304–322. [CrossRef]
38. Bonjean, I. Heterogeneous incentives for innovation adoption: The price effect on segmented markets. *Food Policy* **2019**, *87*, 101741. [CrossRef]
39. Nwafor, C.U.; Ogundeji, A.A.; van der Westhuizen, C. Adoption of ICT-based information sources and market participation among smallholder livestock farmers in South Africa. *Agriculture* **2020**, *10*, 44. [CrossRef]
40. Lusardi, A.; Mitchell, O.S. The Economic Importance of Financial Literacy: Theory and Evidence. *J. Econ. Lit.* **2014**, *52*, 5–44. [CrossRef]
41. Han, L.; Fraser, S.; Storey, D.J. Are good or bad borrowers discouraged from applying for loans? Evidence from US small business credit markets. *J. Bank. Financ.* **2009**, *33*, 415–424. [CrossRef]
42. Watanapongvanich, S.; Khan, M.S.R.; Putthinun, P.; Ono, S.; Kadoya, Y. Financial Literacy and Gambling Behavior in the United States. *J. Gambl. Stud.* **2021**, *38*, 445–463. [CrossRef]
43. Morgan, P.J.; Long, T.Q. Financial literacy, financial inclusion, and savings behavior in Laos. *J. Asian Econ.* **2020**, *68*, 101197. [CrossRef]
44. Ankrah Twumasi, M.; Jiang, Y.; Adhikari, S.; Adu Gyamfi, C.; Asare, I. Financial literacy and its determinants: The case of rural farm households in Ghana. *Agric. Financ. Rev.* **2022**, *82*, 641–656. [CrossRef]
45. Klapper, L.; Lusardi, A.; Van Oudheusden, P. *Financial Literacy around the World: Insights from the S&P Global FinLit Survey*; World Bank: Washington, DC, USA, 2015; pp. 1–27.
46. Atakora, A. Measuring the Effectiveness of Financial Literacy Programs in Ghana. *Int. J. Manag. Bus. Res.* **2016**, *3*, 135–148.
47. Koomson, I.; Villano, R.A.; Hadley, D. The role of financial literacy in households' asset accumulation process: Evidence from Ghana. *Rev. Econ. Househ.* **2022**, 1–24. [CrossRef]
48. Chowa, G.; Despard, M.; Osei-Akoto, I. Financial knowledge and attitudes of youths in Ghana. *YouthSave Res. Br.* **2012**, *12*, 1–7.
49. Ministry of Food and Agriculture. *Facts and Figures*; Ministry of Food and Agriculture: Accra, Ghana, 2015.
50. Andoh, F.K.; Nunoo, J.; Darfor, K.N. Sustaining Small and Medium Enterprises through Financial Service Utilization: Does Financial Literacy Matter? *J. Bus. Entrep. Dev.* **2015**, *5*, 74–94. [CrossRef]
51. Niu, G.; Zhou, Y. Financial literacy and retirement planning: Evidence from China. *Appl. Econ. Lett.* **2018**, *25*, 619–623. [CrossRef]
52. Deng, X.; Zeng, M.; Xu, D.; Wei, F.; Qi, Y. Household Health and Cropland Abandonment in Rural China: Theoretical Mechanism and Empirical Evidence. *Int. J. Environ. Res. Public Health* **2019**, *16*, 3588. [CrossRef]
53. Murendo, C.; Mutsonziwa, K. Financial literacy and savings decisions by adult financial consumers in Zimbabwe. *Int. J. Consum. Stud.* **2017**, *41*, 95–103. [CrossRef]
54. Ma, W.; Zhu, Z. A Note: Reducing Cropland Abandonment in China—Do Agricultural Cooperatives Play a Role? *J. Agric. Econ.* **2020**, *71*, 929–935. [CrossRef]
55. Zhou, X.; Ma, W. Agricultural mechanization and land productivity in China. *Int. J. Sustain. Dev. World Ecol.* **2022**, *29*, 530–542. [CrossRef]
56. Ma, W.; Nie, P.; Zhang, P.; Renwick, A. Impact of Internet use on economic well-being of rural households: Evidence from China. *Rev. Dev. Econ.* **2020**, *24*, 503–523. [CrossRef]
57. Tesfaye, W.; Tirivayi, N. The impacts of postharvest storage innovations on food security and welfare in Ethiopia. *Food Policy* **2018**, *75*, 52–67. [CrossRef]
58. Ankrah Twumasi, M.; Jiang, Y.; Zhou, X.; Addai, B.; Darfor, K.N.; Akaba, S.; Fosu, P. Increasing Ghanaian fish farms' productivity: Does the use of the internet matter? *Mar. Policy* **2021**, *125*, 104385. [CrossRef]
59. Ma, W.; Zhu, Z. Internet use and willingness to participate in garbage classification: An investigation of Chinese residents. *Appl. Econ. Lett.* **2020**, *28*, 788–793. [CrossRef]
60. Lokshin, M.; Sajaia, Z. Maximum Likelihood Estimation of Endogenous Switching Regression Models. *Stata J.* **2004**, *4*, 282–289. [CrossRef]
61. Rees, H.; Maddala, G.S. Limited-Dependent and Qualitative Variables in Econometrics. *Econ. J.* **1985**, *95*, 493. [CrossRef]
62. Greene, W.W.H. *Econometric Analysis*, 7th ed.; Prentice-Hall: Hoboken, NJ, USA, 2012; ISBN 978-0-273-75356-8.
63. Heckman, J.J. Sample Selection Bias as a Specification Error. *Econometrica* **1979**, *47*, 153. [CrossRef]
64. Wooldridge, J.M. Control function methods in applied econometrics. *J. Hum. Resour.* **2015**, *50*, 420–445. [CrossRef]

65. Bucher-Koenen, T.; Lusardi, A.; Alessie, R.; van Rooij, M. How Financially Literate Are Women? An Overview and New Insights. *J. Consum. Aff.* **2017**, *51*, 255–283. [CrossRef]
66. Lusardi, A.; Mitchell, O.S. Financial literacy and planning: Implications for retirement wellbeing. In *Financial Literacy: Implications for Retirement Security and the Financial Marketplace*; NBER: Cambridge, MA, USA, 2013; ISBN 9780199696819.
67. Khanal, A.R.; Mishra, A.K. Financial performance of small farm business households: The role of internet. *China Agric. Econ. Rev.* **2016**, *8*, 553–571. [CrossRef]
68. Ma, W.; Renwick, A.; Nie, P.; Tang, J.; Cai, R. Off-farm work, smartphone use and household income: Evidence from rural China. *China Econ. Rev.* **2018**, *52*, 80–94. [CrossRef]
69. Van Rooij, M.; Lusardi, A.; Alessie, R. Financial literacy and stock market participation. *J. Financ. Econ.* **2011**, *101*, 449–472. [CrossRef]
70. Wooldridge, J.M. *Introductory Econometrics*; Cengage Learning: Mason, OH, USA, 2013; ISBN 9780324581621.
71. Fernandez-Cornejo, J.; Mishra, A.; Nehring, R.; Hendricks, C.; Southern, M.; Gregory, A. *Off-Farm Income, Technology Adoption, and Farm Economic Performance*; Economic Research Report; Economic Research Service, USDA: Washington, DC, USA, 2007.
72. Li, M.; Gan, C.; Ma, W.; Jiang, W. Impact of cash crop cultivation on household income and migration decisions: Evidence from low-income regions in China. *J. Integr. Agric.* **2020**, *19*, 2571–2581. [CrossRef]
73. Saqib, S.E.; Khan, H.; Panezai, S.; Ali, U.; Ali, M. Credit Fungibility and Credit Margin of Investment: The Case of Subsistence Farmers in Khyber Pakhtunkhwa. *Sarhad J. Agric.* **2017**, *33*, 661–667. [CrossRef]

agriculture

MDPI

Article

Re-Measurement of Agriculture Green Total Factor Productivity in China from a Carbon Sink Perspective

Zhuohui Yu [1,2], Qingning Lin [2,*] and Changli Huang [3]

1 College of Economics, Northwest Normal University, No. 967 East Road, Anning District, Lanzhou 730070, China
2 Institute of Agricultural Economics and Development, Chinese Academy of Agricultural Sciences, 12 South Avenue, Zhongguancun, Haidian District, Beijing 100081, China
3 Gansu Academy of Agricultural Sciences, No. 1, New Village of Academy of Agricultural Sciences, Anning District, Lanzhou 730070, China
* Correspondence: linqingning@caas.cn; Tel.: +86-10-8210-6191

Abstract: Accurate measurement of agricultural total factor productivity (AGTFP) is crucial to measure the level of sustainable agricultural development, and agricultural carbon sink is an important element to leverage the development of green transformation. Few studies have incorporated agricultural carbon sink into the measurement framework of AGTFP, and the evolutionary dynamics and related spatial effects of Chinese AGTFP from the perspective of carbon sinks are unclear. On this basis, the paper used a provincial-level agricultural panel data set of China from 2000 to 2019 to measure the provincial indicators of agricultural carbon sinks, CO_2 emissions and agricultural non-point source pollution. Then, we incorporated these environmental factors into the measurement framework of AGTFP and used the SBM-DEA model to calculate the Chinese AGTFP from the perspective of carbon sinks. We further analyzed the spatial and temporal divergence and convergence of AGTFP in China using Moran'I and spatial econometric models. We found that after measuring AGTFP, including agricultural carbon sinks, 28 out of 30 Chinese provinces showed an increased trend, but the development gap between regions was obvious. The spatial econometric model showed a significantly positive spatial correlation between the AGTFP of each province and did not have absolute α-convergence and absolute β-convergence characteristics. After adding the control variables of resource endowment of each province, it showed conditional β-convergence characteristics, and the spatial spillover effect of China's AGTFP was increasing. Finally, the paper proposed policy recommendations for the sustainable and coordinated development of China's agricultural regions in response to the research findings.

Keywords: carbon sink; agriculture green total factor productivity in China; re-measurement

Citation: Yu, Z.; Lin, Q.; Huang, C. Re-Measurement of Agriculture Green Total Factor Productivity in China from a Carbon Sink Perspective. *Agriculture* **2022**, *12*, 2025. https://doi.org/10.3390/agriculture12122025

Academic Editors: Moucheng Liu, Xin Chen and Yuanmei Jiao

Received: 19 October 2022
Accepted: 24 November 2022
Published: 27 November 2022

1. Introduction

Since the reform and opening up, Chinese agriculture has made great progress in ensuring food security and economic stability. However, Chinese agricultural production has long relied on the traditional factor-driven pattern, and the overuse of production factors has contributed to the deterioration of carbon emissions and the increase in agricultural surface pollution while promoting agricultural development [1]. According to the bulletin of the first national pollution source census, the emissions of the three main agricultural water pollutants in China account for a large proportion of total pollution, including chemical oxygen demand (COD) accounting for 43.71%, total nitrogen (TN) accounting for 57.19% and total phosphorus (TP) accounting for 67.27%. COD emissions from agricultural pollution exceed those from the industrial sector, becoming the main source of COD emissions. Energy consumption and CO_2 emissions are increasing year by year [2]. Various phenomena, such as overconsumption of resources and energy, and gradual deterioration of ecological badlands, are seriously limiting the sustainable development of Chinese agriculture, and the changes in production methods around agriculture are imminent. Total factor productivity (TFP) is not only the main tool to study economic growth but also a key method to determine the quality of economic growth [3]. There is a great potential for synergy between TFP and sustainable agricultural development and ecological resilience [4]. The improvement of agricultural green total factor productivity (AGTFP) is a vital indicator to guarantee the green development of agriculture and even economic development [5,6]. Therefore, to clarify how to maintain sustainable agricultural development, exploring the level of green productivity of Chinese agriculture under resource and environmental constraints by measuring AGTFP is crucial.

In addition, global climate problems are becoming increasingly serious, and climate warming threatens global food security by affecting agricultural production [7–10], and climate change has long been a common challenge for people around the world to face. As a major contributor to climate change, the development of agriculture must join the action to cope with the global climate crisis. Agriculture contains not only carbon sources but also the function of the carbon sink in its production process. Therefore, agriculture is a large carbon sink system, and a healthy agroecosystem can offset up to 80% of global greenhouse gas emissions released due to agricultural production processes [11]. Therefore, taking into full consideration the role of agricultural carbon sinks, grasping the development process of low-carbon agriculture, re-measuring the green total factor productivity (AGTFP) of China from the perspective of carbon sinks and releasing the huge potential of agricultural emission reduction will become the keys to promoting the green and sustainable development of agriculture, to achieving China's carbon peaking and carbon neutrality goals and to completing the transformation of the economy to a low-carbon development.

How to improve agricultural productivity has been the focus of scholars' research [12–14], and agricultural total factor productivity (ATFP) is also considered as a measure of agricultural productivity. There are currently three main methods to calculate ATFP. The first method is the growth accounting method, which was used by Fan (1991) [15] to measure ATFP in China, and Wen (1993) [16] also used the Solow residual method and reached similar conclusions as Fan (1991) [15]. The second method is stochastic frontier analysis (SFA), which can construct a frontier surface suitable for the characteristics of agricultural production [17], but it requires a predetermined production function. Coelli et al. (2003) [18] calculated the ATFP of Bangladesh using the SFA approach and found a U-shaped agricultural technology progress. Chen and Gong (2021) [19] estimated four AGTFPs under different forms of production functions. The third method is the data envelope method analysis (DEA), which does not require a predetermined functional form and is used to determine productivity levels by creating a piecewise linear production frontier and comparing it with the optimal frontier surface [20]. DEA is capable of handling multiple inputs and outputs. Po-Chi et al. (2008) [14] used sequential DEA to calculate the output-oriented Malmquist productivity index and its decomposition; they found that the main source of productivity growth is technological progress. In recent years, along with global climate change and ecological deterioration, green growth in agriculture has become an essential element to improve agricultural productivity, and it is the key to sustainable agricultural development [20,21]. Agricultural green total factor productivity (AGTFP) is an objective indicator of sustainable agricultural development [22], revealing the sustainable growth component beyond input factors under environmental pressure. Since SFA is difficult to meet the needs of multiple outputs in agricultural production [1], the advantages of DEA methods, such as measuring multiple inputs and multiple outputs, are widely used in the assessment, especially when incorporating environmental factors into the measurement framework of AGTFP [23–28]. The specific measurement method is to attribute environmental factors, such as carbon source pollution and non-point source pollution, generated from the agricultural production process as non-desired outputs to the output side and then use the DEA method to measure AGTFP [2,20,24,29]. However, the agricultural production process includes not only carbon emissions but also carbon sinks, and a healthy agroecosystem can effectively reduce the CO_2 released from the agricultural production process [30]. Currently, scholars' research focuses on CO_2 and non-point source pollution emissions [2]. Zhang et al. (2017) [31] established a method and estimated the carbon footprint of grain production in China based on life cycle analysis (LCA). The results showed that grain production had a high carbon footprint in 2013. Cheng et al. (2015) [32] also conducted similar studies as Zhang et al. (2017) [31]. Some scholars have estimated and studied carbon sinks. For example, Lin (2018) [33] calculated the green production efficiency of forests based on carbon sinks. Zhang et al. (2022) [34] measured the efficiency of net carbon sinks in 285 Chinese cities from 2012 to 2017. Chen et al. (2021) [35] estimated the carbon sink of crop production systems from four aspects: tree; soil organic carbon; fertilizer application; and no-till management. Chen estimated the carbon footprint of farmers' agricultural production through a multi-system boundary scenario approach and included agricultural carbon sinks in the research framework to judge the contribution of farmers' agricultural production to climate change Chen (et al. (2020)) [36]. There is still a great lack of studies that include carbon sink, carbon emissions and non-point pollution jointly in the measurement framework of AGTFP. Hence, in order to accurately measure China's AGTFP from the perspective of carbon sink, as well as to grasp the sustainable development level of agriculture under environmental constraints, we used the DEA method to add carbon sink to the environmental factors to measure China's AGTFP. On the other hand, another key aspect to assess the sustainable development level of agriculture at this stage is to study the spatial effect of agricultural AGTFP [37]. Wei et al. (2018) [38] studied the factors affecting agriculture using a spatial error model (SEM) and found that factors such as industrial agglomeration and the level of science and technology had positive effects on agricultural green production efficiency. Therefore, to further grasp the level of sustainable agricultural development in China, we used a spatial econometric model to study the relevant spatial effects of AGTFP after completing the measurement of AGTFP that incorporates agricultural carbon sink factors.

Previous studies on the measurement of AGTFP and its spatial effects provide the basis for this paper. However, few studies have included carbon sink factors in the measurement framework of AGTFP, and there is a lack of research on the spatial effects of AGTFP in China from the perspective of carbon sinks. The marginal contributions of this paper are as follows. First, agricultural ecosystems are an essential part of global terrestrial ecosystems, an important source and sink of atmospheric carbon. Agricultural soils have a great carbon sink potential, which has a large impact on mitigating climate change. Relatively few scholars have measured the data on carbon sink as well as net carbon emission indicators. This paper measured agricultural carbon sink, CO_2 emissions and non-point source pollution emissions in Chinese provinces from 2000 to 2019 and included them as non-expected outputs in the calculation framework of AGTFP, which enriches the measurement of AGTFP. Second, based on the previous measured data, we used the global Moran'I index, absolute α convergence, absolute β convergence, conditional β convergence and spatial Durbin model (SDM) to study the spatial autocorrelation, convergence and other spatial effects of China's AGTFP from multiple perspectives to reveal the spatial and temporal convergence of China's AGTFP from a dynamic perspective. Third, this research focused on the agricultural development at the provincial level, and the findings are of great practical significance for promoting the coordinated and high-quality development of regional green agriculture in China. Therefore, the research significance of this paper was demonstrated through the following points. First, we recalculated China's agricultural carbon sinks using the latest carbon equivalent factors from the UN Intergovernmental Panel on Climate Change (IPCC) Sixth Assessment Report to provide a new perspective for developing a more effective CO_2 reduction strategy. Second, we estimated the net carbon emissions of China's agriculture, which can more accurately indicate the actual growth of China's agriculture and provide a reference for decision making to precisely reduce the regional disparity of China's AGTFP. Third, we analyzed the dynamic convergence of AGTFP in the spatial dimension to clarify the dynamic evolutionary characteristics of AGTFP convergence and to reveal the sources of regional disparities in AGTFP growth in China.

2. Materials and Methods

2.1. Measurement of Agriculture Carbon Sinks and Carbon Emissions and Non-Point Source Pollution

At present, there are no relevant statistics on the environmental indicators of agricultural carbon source emissions of CO_2, carbon sinks and agricultural non-point source pollution of CO_2. Therefore, it was necessary to measure the above three indicators and calculate them. Then, the net carbon emissions in the agricultural production process were obtained by subtracting the carbon sequestration by carbon sinks from the agricultural carbon source emissions, which was a good quantitative basis for measuring China's AGTFP from the perspective of carbon sinks in the following context.

2.1.1. Measurement of Agriculture Carbon Sinks

The United Nations Framework Convention on Climate Change (UNFCCC) referred to the concept of carbon sinks as "processes or activities that reduce greenhouse gases in the atmosphere". Crop carbon sequestration referred to the process by which crops convert CO_2 in the air into carbohydrates through photosynthesis, releasing oxygen while fixing the carbon in the crop for its own growth and development. This section draws on the calculations used by Chen et al. (2021) [35] to calculate the carbon sink by crop production systems, including: carbon absorption by trees (CS_{TA}) and soil organic carbon (SOC) increases due to straw, litter, pruning and root residue return (CS_{SR}); manure application (CS_{MA}); and no-tillage management (CS_{NT}).

$$TCS_i = CS_{TA} + CS_{SR} + CS_{MA} + CS_{NT} \tag{1}$$

where TCS_i represents the total carbon sequestration. CS_{TA} represents the carbon absorbed by tea and fruit trees aside from that removed by harvesting, pruning and litter, which was 527.5 (Li, 2012) [39] and 930 (Lv, 2019) [40] kg C ha^{-1} yr^{-1}, respectively. The detailed calculation process and explanation of CS_{SR}, CS_{MA} and CS_{NT} are shown in Appendix A.

2.1.2. Measurement of Agriculture Carbon Emissions

The United Nations Framework Convention on Climate Change (UNFCCC) referred to the concept of carbon sources as "processes or activities that emit greenhouse gases into the atmosphere". Based on the carbon accounting approach of the United Nations Intergovernmental Panel on Climate Change (IPCC), the formula for agricultural carbon emissions was constructed as follows.

$$E_c = \sum E_i = \sum T_i \times \delta_i \tag{2}$$

where E_c represents the total agricultural carbon emission, E_i represents the emission of the i-th category of agricultural carbon source, T_i represents the specific value of the i-th category of agricultural carbon source, and δ_i represents the carbon emission coefficient of each agricultural carbon source. Based on previous studies [3,41], the paper determined the corresponding carbon sources and carbon emission coefficients from agricultural land use, rice and livestock breeding, and the indirect N_2O emissions from in-field nitrogen fertilizer application and straw burning based on the characteristics of agricultural production activities and consideration of data availability. Carbon emissions from agricultural land use covered carbon emissions from fertilizers, pesticides, agricultural films, diesel, tillage, irrigation, etc., in the agricultural production process. In addition, the conversion of carbon dioxide (CO_2), nitrous oxide (N_2O), methane (CH_4), etc., into standard C equivalents and the unification of measurement units facilitated the calculation and subsequent comparison of content. The UN Intergovernmental Panel on Climate Change (IPCC) Sixth Assessment Report stipulated that the conversion C-equivalent standard was that the greenhouse effect caused by 1 t N_2O is equivalent to that caused by 273 t CO_2, and the greenhouse effect caused by 1 t CH_4 is equivalent to the greenhouse effect caused by 27 t CO_2. Because 1 t CO_2 contains 0.272 7 t C, the C contained in 1 t N_2O and 1 t CH_4 is approximately 74.256 t and 7.344 t. The detailed calculation process and explanation of carbon emission are shown in Appendix A.

2.1.3. Measurement of Agriculture Carbon Non-Point Source Pollution

The paper used the idea of the inventory analysis method to account for agricultural non-point source pollution. The method assumed that a certain agricultural activity corresponds to a certain amount of agricultural pollution emissions and integrated a variety of analytical methods to establish the agricultural activity and pollution emissions response relationship, with the unit as the core. The pollutants were mainly COD, TN and TP, and the formula for accounting for agricultural non-point source pollution emissions are as follows [2].

$$E_n = \sum_i EU_i \rho_i (1 - \eta_i) C_i (EU_i, S) = \sum_i PE_i (1 - \eta_i) C_i (EU_i, S) \tag{3}$$

where E_n represents the emission of agricultural non-point source pollution (i.e., CODCR, TN and TP). EU_i represents the indicator statistic of unit i; ρ_i represents the pollution production coefficient of pollutant of unit i; η_i represents the coefficient characterizing the efficiency of relevant resource utilization; PE_i represents the pollution production of pollutant of unit i. This indicator does not take into account the maximum potential pollution caused by comprehensive resource utilization and management factors. C_i represents the emission coefficient of pollutant of unit i, which is determined by the unit characteristics (EU_i) and spatial characteristics (S) and characterizes the combined effects of regional environment, rainfall and various management measures on agricultural non-point source pollution.

The indicators of agricultural non-point source pollutant discharges evaluated in the paper mainly included COD_{CR}, TN and TP remitted to water bodies through surface runoff and farmland drainage, etc. Therefore, based on the characteristics of agricultural production activities, the identified pollution-producing units were pollution discharges from farmland fertilizers, livestock and poultry breeding and farmland solid waste. According to the Class III standard on surface water environmental quality standard (GB3838-2002), the individual pollutant indicators were converted into equivalent emissions. The formula is: Pollutant equivalent emissions = pollutant emissions/pollutant discharge evaluation standard.

2.2. *SBM-DEA Model*

DEA has become a mainstream technique for efficiency evaluation, since it has many advantages, such as not assuming functional relationships, non-subjective weights and the ability to analyze decision unit invalid factors [2]. The DEA method is usually used to evaluate the efficiency of production containing non-desired outputs. Although the traditional directional distance function can better solve the problem of evaluating the efficiency of production containing non-desired outputs, it cannot eliminate the non-efficiency components caused by the input–output slack. To solve the problem of relaxation of variables and the measurement error caused by radial direction, Tone (2001) [42] proposed a non-radial, non-oriented SBM data envelopment analysis model based on relaxation variables, but that model still cannot distinguish and rank multiple equally valid cells. Therefore, Tone (2002) [43] proposed a super-efficient SBM model to solve that problem. Since SBM-DEA takes the input–output slack variables into account, making the efficiency evaluation results more accurate and solving the problem of further comparing and ranking many effective units, it has thus been widely used by scholars [23–28]. In this paper, we applied the method of Tone (2002) [43] to measure the AGTFP of China from the perspective of carbon sink.

We supposed there existed M decision-making units (DMUs). $P(x)$ represents the set of production possibilities; x represents the production input; y represents the economic output; and b represents the undesired output—all of which can be freely disposed of for input factor x and economic output y. Therefore, if $(y, b) \in P(x)$ and $y' \leq y, x' \geq x$, then $(y', b) \in P(x)$ or $P(x') \in P(x)$. Similarly, when the environmental output also satisfies free disposability, the environmental output indicator will also satisfy the above axioms. When agriculture does not have to pay the corresponding economic cost for the environmental pollution generated during the production process, the production possibility set will take the following form.

$$P(x) = \left\{ \begin{array}{c} (x, y, b) : \sum_{m=1}^{M} z_m x_m \leq x; \sum_{m=1}^{M} z_m y_m \geq y; \\ \sum_{m=1}^{M} z_m b_m \leq b, z_m \geq 0, m = 1, .., M \end{array} \right. \tag{4}$$

When the environmental output is weakly disposable, the environmental output b will satisfy the following axiom: if $(y, b) \in P(x)$ and $0 < \theta < 1$, then $(\theta y, \theta b) \in P(x)$. This axiom states that each unit of emission reduction will cause an equally proportional reduction in economic output. That is, it is the economic cost of the agricultural production process due to emissions, just as the non-point source pollution emission rights and carbon emission rights gradually established in China are the economic costs due to emissions. In this case, the production may take the form of:

$$P(x) = \left\{ \begin{array}{c} (x, y, b) : \sum_{m=1}^{M} z_m x_m \leq x; \sum_{m=1}^{M} z_m y_m \leq y; \\ \sum_{m=1}^{M} z_m b_m = b, z_m \geq 0, m = 1, .., M \end{array} \right. \tag{5}$$

The specific expression of the super-efficiency model constructed by Tone (2002) [43] is as follows.

$$\rho = min \frac{\frac{1}{m} \sum_{i=1}^{M} \frac{\overline{x}}{x_{ik}}}{\frac{1}{s_1 + s_2} \left(\sum_{i=1}^{s_1} \frac{\overline{y_l^d}}{y_{l_0}^d} + \sum_{k=1}^{s_2} \frac{\overline{y_k^d}}{y_{k_0}^d} \right)}$$
$$s.t. \left\{ \begin{array}{c} \overline{x} \geq \sum_{j=1, j \neq j_0}^{n} x_{ij} \lambda_j; \overline{y^d} \leq \sum_{j=1, j \neq j_0}^{n} y_{lj}^d \lambda_j; \overline{y^u} \leq \sum_{j=1, j \neq j_0}^{n} y_{kj}^d \lambda_j \\ \overline{y^d} \leq y_{lj}^d; \overline{y^u} \leq y_{kj}^d \\ \lambda_j \geq 0, i = 1, \ldots, m; j = 1, \ldots, n; l = 1, \ldots, s_1; k = 1, \ldots, s_2 \end{array} \right. \tag{6}$$

where n denotes the number of decision units, which is the number of provinces in this study. Each DMU consists of input m, desired output s_1 and non-desired output s_2. x denotes the elements in the input matrix; y^d denotes the elements in the desired output matrix; y^u denotes the data in the non-desired output matrix; and ρ denotes the efficiency value of the DMU.

The green production efficiency values measured by the SBM model are static, and the Malmquist model complements the SBM model well by analyzing dynamically the changes in efficiency values between the two preceding and following years. Therefore, the global reference Malmquist model (GML model), which uses the sum of the periods as a possible reference set, is used to calculate the production efficiency values.

$$s^g = s^1 \cup s^2 \cup \ldots \cup s^p = \left\{ \left(x_j^1, y_j^1\right) \cup \left(x_j^2, y_j^2\right) \cup \ldots \cup \left(x_j^p, y_j^p\right) \right\} \tag{7}$$

The index formula for GML is as follows:

$$M_g\left(x^{t+1}, y^{t+1}, x^t, y^t\right) = \frac{E^g\left(x^{t+1}, y^{t+1}\right)}{E^g\left(x^t, y^t\right)} \tag{8}$$

The same global frontier is referenced in the calculation of the Malmquist index for the two adjacent periods, but the calculation of the efficiency change still uses the respective frontier, so that the efficiency change (EC) is expressed as

$$EC = \frac{E^{t+1}\left(x^{t+1}, y^{t+1}\right)}{E^t\left(x^t, y^t\right)} \tag{9}$$

where the degree to which frontier $t + 1$ is close to the global frontier is represented by $\frac{E^g(x^{t+1},y^{t+1})}{E^{t+1}(x^{t+1},y^{t+1})}$, and a larger ratio indicates that frontier t + 1 is closer to the global frontier, and the degree to which frontier t is close to the global frontier is represented by $\frac{E^g(x^t,y^t)}{E^t(x^t,y^t)}$, with a larger ratio indicating that the frontier t is closer to the global frontier. The variation of efficiency can be obtained by dividing the above two values.

$$TC_g = \frac{E^g(x^{t+1},y^{t+1})/E^{t+1}(x^{t+1},y^{t+1})}{E^g(x^t,y^t)/E^t(x^t,y^t)} = \frac{E^g(x^{t+1},y^{t+1})}{E^{t+1}(x^{t+1},y^{t+1})} \times \frac{E^t(x^t,y^t)}{E^g(x^t,y^t)} \quad (10)$$

Thus, the Malmquist index can be decomposed into efficiency changes and technological changes.

$$M_g\left(x^{t+1},y^{t+1},x^t,y^t\right) = \frac{E^g(x^{t+1},y^{t+1})}{E^g(x^t,y^t)} = \frac{E^{t+1}(x^{t+1},y^{t+1})}{E^t(x^t,y^t)}\left(\frac{E^g(x^{t+1},y^{t+1})}{E^{t+1}(x^{t+1},y^{t+1})} \times \frac{E^t(x^t,y^t)}{E^g(x^t,y^t)}\right) = EC \times TC \quad (11)$$

If ML > 1, it means that AGTFP is increasing; conversely, if ML < 1, it means that AGTFP is decreasing. EC > 1 indicates that the DMU moved to the best practice frontier; TC measures the movement of the best practice frontier caused by technological progress.

2.3. *Spatial Effect Model*

2.3.1. Method of Spatial Autocorrelation

The study of spatial autocorrelation is a crucial concept to reveal the distribution of spatial data, and the calculation of the degree of correlation in spatial autocorrelation is the primary method to study spatial autocorrelation [34]. The autocorrelation test of AGTFP is the first step in constructing the spatial econometric model. We applied RSDA to test the spatial correlation and selected the global spatial correlation as well as the local spatial correlation in ESDA analysis tool to test the spatial correlation of AGTFP.

The global spatial correlation can be used to analyze the spatial agglomeration state of AGTFP, and the Greary'C coefficient and Moran'I index are used in most cases. Since the global Moran'I index can more closely reflect the degree of similarity between neighboring regions, we chose the global Moran'I index to test the spatial correlation of AGTFP. The formula for constructing the global Moran'I index is as follows.

$$I = \frac{n\sum_{i=1}^{n}\sum_{j=1}^{N} w_{ij}(y_i - \overline{y})(y_j - \overline{y})}{S_0\sum_{i=1}^{N}(y_i - \overline{y})^2} \quad (12)$$

where y_i and y_j represent the AGTFP of the i-th and j-th provinces, respectively; $n = 1, 2, \ldots, 30$ represents the number of provinces that we studied; $\overline{y}$ represents the mean value of AGTFP of the 30 provinces; w_{ij} is the spatial adjacency weight matrix; $S_0 = \sum_{i=1}^{N}\sum_{j=1}^{N} w_{ij}$ represents the spatial weight aggregation; and the Moran'I $\in [-1,1]$. The larger the value of Moran'I index, the higher the degree of spatial correlation between regions. If the Moran'I index is significantly greater than 0, it means that there is a positive spatial correlation between regions, which is expressed as "high-high" or "low-low" spatial clustering. If the Moran'I index is significantly less than 0, it means that there is a negative spatial correlation between regions, which is expressed as "high-low" or "low-high" spatial clustering. If the Moran'I index is 0, it means that there is no spatial correlation between regions, and the AGTFP of each province is independently distributed. After the Moran'I index is obtained, its significance needs to be tested. In this section, we will use the Z statistic test, which is calculated by the following formula.

$$Z(I) = \frac{I - R(I)}{\sqrt{VAR(I)}} \quad (13)$$

where $R(I) = \frac{-1}{n-1}$, $VAR(I) = \left[\frac{1}{w_0^2(n^2-1)}\left(n^2w_1 + nw_2 + 3w_0^2\right)\right] - R^2(I)$, $w_0 = \sum_{i=1}^{N}\sum_{j-1}^{N} w_{ij}$, $w_1 = \frac{1}{2}\sum_{i=1}^{N}\sum_{j=1}^{N}\left(w_{ij} + w_{ji}\right)^2$, $w_2 = \sum_{i=1}^{N}\left(w_{i\cdot} + w_{\cdot j}\right)^2$. $w_{i\cdot}$ and $w_{\cdot j}$ are the sum of the *i*-th row and *j*-th column in the spatial weight matrix. If the value of Z(I) is greater than zero, it means that there is a spatially positive correlation of AGTFP between provinces; if the value of Z(I) is less than zero, it means that there is a spatially negative correlation of AGTFP between provinces; if the value of Z(I) is equal to zero, it means that there is a spatially independent distribution of AGTFP between provinces.

2.3.2. Method of Spatial Convergence Analysis

Since the convergence analysis can visualize the performance of an algorithm and evaluate an algorithm scientifically from a theoretical point of view, convergence analysis is widely applied by scholars [44]. The methods for studying spatial convergence are absolute α convergence, absolute β convergence and conditional β convergence. Absolute α convergence refers to the fact that the gap between different regions will gradually decrease and eventually converge with time. Absolute β convergence assumes that the marginal factor rewards are decreasing. Under this premise, the regions will eventually reach the same steady-state level as time elapses. Conditional β convergence indicates that the resource endowment conditions of different regions are different and closely related to economic growth, making it difficult to achieve a consistent steady-state level among regions. Previous econometric models have led to biased convergence conclusions due to often ignoring the correlation with geographic location [45]. Therefore, we incorporated spatial factors into previous econometric models to examine the regional convergence differences of AGTFP in China from a spatial perspective.

(1) Absolute α convergence analysis

When absolute α convergence is tested for the dispersion of AGTFP in China, if α shows a decreasing trend, there is a convergence trend among provinces. Different tests have different sensitivities to the data, so the α coefficient and the coefficient of variation will be used to jointly test the convergence characteristics of AGTFP among provinces to ensure the robustness of the test results. The equations of each test method are as follows.

$$\alpha = \sqrt{\frac{\sum_{i=1}^{N}\left(lny_{it} - \overline{lny_t}\right)^2}{n}} \tag{14}$$

$$CV = \frac{S}{\overline{y_t}} \tag{15}$$

where y_{it} is the AGTFP of the *i*-th province in *t*-th year, and $\overline{y_t}$ is the mean value of the AGTFP of the provinces in *t*-th year.

(2) Absolute β convergence analysis

Absolute β convergence can test whether provinces that started with lower AGTFP can catch up with provinces that started with higher AGTFP through higher growth rates. Based on the method of Barro et al. (1995) [45], we used an absolute β convergence model. for the test, and the model equation is as follows.

$$\frac{ln(y_{it}/y_{i0})}{T} = \alpha + \beta y_{i0} + \mu_{it} \tag{16}$$

where y_{it} and y_{i0} are the AGTFP of the *i*-th province in *t*-th year. T represents the average annual growth rate of AGTFP of province *i* from 2000 to 2019; α and β are parameters to be estimated; μ_{it} is the random error term. If the parameter β is significantly negative, it means that the AGTFP among Chinese provinces has an absolute β-convergence trend.

(3) Conditional β convergence analysis

Conditional β convergence refers to the fact that the steady-state level of AGTFP in each province is associated with some resource endowment conditions. It is difficult to reach the same steady-state level in all provinces. In order to consider the influence of external environment on the steady-state level of AGTFP in each province, we added control variables to the model when conducting the conditional β convergence test. If the estimation result of β remains significantly negative, it indicates the existence of conditional β convergence among provinces. Based on previous studies, we selected the level of economic development (GDP), agricultural industrial restructuring (AIR), agricultural infrastructure (AID), energy consumption (EC), effective irrigation rate (EI) and disaster incidence rate (DOR) as the control variables for each province. GDP was expressed as gross output per capita. AIR was expressed as the ratio of total plantation output to total agricultural output. AID was expressed as the ratio of road mileage to provincial and district administrative area. EC was expressed as rural electricity consumption. EI was expressed as the ratio of irrigated area to total sown area of crops. DOR was expressed as the ratio of disaster area to total sown area of crops. The conditional β convergence test model for AGTFP in China is as follows.

$$d(lny_{it}) = lny_{it} - lny_{i(t-1)} = \alpha + \beta lny_{i(t-1)} + \gamma x_{it} + \mu_{it} \tag{17}$$

where lny_{it} represents the AGTFP of the i-th province in t-th year. x_{it} is the control variable mentioned above. α, β and γ are the parameters to be estimated. Additionally, μ_{it} is the random error term.

(4) Spatial Econometric Model

Traditional econometric models largely ignore the geographical correlation between regions, thus yielding biased spatial convergence results [45]. The inclusion of spatial factors can not only avoid the endogeneity of spatial spillover effects but also study the direction of spatial spillover effects. Therefore, spatial econometric models are mostly used by scholars to study spatial characteristics [41,46]. Currently, scholars often apply SEM, SDM and SLM [47,48] to introduce geographic features to construct models, and the spatial lag model (SAR) and spatial error model (SEM) can reflect the correlation between different regions. Based on the studies of scholars such as Yu et al. (2012) [46] and Elhorst (2012) [41], we combined spatial factors to construct a convergence model and consider a spatial perspective to study the convergence of regional differences in AGTFP. The y_t with one period lag is set as the explanatory variable in the β convergence model to construct the dynamic space (SDM) conditional β convergence model, dynamic space (SAR) conditional β convergence model and dynamic space (SEM) conditional β convergence model of AGTFP in each province of China. The specific models are as follows.

$$\begin{gathered} ln\frac{y_{it}}{y_{it-p}} = \alpha + \beta lny_{it-p} + \rho wln\frac{y_{it}}{y_{it-p}} + \gamma x_{it} + \varepsilon_{it}, \\ \varepsilon_{it} = \lambda w\varepsilon_{it} + \mu_{it}, \mu_{it} \sim N(0,\sigma^2) \end{gathered} \tag{18}$$

SAR conditional β convergence model

$$ln\frac{y_{it}}{y_{it-p}} = \alpha + \beta lny_{it-p} + \rho wln\frac{y_{it}}{y_{it-p}} + \gamma x_{it} + \varepsilon_{it}, \varepsilon_{it} \sim N\left(0,\sigma^2\right) \tag{19}$$

SEM conditional β convergence model

$$ln\frac{y_{it}}{y_{it-p}} = \alpha + \beta lny_{it-p} + \gamma x_{it} + \varepsilon_{it}, \varepsilon_{it} = \lambda w\varepsilon_{it} + \mu_{it}, \mu_{it} \sim N\left(0,\sigma^2\right) \tag{20}$$

where y_{it} and y_{it-p} are the values of AGTFP for each Chinese province in t-th year and $(t-p)$-th year; w is the spatial weight matrix; α, β and γ are the parameters to be estimated; λ and ρ represent the spatial correlation coefficients, which are a reflection of the relationship between AGTFP interactions among provinces; ε_{it} and μ_{it} are both random error terms obeying independent identical distribution; x_{it} represents the control variables. If β is significantly negative, it indicates that the AGTFP in each province showed dynamic spatial convergence. We selected the number of lags as one period.

2.4. Variable Selection and Data Source

Based on the production characteristics of agriculture, the paper selected the input–output data of 30 provinces (excluding Tibet) in mainland China from 2000 to 2019 to calculate the AGTFP. The input indicators included land, labor, machinery and fertilizer. The output indicators included the desired output and non-desired output, and non-desired output included agricultural non-point pollution (NP) and agricultural net carbon emissions (NCE). According to the class III surface water environmental quality standard (GB3838-2002), the calculation of agricultural non-point source pollution was converted to the three types of agricultural non-point source pollution emissions in agricultural pollution loads. In addition, when calculating the net agricultural carbon emissions, greenhouse gases, such as CO_2, N_2O and CH_4, emitted into the atmosphere during the production process were uniformly converted into standard carbon (C) equivalents, thus unifying the measurement units and subtracting them from agricultural carbon sequestration to obtain the net agricultural carbon emissions. Additionally, when calculating the spatial econometric model, we selected the level of economic development (GDP), agricultural industrial restructuring (AIR), agricultural infrastructure (AID), energy consumption (EC), effective irrigation rate (EI), disaster occurrence rate (DOR), financial support for agriculture (FS) and major grain producing areas (MGP) of each province as the control variables. The specific index selection and data sources are shown in Table 1.

Table 1. AGTFP assessment indicators and data sources.

	Assessment Indicators	Indicators' Explanation	Unit	Source	Reference
Input	Land	the total sown area of crops	10^4 hectares	"China Agricultural Statistics" and "China Rural Statistical Yearbook"	Gong (2020) [17] Chen et al. (2021) [19]
	Labor	employees in the primary industry	10^4 People	"China Statistical Yearbook"	Gong (2020) [17] Chen et al. (2021) [19]
	Machinery	the total power of agricultural machinery	10^4 Tons	"China Agricultural Statistics" and "China Rural Statistical Yearbook"	Gong (2020) [17] Chen et al. (2021) [19]
	Fertilizer	the amount of chemical fertilizer actually used in agricultural production calculated by the pure method	10^4 kilowatts	"China Agricultural Statistics" and "China Rural Statistical Yearbook"	Gong (2020) [17] Chen et al. (2021) [19]
Output	GVAO (Expected output)	the total output value of agriculture, forestry, animal husbandry and fishery at constant prices in 2000	10^8 CNY	"China Agricultural Statis-tics" and "China Rural Statistical Yearbook"	Gong (2020) [17] Chen et al. (2021) [19]
	NP (Non-expected output)	the pollution of chemical oxygen demand, total nitrogen and total phosphorus caused by pollutants entering the water body through surface runoff and farmland drainage	10^4 Tons	Calculated results	Yu et al. (2022) [11] Shen et al. (2018) [20]
	NCE (Non-expected output)	the value that uses agricultural carbon emissions minus agricultural carbon sinks	10^4 Tons	Calculated results	Yu et al. (2022) [11]

Table 1. *Cont.*

	Assessment Indicators	Indicators' Explanation	Unit	Source	Reference
control variables	GDP	GDP per capita	10^4 CNY	China Statistical Yearbook	Liu et al. (2021) [5]
	AIR	the total output value of planting industry/total agricultural output value	-	"China Agricultural Statistics" and "China Rural Statistical Yearbook"	Yu et al. (2022) [11] Liu et al. (2021) [5]
	AID	number of road miles/administrative area	-	China Regional Economic Statistics Yearbook	Wang et al. (2021) [22]
	EC	rural electricity consumption	10^8 kW/h	China Agricultural Statistics	Reza et al. (2016) [49]
	EI	the effective irrigated area/total sown area of crops	-	"China Agricultural Statistics" and "China Rural Statistical Yearbook"	Kumar et al. (2008) [50]
	DOR	agricultural disaster area/total sown area of crops	-	"China Agricultural Statistics" and "China Rural Statistical Yearbook"	Nwaiwu et al. (2015) [51]
	FS	local financial expenditure on agriculture, forestry and water affairs	10^8 CNY	National Bureau of Statistics	Gong (2020) [17]
	MGP	MGP is a dummy variable, if a province belongs to the major grain producing area, MGP = 1, otherwise MGP = 0	-	Ministry of Agriculture and Rural Affairs of the People's Republic of China	Li et al. (2022) [28]

Note: GVAO represents gross value of agriculture, NP represents agricultural non-point pollution, NCP represents agricultural net carbon emissions, GDP represents GDP per capita, AIR represents agricultural industrial restructuring, AID represents agricultural infrastructure, EC represents energy consumption, EI represents effective irrigation rate, DOR represents disaster occurrence rate, FS represents financial support for agriculture, and MGP represents major grain producing areas.

3. Results and Analysis

3.1. Calculation Results of Agricultural Net Carbon Emissions

Based on the measurement methods introduced in Section 2.1, we calculated the agricultural carbon source emissions, carbon sinks and net carbon emissions for each province in China. The following Table 2 lists the mean values of carbon emissions, carbon sinks and net carbon emissions for 2000–2019. Beijing had the smallest agricultural carbon emission, with a mean value of 92.563×10^4 tons. Henan had the largest agricultural carbon emission, with a mean value of 2414.393×10^4 tons. Guangdong had the largest agricultural carbon sink, with a mean value of 1034.076×10^4 tons. Henan had the highest mean value of agricultural net carbon emissions, while Beijing had the lowest. Agricultural land use carbon emissions were highest in Henan, Shandong and Hebei. Rice fields carbon emissions were highest in Jiangxi, Jiangsu and Hunan. Shandong, Henan and Chongqing had the highest livestock and poultry farming carbon emissions. The highest carbon sinks by crop production systems and soil organic carbon occurred in Guangdong, Guangxi and Shanxi.

Table 2. Total agricultural net carbon emissions in China's provinces in 2000–2019. Unit: 10,000 t.

Area	CE_{LU}	CE_{RF}	CE_{LP}	INE	TCE	CS_{TA}	CS_{SR}	CS_{MA}	CS_{NT}	TCS	NCE
Beijing	29.737	0.150	56.694	5.983	92.563	62.515	3.599	0.006	4.723×10^{-5}	66.120	28.168
Tianjin	49.472	1.539	64.482	22.896	138.389	32.452	1.872	0.005	1.821×10^{-5}	34.329	104.060
Hebei	829.965	9.111	738.539	43.523	1621.138	898.696	51.783	0.061	2.701×10^{-5}	950.540	670.598
Shanxi	263.153	0.084	190.409	28.587	482.233	291.346	16.791	0.014	5.127×10^{-6}	308.150	174.083
Inner Mongolia	447.113	5.962	750.333	29.127	1232.535	58.209	3.382	0.046	1.271×10^{-5}	61.636	1170.898
Liaoning	368.994	34.822	484.129	44.610	932.554	327.118	18.826	0.046	3.960×10^{-6}	345.990	586.564
Jilin	407.043	26.695	405.600	35.111	874.449	59.334	3.449	0.031	6.423×10^{-6}	62.814	811.634
Heilongjiang	688.669	159.465	462.716	16.086	1326.936	34.033	2.079	0.029	3.008×10^{-6}	36.141	1290.794
Shanghai	45.449	43.035	35.679	5.859	130.021	17.327	1.000	0.004	1.545×10^{-6}	18.330	111.691
Jiangsu	686.999	794.893	375.799	32.792	1890.483	195.794	11.862	0.043	8.459×10^{-6}	207.700	1682.784
Zhejiang	332.842	291.746	165.505	26.156	816.249	377.078	24.871	0.018	2.616×10^{-7}	401.967	414.282
Anhui	691.652	696.298	444.793	34.173	1866.916	179.744	12.967	0.045	7.065×10^{-6}	192.756	1674.160
Fujian	280.960	215.281	218.843	29.939	745.022	563.643	35.886	0.024	3.186×10^{-9}	599.553	145.469
Jiangxi	368.187	834.582	378.977	50.660	1632.405	359.088	21.829	0.035	5.401×10^{-7}	380.952	1251.453
Shandong	1095.156	18.440	1054.839	92.867	2261.302	621.067	36.179	0.100	2.606×10^{-5}	657.347	1603.955
Henan	1177.588	69.285	1096.862	70.658	2414.393	433.098	26.304	0.088	1.539×10^{-5}	459.490	1954.902
Hubei	643.983	726.931	459.902	55.751	1886.568	421.412	28.265	0.045	7.556×10^{-7}	449.722	1436.845
Hunan	573.306	915.444	665.073	62.096	2215.920	497.848	30.605	0.061	2.155×10^{-7}	528.514	1687.405
Guangdong	463.018	463.049	496.962	58.187	1481.216	976.971	57.046	0.058	1.100×10^{-8}	1034.076	447.140
Guangxi	475.255	459.621	510.989	36.565	1482.430	937.841	54.912	0.050	1.725×10^{-6}	992.803	496.453
Hainan	97.717	68.554	97.052	11.756	275.080	155.198	8.960	0.009	4.078×10^{-8}	164.167	110.913
Sichuan	224.263	121.680	238.992	55.059	639.993	230.944	13.879	0.022	1.147×10^{-8}	244.846	395.147
Chongqing	622.403	350.763	1081.456	86.247	2140.868	623.061	39.998	0.088	2.333×10^{-7}	663.147	1477.722
Guizhou	265.837	105.476	384.729	45.753	801.795	316.310	22.073	0.026	4.982×10^{-8}	338.410	488.322
Yunnan	466.099	48.739	651.605	51.209	1217.651	505.417	35.133	0.046	4.325×10^{-8}	540.596	677.054
Shanxi	14.300	10.571	272.830	26.585	662.767	953.935	56.480	0.016	1.201×10^{-5}	101.043	561.724
Gansu	363.115	0.237	222.148	26.094	554.953	347.576	20.155	0.022	1.779×10^{-6}	36.775	518.178
Qinghai	306.711	0.000	359.654	14.022	414.105	5.399	0.311	0.014	3.715×10^{-6}	5.724	408.381
Ningxia	36.622	3.808	306.409	17.780	422.686	80.036	4.608	0.005	2.265×10^{-6}	84.649	338.037
Xinjiang	93.364	5.134	98.497	1.756	527.076	657.896	37.864	0.031	1.765×10^{-6}	69.579	457.497

Note: CE_{LU} represents the agricultural land use carbon emissions; CE_{RF} represents the rice fields carbon emissions; CE_{LP} represents the livestock and poultry farming carbon emissions; INE represents the indirect N_2O emissions; TCE represents the total carbon emissions; CS_{TA} represents the carbon sinks by crop production systems, including carbon absorption by trees; CS_{SR} represents the soil organic carbon increases due to straw, litter, pruning and root residue return; CS_{MA} represents carbon sink by manure application; CS_{NT} represents the carbon sink by no-tillage management; TCS represents the total carbon sinks; NCE represents the net carbon emission.

3.2. Empirical Results and Analysis of China's AGTFP

Based on MAXDEA software, we separated technical progress and technical efficiency to obtain the average annual growth rates for the MI index, EC index and TC index in each province from the carbon sink perspective (Table 3). Except for Heilongjiang and Ningxia, whose AGTFP was decreasing, the AGTFPs of all the other 28 provinces were increasing. Among them, Beijing, Tianjin and Chongqing had an average annual growth rate of MI over 1. Beijing had the highest AGTFP growth rate of 2.068%, and Heilongjiang had the lowest AGTFP growth rate of −0.094%. The EC growth rates in Shanxi, Inner Mongolia, Jilin, Heilongjiang, Jiangsu, Zhejiang, Fujian, Shandong, Guangxi, Guizhou, Ningxia and Xinjiang were negative, while Fujian had the lowest EC growth rate of −0.162%. The remaining provinces had a positive average annual growth rate of EC. The average annual growth rate of EC in Beijing and Shanghai exceeded 1. Beijing had the highest TC growth rates. The average annual growth rate of TC in Beijing was 3.030%. The negative average annual growth rate of TC in Tianjin, Shanghai and Qinghai indicated that the technological progress showed a decreasing trend.

Table 3. Average annual growth rates of MI, EC and TC from 2000 to 2019 from the carbon sink perspective (%).

Province	MI	EC	TC	Province	MI	EC	TC
Beijing	2.068	3.517	3.030	Henan	0.183	0.010	0.173
Tianjin	0.613	0.341	−0.271	Hubei	0.147	0.129	0.276
Hebei	0.045	0.538	0.586	Hunan	0.347	0.102	0.245
Shanxi	0.201	−0.119	0.082	Guangdong	1.066	0.125	0.941
Inner Mongolia	0.183	−0.046	0.229	Guangxi	0.188	−0.195	0.384
Liaoning	0.168	0.003	0.171	Hainan	0.397	0.866	0.466
Jilin	0.368	−0.110	0.258	Sichuan	0.380	0.068	0.311
Heilongjiang	−0.094	−0.143	0.049	Chongqing	1.844	0.977	0.858
Shanghai	0.006	5.363	−0.085	Guizhou	0.157	−0.273	0.431
Jiangsu	0.500	−0.045	0.545	Yunnan	0.366	0.215	0.150
Zhejiang	0.639	−0.060	0.699	Shanxi	0.214	0.102	0.316
Anhui	0.127	0.043	0.170	Gansu	0.456	0.109	0.347
Fujian	0.723	−0.162	0.561	Qinghai	0.578	0.683	−0.104
Jiangxi	0.275	0.046	0.229	Ningxia	−0.035	−0.233	0.269
Shandong	0.390	−0.060	0.450	Xinjiang	0.149	−0.027	0.176

Note: MI represents total factor productivity, EC represents efficiency changes, TC represents technological progress.

3.3. Spatial Effect Analysis

On the basis of measuring China's AGTFP, we further analyzed its distribution pattern, the spatial effects, the power source, spatial and temporal divergence and convergence of China's AGTFP growth. We explained the spatial and temporal convergence of China's AGTFP in a panoramic manner from the perspective of spatial and temporal dynamics.

3.3.1. Empirical Results and Analysis of Spatial Autocorrelation

According to the calculation methods of spatial autocorrelation, we conducted a test on the mean value of China's AGTFP, and the test results are shown in Table 4. Table 4 shows the results of the global autocorrelation test of AGTFP in China, where the Moran'I index of AGTFP was greater than 0 and passed the 1% significance level test in 2000 and 2018, while it passed the 10% significance level test in 2007, 2010, 2016, and in the remaining years, it passed the 5% significance level test. Overall, there was a significantly positive spatial correlation between the AGTFP of each province in China. In addition, a larger Moran'I value indicates a stronger spatial correlation; a maximum value of 0.243 in 2016 indicates the strongest spatial correlation. The Moran'I index had fluctuated during 2000–2019, but the overall trend was upward, rising from 0.103 in 2000 to 0.153 in 2019. This indicates that there was a presence of agricultural green technology diffusion and technology exchange among neighboring provinces, with an overall increasing trend of diffusion and exchange, as indicated by the spatial spillover effects. Resource endowment and natural location conditions were inextricably linked to agricultural green production, and the convergence of agricultural green technology conditions was higher in neighboring or closer provinces. With the diffusion and exchange of knowledge and green technology, the AGTFP in neighboring or closer provinces was spatially correlated.

Table 4. Global correlation test results of AGTFP in China.

Year	AGTFP		
	Moran'I	**Z Value**	***p* Value**
2000	0.103	0.305	0.000 ***
2001	0.024	0.576	0.038 **
2002	0.126	0.888	0.028 **
2003	0.215	2.223	0.018 **
2004	0.057	0.828	0.013 **
2005	0.123	0.809	0.020 **
2006	0.064	0.879	0.021 **
2007	0.114	1.358	0.019 **
2008	0.182	2.287	0.087 *
2009	0.118	1.382	0.011 **
2010	0.143	0.967	0.084 *
2011	0.094	1.139	0.016 **
2012	0.085	0.383	0.012 **
2013	0.033	0.020	0.035 **
2014	0.216	2.042	0.049 **
2015	0.091	1.208	0.021 **
2016	0.234	1.842	0.083 *
2017	0.181	2.453	0.033 **
2018	0.150	0.154	0.001 ***
2019	0.153	0.305	0.044 **

Robust standard errors in parentheses *** $p < 0.01$, ** $p < 0.05$, * $p < 0.1$.

3.3.2. Empirical Results and Analysis of Spatial Convergence

Based on previous studies [31,52], we used convergence methods such as absolute α convergence, absolute β convergence and conditional β convergence to analyze the convergence of AGTFP in China.

(1) Empirical Results and Analysis of Absolute α Convergence

According to the calculation methods of the α coefficient and coefficient of variation, we performed a test on the mean value of China's AGTFP, and the test results are shown in Figure 1. Figure 1 shows that the test results of each method had different values, but the trend was relatively smooth and had a small upward trend, which indicated that China's AGTFP will not have an absolute alpha convergence trend in a certain period of time. The reason for such a situation may be that the paper involved green technologies, such as environmental pollution and resource saving. However, the current lack of motivation to promote related technologies makes it difficult for Chinese agricultural green technologies to diffuse. Additionally, the provinces with higher AGTFP in the initial year maintained higher efficiency levels, while the provinces with lower AGTFP in the initial year had difficulty in imitating and learning quickly. This made it difficult for absolute α convergence trends to occur within a certain period of time.

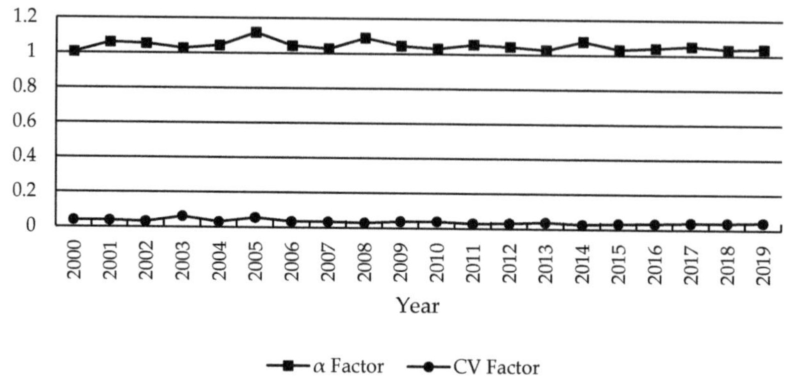

Figure 1. Trend of α convergence of AGTFP in China. Note: CV represents coefficient of variation.

(2) Empirical Results and Analysis of Absolute β Convergence

Based on the calculation methods of absolute β convergence, we performed a test on the mean value of China's AGTFP, and the test results are shown in Table 5. Table 5 shows that the results for absolute β convergence of AGTFP and the β coefficients of the eastern, central, western regions and the national average were significantly negative at the 1% level. This indicates that the AGTFP of the national region, the eastern region, the central region, the western region were characterized by absolute β convergence. In addition, the β coefficients were significantly positive at the 1% level for all three time periods within the period 2000–2019, except for 2000–2004, where all β coefficients were significantly negative at the 1% level, indicating that China's AGTFP was characterized by non-absolute β convergence.

Table 5. Absolute β convergence results for AGTFP.

Factor	Nationwide	Sub-Region			Sub-Time			
		East	Central	West	2000–2004	2005–2009	2010–2014	2014–2019
β	−0.941 ***	−0.904 ***	−0.989 ***	−0.911 ***	−1.313 ***	1.100 ***	1.126 ***	1.171 ***
	(0.043)	(0.069)	(0.084)	(0.075)	(0.097)	(0.124)	(0.098)	(0.083)
α	0.958 ***	0.922 ***	1.008 ***	0.927 ***	1.331 ***	1.127 ***	1.142 ***	1.093 ***
	(0.044)	(0.071)	(0.083)	(0.077)	(0.099)	(0.127)	(0.099)	(0.085)
R^2	0.471	0.487	0.478	0.450	0.646	0.469	0.467	0.576

Note: Robust standard errors in parentheses *** $p < 0.01$.

(3) Empirical Results and Analysis of Conditional β Convergence Analysis

Based on the calculation methods of conditional β convergence, we performed a test on the mean value of China's AGTFP, and the test results are shown in Table 6. Table 6 shows the results of the conditional β convergence of AGTFP in China. First, from the time perspective, the β coefficients of the national, eastern, central and western regions were significantly negative, and the national, eastern and western regions all pass the test at the 1% significance level. This indicates that the AGTFP of each region in China had a conditional β convergence posture. Second, from the time perspective, the β coefficients of China's AGTFP are significantly negative for the period 2000–2019, and all four periods in Table 5 pass the 1% significance level test. This indicates that the AGTFP of China had a conditional beta convergence posture. Overall, the AGTFP of China in the national, eastern, central and western regions had significant conditional β convergence characteristics. Because of the differences in resource endowment of different provinces, the AGTFP in different provinces converges to its own steady-state level at different rates.

Table 6. Conditional β convergence results for AGTFP.

Factor	Nationwide	Sub-Region			Sub-Time			
		East	Central	West	2000–2004	2005–2009	2010–2014	2014–2019
β	−0.911 ***	−0.887 ***	−0.961 ***	−0.905 **	−1.093 ***	−0.882 ***	−0.891 ***	−0.984 ***
	(0.042)	(0.068)	(0.081)	(0.076)	(0.091)	(0.108)	(0.086)	(0.072)
α	0.930 ***	0.895 ***	0.917 ***	0.916 ***	1.028 ***	0.956 ***	0.896 ***	1.081 ***
	(0.044)	(0.078)	(0.107)	(0.089)	(0.099)	(0.130)	(0.088)	(0.087)
GDP	0.001 ***	0.002 ***	0.003 ***	0.004 ***	0.007 ***	0.005 ***	0.001 ***	0.001 ***
	(0.001)	(0.002)	(0.006)	(0.004)	(0.007)	(0.006)	(0.002)	(0.00)
AIR	0.001	0.007	0.100	−0.009	0.068	−0.029	0.013	−0.053
	(0.032)	(0.066)	(0.118)	(0.68)	(0.078)	(0.101)	(0.049)	(0.055)
AID	0.002	0.001	0.032	−0.001	0.005	−0.003	−0.004 **	0.003 *
	(0.001)	(0.002)	(0.022)	(0.002)	(0.004)	(0.006)	(0.001)	(0.02)
EC	−0.001	−0.001	−0.022	0.003	0.00	0.004	−0.002	0.001
	(0.001)	(0.001)	(0.015)	(0.014)	(0.004)	(0.003)	(0.001)	(0.001)
EI	−0.005	0.002	0.042	−0.007	0.009	0.006	0.008	−0.022 **
	(0.007)	(0.038)	(0.101)	(0.012)	(0.018)	(0.026)	(0.011)	(0.01)
DOR	0.001	0.051	−0.049	−0.012	0.035	−0.017	0.002	0.018 *
	(0.0019)	(0.031)	(0.044)	(0.036)	(0.045)	(0.049)	(0.034)	(0.034)
FS	−0.001	−0.00	−0.003	−0.004	00.046	−0.024	0.001	−0.006 ***
	(0.001)	(0.002)	(0.005)	(0.003)	(0.044)	(0.015)	(0.003)	(0.002)
MGP	−0.004	−0.015	−0.011	0.014	−0.001	−0.009	0.010	−0.013
	(0.005)	(0.011)	(0.023)	(0.014)	(0.012)	(0.016)	(0.008)	(0.009)
R^2	0.474	0.503	0.492	0.460	0.661	0.469	0.510	0.634

Note: Robust standard errors in parentheses *** $p < 0.01$, ** $p < 0.05$, * $p < 0.1$. GDP represents GDP per capita, AIR represents agricultural industrial restructuring, AID represents agricultural infrastructure, EC represents energy consumption, EI represents effective irrigation rate, DOR represents disaster occurrence rate, FS represents financial support for agriculture, and MGP represents major grain producing areas.

3.3.3. Empirical Results and Analysis of SDM Model

Before the spatial analysis, we proceeded with some preliminary statistical tests (Table 7). The results of the LM test showed significant spatial error and spatial lag; therefore, a spatial model should be used instead of a mixed regression model. The fixed effects model was determined by the Hausman test. The likelihood ratio (LR) and the Wald test showed that SDM cannot be degraded to SAR and SEM models; therefore, we used the dynamic spatial model (SDM) to study the dynamic spatial change dynamics of AGTFP in China.

Table 7. Statistical tests of the spatial econometric model.

		Statistic	*p* Value
LM	Spatial error	14.236	0.000 ***
	Spatial lag	33.587	0.000 ***
Hausman	-	20.040	0.000 ***
LR	SDM-SAR	43.254	0.000 ***
	SDM-SEM	13.187	0.001 ***
Wald	SDM-SAR	12.041	0.004 ***
	SDM-SEM	14.012	0.001 ***

Note: *** $p < 0.01$.

Table 8 shows the results of the conditional β convergence test for the dynamic spatial SDM of AGTFP. Table 6 illustrated that after incorporating the spatial factors and lagged variables of China's AGTFP, the β coefficient was still significantly negative at the 1% statistical level. This indicates that the regional convergence characteristics of China's AGTFP were still evident after considering the endowment conditions of each province's GDP, AIR, AID, EC, EI and DOR. Therefore, the potential factors, such as inter-regional agricultural production factor flows and institutional environment, also played a non-negligible role in regional disparities. In addition, the spatial correlation coefficient ρ passed the 1% significance level test and was positive, indicating that the spatial spillover effect of AGTFP in China was increasing, and it was necessary to further promote the exchange of agricultural-related green production activities among provinces, and the regions with higher AGTFP played a demonstrative role in driving other Chinese provinces with lower AGTFP to improve continuously.

Table 8. Results of the conditional β convergence test for the dynamic spatial SDM of AGTFP.

Variable	SDM	Variable	SDM
β	−0.942 *** (0.042)	DOR	0.002 *** (0.021)
α	0.604 (0.109)	FS	0.002 ** (0.00)
GDP	0.003 *** (0.002)	MGP	0.002 (0.049)
AID	−0.001 (0.003)	ρ	15.009 ** (1.015)
AIR	0.018 (0.081)	σ^2	0.003 *** (0.001)
EC	0.001 (0.002)	R^2	0.466
EI	0.0139 *** (0.044)	Log-likelihood	816.547

Note: Robust standard errors in parentheses *** $p < 0.01$, ** $p < 0.05$. GDP represents GDP per capita, AIR represents agricultural industrial restructuring, AID represents agricultural infrastructure, EC represents energy consumption, EI represents effective irrigation rate, DOR represents disaster occurrence rate, FS represents financial support for agriculture, and MGP represents major grain producing areas.

4. Discussion

Over the period of 2000–2019, the AGTFP in most Chinese provinces showed an upward trend, which is similar to the growth trend of AGTFP measured by scholars such as Chen et al. (2021) [2], Huang et al. (2022) [53] and Yang et al. (2022) [54]. However, the AGTFP of each province differed from these studies. The reason is that we put carbon sinks into the measurement framework of AGTFP, which can effectively reduce CO_2 emissions. Additionally, Lin (2018) [31] and Chen et al. (2021) [35] came to the same conclusion. Chen et al. (2021) [35] studied the carbon sequestration and carbon footprint of 16 crop production systems in China from 2001 to 2018, and they found that the crop system can effectively alleviate its own carbon emission. Additionally, other scholars [36,55–58] have also calculated the agricultural carbon sink by crop production systems, including carbon absorption by trees and soil organic carbon, manure application and no-tillage management, and they came to similar conclusions. Therefore, there is a minor difference from the results of AGTFP measurement without considering carbon sinks. The significant increase in China's AGTFP indicates that after China's economy entered a medium- to high-speed development stage, China has focused great attention on the transformation of the economy to a high-quality development model over the past decades. A series of material input reduction and various comprehensive management measures have gradually taken effect and successfully put the economy and the environment on a harmonious development track. However, the decomposition indicators of AGTFP in each province were not promising, with 11 provinces showing a decreasing trend in technical efficiency to varying degrees, similar to the findings of Sun et al. (2020) [59], who found a significant increase in AGTFP in China, and 25 provinces showed a decreasing trend in the decomposition indicators of the AGTFP trend. Although all provinces are trying to innovate their economic development models and have accomplished great achievements in stabilizing the economy, adjusting the structure and promoting development, the gap between the advanced and backward provinces still exists. Guo et al. (2021) [60] had similar findings on this point. The main reason for the occurrence of the above situation may be the obvious difference in economic development between different regions, with different resource endowments and industrial advantages, and distinct degrees of green and low-carbon development in agriculture.

In response to the forms of agricultural development in different regions, applying local policies will become one of the effective paths to promote green agricultural development.

In addition, spatial factors had a positive contribution to AGTFP growth. Spatial proximity can promote the dissemination of agricultural green technology and knowledge. The neighboring regions can share high-quality agricultural resource elements. The results of the study through the spatial econometric model indicated that the Moran'I index of AGTFP in each province was significantly positive, showing that the green development between different provinces was spatially interconnected, and cross-regional cooperation and agriculture promotion were of great practical importance. Chen et al. (2022) [61] also argued that the exhibition of cross-regional cooperation targeted the policies. On the other hand, the convergence test showed that the Chinese AGTFP did not have an absolute σ and β convergence trend, and the gap between the regions will not be reduced, which is also consistent with the findings of Guo et al. (2021) [60]. The possible reasons for this result are that the relevant green technologies are currently not accessible, technology promotion is more sluggish, and green technologies are difficult to diffuse. Higher AGTFP efficiency zones maintain higher levels of efficiency, and lower efficiency zones find it difficult to imitate them. The spatial econometric model in this paper showed that the AGTFP had a conditional β convergence posture and had a dynamic spatial conditional β convergence state, while Xu et al. (2022) [62] concluded that the AGTFP did not have a dynamic spatial conditional β convergence state, which is inconsistent with the findings of this paper. The reason for the occurrence of the above may be the inconsistency of the conditional resource endowment of the selected provinces, which can lead to different study results.

5. Conclusions and Recommendations

5.1. Conclusions

From the perspective of agricultural carbon sink, the paper took the agricultural net carbon emissions and agricultural non-point source pollution as unexpected outputs and incorporated them into the calculation framework of AGTFP. We used the super-efficiency productivity index model SBM-DEA to calculate and evaluate the AGTFP in 30 provinces of China from 2000 to 2019. Then, we used the global Moran'I index to analyze the spatial concentration of AGTFP in various provinces of China and studied the convergence trend of China's AGTFP through the absolute α convergence, absolute β convergence and conditional β convergence. Finally, we used the dynamic spatial SDM model to explore the spatiotemporal differentiation and dynamic spatial convergence characteristics of China's AGTFP growth. Our findings can provide a reference for proposing an optimal pathway to improve the AGTFP from the perspective of agricultural carbon sinks, and they are useful for identifying the sources of regional differences in China's green agricultural development, narrowing the regional differences and providing the theoretical support and decision-making basis for regional green agricultural development. Our research also contributes to a well-balanced institutional mechanism for coordinated regional development at the level of green agricultural development. The main research conclusions are as follows:

(1) From the perspective of agricultural carbon sink, the AGTFPs of 28 out of 30 provinces in China were growing, while that of Heilongjiang and Ningxia was decreasing. Among them, the average annual growth rate of MI in Beijing, Guangdong and Chongqing exceeded 1. The average annual growth rate of AGTFP in Beijing was the highest, reaching 2.068%, while that in Heilongjiang was the lowest, reaching −0.094%. In addition, the growth of AGTFP in most provinces was attributed to the improvement of technological progress.

(2) Overall, there was a significantly positive spatial correlation between the AGTFPs in various provinces of China. The Moran'I index of the AGTFP showed an upward trend of fluctuation during the study period, rising from 0.103 in 2000 to 0.153 in 2019, among which the maximum value was 0.243 in 2016. This indicated the presence of diffusion and technology exchange between neighboring provinces regarding agricultural green

technology. With the diffusion and exchange of knowledge and green technology, the AGTFP in neighboring provinces or closer provinces had spatial relevance.

(3) The AGTFP in China did not have absolute α convergence and absolute β convergence characteristics; provinces with higher AGTFP in the initial year maintained higher efficiency levels, while low-AGTFP regions found it difficult to quickly imitate and learn. However, after controlling for the control variable of resource endowment of each province, the conditional β convergence characteristics showed that the convergence characteristics of different provinces were closely related to different resource endowments. Additionally, there were still obvious conditional β convergence characteristics after the spatial factors were considered. The spatial correlation coefficient ρ was positive at the significance level of 1%, which indicated that the spatial spillover effect of AGTFP in China was constantly increasing.

5.2. Recommendations

Based on the above research conclusions, we proposed corresponding countermeasures and suggestions:

(1) According to the development of agriculture in different provinces, local policies will become one of the effective ways to promote sustainable agricultural development. First, for provinces with high agricultural land use carbon emissions, such as Henan, Shandong and Hebei, local governments should increase efforts to return farmland to forests, reduce the use of chemical fertilizers and pesticides, pay attention to conservation tillage systems, reasonably carry out tillage and crop rotation to enhance the carbon sink function of the region, offset the higher carbon emissions and improve the ecological environment of farmland. Second, in provinces where rice fields emit a high amount of carbon dioxide, such as Jiangxi, Jiangsu and Hunan, the government should strengthen the management of rice fields, promote over-belly return, develop biogas and strictly prohibit burning in situ to inhibit the spread of greenhouse gases and cultivate soil fertility. Third, for Shandong, Henan, Chongqing and other provinces with high carbon emissions from livestock and poultry breeding, it is essential to reasonably plan the livestock industry, reasonably treat livestock and poultry manure using modern composting processes, vigorously promote biogas projects and implement carbon reduction policies, i.e., using clean energy instead of traditional energy.

(2) The government should focus on developing a series of policies to enhance the carbon sink capacity of agricultural land in order to reduce the concentration of greenhouse gases in the atmosphere, mainly from the following three aspects. The government should adopt a protective farming system, reduce the use of chemical fertilizers and pesticides, decrease straw burning and promote straw return to the fields according to the resource endowment conditions of different regions through government subsidies in order to enhance the carbon sink capacity of farmland, increase the carbon sink capacity of grasslands through rational planning of livestock farming, implementation of grazing pause or even grazing ban and returning grazing to grass. Afforestation and reforestation in eligible areas can significantly improve the vegetation cover of land, and the carbon sink capacity of agricultural land can be increased.

(3) Policy makers should develop AGTFP growth strategies for different provinces according to the spatial characteristics of China's AGTFP and local conditions. The endowment conditions, such as geographic and natural conditions, vary significantly among the regions in China, but the AGTFPs among different provinces have obvious spatial correlation. Therefore, policy makers should consider each province's factor endowment advantages, as well as resource and environmental carrying capacity, tapping the potential of the carbon sink market and formulating relevant measures to reduce emissions and increase sinks, as well as prevent and treat pollution to improve the ecological environment. With rich carbon sinks, Guangdong, Guangxi and Shanxi should further maximize the spatial spillover effect, improve the radiation demonstration role and realize the docking of green technology and green growth through management experience and technology

exchange. To further enhance the AGTFP, agricultural carbon sinks should be increased, and the agricultural ecological environment should be optimized.

(4) Local governments should combine their own agricultural development to promote a coordinated development of AGTFP in each province at multiple levels, so as to achieve high-quality development of agriculture. First, the government should increase financial support for green agricultural development and enhance the conservation of agricultural resources while improving the efficiency of agricultural resource utilization. Second, the eastern and central provinces should further improve the efficiency of effective irrigation, actively develop water-saving agriculture, promote dry-farming and water-saving agricultural technologies and improve the efficiency of water resources utilization. Gansu, Xinjiang and other western provinces should further strengthen environmental management, optimize the agricultural industrial structure, formulate policies to reduce energy consumption and improve energy use efficiency, and develop effective strategies to deal with major disasters to reduce the negative impact of disasters on ecological and agricultural production activities.

Author Contributions: Conceptualization, Z.Y.; Data curation, Z.Y. and C.H.; Investigation, C.H.; Methodology, Z.Y.; Software, Q.L.; Validation, Q.L.; Writing—original draft, Z.Y., Q.L. and C.H.; Writing—review and editing, Q.L. All authors have read and agreed to the published version of the manuscript.

Funding: This research was funded by Northwest Normal University Young Teachers' Research Ability Enhancement Program (NWNU-SKQN2022-33).

Institutional Review Board Statement: Not applicable.

Data Availability Statement: The datasets used and analyzed in the current study are available from the corresponding author on reasonable request.

Conflicts of Interest: The authors declare no conflict of interest.

Appendix A

The detailed calculation methods for CS_{SR}, CS_{MA} and CS_{NT} are as follows:

$$CS_{SR} = \frac{SR_i + RB_i}{1000} \times 29.025 + 272.33 \quad \text{(A1)}$$

where the tree body does not consider the root residue, and the litter and pruning are equivalent to straw return. The biomasses of litter and pruning for tea and fruit trees (take citrus, for example) are 1682 (You, 2008) [55] and 1843 (Wu et al., 2010) [56] kg ha^{-1}, respectively.

$$CS_{MA} = M_{i,c} \times 19.1\% \quad \text{(A2)}$$

where Mc refers to the carbon input due to manure application. This value can be calculated by Equation (5). The 19.1% refers to the percentage of input carbon converted into soil organic carbon (Wang et al., 2015) [57].

$$CS_{NT} = 120 \times NTR \quad \text{(A3)}$$

where 120 refers to no-tillage management, which can increase SOC by 120 kg ha^{-1} (Luo et al., 2010) [58]; NTR refers to the proportion of no-tillage area to total area.

According to the existing research literature, the emission coefficients and reference sources of various carbon sources are summarized as follows (Table A1).

Table A1. Agricultural land use carbon emission sources, carbon emission coefficients and reference sources.

Carbon Source	Carbon Emission Coefficient	Reference Source
Fertilizer	0.8965 kgC·kg^{-1}	West and Marland (2002) [63]
Pesticide	4.9341 kgC·kg^{-1}	West and Marland (2002) [63]
Agricultural Film	5.18 kgC·kg^{-1}	Wang and Zhang (2016) [64]
Diesel Fuel	0.5927 kgC·kg^{-1}	IPCC (2007) [65]
Plowing	312.6 kgC·hm^{-2}	Wu and Li (2007) [66]
Agricultural Irrigation	25 kgC·ha^{-1}	Dubey and Lal (2009) [67]

The CH_4 emissions produced by rice planting not only account for most of the CH_4 emissions in China but also have a heavy impact on the global atmospheric CH_4 emissions. Therefore, when considering the carbon emission coefficient of rice production, it needs to be considered by varieties and regions. On the basis of Min and Hu (2012) [68], the obtained C emission coefficients of rice by variety and region were transformed into the C emission coefficients, and the C emission coefficients of rice varieties (early rice, mid-season rice and late rice) were obtained by province (Table A2).

Table A2. Rice carbon emission coefficients in each province. Unit: kg·hm^{-1}.

Area	Early Rice (Single Cropping Rice)	Mid-Season Rice (Single Cropping Late Rice, Winter Paddy Field and Wheat Stubble Rice)	Double-Cropping Late Rice
Beijing	0	901.96	0
Tianjin	0	773.11	0
Hebei	0	1045.12	0
Shanxi	0	451.32	0
Inner Mongolia	0	608.80	0
Liaoning	0	629.94	0
Jilin	0	379.74	0
Heilongjiang	0	566.54	0
Shanghai	846.05	3672.59	1874.81
Jiangsu	1095.57	3650.70	1881.63
Zhejiang	979.68	3951.42	2352.04
Anhui	1141.93	3493.29	1881.63
Fujian	527.68	2963.57	3586.01
Jiangxi	1054.67	4460.01	3122.42
Shandong	0	1431.68	0
Henan	0	1216.92	0
Hubei	1193.74	3065.74	2658.83
Hunan	1002.85	3836.89	2324.77
Guangdong	1026.03	3887.34	3517.83
Guangxi	846.05	3257.40	3347.39
Hainan	915.59	3564.87	3367.85
Sichuan	446.54	1754.14	1261.24
Chongqing	446.54	1754.14	1261.24
Guizhou	347.70	1503.26	1431.68
Yunnan	162.26	494.27	518.13
Shanxi	0	852.87	0
Gansu	0	465.6	0
Qinghai	0	0	0
Ningxia	0	501.08	0
Xinjiang	0	715.83	0

Livestock and poultry farming is an important emission source of CH_4 and N_2O emissions. The CH_4 emission coefficient comprises the CH_4 emission coefficient of gastrointestinal fermentation of livestock and poultry and the CH_4 emission coefficient of livestock and poultry excrement. The N_2O emission coefficient is the N_2O emission coefficient of livestock and poultry excrement. According to the development of animal husbandry in China, the research objects are mainly CH_4 emissions and excrement caused by gastrointestinal fermentation of cattle (dairy cows, cattle, buffalo), sheep, pigs, horses, donkeys, mules, camels, rabbits and other poultry during the breeding process. Emissions of CH_4 and N_2O are generated during processing. Based on the research of Min and Hu (2012) [68], the emission coefficients of various carbon sources were summarized and converted into C exclusion coefficients (Table A3).

Table A3. Carbon emission coefficients of various livestock and poultry breeds. Unit: kg·$head^{-1}$·a^{-1}.

Livestock and Poultry Breeds	CH_4 Emission Coefficient		N_2O Emission Coefficient	C Emission Coefficient
	Gastrointestinal Fermentation	Fecal Discharge	Fecal Discharge	
Cows	68	16	1	653.9346
Cattle	47.8	1	1.39	445.9218
Buffalo	55	2	1.34	497.4921
Sheep	5	0.16	0.33	61.9956
Pig	1	3.5	0.53	43.4790
Horse	18	1.64	1.39	246.8535
Donkey	10	0.9	1.39	187.2686
Mule	10	0.9	1.39	187.2686
Camel	46	1.92	1.39	439.6524
Rabbit	0.254	0.08	0.02	3.9023
Birds	-	0.02	0.02	1.7616

Note: Since the amount of CH_4 produced by the gastrointestinal fermentation of poultry is small, the emission of CH_4 caused by the gastrointestinal fermentation of poultry is not considered.

The indirect N_2O emissions from in-field nitrogen fertilizer application and straw burning should also be taken into account in carbon emissions. The indirect N_2O emissions are estimated using the following equations.

$$INE_i = (N_2O_{i,ATD-N} + N_2O_{i,L-N} + N_2O_{i,SB}) \times \frac{44}{28} \times 265 \quad \text{(A4)}$$

where INE_i represents the total indirect N_2O emissions; $N_2O_{i,ATD-N}$ represents the N_2O emissions from the atmospheric deposition of volatility; $N_2O_{i,L-N}$ represents the N_2O emissions from leaching and runoff; $N_2O_{i,SB}$ represents the total N_2O emissions from crop straw burning; 44/28 is the molecular conversion factor of N_2 to N_2O; and 265 is the global warming potential of N_2O for a 100-year period.

$$N_2O_{i,ATD-N} = (F_{i,SN} \times EF_{SN-ATD} + F_{i,ON} \times EF_{ON-ATD}) \times 1\% \quad \text{(A5)}$$

$N_2O_{i,ATD-N}$ represents the N_2O emission from atmospheric deposition of N volatility; $F_{i,SN}$ represents the annual amount of synthetic fertilizer N applied to soils; $F_{i,ON}$ represents the amount of manure, compost and other organic N applied to soils; $F\ EF_{SN-ATD}$ represents the fraction of synthetic fertilizer N that volatilizes as NH_3 and NO_x, equal to 11% (IPCC, 2019) [69]; EF_{ON-ATD} represents the fraction of applied organic N fertilizer material that volatilizes as NH_3 and NO_x, equal to 21% (IPCC, 2019) [69]; 1% represents the emission factor for N_2O emissions from atmospheric deposition of N on soils and water surfaces (IPCC, 2019) [69].

$$N_2O_{i,l-N} = (F_{i,SN} + F_{i,ON} + F_{i,CRN}) \times EF_{L-N} \times 1.1\% \quad \text{(A6)}$$

$N_2O_{i,l-N}$ represents the N_2O emission from leaching and runoff; $F_{i,SN}$ represents the annual amount of crop residues' return to soils; EF_{L-N} represents the fraction of all N added to/mineralized in soils in regions where leaching/runoff occurs, equal to 24% (IPCC,

2019) [52]; 1.1% represents the emission factor for N_2O emissions from N leaching and runoff (IPCC, 2019) [52].

$$\begin{aligned} N_2O_{i,SB} &= (SB_i \times EF_{SB-D} + SB_i \times EF_{SB-ATD}) \times 1\% \\ SB_i &= Y_i \times RSY_i \times PSB_i \end{aligned} \tag{A7}$$

$N_2O_{i,SB}$ represents the total N_2O emission from crop straw burning; RSY_i represents the ratio of straw to yield (Table 4); SB_i represents the dry matter quality (moisture content is 22%) of straw burned; EF_{SB-D} represents direct N_2O released from straw burning (See Table A4); EF_{SB-ATD} represents the fraction of NH_3 and NO_x released from straw burning (See Table A4); PSB_i represents the proportion of straw burned as part of the total straw biomass (Table A5).

Table A4. Greenhouse gas emissions from straw burning by crops per unit weight. Unit: kg kg^{-1}.

Crop	EF_{SB-D}	EF_{SB-ATD}	CH_4
Rice	0.0008	0.0023	0.0025
Wheat	0.0003	0.0021	0.0025
Maize	0.0004	0.0022	0.0025
Beans	0.0007	0.0027	0.0025
Potato	0.0007	0.0027	0.0025
Rape seed	0.0007	0.0027	0.0025
Vegetables	0.0007	0.0027	0.0025
Fruits	-	-	-

Note: EF_{SB-D} represents direct N_2O released from straw burning; EF_{SB-ATD} represents the fraction of NH_3 and NO_x released from straw burning. All the data in this table are summarized from Chen et al. 2021 [35].

Table A5. Ratio of straw biomass to yield, the nitrogen concentration in crop straw (root) and return part to the total straw biomass.

Crop	RSY	Straw and Root N Concentration (%)	PSB (%) 2001–2005	PSB (%) 2012–2018
Rice	1.1	0.91	41.9	11.9
Wheat	0.9	0.65	30.6	12.0
Maize	0.8	0.92	44.1	30.2
Beans	1.0	1.81	33.9	16.3
Potato	2.0	2.37	15.7	19.7
Rape seed	0.4	0.87	41.3	42.3
Vegetables	5.9	2.98	28.9	18.2
Fruits	-	2.6	-	-

Note: RSY_i represents the ratio of straw to yield; RAR refers to the ratio of above-ground biomass to root biomass; PSB_i represents the proportion of straw burned as part of the total straw biomass. All the data in this table are summarized from Chen et al. 2021 [35].

References

1. Liu, Y.; Feng, C. What drives the fluctuations of "green" productivity in China's agricultural sector? A weighted Russell directional distance approach. *Resour. Conserv. Recycl.* **2019**, *147*, 201–213. [CrossRef]
2. Chen, Y.; Miao, J.; Zhu, Z. Measuring green total factor productivity of China's agricultural sector: A three-stage SBM-DEA model with non-point source pollution and CO_2 emissions. *J. Clean. Prod.* **2021**, *318*, 128543. [CrossRef]
3. Cheng, K.; Pan, G.; Smith, P.; Luo, T.; Li, L.; Zheng, J.; Zhang, X.; Han, X.; Yan, M. Carbon footprint of China's crop production—An estimation using agro-statistics data over 1993–2007. *Agric. Ecosyst. Environ.* **2011**, *142*, 231–237. [CrossRef]
4. Coomes, O.T.; Barham, B.L.; MacDonald, G.K.; Ramankutty, N.; Chavas, J.P. Leveraging total factor productivity growth for sustainable and resilient farming. *Nat. Sustain.* **2019**, *1*, 22–28. [CrossRef]
5. Liu, D.; Zhu, X.; Wang, Y. China's agricultural green total factor productivity based on carbon emission: An analysis of evolution trend and influencing factors. *J. Clean. Prod.* **2021**, *278*, 123692. [CrossRef]
6. Wang, Y.; Xie, L.; Zhang, Y.; Wang, C.; Yu, K. Does FDI Promote or Inhibit the High-Quality Development of Agriculture in China? An Agricultural GTFP Perspective. *Sustainability* **2019**, *11*, 4620. [CrossRef]
7. Wheeler, T.; Von Braun, J. Climate Change Impacts on Global Food Security. *Science* **2013**, *341*, 508–513. [CrossRef]

8. Stevanović, M.; Popp, A.; Lotze-Campen, H.; Dietrich, J.P.; Müller, C.; Bonsch, M.; Schmitz, C.; Bodirsky, B.L.; Humpenöder, F.; Weindl, I. The impact of high-end climate change on agricultural welfare. *Sci. Adv.* **2016**, *2*, e1501452. [CrossRef]
9. Moore, F.C.; Baldos, U.L.C.; Hertel, T. Economic impacts of climate change on agriculture: A comparison of process-based and statistical yield models. *Environ. Res. Lett.* **2017**, *12*, 065008. [CrossRef]
10. Schlenker, W.; Lobell, D.B. Robust negative impacts of climate change on African agriculture. *Environ. Res. Lett.* **2010**, *5*, 014010. [CrossRef]
11. Yu, Z.H.; Mao, S.P.; Lin, Q.N. Has China's Carbon Emissions Trading Pilot Policy Improved Agricultural Green Total Factor Productivity? *Agriculture* **2022**, *19*, 1444. [CrossRef]
12. McMillan, J.; Whalley, J.; Zhu, L. The Impact of China's Economic Reforms on Agricultural Productivity Growth. *J. Political Econ.* **1989**, *97*, 781–807. [CrossRef]
13. Eroğlu, N.A.; Bozoğlu, M. Impacts of the support policies on agricultural efficiency and total factor productivity in Turkey. *Anadolu J. Agric. Sci.* **2017**, *32*, 35. [CrossRef]
14. Po-Chi, C.H.E.N.; Ming-Miin, Y.U.; Chang, C.C.; Shih-Hsun, H.S.U. Total factor productivity growth in China's agricultural sector. *China Econ. Rev.* **2008**, *19*, 580–593.
15. Fan, S. Effects of Technological Change and Institutional Reform on Production Growth in Chinese Agriculture. *Am. J. Agric. Econ.* **1991**, *73*, 266–275. [CrossRef]
16. Wen, G.J. Total Factor Productivity Change in China's Farming Sector: 1952–1989. *Econ. Dev. Cult. Chang.* **1993**, *42*, 1–41. [CrossRef]
17. Gong, B. Agricultural productivity convergence in China. *China Econ. Rev.* **2020**, *60*, 101423. [CrossRef]
18. Coelli, T.; Rahman, S.; Thirtle, C. A stochastic frontier approach to total factor productivity measurement in Bangladesh crop agriculture, 1961–1992. *J. Int. Dev.* **2003**, *15*, 321–333. [CrossRef]
19. Chen, S.; Gong, B. Response and adaptation of agriculture to climate change: Evidence from China. *J. Dev. Econ.* **2021**, *148*, 102557. [CrossRef]
20. Shen, Z.; Baležentis, T.; Chen, X.; Valdmanis, V. Green growth and structural change in Chinese agricultural sector during 1997–2014. *China Econ. Rev.* **2018**, *51*, 83–96. [CrossRef]
21. Liu, Z.; Guan, D.; Crawford-Brown, D.; Zhang, Q.; He, K.; Liu, J. Energy policy: A low-carbon road map for China. *Nature* **2013**, *500*, 143. [CrossRef]
22. Wang, Q.; Wang, H.; Chen, H. Research on the Change of Green Total Factor Productivity in China's Agriculture:1992-2010. *Econ. Rev.* **2012**, *5*, 24–33. (In Chinese)
23. Emrouznejad, A.; Yang, G.-L. A survey and analysis of the first 40 years of scholarly literature in DEA: 1978–2016. *Socio-Econ. Plan. Sci.* **2018**, *61*, 4–8. [CrossRef]
24. Coelli, T.J.; Rao, D.S.P. Total factor productivity growth in agriculture: A Malmquist index analysis of 93 countries, 1980–2000. *Agric. Econ.* **2005**, *32*, 115–134. [CrossRef]
25. Toma, P.; Miglietta, P.P.; Zurlini, G.; Valente, D.; Petrosillo, I. A non-parametric bootstrap-data envelopment analysis approach for environmental policy planning and management of agricultural efficiency in EU countries. *Ecol. Indic.* **2017**, *83*, 132–143. [CrossRef]
26. Liu, S.; Zhang, S.; He, X.; Li, J. Efficiency change in North-East China agricultural sector: A DEA approach. *Agric. Econ.* **2016**, *61*, 522–532. [CrossRef]
27. Diao, P.; Zhang, Z.; Jin, Z. Dynamic and static analysis of agricultural productivity in China. *China Agric. Econ. Rev.* **2018**, *10*, 293–312. [CrossRef]
28. Li, J.; Lin, Q. Can the Adjustment of China's Grain Purchase and Storage Policy Improve Its Green Productivity? *Int. J. Environ. Res. Public Health* **2022**, *19*, 6310. [CrossRef]
29. Dong, G.; Wang, Z.; Mao, X. Production efficiency and GHG emissions reduction potential evaluation in the crop production system based on emergy synthesis and nonseparable undesirable output DEA: A case study in Zhejiang Province, China. *PLoS ONE* **2018**, *13*, e0206680. [CrossRef]
30. Jin, S.; Lin, Y.; Niu, K. Low-carbon driven green transformation of agriculture: Characteristics of China's agricultural carbon emissions and its emission reduction path. *Reform* **2021**, *5*, 29–37. (In Chinese)
31. Zhang, D.; Shen, J.; Zhang, F.; Li, Y.; Zhang, W. Carbon footprint of grain production in China. *Sci. Rep.* **2017**, *29*, 4126. [CrossRef] [PubMed]
32. Cheng, K.; Yan, M.; Nayak, D.; Pan, G.X.; Smith, P.; Zheng, J.F. Carbon footprint of crop production in China: An analysis of National Statistics data. *J. Agric. Sci.* **2015**, *153*, 422–431. [CrossRef]
33. Lin, B. Carbon sinks and output of China's forestry sector: An ecological economic development perspective. *Sci. Total Environ.* **2018**, *655*, 219. [CrossRef]
34. Zhang, A.; Deng, R. Spatial-temporal evolution and influencing factors of net carbon sink efficiency in Chinese cities under the background of carbon neutrality. *J. Clean. Prod.* **2022**, *365*, 132547. [CrossRef]
35. Chen, X.; Ma, C.; Zhou, H.; Liu, Y.; Huang, X.; Wang, M.; Cai, Y.; Su, D.; Muneer, M.A.; Guo, M.; et al. Identifying the main crops and key factors determining the carbon footprint of crop production in China, 2001–2018. *Resour. Conserv. Recycl.* **2021**, *172*, 105661. [CrossRef]

36. Chen, R.; Zhang, R.; Han, H.; Jiang, Z. Is farmers' agricultural production a carbon sink or source?—Variable system boundary and household survey data. *J. Clean. Prod.* **2020**, *266*, 122108. [CrossRef]
37. Altarhouni, A.; Danju, D.; Samour, A. Insurance Market Development, Energy Consumption, and Turkey's CO_2 Emissions. New Perspectives from a Bootstrap ARDL Test. *Energies* **2021**, *14*, 7830. [CrossRef]
38. Wei, Q.; Zhang, B.; Jin, S. A Study on Construction and Regional Comparison of Agricultural Green Development Index in China. *Issues Agric. Econ.* **2018**, *11*, 11–20.
39. Li, Y.S. *Study Under the Condition of Field Grown Tea Tree Biomass and Nutrient Distribution and the Characteristics of Root Growth*; College of Agriculture. Sichuan Agricultural University: Ya'an, China, 2012.
40. Lv, J.L. *Study on Biomass and Carbon Absorption of Apple Economic Forest in the Residual Gully Area of Shanxi Province*; College of Forestry. Inner Mongolia Agricultural University: Hohhot, China, 2019.
41. Elhorst, J.P. Dynamic Spatial Panels: Models, Methods, and Inferences. *J. Geogr. Syst.* **2012**, *14*, 5–28. [CrossRef]
42. Tone, K. A slacks-based measure of efficiency in data envelopment analysis. *Eur. J. Oper. Res.* **2001**, *130*, 498–509. [CrossRef]
43. Tone, K. A slacks-based measure of super-efficiency in data envelopment analysis. *Eur. J. Oper. Res.* **2002**, *143*, 32–41. [CrossRef]
44. Tobler, W.R. A Computer Movie Simulating Urban Growth in the Detroit Region. *Econ. Geogr.* **1970**, *46*, 234–240. [CrossRef]
45. Barro, R.; Sala-I-Martin, X. *Economic Growth*; Mc Graw Hill: New York, NY, USA, 1995.
46. Yu, J.; Lee, L.-F. CONVERGENCE: A SPATIAL DYNAMIC PANEL DATA APPROACH. *Glob. J. Econ.* **2012**, *1*, 1250006. [CrossRef]
47. Yu, X.R. Promoting Agriculture Green Development to Realize the Great Rejuvenation of the Chinese Nation. *Front. Agric. Sci. Eng.* **2020**, *7*, 11233–12113.
48. Chi, Y.; Xu, Y.; Wang, X.; Jin, F.; Li, J. A Win-Win Scenario for Agricultural Green Development and Farmers' Agricultural Income: An Empirical Analysis Based on the EKC Hypothesis. *Sustainability* **2021**, *13*, 8278. [CrossRef]
49. Reza, M.; Amene, A.P. Energy consumption and total factor productivity growth in Iranian agriculture. *Energy Rep.* **2016**, *2*, 218–220.
50. Kumar, P.; Mittal, S.; Hossain, M. Agricultural Growth Accounting and Total Factor Productivity in South Asia: A Review and Policy Implications. *Agric. Econ. Res. Rev.* **2008**, *21*, 145–172.
51. Nwaiwu, I.O.U.; Asiabaka, C.C.; Ohajianya, D.O. Green Economy-A Panacea for the Devastating Effects of Climate Change on Agricultural Productivity in Southeast Nigeria. *Agric. Econ. Res. Rev.* **2015**, *4*, 204226.
52. Anselin, L. *Spatial Econometrics: Methods and Models*; Kluwer Academic Publishers: Dordrecht, The Netherlands, 1988.
53. Huang, X.; Feng, C.; Qin, J.; Wang, X.; Zhang, T. Measuring China's agricultural green total factor productivity and its drivers during 1998–2019. *Sci. Total Environ.* **2022**, *829*, 154477. [CrossRef]
54. Yang, Y.; Ma, H.; Wu, G. Agricultural Green Total Factor Productivity under the Distortion of the Factor Market in China. *Sustainability* **2022**, *14*, 9309. [CrossRef]
55. You, X.Q. *Requirement on Nutrient and Accumulation on Biomass by Up-Ground Parts of Tea Plants under Field Conditions, Tea Research Institute*; Chinese Academy of Agricultural Sciences: Beijing, China, 2008.
56. Wu, Z.D.; Wang, Y.X.; Cai, Z.J.; You, Z.M.; Zhang, W.J.; Weng, B.Q. Amount and decomposition characteristics of litters in citrus orchard in Fuzhou, China. *J. Ecol. Rural Environ.* **2010**, *26*, 231–234.
57. Wang, Y.; Hu, N.; Xu, M.; Li, Z.; Lou, Y.; Chen, Y.; Wu, C.; Wang, Z.-L. 23-year manure and fertilizer application increases soil organic carbon sequestration of a rice–barley cropping system. *Biol. Fertil. Soils* **2015**, *51*, 583–591. [CrossRef]
58. Luo, Z.; Wang, E.; Sun, O.J. Can no-tillage stimulate carbon sequestration in agricultural soils? A meta-analysis of paired experiments. *Agric. Ecosyst. Environ.* **2010**, *139*, 224–231. [CrossRef]
59. Sun, Y.N.; Yang, M.N. A Study of Club Convergence and Sources of Regional Gaps in Green Total Factor Productivity in China. *Quant. Econ. Tech. Econ. Res.* **2020**, *6*, 47–68. (In Chinese)
60. Guo, H.H.; Liu, X.M. Green Total Factor Productivity in Chinese Agriculture Spatial and Temporal Divergence and Convergence. *Quant. Econ. Tech. Econ. Res.* **2021**, *10*, 65–84. (In Chinese)
61. Chen, H.; Zhu, S.; Sun, J.; Zhong, K.; Shen, M.; Wang, X. A Study of the Spatial Structure and Regional Interaction of Agricultural Green Total Factor Productivity in China Based on SNA and VAR Methods. *Sustainability* **2022**, *14*, 7508. [CrossRef]
62. Xu, P.; Jin, Z.; Ye, X.; Wang, C. Efficiency Measurement and Spatial Spillover Effect of Green Agricultural Development in China. *Front. Environ. Sci.* **2022**, *10*, 909321. [CrossRef]
63. West, T.O.; Marland, G. A synthesis of carbon sequestration, carbon emissions, and net carbon flux in agriculture: Comparing tillage practices in the United States. *Agric. Ecosyst. Environ.* **2002**, *91*, 217–232. [CrossRef]
64. Wang, B.; Zhang, W. Study on the measurement of agricultural eco-efficiency and spatial and temporal differences in China. *China Popul. Resour. Environ.* **2016**, *26*, 11–19. (In Chinese) [CrossRef]
65. United Nations Intergovernmental Panel on Climate Change (IPCC). *Climate Change*; Cambridge University Press: Cambridge, UK; New York, NY, USA, 2007.
66. Wu, F.; Li, L.; Zhang, H. Effects of conservation tillage on net carbon release from farmland ecosystems. *J. Ecol.* **2007**, *26*, 2035–2039. (In Chinese)
67. Dubey, A.; Lal, R. Carbon footprint and sustainability of agricultural production systems in Punjab, India, and Ohio, USA. *J. Crop Improv.* **2009**, *23*, 332–350. [CrossRef]

68. Min, J.; Hu, H. Measurement of greenhouse gas emissions from agricultural production in China. *China Popul. Resour. Environ.* **2012**, *22*, 21–27. (In Chinese)
69. United Nations Intergovernmental Panel on Climate Change (IPCC). *N_2O Emissions from Managed Soils, and CO_2 Emissions from Lime and Urea Application*; Cambridge University Press: Cambridge, UK; New York, NY, USA, 2019.

agriculture

MDPI

Article

Degradation Pattern of Five Biodegradable, Potentially Low-Environmental-Impact Mulches under Laboratory Conditions

Jaime Villena [1], Marta M. Moreno [1,*], Sara González-Mora [2], Jesús A. López-Perales [1], Pablo A. Morales-Rodríguez [1] and Carmen Moreno [1]

[1] Higher Technical School of Agricultural Engineering in Ciudad Real, University of Castilla-La Mancha, Ronda de Calatrava 7, 13071 Ciudad Real, Spain
[2] Council of Agriculture, Water and Rural Development, Junta de Comunidades de Castilla-La Mancha, 13270 Almagro (Ciudad Real), Spain
* Correspondence: martamaria.moreno@uclm.es

Abstract: The use of biodegradable (BD) plastic mulch materials as alternatives to the widely used low-density polyethylene (PE) is increasing nowadays, mainly for environmental reasons. However, the success of these materials depends, in addition to fulfilling their function, on completely degrading in the short term, which depends on both their composition and environmental conditions. This study focused on the degradation pattern of five BD plastic materials of different composition (i.e., corn and potato starch, and polylactic acid plastic (PLA) films, blended with different copolyesters during their manufacture), in two soils with different granulometry (Soil 1 has less clay content than Soil 2), taken from organic vegetable fields under controlled laboratory conditions. Conventional PE was used as a reference. The degree of degradation was evaluated through the number of fragments, weight loss, and surface area loss until their total disappearance. The degradation trend of the BD materials was similar in both soils, although much faster in Soil 2. Their total visible disappearance was in the following ranges: potato starch, 225–250 days in Soil 1, 150–200 days in Soil 2; corn starch, 550 days in Soil 1, 300 days in Soil 2; PLA, 1000–1050 days in Soil 1, 350–475 days in Soil 2. PE remained practically intact in both trials. The degradation model of potato starch materials fitted a decreasing exponential model in both soils, while the other bioplastics followed a decreasing Gompertz model, in all cases with steeper slopes in Soil 2. The curves of the degradation models indicated how the same material can degrade differently depending on the type of soil, information that could be useful for users and manufacturers in the framework of a sustainable agriculture.

Keywords: biodegradable plastic mulch; polyethylene; starch; polylactic acid; surface area—weight ratio; degradation model

Citation: Villena, J.; Moreno, M.M.; González-Mora, S.; López-Perales, J.A.; Morales-Rodríguez, P.A.; Moreno, C. Degradation Pattern of Five Biodegradable, Potentially Low-Environmental-Impact Mulches under Laboratory Conditions. *Agriculture* **2022**, *12*, 1910. https://doi.org/10.3390/agriculture12111910

Academic Editors: Moucheng Liu, Xin Chen and Yuanmei Jiao

Received: 16 October 2022
Accepted: 8 November 2022
Published: 13 November 2022

1. Introduction

Mulching is used worldwide in agriculture for several reasons, which can be summarized as follows: increasing and stabilizing soil temperature, weed control, improving crop yields and quality, reducing soil evaporation and erosion, increasing soil water-holding capacity, and improving the efficiency of fertilizers and water, among other benefits [1–5]. For this purpose, polyethylene (PE), a petroleum-based polymer, is the most commonly used, mainly due to its ease of installation and maintenance, high durability, reasonably low price, and its positive effects on crop yields [6,7]. Among PE types, low-density PE (LDPE) is the most commonly used due to its high puncture resistance, impermeability to water, and mechanical stretch properties [8,9]. However, the excessive use of PE for mulching in recent decades has undoubtedly had a negative impact on the environment, as a result of its long degradation period, estimated at about 100 years [10], due to its high molecular weight and chemical stability [11]. Even as microplastics (<5 μm), it can threaten both aquatic [12] and terrestrial life, especially if these particles have adsorbed pesticides [13]. A detailed review of the environmental risk, toxicity, and biodegradation of polyethylene, as well as the positive and negative effects of mulching, can be found in previous studies [13–15].

With the aim of reducing these environmental problems, alternative materials to be used as mulch have increased in recent decades, especially biodegradable (BD) plastic films from renewable resources. In the field, these materials can experience photodegradation, for those parts exposed to solar radiation, and biodegradation as a result of the action of soil microorganisms [16]. The review compiled by Maisara and Mariatti [7] summarized the ideal characteristics a BD mulch film should have, according to the current international standards [17,18], as soil biodegradability, high tensile strength, low cost, good barrier properties, low water permeability, high elongation, and low photosynthetic active radiation transmittance. With the aim of approaching these ideal characteristics as closely as possible, additives and/or other polymers need to be included in their formulation [7,19,20]. In summary, BD plastic mulches can first be classified into synthetic and natural materials. Synthetic polymers include polyhydroxyalkanoates (PHA), polylactic acid (PLA), polybutylene succinate (PBS), and polybutylene adipate terephthalate (PBAT), while natural polymers include starch, lignin, and cellulose, among others. They may all exist as a single polymer, or they may be blended and made up of different polymers [21]. A detailed description of all these components can be found in Manzano et al. [22], Merino et al. [23], and Maisara and Mariatti [7].

Research studies into aspects of the degradation of mulch materials have increased considerably in the last 10 years, due to concerns about the previously mentioned problems [7]. The different methods used for estimating the processes related to mulch degradation in laboratory conditions, summarized as changes in the mechanical and optical properties, the CO_2 evolution/O_2 consumption ratio, the amount of carbon assimilated by the microbial community, or by soil enzymatic measurements, can be seen in Moreno et al. [24].

The present study is linked with that carried out by Moreno et al. [24], analyzing the deterioration pattern of six BD mulch materials (five of which are used in the present study) under field conditions, as well as their deterioration rate after incorporation into the soil. The current study therefore analyzed the degradation pattern of these same five mulch materials, until their total visual disappearance, through the evolution of weight, surface area, and number of fragments over time, in controlled laboratory conditions. The trials were carried out in the same soil used in the previous study, and additionally, the experiment was duplicated in another type of soil with different characteristics, mainly related to the kind of texture. This would allow the importance of the soil in the degradation process of the materials to be highlighted according to their nature and formulation.

2. Materials and Methods

2.1. Experimental Design

A laboratory experiment was conducted at the Sustainable Agriculture Laboratory of the Higher Technical School of Agricultural Engineering in Ciudad Real (University of Castilla-La Mancha, Ciudad Real, Spain).

Five BD films of different compositions used as mulch in agriculture were selected: Mater-Bi® (MB); Sphere 4 (Sp4); Sphere 6 (Sp6); Bioflex® (BFx); Ecovio® (Eco). A conventional standard linear low-density polyethylene (PE) was used as a control. All of these films were black in color and 15 μm in thickness (data provided by the suppliers). The main components and manufacturers of these materials are shown in Table 1.

Table 1. Main characteristics of the mulch materials tested.

Mulch Material	Composition	Manufacturer
Mater-Bi® (MB)	Corn thermoplastic starch, PBAT, vegetable oils	Novamont S.p.A., Italy
Sphere 4 (Sp4), Sphere 6 (Sp6)	Potato thermoplastic starch and biodegradable recycled polymers bioplastic (with a different proportion of its components)	Sphere Group Spain S.L., Spain
Bioflex® (BFx)	PLA, PBS	Fkur-Oerlemans Plastics, Germany
Ecovio® (Eco)	PLA, ecoflex (PBAT)	BASF, Germany
Polyethylene (PE)	Conventional standard linear low density polyethylene	Siberline, Spain

As previously argued [24], all the BD materials used are susceptible to degradation due to different factors (microorganisms, temperature, humidity, and light). The starch-based materials are especially sensitive to humidity, while the PLA-based materials need, in addition to high humidity, higher temperatures for a fast degradation process. In this sense, PBAT as mulch is limited due to the excessive degradation rate, and for this reason several methods have been proposed to delay their degradation time, such as adding an ADR chain extender, UV absorber, and antihydrolytic agent, among others, as compiled by Quiao et al. [25,26].

Soil samples from two different types of soil (Soil 1, Soil 2) were collected from experimental organically managed vegetable fields (EC 848/2018) at the Agrarian Research Centre "El Chaparrillo" (39°0′ N–3°56′ W, altitude 640 m) (Regional Institute of Research and Food Industry and Forestry Development, IRIAF), Ciudad Real, Spain, at the end of June. Soil samples from Soil 1 correspond to the plot used in the field trials described in Moreno et al. [24].

In the laboratory, soil samples were spread out and air dried at room temperature for 72 h. They were then sieved through a 2 mm mesh sieve and analyzed for physical and chemical properties (Table 2). The differences between the two soils were mainly textural in nature, especially because of the clay content (Soil 1: 8.8%, sandy-loam; Soil 2: 29.0%, clay-loam). Additionally, in order to estimate the microbiological status of both soils at the beginning of the trial, enzyme activity was estimated through dehydrogenase activity (DHA), widely used as a good indicator of oxidative status in soils, according to Casida et al. [27,28] with further modifications [29,30]. Similar values were found in both soils, around 148 μg g^{-1} $soil^{-1}$ day^{-1}.

Table 2. Physical–chemical properties of soils.

Soil Parameter	Soil 1	Soil 2
entry 1	data	data
Sand (2–0.05 mm) (%)	55.2	45.0
Silt (0.05–0.002 mm) (%)	36.0	26.0
Clay (<0.002 mm) (%)	8.8	29.0
Soil textural class (USDA)	Sandy loam	Clay loam
Wilting point (m^3 m^{-3})	0.100	0.160
Field capacity (m^3 m^{-3})	0.230	0.350
pH (1:2.5 soil:water)	8.2	8.0
EC (1:5 soil:water) (dS m^{-1})	0.76	0.65
Organic matter (Walkley-Black) (%)	1.6	1.7
Total carbonates (%)	6.0	7.5
Total nitrogen (%)	0.09	0.08
C/N ratio	7.9	9.5
Assimilable phosphorus concentration (g kg^{-1})	0.017	0.020
Exchangeable potassium concentration (g kg^{-1})	0.351	0.409
Exchangeable calcium concentration (g kg^{-1})	2.324	2.480
Exchangeable magnesium concentration (g kg^{-1})	0.216	0.254
Exchangeable sodium concentration (g kg^{-1})	0.008	0.006

In preparing the trial, the procedure adopted was, in general terms, that established by Barragán et al. [16] but with the following modifications.

From each type of mulch material, 180 pieces of 8 × 8 cm^2 were cut (180 samples × 6 materials = 1080 samples in total) and weighed individually on a precision balance (mod. Crystal, 0.1 mg precision). Next, the samples were individually placed in non-biodegradable plastic containers (polyethylene terephthalate, PET) with a capacity of 500 mL, perforated at the top and on the sides, previously filled with 400 mL of soil from each of the two soils tested. Plastic samples were carefully buried in the central part of the containers, fully extended, leaving the same distance between the top and the bottom. Therefore, 90 samples of each material were placed in each soil type (90 samples × 6 materials = 540 samples for each soil). Distilled water was then added to each container, to adjust the water soil content up to 50% of water-holding capacity. This value was determined using the methodology proposed by Jarrel et al. [31]. The containers were then transferred to an environmental chamber at a constant temperature of 25 °C and in dark conditions. Throughout the trial period, the soil humidity was checked weekly by randomly weighing 10 containers of each soil and correcting the weight loss by adding water.

2.2. Laboratory Measurements of Mulch Materials

From the beginning of the trial, the film samples were periodically extracted, up to their total degradation, at variable intervals according to their state (25 sampling dates maximum). To determine these dates, the film samples were visually inspected approximately every 10 days. On each sampling date, 36 film samples were extracted, corresponding to the material that had not yet been completely degraded (6 materials × 3 repetitions of each material × 2 soils), in order to determine the weight and surface area of each buried material. Next, the remnants of the films were carefully separated from the soil, cleaned with distilled water and cotton, dried at room temperature to constant weight, and weighed on a precision balance (±0.1 mg). The weights of the samples were expressed in grams, and as a percentage of the initial weight (new material), calculated as follows:

$$\text{Weight } (\%) = (Wn/Wo) \times 100 \quad (1)$$

where Wo is the initial weight of the sample before starting the test and Wn the weight of the material on date n during the test.

Once weighed, the material remnants were photographed in order to determine their surface area. This was done by placing the samples in a glass support with a white background, illuminated from below, and photographed with a digital compact camera with built-in lens and optical viewfinder, model Canon PowerShort G11, 35 mm. The photographs were analyzed with the ImageJ® program. The scale for each image was established by drawing a line between two known points, 100 mm apart, using the program's Set Scale function. Thus, the number of fragments and the surface area of each were obtained directly. As for weight, surface area data was expressed in cm^2, or as a percentage of the initial value, calculated as follows:

$$\text{Area } (\%) = (An/Ao) \times 100 \quad (2)$$

where Ao is the initial surface area of the sample before starting the test (64 cm^2) and An the surface area of the material on date n during the test.

Additionally, the changes in the morphology of the surface areas of the films PE, MB, Sp6 and BFx during the degradation process were examined 100 days after the start of the trial, through micro-photographs taken by scanning electron microscopy (SEM, Scanning Electron Microscope, mod. JEOL JSM-6610LV) and compared with new materials. The scanning of the samples was performed at 500 and 1000 magnifications.

2.3. Statistical Analysis

The analysis of the data, both the descriptive study and the corresponding Analysis of Variance (ANOVA) and Linear and Non-Linear Regression for the establishment of models, was carried out with Infostat v. 2015 professional, with connectivity to the statistical package R (https://cran.r-project.org) (accessed on 15 June 2022), taking a significance level of 0.05.

When statistically significant differences were found, multiple range comparisons (Duncan's test) and the Least Significant Difference (Fisher's LSD) test were performed, incorporating the value of the latter through segments in the corresponding graphs.

In the choice of the best degradation model among those tested, the respective values of the AIC index (Akaike Information Criterion) were compared. The AIC is an estimator based on Information Theory, a measurement of the relative quality of the different statistical models that represent a data set [32]. AIC is calculated as follows:

$$\mathrm{AIC} = 2\,K - 2\ln(L) \qquad (3)$$

where K is the number of independent variables used en the models and L is the log-likelihood estimate. The best-fit model according to AIC is the one that explains the greatest amount of variation using the fewest possible independent variables. Thus, the model with the lowest AIC index value was taken as appropriate for modelling.

3. Results and Discussion

3.1. Evolution over Time of Weight, Surface Area and Number of Fragments

The initial weights of all the materials tested were in the range 0.111 (BFx) to 0.130 (Sp6) grams per 8×8 cm^2 of surface area. However, from the first sampling date (15 days), significant differences both in weight and in surface area were established among materials in both soils, as shown in Figure 1 (values expressed as percentage in relation to the initial values: new materials, 0 days). Thus, the evolution of the weight and surface area of the materials in each soil was broadly similar. From 50 days in Soil 1 (Figure 1a,b), Sp4 and Sp6 stood out as the most degraded, a circumstance that was maintained until their total disappearance (at 225 days in Sp4 and 250 days in Sp6). Similar behavior was observed in Soil 2 (Figure 1c,d), although in this case the total degradation of these materials occurred somewhat earlier (at 150 days in Sp4 and 200 days in Sp6). MB totally disappeared at 550 days in Soil 1 and at 300 days in Soil 2. As expected, PE remained practically intact in both soils, while the PLA-based materials (BFx and Eco) showed intermediate behaviors; thus, Eco completely disappeared at 1000 days in Soil 1 and at 350 days in Soil 2, while visible remnants of BFx remained up to 1050 and 475 days in Soils 1 and 2, respectively.

In the field study previously carried out in Soil 1 by Moreno et al. [24], a similar trend was also observed in the degradation of the buried part of the tested mulch materials according to their nature (starch > PLA > PE) at the end of the crop cycle (145 days after transplanting). Although that timeframe (145 days) was not enough to achieve a significant deterioration of the buried part of the PLA mulches in the field (~10% as average), the starch-based materials did show it, especially Sp4 and MB (~40%), although this was not as pronounced as under laboratory conditions, in agreement with previous works [33].

It is noteworthy that the total variability of the current trial increased as it progressed, reaching a coefficient of variation of 400% (data not shown) at the end of the experiment in each soil (1050 days in Soil 1 and 475 days in Soil 2).

These trends can also be seen in Figures 2 and 3, which show a photographic sequence of the evolution of the different materials (one of the three repetitions of each one) until the end of the tests in both soils.

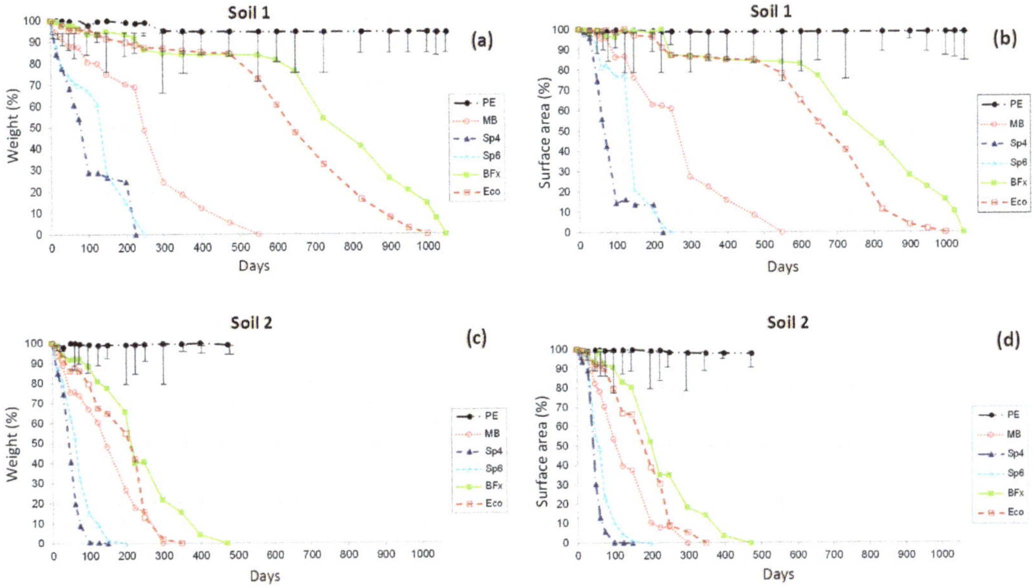

Figure 1. Variation of the weight (**a**,**c**) and surface area (**b**,**d**) of the mulch materials in Soils 1 and 2. Data expressed as percentages of the initial values. Vertical bars at each date represent the least significant difference (LSD) at a significance level of 0.05 among treatment means.

In the comparison of soils, with the exception of PE (which remained practically intact in both), the loss of weight and surface area of the materials was more pronounced in Soil 2 than in Soil 1 (therefore, degradation in Soil 2 was faster). This may have been a consequence of the larger soil–material contact surface in soils with a greater clay component (Soil 2), since the enzymatic activity (ADH) was similar at the beginning of the experiment in both soils. Regarding the materials, in general the degradation depended on their nature, being faster in those based on starch (especially potato (Sp4 and Sp6) compared with corn (MB)) than in PLA compounds (BFx and Eco).

These results differ from those obtained by Barragán et al. [16] under laboratory conditions similar to those of the present study. The former study was carried out in a clay-loam soil with a slightly basic pH, non-saline, with 3.13% organic matter and a high carbonate content (29%), and in that case, MB and BFx practically disappeared at 180 days and a film based on potato starch at 160 days. An explanation for the lower film degradation in our trial could be the lower organic matter content of the soil (1.6% and 1.7% in Soils 1 and 2, respectively), because a high organic matter content favors soil microorganism activity and therefore the process of biodegradation [16,34].

Although Sp4, Sp6, and MB are formulated with a biodegradable base (potato or corn starch), they were blended with different types and amounts of copolyesters during their manufacture. This could well be the cause of the variations in their degradation processes. Likewise, the additives and the presence of vegetable oils in the MB blend with starch could lead to slower degradation compared with potato starch compounds [16]. In this sense, according to Vázquez et al. [35], the presence of amylose–lipid complexes has a negative effect on the enzymatic digestibility of starch, which could also explain the behavior of MB.

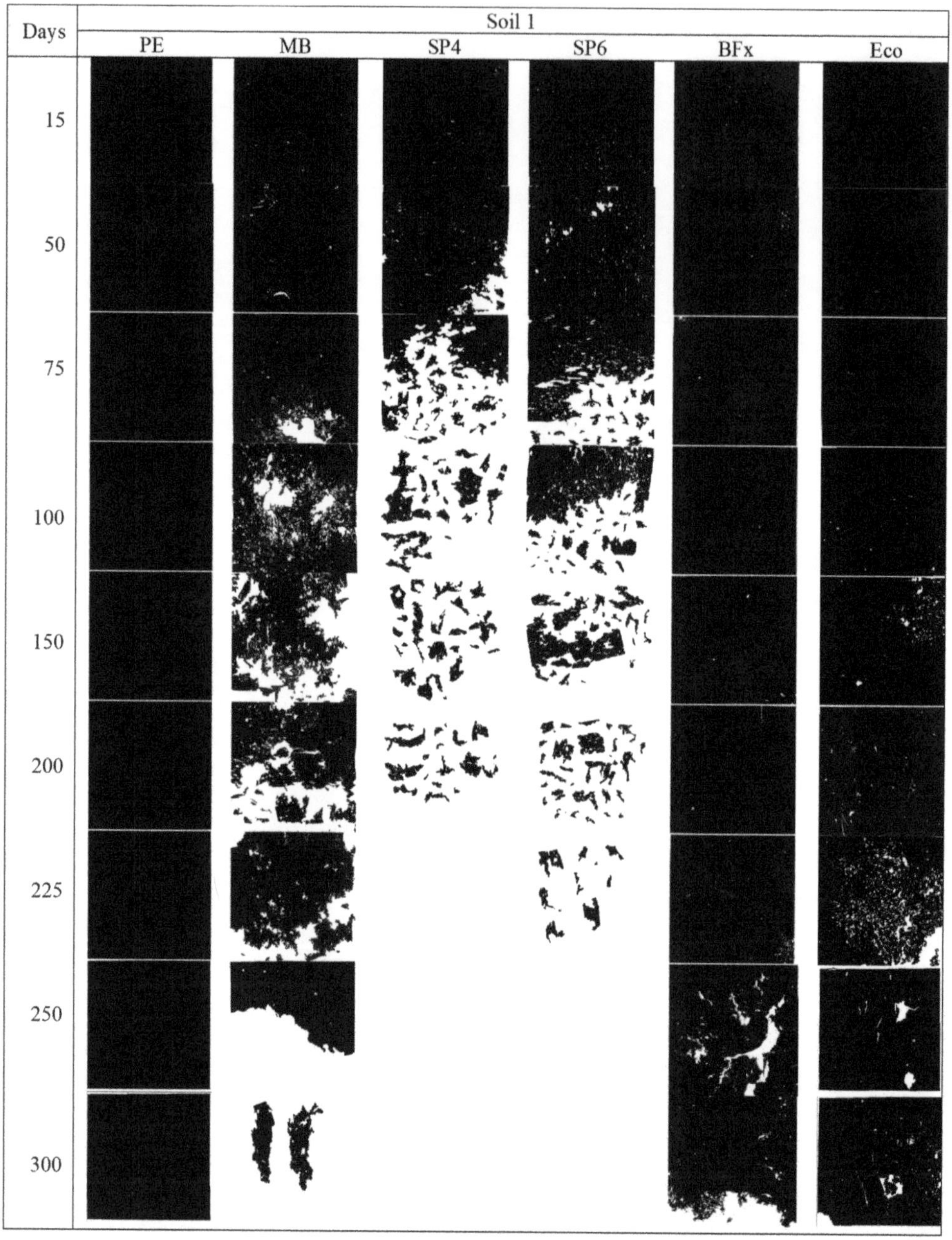

Figure 2. *Cont.*

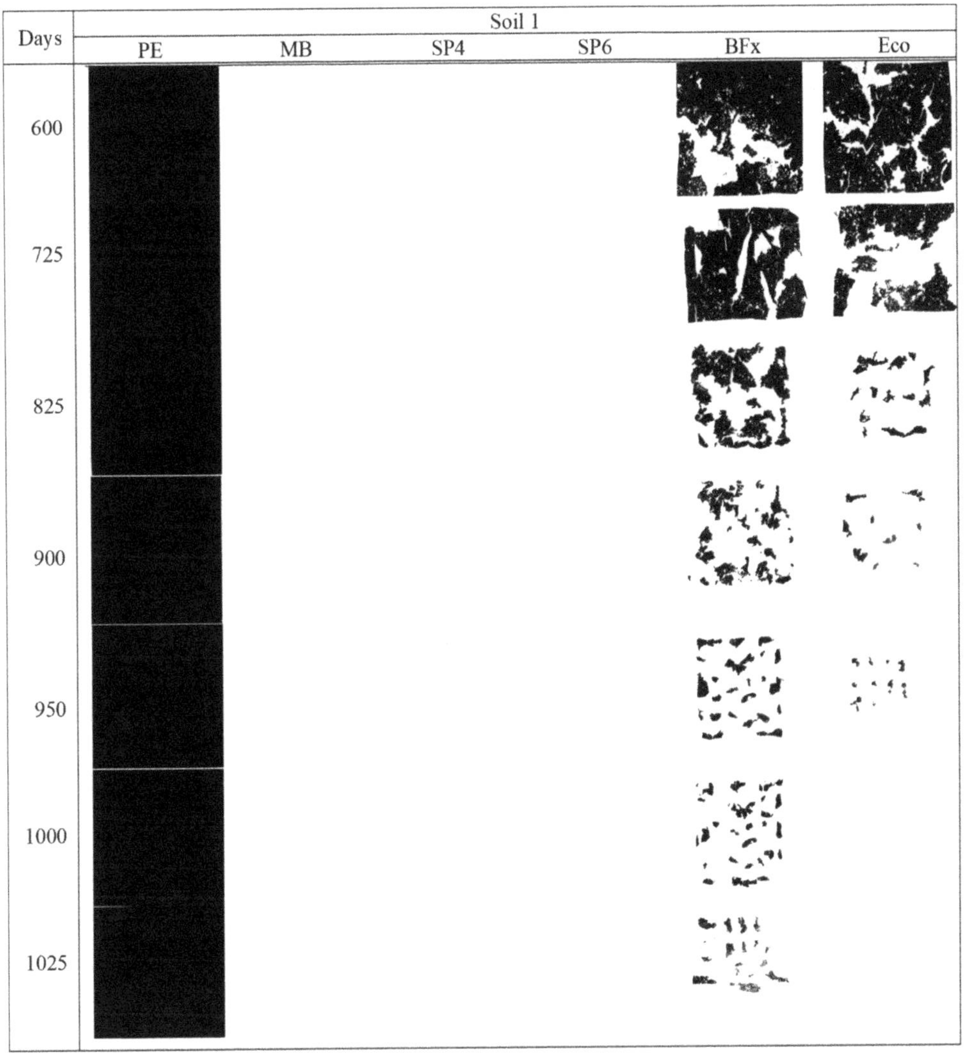

Figure 2. Evolution of the mulch materials in Soil 1 (reference area of 8 × 8 cm^2). Photographs correspond to one of the three replications.

In a similar study, Mostafa et al. [36] found that pieces of MB buried in a sandy-loam soil at 25 °C degraded by up to 70% in five months. In our trial, however, the degradation of MB in the soil of a similar textural class (Soil 1) was much lower at that date (around 20%), which could be attributable, as in the previous comparison with Barragán et al. [16], to the lower organic matter content in our study.

In the case of the PLA materials, their slower degradation in comparison with starch-based materials could be explained by the temperatures maintained during the whole trial (≈25 °C), which were lower than the minimum threshold values indicated previously [37] for PLA degradation (≥30 °C). Other authors also support this [16,38,39], arguing that the slow degradation of PLA materials is a result of the normally low temperature of the soil and the limited hydrolysis of PLA in the soil environment.

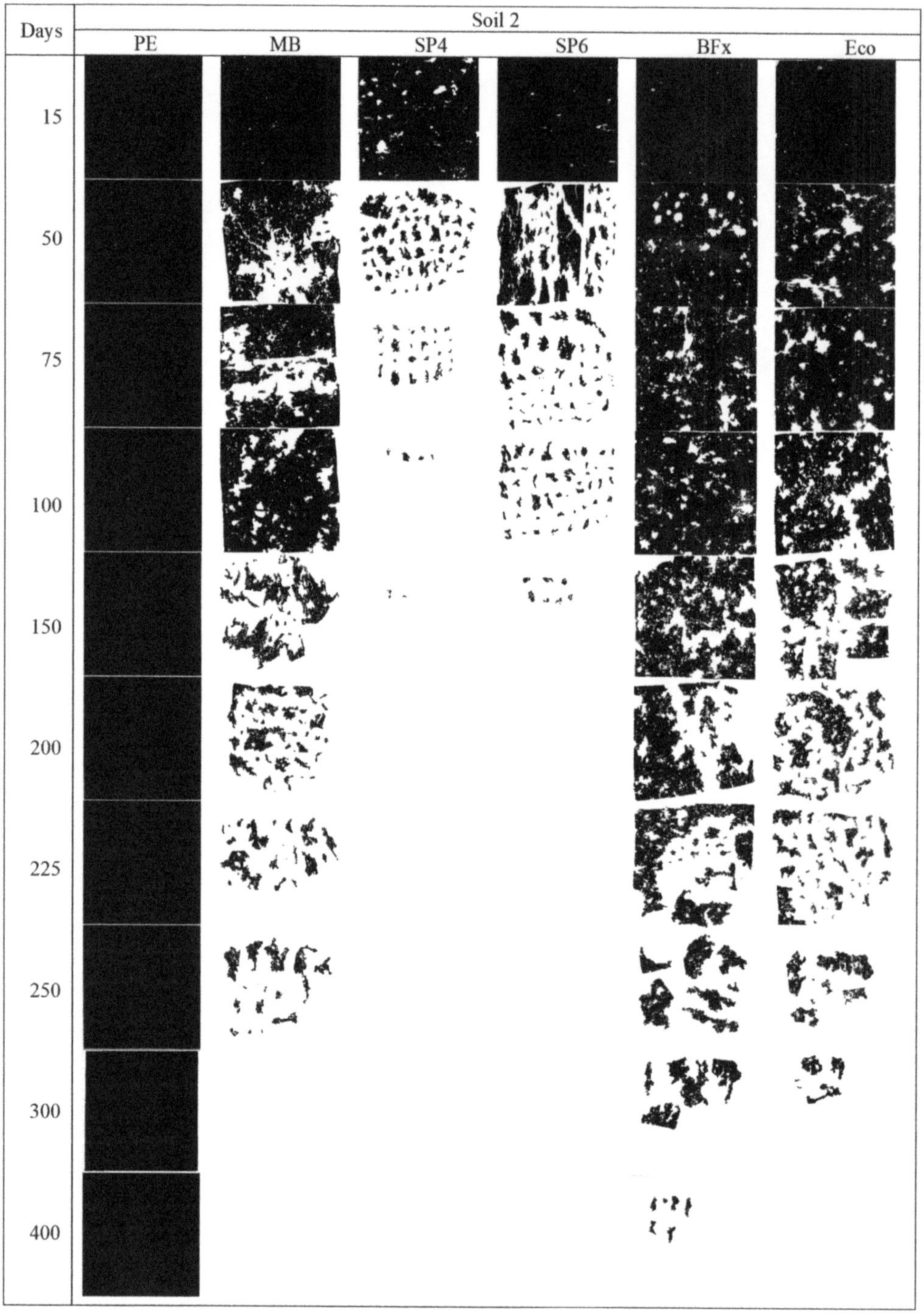

Figure 3. Evolution of the mulch materials in Soil 2 (reference area of 8 × 8 cm^2). Photographs corresponding to one of the three replications.

Comparing both PLA materials, the higher degradation rate of Eco compared with BFx could be caused by the copolyester component added in its manufacture (PBAT in Eco and PBS in BFx) [16,36].

When discussing the different behavior of the materials used in the trials, it should also be noticed that the thickness of the films affects the disintegration process, as specifically discussed in previous works [40]. In our study, all the films used theoretically had the same thickness, although this information was given by the supplier companies.

With regard to the number of fragments (Supplementary Tables S1 and S2), the disintegration of the materials was higher in Soil 2 than in Soil 1, according to weight and surface area behavior. As examples, Sp4 and Sp6 in Soil 2 were in 31 and 25 fragments, respectively, at 50 days, while in Soil 1, 10 and 1 pieces were registered. MB in Soil 2 had 12 fragments at 75 days, while in Soil 1 the material was still practically intact. In the PLA materials, the slower fragmentation is very striking; in particular, Eco remained practically intact up to 475 days in Soil 1 and 100 days in Soil 2. As expected, PE did not undergo any disintegration process. All these circumstances led to great variability (measured through the coefficient of variation) in the number of fragments observed in the trial, which reached 100% after 50 days in practically all the sampling dates (Supplementary Tables S1 and S2).

3.2. SEM Microphotographs

Microphotographs of PE, MB, Sp6, and BFx taken by SEM in new materials and after 100 days in Soils 1 and 2 (500 and 1000 magnifications, Figures 4 and 5, respectively) showed that the surface of the new materials was smooth, without cracks or roughness. PE presented a very homogeneous surface due to the uniformity of the mixture of the granules it is made from. The surface of the BD materials revealed a fairly uniform dispersal of the starch particles for MB and Sp6, and of polylactic acid for BFx. These particles are embedded in a continuous matrix made up of the synthetic polymeric component used for the formulation [1]. At 100 days, a greater degree of degradation of the materials (except for PE, which remains intact) was observed, with differences between the two soils clearly visible. Thus, MB in Soil 2 presented a greater number of cracks than in Soil 1, although they were very uniformly distributed across the surface (Figures 4f and 5f). Sp6 in both soils presented larger whitish granules in comparison with new material, which could be caused by the swelling of the starch particles due to the effect of humidity. Likewise, in Soil 1 (Figure 4g), some holes of a similar size to those of the original starch particles can be observed (see arrows on the figure). This finding could be explained by the fact that, when biodegradable mulch films are in contact with the soil, the microorganisms present feed on the original starch particles [33,41]. In Soil 2, a greater roughness was observed, probably due to a higher concentration of starch particles in these areas. It is worth noting the shape of the cracks in this material, in which distortion of the filaments is observed when the material is degrading (Figures 4k and 5k). In BFx there were no clear signs of degradation (no cracks or holes); however, at 100 days, the granules observed in the new material became smaller in Soil 2 than in Soil 1 (Figure 4d,h,l and Figure 5d,h,l).

3.3. Surface Area—Weight Ratio

The optimal model that adjusted to the surface area–weight ratio for each BD film in both soils was the Gompertz model with Equation (4):

$$y = \alpha e^{-\beta e^{-\gamma x}} \tag{4}$$

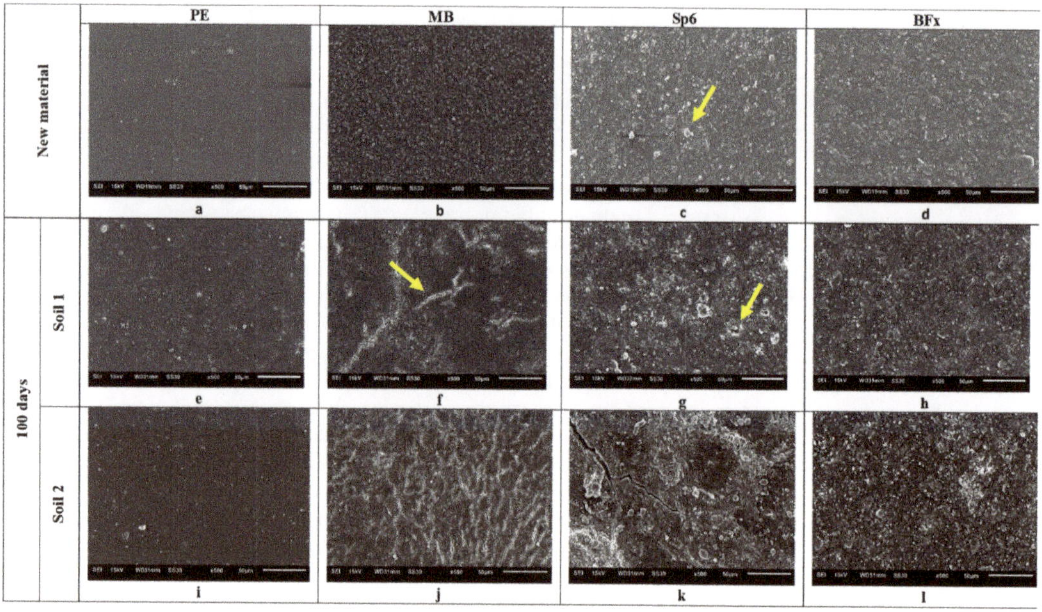

Figure 4. SEM micrographs corresponding to PE, MB, Sp6, and BFx in new materials and after 100 days in Soils 1 and 2 (500× magnification). The yellow arrows point out different cracks, holes or granules in the materials.

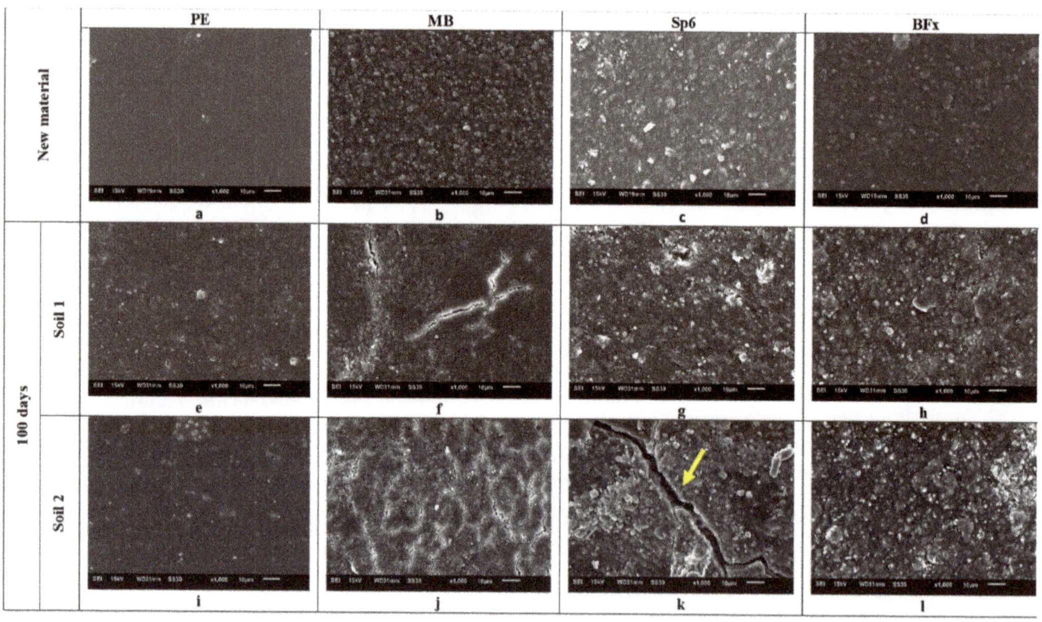

Figure 5. SEM micrographs corresponding to PE, MB, Sp6, and BFx in new materials and after 100 days in Soils 1 and 2 (1000× magnification). The yellow arrows point out different cracks, holes or granules in the materials.

where y represents the surface area, x the weight (percentage values) and α, β, γ, the corresponding parameters of the model (see equations for the sigmoid curves in Figure 6).

The small plateau of points in the sigmoid curves (on the right side of most of the figures, especially pronounced in Sp4 and Sp6) corresponds to the first stages of degradation. At this stage there was a reduction in the weight of the materials as a result of a decrease in their thickness (refining), but there was as yet no breaking (no cracks or holes), thus keeping their surface areas practically intact. As an example, we can see that the breaking threshold for the surface considered (8×8 cm^2) would be reached when the weight dropped to 80% in Sp6 in Soil 2 or to 90% in PLA materials in Soil 1. In the case of PE, however, there was only a slight variation in the weight of the samples in both soils, although the surface area remained intact.

3.4. Degradation Models of the Mulch Materials

In constructing the corresponding degradation models of MB, Sp4, Sp6, BFx, and Eco, a sufficient range was considered in the regressor variable (time) to ensure their total practical disappearance in both soils. This time-frame exceeds the length of the crop cycles of the mulched annual vegetable crops. Thus, two time limits were established: 300 days (Sp4, Sp6) and 1000 days (MB, BFx, Eco).

With the data corresponding to the remaining weights and surface areas of each material in each soil, different models were tested, showing their degradation process over time, and those with the lowest AIC were chosen. Thus, exponential models were adopted for Sp4 and Sp6, and Gompertz models for MB, BFx, and Eco.

3.4.1. Sp4 and Sp6: Exponential Model of Degradation

The exponential model satisfies the following condition: the rate of change of the response variable y (weight, surface area) with respect to time t is proportional to y, where b is the constant of proportionality interpreted as the growth rate (decrease) in Equation (5), which solves this model:

$$y = ae^{bt} \tag{5}$$

The values obtained from the parameters of the exponential model (5), adjusted for the remaining weight and surface area at instant t, corresponding to the different materials and soils, are presented in Table 3.

Table 3. Coefficients of the exponential model * of degradation corresponding to the remaining weight and surface area of Sp4 and Sp6 mulch materials (300 days, complete degradation in both soils).

	Material	Soil 1			Soil 2		
		a	b	AIC	a	b	AIC
Weight (g)	Sp4	0.135	−0.020	−179.9	0.140	−0.035	−129.4
	Sp6	0.140	−0.015	−471.8	0.150	−0.020	−488.0
Surface area (cm^2)	Sp4	66.30	−0.018	217.5	64.10	−0.028	186.1
	Sp6	68.01	−0.015	272.5	66.23	−0.020	279.9

* $y = ae^{bt}$. AIC: Akaike Information Criterion. y: remaining weight/surface area.

Both in Sp4 and Sp6, the curves of the exponential degradation models relative to weight and surface area in Soils 1 and 2 were similar to each other (Figure 7): in both materials, the curve relative to Soil 2 always appeared below that corresponding to Soil 1, which shows a faster decrease in weight and surface area over time in Soil 2, as they are curves with a steeper slope (b values more negative in Soil 2, Table 3).

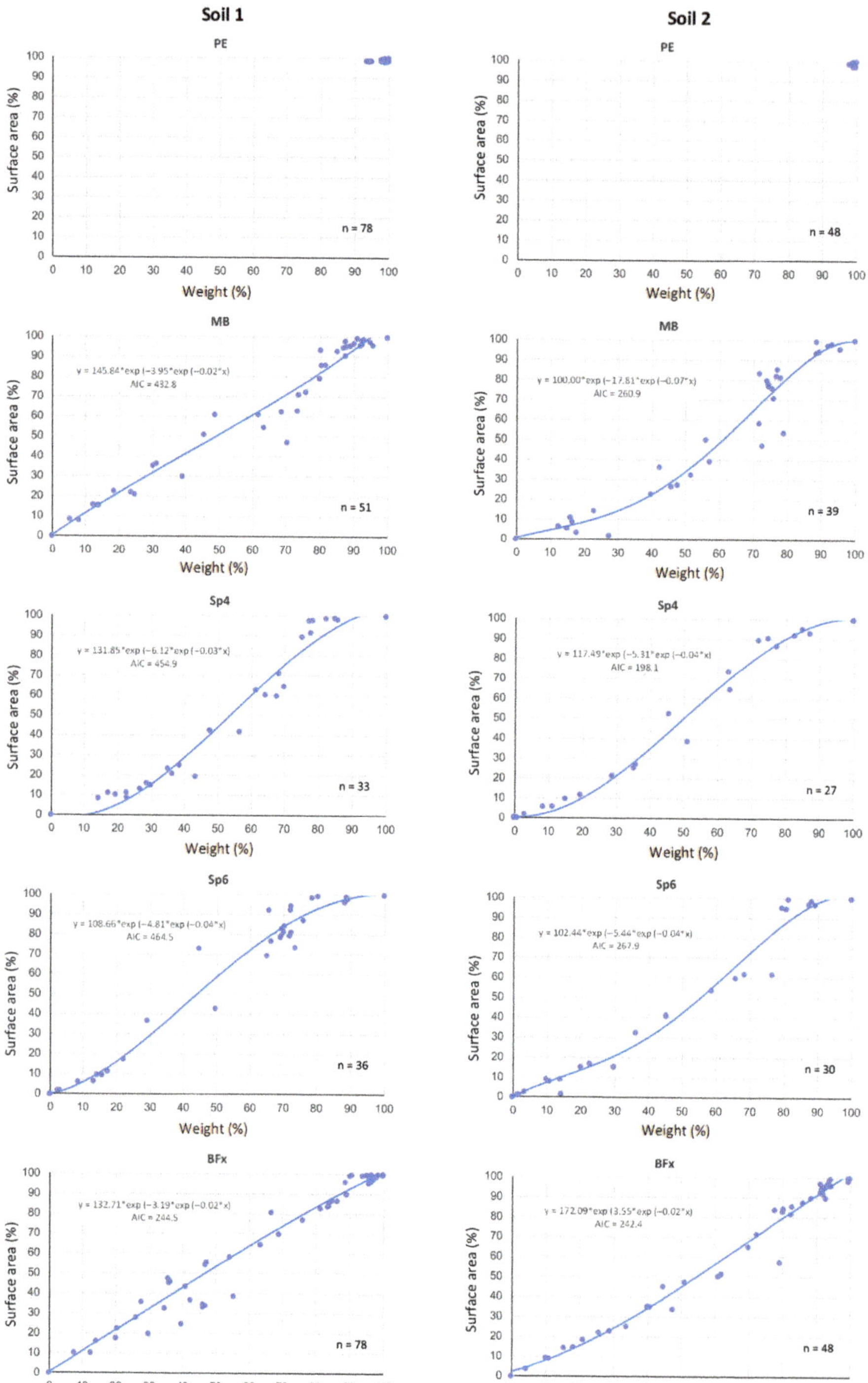

Figure 6. *Cont.*

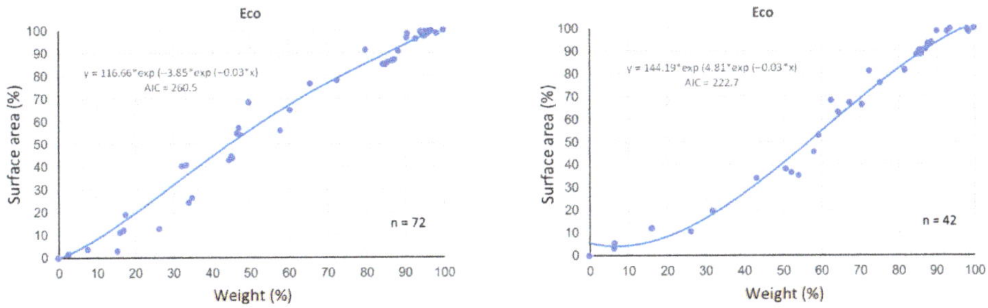

Figure 6. Surface area–weight ratio of the different mulch materials in Soils 1 and 2. Data expressed as percentages of the initial values.

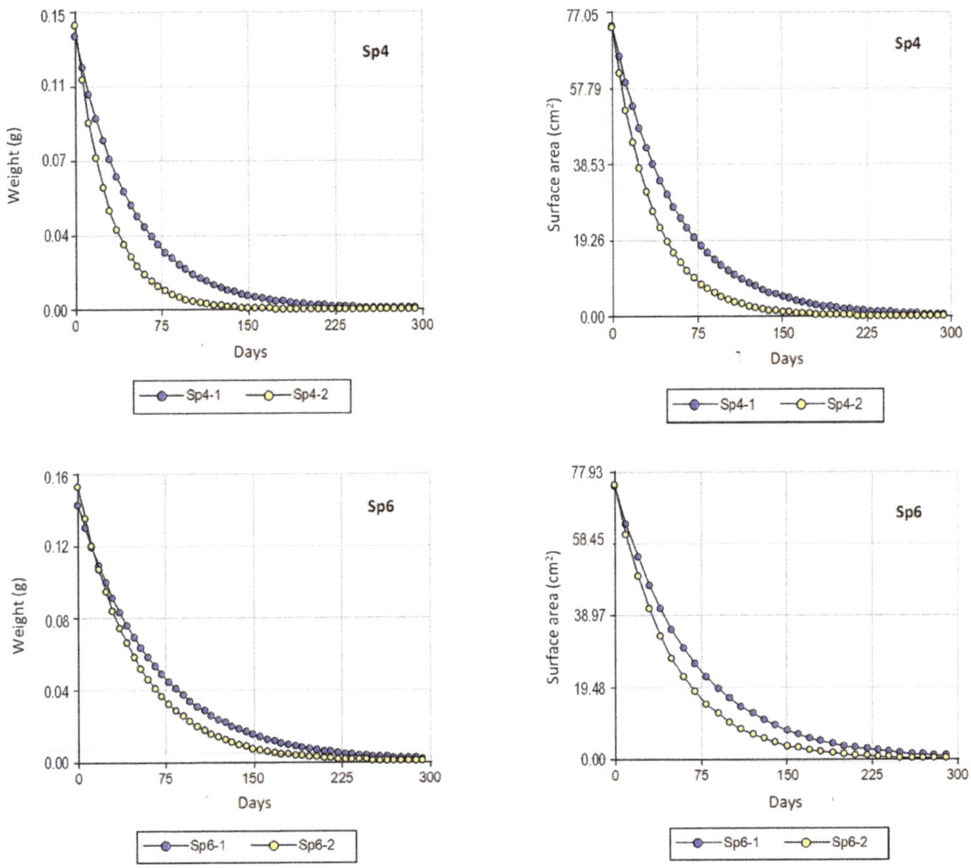

Figure 7. Degradation curves relative to weight and surface area variation of Sp4 and Sp6 in Soils 1 (blue) and 2 (yellow). Exponential model (300 days).

Comparing both materials, the curves related to Sp4 had more negative slopes than those corresponding to Sp6 (Figure 7 and b values in Table 3). This would indicate a higher rate of degradation (more negative) in Sp4 and therefore, an earlier disappearance than Sp6, especially in Soil 1 (Sp4: ≈225 days, Soil 1, ≈150 days, Soil 2; Sp6: ≈250 days, Soil 1, ≈200 days, Soil 2).

3.4.2. MB, BFx, and Eco: Gompertz Model of Degradation

In these materials, the Gompertz Equation (6) of parameters α, β, γ, where this last is considered the (negative) growth rate, satisfactorily modeled the decay of the response variable y (weight, surface area) with respect to time t,

$$y = \alpha e^{-\beta e^{-\gamma x}} \tag{6}$$

The values obtained from the parameters of the Gompertz model (6), adjusted for the remaining weight and surface area at instant t, corresponding to the different materials and soils, are shown in Table 4.

Table 4. Coefficients of the Gompertz model * of degradation corresponding to the remaining weight and surface area of MB, BFx and Eco mulch materials (1.000 days, complete degradation in both soils).

	Material	Soil 1				Soil 2	
		α	β	γ	AIC	α	β
Weight (g)	MB	0.120	0.060	-0.008	-465.0	0.160	0.300
	BFx	0.110	0.010	-0.007	-363.0	0.110	0.070
	Eco	0.130	0.010	-0.006	-360.9	0.130	0.010
Surface area (cm^2)	MB	75.00	0.170	-0.007	411.9	100.0	0.43
	BFx	67.30	0.010	-0.007	253.2	70.70	0.080
	Eco	64.00	0.010	-0.007	194.9	69.90	0.080

* $y = \alpha e^{-\beta e^{-\gamma t}}$. AIC: Akaike Information Criterion. y: remaining weight/surface area.

As in the exponential case for Sp4 and Sp6, the Gompertz models for the weight and surface area of MB, BFx,h and Eco also showed great similarity to each other, and the degradation curves relative to Soil 2 were always below those corresponding to Soil 1. This indicates that the (negative) growth rate is higher in Soil 2 than in Soil 1 (parameter γ and the slopes of the curves more negative in Soil 2) (Figure 8, Table 4). This is especially noteworthy in Eco, with total disappearance in Soil 2 around 350 days compared with 1000 days in Soil 1.

As expected, in all the materials and soils of both models, the corresponding growth rates (b and γ parameters in the exponential and Gompertz models, respectively) were negative, corroborating the decreasing behavior of these curves. This corresponds logically to the degradation curves in which the variation of the weight and surface area of the materials over time has been modeled. Likewise, in all cases, a faster disappearance of the materials was observed in Soil 2 (clay-loam) than in Soil 1 (sandy-loam), being more pronounced in the films of the second model, especially in Eco, as previously discussed.

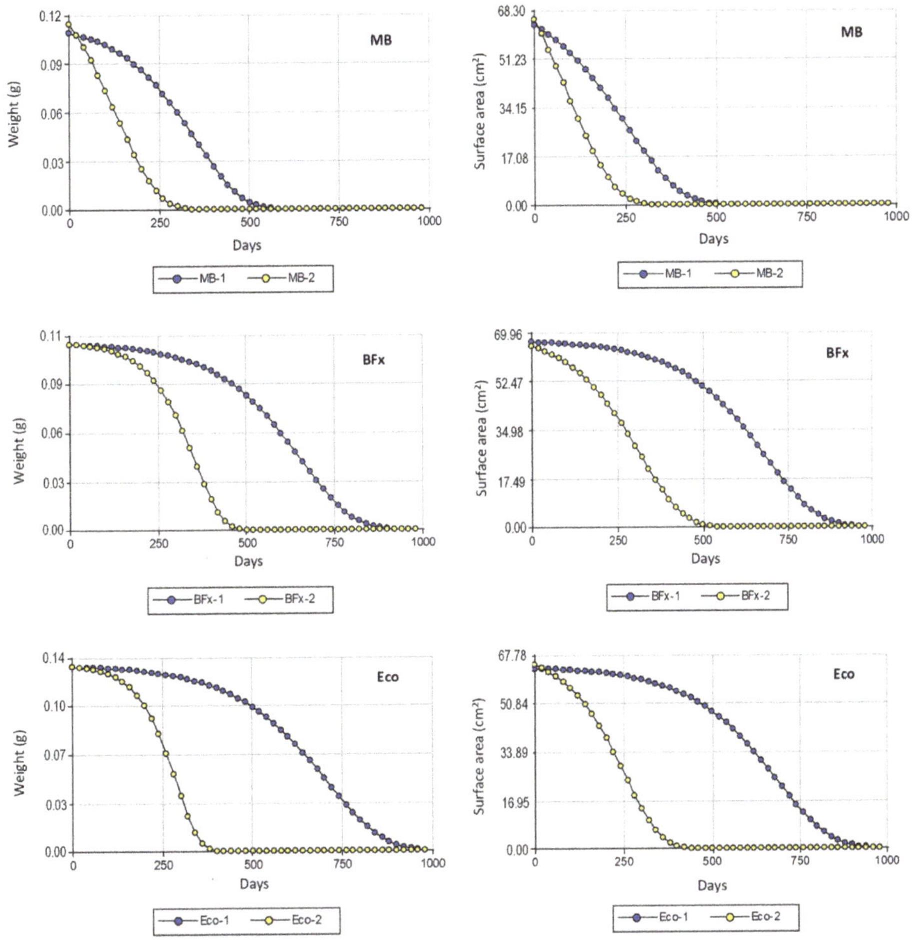

Figure 8. Degradation curves relative to weight and surface area variation of MB, BFx, and Eco in Soils 1 (blue) and 2 (yellow). Gompertz model (1000 days).

4. Conclusions

An important concern with regard to biodegradable (BD) mulch materials used as an alternative to the conventional PE in farms is to deepen the knowledge of their physical degradation process under different environments. In this study, using five BD plastics in laboratory conditions, and two types of soils, it was highlighted that (i) Biodegradable plastics degrade faster (based on weight and surface area loss, and greater disintegration) in soils with a higher clay content; (ii) the degradation of starch-based materials is faster than in those made from polylactic acid, especially those made from potato starch; (iii) the degradation model of potato-starch materials fits a decreasing exponential model in both soils, while corn-starch and polylactic acid mulches fit a decreasing Gompertz model, in all cases with steeper slopes in the soil with a higher clay content; And (iv) degradation curves based on surface area and weight indicate how the same material can degrade differently depending on the soil granulometry. The different behavior of the BD materials depending on both their composition and the type of soil where there are to be used would provide interesting complementary information to field trials to be taken into consideration by both manufacturers and users through accurate and sustainable tools.

Supplementary Materials: The following supporting information can be downloaded at: https://www.mdpi.com/article/10.3390/agriculture12111910/s1, Table S1: Number of fragments of the mulch materials in the different sampling dates. Soil 1; Table S2: Number of fragments of the mulch materials in the different sampling dates. Soil 2.

Author Contributions: Conceptualization, M.M.M. and C.M.; software, C.M., J.V. and P.A.M.-R.; methodology, J.V., C.M., P.A.M.-R. and J.A.L.-P.; formal analysis, J.V., C.M. and S.G.-M.; investigation, M.M.M., C.M. and J.A.L.-P., data curation, J.V., S.G.-M. and P.A.M.-R., writing—original draft, J.V., J.A.L.-P. and C.M., writing—review and editing, J.V., M.M.M. and C.M., funding acquisition, M.M.M. All authors have read and agreed to the published version of the manuscript.

Funding: This research was funded by the National Institute for Agricultural and Food Research and Technology (INIA), Ministry of Economy and Competitiveness (grant number: RTA2011-00104-C04-03), Spain.

Institutional Review Board Statement: Not applicable.

Data Availability Statement: Data will be made available on request.

Conflicts of Interest: The authors declare no conflict of interest.

References

1. Scarascia-Mugnozza, G.; Schettini, E.; Vox, G.; Malinconico, M.; Immirzi, B.; Pagliara, S. Mechanical properties decay and morphological behaviour of biodegradable films for agricultural mulching in real scale experiment. *Polym. Degrad. Stab.* **2006**, *91*, 2801–2808. [CrossRef]
2. Moreno, M.M.; Moreno, A. Effect of different biodegradable and polyethylene mulches on soil properties and production in a tomato crop. *Sci. Hortic.* **2008**, *116*, 256–263. [CrossRef]
3. Moreno, M.M.; Cirujeda, A.; Aibar, J.; Moreno, C. Soil thermal and productive responses of biodegradable mulch materials in a processing tomato (*Lycopersicon esculentum* Mill.) crop. *CSIRO Publ. Soil Res.* **2016**, *54*, 207–215. [CrossRef]
4. Nishigaki, T.; Shibata, M.; Sugihara, S.; Mvondo-Ze, A.D.; Araki, S.; Funakawa, S. Effect of mulching with vegetative residues on soil water erosion and water balance in an oxisol cropped by cassava in East Cameroon. *Land Degrad. Dev.* **2016**, *28*, 682–690. [CrossRef]
5. Prosdocimi, M.; Tarolli, P.; Cerda, A. Mulching practices for reducing soil water erosion: A review. *Earth Sci. Rev.* **2016**, *161*, 191–203. [CrossRef]
6. Haapala, T.; Palonen, P.; Korpela, A.; Ahokas, J. Feasibility of paper mulches in crop production—A review. *Agric. Food Sci.* **2014**, *23*, 60–79. [CrossRef]
7. Maisara, A.M.A.; Mariatti, M. Formulation of biodegradable plastics mulch film for agriculture crop protection: A review. *Polym. Rev.* **2022**, *62*, 890–918.
8. Kasirajan, S.; Ngouajio, M. Polyethylene and biodegradable mulches for agricultural applications: A review. *Agron. Sustain. Dev.* **2012**, *32*, 501–529. [CrossRef]
9. Serrano-Ruiz, H.; Martin-Closas, L.; Pelacho, A.M. Biodegradable plastic mulches: Impact on the agricultural biotic environment. *Sci. Total Environ.* **2021**, *750*, 141228. [CrossRef]
10. Bilck, A.P.; Grossmann, M.V.E.; Yamashita, F. Biodegradable mulch films for strawberry production. *Polym. Test.* **2010**, *29*, 471–476. [CrossRef]
11. Ghatge, S.; Yang, Y.; Ahn, J.H.; Hur, H.G. Biodegradation of polyethylene: A brief review. *Appl. Biol. Chem.* **2020**, *63*, 27. [CrossRef]
12. Duncan, E.M.; Arrowsmith, J.; Bain, C.; Broderick, A.C.; Lee, J.; Metcalfe, K.; Pikesley, S.K.; Snape, R.T.E.; Van Sebille, E.; Godley, B.J. The true depth of the Mediterranean plastic problem: Extreme microplastic pollution on marine turtle nesting beaches in Cyprus. *Mar. Pollut.* **2018**, *136*, 334–340. [CrossRef] [PubMed]
13. Gao, H.H.; Yan, C.R.; Liu, Q.; Ding, W.L.; Chen, B.Q.; Li, Z. Effects of plasticmulching and plastic residue on agricultural production: A meta-analysis. *Sci. Total Environ.* **2019**, *651*, 484–492. [CrossRef] [PubMed]
14. El-Sherif, D.M.; Eloffy, M.G.; Emesery, A.; Abouzid, M.; Gad, M.; El-Seedi, H.R.; Brinkmann, M.; Wang, K.; Naggar, Y.A. Environmental risk, toxicity, and biodegradation of polyethylene: A review. *Environ. Sci. Pollut. Res.* **2022**, *29*, 81166–81182, Epub ahead of print. [CrossRef] [PubMed]
15. Kader, M.A.; Senge, M.; Mojid, M.A.; Ito, K. Recent advances in mulching materials and methods for modifying soil environment. *Soil Tillage Res.* **2017**, *168*, 155–166. [CrossRef]
16. Barragán, H.; Pelacho, A.M.; Martín-Closas, L. Degradation of agricultural biodegradable plastics in the soil under laboratory conditions. *Soil Res.* **2016**, *54*, 216–224. [CrossRef]
17. *ASTM D5988–12*; Standard Test Method for Determining Aerobic Biodegradation of Plastic Materials in Soil. ASTM International: West Conshohocken, PA, USA, 2012.
18. *EN 17033*; Plastics-Biodegradable Mulch Films for Use in Agriculture and Horticulture-Requirements and Test Methods. European Standard; European Committee for Standardization: Brussels, Belgium, 2018.

19. Brodhagen, M.; Goldbergerb, J.R.; Hayes, D.G.; Inglis, D.A.; Marsh, T.L.; Miles, C. Policy considerations for limiting unintended residual plastic in agricultural soils. *Environ. Sci. Policy* **2017**, *69*, 81–84. [CrossRef]
20. Hayes, D.G.; Anunciado, M.B.; Debruyn, J.M.; Bandopadhyay, S.; Schaeffer, S.; English, M.; Ghimire, S.; Miles, C.; Flury, M.; Sintim, H.Y. Biodegradable plastic mulch films for sustainable specialty crop production. In *Polymers for Agri-Food Applications*; Gutiérrez, T.J., Ed.; Springer: Cham, Switzerland, 2019; pp. 183–213.
21. Yang, Y.; Li, P.; Jiao, J.; Yang, Z.; Lv, M.; Li, Y.; Zhou, C.; Wang, C.; He, Z.; Liu, Y.; et al. Renewable sourced biodegradable mulches and their environment impact. *Sci. Hortic.* **2020**, *268*, 109375. [CrossRef]
22. Manzano, V.; García, N.L.; Rodríguez Ramírez, C.; D'Accorso, N.; Goyanes, S. Mulch plastic systems: Recent advances and applications. In *Polymers for Agri-Food Applications*; Gutiérrez, T.J., Ed.; Springer: Cham, Switzerland, 2019; pp. 265–290.
23. Merino, V.; Mansilla, A.Y.; Casalongué, C.A.; Alvarez, V.A. Performance of bio-based polymeric agricultural mulch films. In *Polymers for Agri-Food Applications*; Gutiérrez, T.J., Ed.; Springer: Cham, Switzerland, 2019; pp. 215–240.
24. Moreno, M.M.; Gonzalez-Mora, S.; Villena, J.; Campos, J.A.; Moreno, C. Deterioration pattern of six biodegradable, potentially low-environmental impact mulches in field conditions. *J. Environ. Manag.* **2017**, *200*, 490–501. [CrossRef]
25. Quiao, R.; Wang, X.; Qin, G.; Liu, Q.; Liu, J.; He, W. Preparation of organic crystal seed and its application in improving the functional period of biodegradable agricultural film. *Crystals* **2021**, *11*, 826. [CrossRef]
26. Quiao, R.; Zhao, C.-P.; Liu, J.-L.; Zhang, M.-L.; He, W.-Q. Synthesis of novel ultraviolet absorbers and preparation and field application of anti-iltraviolet aging PBAT/UVA films. *Polymers* **2022**, *14*, 1434. [CrossRef] [PubMed]
27. Casida, L.E.; Klein, D.A.; Santoro, T. Soil dehydrogenase activity. *Soil Sci.* **1964**, *98*, 371–376. [CrossRef]
28. Casida, L.E. Microbial metabolic activity in soil as measured by dehydrogenase determinations. *Appl. Environ. Microbiol.* **1977**, *34*, 630–636. [CrossRef] [PubMed]
29. Barajas, M. Ensayos de metabolismo microbiano en suelo: Actividad deshidrogenasa y tasa de mineralización del nitrógeno. In *Ensayos Toxicológicos Para la Evaluación de Sustancias Químicas en Agua y Suelo: La Experiencia en México*; Ramírez, P., Mendoza, A., Eds.; Instituto Nacional de Ecología: Mexico City, Mexico, 2008.
30. Montejo, M.; Torres, C.P.; Martínez, A.; Tenorio, J.A.; Cruz, M.R.; Ramos, F.R.; Cuevas, M.C. Técnicas para el análisis de actividad enzimática en suelos. In *Métodos Ecotoxicológicos Para la Evaluación de Suelos Contaminados con Hidrocarburos*; Cuevas, M.C., Espinosa, G., Ilizaliturri, C., Mendoza, A., Eds.; INECC: Mexico City, Mexico, 2012.
31. Jarrell, W.M.; Armstrong, D.E.; Grigal, D.F.; Kelly, E.F.; Monger, H.C.; Wedin, D.A. Calculating gravimetric water content and water holding capacity. In *Standard Soil Methods for Long-Term Ecological Research*; Robertson, G.P., Coleman, D.C., Bledsoe, C.S., Sollins, P., Eds.; Soil Methods; Oxford University Press: Oxford, UK, 1999; pp. 55–73.
32. Akaike, H. A new look at the statistical model identification. *IEEE Trans. Automat. Control* **1974**, *19*, 716–723. [CrossRef]
33. Griffin-LaHue, D.; Ghimire, S.; Yu, Y.; Scheenstra, E.J.; Miles, C.A.; Flury, M. In-field degradation of soil-biodegradable plastic mulch films in a Mediterranean climate. *Sci. Total Environ.* **2022**, *806*, 150238. [CrossRef]
34. Borrowman, C.K.; Johnston, P.; Adhikari, R.; Saito, K.; Patti, A.F. Environmental degradation and efficacy of a sprayable, biodegradable polymeric mulch. *Polym. Degrad. Stab.* **2020**, *175*, 109126. [CrossRef]
35. Vázquez, A.; Foresty, M.L.; Cyras, V. Production, chemistry and degradation of starch-bases polymers. In *Biopolymers—New Materials for Sustainable Films and Coatings*; Plackett, D., Ed.; John Wiley & Sons Ltd.: Hoboken, NJ, USA, 2011; pp. 277–299.
36. Mostafa, H.; Sourell, H.; Bockisch, F. Mechanical properties of some bioplastics under different soil types used as biodegradable drip tubes. *Agric. Eng. Int. CIGR J.* **2010**, *12*, 12–21.
37. Rudnik, E.; Briassoulis, D. Comparative biodegradation in soil behaviour of two biodegradable polymers based on renewable resources. *J. Polym. Environ.* **2011**, *19*, 18–39. [CrossRef]
38. Ho, K.L.G.; Pometto, A.L. Temperature effects on soil mineralization of polylactic adic platic in laboratory respirometers. *J. Environ. Polym. Degrad.* **1999**, *7*, 101–108. [CrossRef]
39. Ho, K.L.G.; Pometto, A.L.; Hinz, P.N. Effects of temperature and relative humidity on polylactic acid plastic degradation. *J. Environ. Polym. Degrad.* **1999**, *7*, 83–92. [CrossRef]
40. Liu, Q.; Wang, Y.; Liu, J.; Liu, X.; Dong, Y.; Huang, X.; Zhen, Z.; Lv, J.; He, W. Degradability and properties of PBAT-based biodegradable mulch films in field and their effects on cotton planting. *Polymers* **2022**, *14*, 3157. [CrossRef] [PubMed]
41. Liu, M.; Huang, Z.; Yang, Y. Analysis of biodegradability of three biodegradable mulching films. *J. Polym. Environ.* **2010**, *18*, 148–154. [CrossRef]

agriculture

MDPI

Article

The Influence of Converting Food Crops to Forage Crops Policy Implementation on Herbivorous Livestock Husbandry Development—Based on Policy Pilot Counties in Hebei, China

Huanhuan Zhang, Guogang Wang *, Jinge Liu, Shuai Hao and Shengnan Huang

Institute of Agricultural Economics and Development, Chinese Academy of Agricultural Sciences, Beijing 100081, China
* Correspondence: wangguogang@caas.cn; Tel.: +86-150-1035-6136

Abstract: In the context of increasing consumption of herbivorous livestock products, competition between humans and animals for food, and increasing environmental constraints, it is necessary to solve the problem of sustainable development of China's livestock industry and increase the protection and development of the grassland livestock industry while making good use of production resources in agricultural areas in order to explore the development potential of the herbivorous livestock industry in agricultural areas. The Converting Food Crops to Forage Crops Policy (CFFP), as an important measure of agricultural supply-side structural reform, aims to develop a high-quality forage industry and a high-quality herbivorous livestock industry. However, over the years of policy implementation, few studies have examined the impact effects of the policy on the development of the regional herbivorous livestock industry. To fill this research gap and provide theoretical support for subsequent policy implementation, the study used the synthetic control method to examine the impact of policy implementation on the development of herbivorous livestock production in the pilot counties in Hebei Province from 2010 to 2020. The study discovered that the policy's implementation encouraged the expansion of herbivorous livestock production in the pilot counties, but the policy's effects on various regions and livestock species varied due to the influence of local production bases and resource endowments.

Keywords: Converting Food Crops to Forage Crops Policy (CFFP); policy effect; herbivorous livestock husbandry

Citation: Zhang, H.; Wang, G.; Liu, J.; Hao, S.; Huang, S. The Influence of Converting Food Crops to Forage Crops Policy Implementation on Herbivorous Livestock Husbandry Development—Based on Policy Pilot Counties in Hebei, China. *Agriculture* **2022**, *12*, 1872. https://doi.org/10.3390/agriculture12111872

Academic Editor: Christos Karelakis

Received: 8 October 2022
Accepted: 4 November 2022
Published: 8 November 2022

1. Introduction

Advancing the sustainable development of the livestock industry is an important part of advancing the sustainable development of China's agriculture. Although the expansion of China's livestock industry has played an important role in meeting the consumer demand of residents and promoting the income of farmers and herders, under the requirements of tightening resource constraints, upgrading residents' consumption, and green ecology, how to enhance the productivity of animal husbandry to meet the increasing consumer demand of residents is still an important goal for the development of animal husbandry [1,2]. With the growth of the population, urbanization, and improvement of living standards driving the growth of global demand for animal protein, China's per capita consumption of beef and lamb increased by 84.58% and 54.30%, respectively, compared with 2000. The growing consumer demand for the development of herbivorous livestock husbandry, as an important part of animal husbandry, puts forward new requirements [3,4]. According to the OECD-FAO Agricultural Outlook 2021–2030, China's per capita beef consumption will reach 3.99 kg/person in 2029, and milk consumption continues to rise. With the backdrop of sustained growth in global demand for animal protein, meeting China's demand for livestock products also requires expanding domestic production capacity [5].

The majority of China's herbivorous livestock products and forage resources required for the development of the herbivorous livestock industry have always come from pasture areas, but against the background of increasing demand for herbivorous livestock products and degradation of grassland productivity, the contradiction between the rapid growth of residents' demand for herbivorous livestock products and the insufficient supply of high-quality forage has sharpened [6,7]. While protecting grasslands and supporting the development of grassland animal husbandry, it has become an important trend in the development of herbivorous animal husbandry in agricultural areas by utilizing the resources of agricultural areas and tapping the potential of herbivorous animal husbandry in agricultural areas [8]. The high percentage of livestock that eat grains, such as pigs and poultry, and the slow growth rates of livestock that are fed on grass, such as cattle and sheep, combined with the traditional idea of valuing grain production, have prevented China from developing its herbivorous livestock industry which feeds on high-quality forage. Many studies have been conducted to prove the importance of forage feeding to cattle, sheep, and other herbivorous livestock in improving production efficiency, upgrading quality, and ensuring product safety, and the role of forage in the development of herbivorous livestock has been widely recognized [9,10]. The Chinese government and scholars have also begun to realize the important role of forage in the transformation and upgrading of herbivorous livestock husbandry and have begun to pay attention to the importance of herbivorous livestock husbandry development in agricultural areas in the sustainable development of livestock husbandry while protecting grassland ecology and developing grassland livestock husbandry [11–13]. In 2015, in order to promote the structural adjustment of the plantation industry and the transformation and upgrading of the grassland livestock husbandry industry, the Chinese government began to arrange financial funds to support the development of Converting Food Crops to Forage Crops Policy (CFFP) pilots, which provides new ideas for the development of a green and sustainable modern livestock husbandry industry. However, compared with related fields such as grassland livestock husbandry and traditional farming, there is a lack of research on CFFP [14–16]. To fill this gap, the study will focus on analyzing the impact effects of policy implementation on the development of herbivorous livestock farming in the pilot counties based on the existing literature, focusing on policy implementation ideas and objectives. The study will further analyze whether there are differences in policy effects among different types of regions and different livestock species and how to explain such differences.

The remainder of the study is structured as follows. The study area, data sources, policy introduction, and model selection are included in Section 2. The empirical findings are reported in Section 3 together with examinations of their robustness. The results are discussed in Section 4. The conclusions are presented in Section 5.

2. Materials and Methods

2.1. The Policy

As the constraints on the development of herbivorous livestock husbandry in traditional pasture areas increase, the development of modern livestock husbandry also places new requirements on the structure and mode of herbivorous livestock husbandry. CFFP, as an important measure to adjust structure and change mode and promote structural reform on the supply side of agriculture in China, aims to play a leading role in financial funds, mobilizing farmers' enthusiasm for forage cultivation through market mechanisms, building a new agricultural and livestock husbandry structure combining farming and raising animals, and promoting the development of herbivorous livestock husbandry (the policy implementation framework is shown in Figure 1). In 2015, the central government allocated special funds to begin the policy in 30 counties in the 10 regions of Liaoning, Heilongjiang, Jilin, Inner Mongolia, Hebei, Shanxi, Gansu, Qinghai, and Ningxia; at present, the policy has been implemented in more than 900 counties.

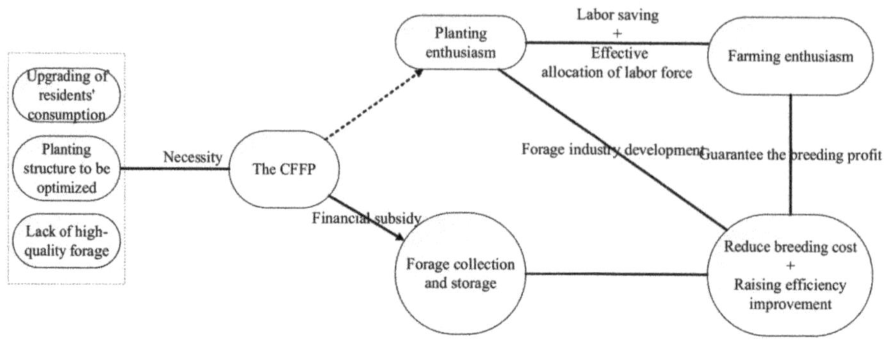

Figure 1. The implementation framework for the CFFP.

CFFP's subsidy funds are used primarily for high-quality forage storage. Policy subsidies are primarily for large-scale herbivorous livestock farms (households) with forage storage and use capacity or professional harvesting enterprises (cooperatives) with consistent forage supply and marketing orders. In terms of the essence of CFFP, government subsidies linked to forage harvest are the means; increasing forage supply is the channel; and developing a quality forage industry and promoting herbivorous livestock development are the main objectives. The development of the forage industry is the basis of the transformation and upgrading of the herbivorous livestock industry, and the development of the herbivorous livestock industry is the driving force behind the rapid development of the forage industry. On this basis, the policy is to promote the development of the forage industry and the transformation and upgrading of the herbivorous livestock industry in accordance with the circular development concept of "planting to drive breeding, breeding to promote planting."

2.2. The Study Area

The implementation area of CFFP focuses on two types of areas: agricultural and semi-agricultural and semi-pastoral areas, and has been expanded from the initial 30 pilot counties to more than 900 pilot counties at present. On the basis of the comprehensive consideration of regional characteristics and data feasibility and taking into account representativeness and practicality, the study selects the first batch of pilot counties for CFFP—Xingtang County and Weichang County, Hebei Province—as the study area to analyze and explore the impact of CFFP on the development of the regional herbivorous livestock industry. Xingtang and Weichang counties belong to different types of regions, and the comparative analysis of the effect of grain on feed policy in the two regions can further explore the regional differences in the policy effect while analyzing the policy effect.

Hebei Province, which is rich in production resources, is bounded by 36°05′ and 42°40′ N latitude and 113°27′ and 119°50′ E longitude. According to the data of the Third National Land Survey, Hebei Province has 6520 thousand hectares of arable land (5.1% of the national arable land area) and 1947.27 thousand hectares of grassland. In 2021, Hebei's share of grain production in the country was 5.6%, and the total protein of livestock products in the country was 7.28%, of which the total protein of major herbivorous livestock products in the country was 9.65%, which is much higher than the average of all provinces (municipalities and regions). As an important area for the supply of agricultural products in Beijing, Tianjin, and Hebei, the development of its livestock industry plays an important role in meeting the growing demand for herbivorous livestock products. Xingtang County belongs to Shijiazhuang City, Hebei Province, with 47.53 thousand hectares of arable land, planted mainly with corn, wheat, peanuts, and other food crops and cash crops, with a small proportion of grassland area, which is a typical agricultural county. According to the policy tracking data, the milk production of Xingtang County in 2021 was 267,800 tons, and the annual slaughter of beef cattle and sheep reached 25,800 and 75,500 heads, respectively.

Weichang County belongs to Chengde City, Hebei Province, with 112.04 thousand hectares of arable land and 111.65 thousand hectares of grassland. It is a semi-agricultural and semi-pastoral county in transition from pastoral to agricultural areas. In 2021, milk production in Weichang County was 29,500 tons, and the annual slaughter of beef cattle and meat sheep reached 170,000 and 250,000 heads, respectively.

2.3. Data

The Hebei province county panel data used in the study are primarily from the Hebei Rural Statistical Yearbook, with missing data supplemented by regional government work reports.

Focusing on the research objective of "the impact of policy implementation on the development of the herbivorous livestock industry in pilot counties", two indicators, herbivorous livestock production level and herbivorous livestock production concentration index (HPCI), were selected as predictor variables to analyze the level of herbivorous livestock production in pilot counties and the contribution of pilot counties to the overall herbivorous livestock production level in Hebei Province based on county data. Herbivorous livestock farming is a production system that uses forage to feed herbivorous animals such as cattle, sheep, horses, and rabbits to obtain livestock products, of which cattle and sheep are the two main types of herbivorous livestock. On the basis of full consideration of the actual situation and data availability, the study focused on two major types of herbivorous livestock, cattle and sheep, and explored the impact of the CFFP implementation on the development of major herbivorous livestock farming.

On the basis of the existing research results [14,17], we take the annual slaughter volume of livestock as a measure of the regional herbivorous livestock production level and refer to the existing standards (the "one cow is equal to five sheep units" standard, according to the "Inner Mongolia Autonomous Region basic grassland protection regulations") to unify the slaughter volume of cattle and sheep into sheep units to facilitate the analysis and calculation of the overall development level of the regional herbivorous livestock industry.

The contribution of a part to the overall production total can usually be expressed as a production concentration index. In this study, this index is expressed as the proportion of the pilot counties' herbivorous livestock production levels to the overall production levels in Hebei Province, i.e., the herbivorous livestock production concentration index (HPCI). In addition, the index can be used to indicate changes in the layout of regional livestock production and is widely used in studies related to industrial, agricultural, and other industrial development [18]. HPCI can be calculated by the equation $\omega_{it} = \frac{w_{it}}{W_T}(i = 1, 2 \ldots, N; t = 1, 2 \ldots, T)$, where w_{it} denotes the level of herbivorous livestock production in pilot counties in period 't', and W_T refers to the total level of overall herbivorous livestock production in Hebei Province in period 't'.

In addition, in order to consider the fitting effect of synthetic control objects in the empirical analysis and the robustness of the results, we used important factors affecting the development of herbivorous livestock farming as predictor control variables. The study considered the impact of population growth, economic development, and regional agricultural production resources on the development of livestock farming and used gross regional domestic product (in the analysis, the gross regional domestic product is deflated to obtain the real gross regional domestic product with 2010 as the base period), population, grain production, and predictor variables with a three-period lag as predictor control variables [19]. Finally, in order to eliminate the effect of magnitude, the empirical analysis part of the study logarized the variables of livestock slaughter, industry concentration, gross regional product, population size, and grain production before conducting the analysis.

2.4. Econometric Method

The study examines the effects of the CFFP implementation on the level of growth of herbivorous livestock husbandry in the region using a synthetic control method, with reference to prior research [20]. The study selected Xingtang and Weichang counties in

Hebei province as the treatment groups, and 2015 was used as the time point of policy intervention. In the control group, counties in Hebei province that were classified as pilot counties for CFFP conversion after 2015 were removed, and counties that had not implemented the policy since the CFFP was implemented were used as the control group.

The "counterfactual" reference group for each treatment group was constructed through the weighted average of the control groups to simulate the development of herbivorous livestock in the region without the implementation of the policy. The comparison of the production level of herbivorous livestock in the region with and without the implementation of the policy is the policy effect of policy implementation on the development level of herbivorous livestock in the region.

Suppose there are N + 1 regions. Region 'i' begins implementing the CFFP in period T_0, and the other N regions do not implement the policy. The potential outcome of region 'i' implementing the policy in period 't' is denoted by Y_{1it}, the potential outcome of region 'i' not implementing the policy in period 't' is denoted by Y_{0it}, and the causal effect of region implementing the policy is denoted by $\tau_{it} = Y_{\ 1it} - Y_{\ 0it}$, where $i = 1,\ 2, \ldots, T$.

The result of observed herbivorous livestock production in region 'i' in period 't' is $Y_{it} = D_{it}Y_{1it} - (1 - D_{it})Y_{0it} = Y_{0it} + \tau_{it}D_{it}$, where D_{it} denotes the policy implementation status of region 'i' in period 't'. If region 'i' is subject to policy intervention in period 't', the value is 1, otherwise the value is 0. Assume that region 'i' is subject to policy intervention after period T_0, while the other N regions have never been subject to policy intervention in all periods. For $t > T_0$, the policy effect can be written as $\tau_{it} = Y_{1it} - Y_{0it}$, where Y_{1it} is observable owing to the policy intervention in region 'i' after period 't', while Y_{0it} is not observable. This is assuming that the other N areas are never subject to policy intervention in all times. The following model can be used to estimate the counterfactual result for region 'i':

$$Y_{0it} = \delta_t + \theta_t Z_i + \gamma_t \mu_i + \varepsilon_{it} \tag{1}$$

where δ_t stands for time fixed effects; Z_i are the (K × 1)-dimensional observable covariates; θ_t is the (1 × K)-dimensional vector of unknown parameters; γ_t is the (1 × F)-dimensional vector of unobservable common factors; μ_i is the (F × 1)-dimensional vector of coefficients; and ε_{it} are the unobservable short-term shocks in each region, which are supposed to have a mean value of '0' at the region level.

Equation (1) is an extension of the traditional Differences-in-Differences (DID). The traditional DID model allows the presence of unobservable factors to limit the effects by transforming the effects of these factors into constants in time. γ_t in Equation (1) is not constant, allowing the unobservable factor effects to vary in time.

Assume that only the first region '1' (i = 1) has implemented CFFP and that none of the other regions have done so. Consider an (N × 1)-dimensional weight vector $W = (w_2, \ldots, w_{N+1})$ that satisfies $w_j \geq 0, j = 2, \ldots, N+1$ and $w_2 + \ldots + w_{N+1} = 1$ in order to determine Y_{01t}. A synthetic control group is represented by each vector W. Each control group region's outcome variable values are weighted to produce:

$$\sum_{j=2}^{N+1} w_j Y_{jt} = \delta_t + \theta_t \sum_{j=2}^{N+1} w_j Z_j + \gamma_t \sum_{j=2}^{N+1} w_j \mu_j + \sum_{j=2}^{N+1} w_j \varepsilon_{jt} \tag{2}$$

Suppose there exists a weight $W^* = w_2^*, \ldots, w_{N+1}^*$ such that:

$$\sum_{j=2}^{N+1} w_j^* Y_{j1} = Y_{11}, \sum_{j=2}^{N+1} w_j^* Y_{j2} = Y_{12}, \ldots, \sum_{j=2}^{N+1} w_j^* Y_{jT_0} = Y_{1T_0}, \sum_{j=2}^{N+1} w_j^* Z_j = Z_1 \tag{3}$$

Abadie [21] proves that if $\sum_{t=1}^{T_0} \gamma_t' \gamma_t$ is a non-singular square matrix, then we have:

$$Y_{01t} - \sum_{j=2}^{N+1} w_j^* Y_{kt} = \sum_{j=2}^{N+1} w_j^* \sum_{s=1}^{T_0} \gamma_n \left(\sum_{n=1}^{T_0} \gamma_n' \gamma_n \right)^{-1} \gamma_s' (\varepsilon_{js} - \varepsilon_{1s}) - \sum_{j=2}^{N+1} w_j^* (\varepsilon_{jt} - \varepsilon_{1t}) \quad (4)$$

Under general conditions, Equation (4) converges to '0'. For $T_0 < t \leq T$, the counterfactual results for region '1' can be approximated by a synthetic control group, $\hat{Y}_{01t} = \sum_{j=2}^{N+1} w_j^* Y_{jt}$, which yields an estimate of the policy effect:

$$\hat{\tau}_{1t} = Y_{1t} - \sum_{j=2}^{N+1} w_j^* Y_{jt}, t \in [T_0 + 1, \ldots, T] \quad (5)$$

The secret to finding $\hat{\tau}_{1t}$ is to identify the proper weight W^* that ensures the validity of Equation (3). On the basis of the Abadie development program, a synthetic pilot county that approximates the actual pilot county development trend can be obtained. The development trend of herbivorous animal husbandry in the synthetic pilot counties, obtained by weighting, actually simulates the development trend of the pilot counties without the policy, and the level difference between them is the policy effect of the CFFP.

3. Results

This section assesses the impact of the CFFP implementation on herbivorous livestock development in the pilot counties based on the synthetic control method, and the policy effects are captured by the differences in the predictor variables after the policy implementation. The synthetic control method as a data-driven method, creating the synthetic area approximation fitting the pre-policy implementation development trend in the pilot area, is the basis for an accurate assessment of the policy implementation effect. Due to the high production level of the cattle industry in Weichang County, when it is used as a predictor variable to assess the policy effect, it is not possible to find suitable weights to fit the change trend before the policy implementation. In this case, the synthetic control method is no longer applicable to assess the impact of the CFFP implementation on the beef cattle production level in Weichang County. The study refers to the existing studies [20,22] and compensates for this deficiency by an alternative method—Differences-in-Differences (DID).

3.1. Impacts on Production Levels

3.1.1. The Effect of CFFP Implementation on the Production Level of Herbivorous Livestock

Figure 2 illustrates the fitting of the production level of herbivorous animal husbandry in actual pilot counties and artificial pilot counties from 2010 to 2020. The CFFP's implementation year is indicated by the location of the vertical dotted line. From Figure 2a, it can be seen that before the policy implementation, synthetic Xingtang County and Xingtang County were very close in the change trend, indicating that synthetic Xingtang County better fit the change trend of the herbivorous livestock production level in Xingtang County; after the policy implementation, the herbivorous livestock production level in Xingtang County was higher than synthetic Xingtang County, and the difference between the two represents the policy effect, indicating that the implementation of the CFFP promoted the herbivorous livestock production level in Xingtang County. Similarly, as can be seen from Figure 2b, the fitted polder counties better fit the trend of the actual polder counties, and the difference between the trend of the actual polder counties and the synthetic polder counties since the year of policy implementation was positive, indicating that the CFFP implementation had the same positive effect on the improvement of the level of herbivorous livestock production in polder counties. Whether the level of herbivorous livestock production in Xingtang or Weichang counties was used as a predictor variable, the policy had a catalytic effect on the level of herbivorous livestock production in the pilot counties.

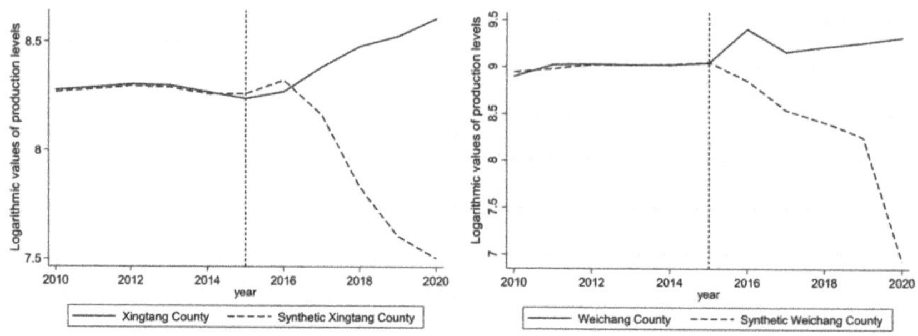

(a) (b)

Figure 2. Effect of the CFFP on the production level of herbivorous livestock, (**a**) the trend of Xingtang County and synthetic Xingtang County; (**b**) the trend of Weichang County and synthetic Weichang County.

From the viewpoint of the policy action path, CFFP has two main impacts on the development of the herbivorous livestock industry. Firstly, the policy encourages the development of the forage industry, which provides high-quality forage for the herbivorous livestock industry, promotes the optimization of the diet structure of the herbivorous livestock industry, and improves breeding efficiency. Secondly, the policy subsidizes the forage storage link (most of the large-scale farms with forage demand meet the storage conditions), which in turn reduces the breeding cost to a certain extent and maintains the enthusiasm of herbivorous livestock farmers. However, at the early stage of policy implementation, agricultural operators are more willing to adopt a wait-and-see attitude due to a lack of understanding of the policy content and objectives [23].

From the perspective of planting, the policy is initially influenced by the implementation efforts and farmers' perceptions. Rational farmers tend to have reservations about planting forage crops with unfamiliar production technology and low levels of social service development, and this influence will in turn spread through the peer effect in the farmers' group species, which will then evolve into group behavior [24]. On the other hand, considering that wheat, corn crops, and other food crops are the main competitive crops of forage crops, under the influence of food support policy and planting habits, farmers have a certain preference for traditional crop planting [25,26]. This is coupled with the lack of direct guiding effect of policy on farmers' planting structure adjustment, causing the policy in the early pilot areas of the forage industry development drive to be limited. As the basis for the development of herbivorous livestock industry, the slow development of the forage industry will have a direct impact on the back end of the breeding chain.

From the perspective of breeding, herbivorous livestock breeding has long-cycle and high-cost characteristics, and the breeding body will not substantially adjust the planting scale. Although the subsidy of CFFP projects can reduce the cost of breeding to a certain extent compared with the universal policy of grain subsidy, the target and standard of CFFP subsidy have a certain threshold. For example, project funds in Xingtang County, Hebei Province, are used to subsidize large-scale farms that harvest more than 33.33 hectares of whole-plant silage corn, and small-scale subjects do not directly benefit from the policy implementation. In the case of uncertainty about expected returns, the farming body will not easily change the scale of existing herbivorous livestock operations.

The improvement of policy content, the increase of publicity and the stability of policy support, the deepening of farmers' policy perception, the increase of farmers' willingness to participate, the orderly formation of the industrial development environment, the forage industry, and herbivorous livestock development under the guidance of the policy gradually formed a food cycle, promoting the implementation of the policy and the realization of policy objectives [27,28].

3.1.2. The Effect of CFFP Implementation on the Levels of Production of Various Livestock Species

From the actual change trend in Xingtang County in Figure 3, it can be seen that the production level of the cattle industry in Xingtang County has been in a stable growth trend, especially after the implementation of the policy. The production level of the cattle industry has increased significantly. From the change trend of policy effect, the policy effect of the production level of the cattle industry in Xingtang County is significantly positive and continuously increasing. Compared with the cattle industry, on the one hand, the actual sheep industry production level in Xingtang County fluctuates slightly around a certain level and does not show a significant increase. On the other hand, the policy effect of the sheep industry production level is not stable and even had a significant negative effect at the start of policy implementation.

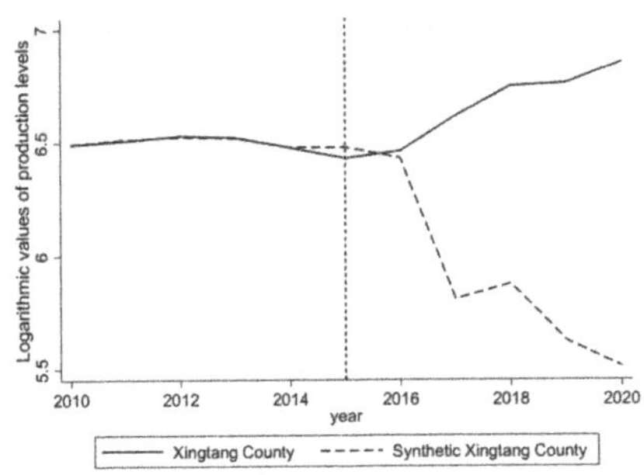

Figure 3. Effect of the CFFP on the production level of cattle industry.

From Figure 4b, it can be seen that the policy had a negative effect on the production level of the sheep industry in Weichang County in the first two years of implementation, and this negative effect weakened and changed to a positive effect with the implementation of the policy. In addition, since the level of cattle production in Weichang County is generally higher than that in other regions, it is not possible to find suitable weights to fit the trend before the implementation of the policy, so it is not analyzed here.

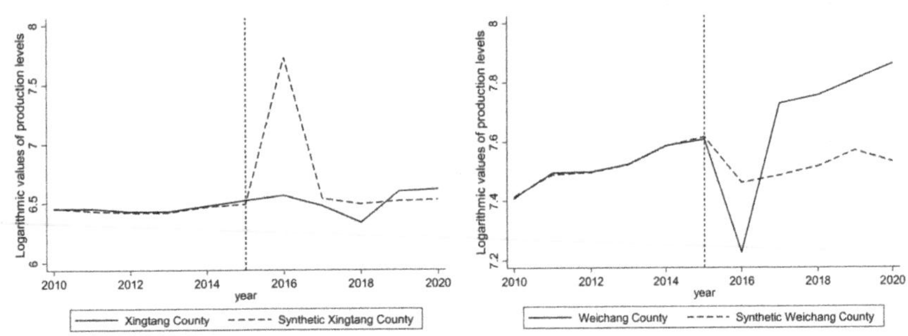

(a) (b)

Figure 4. Effect of the CFFP on the production level of sheep industry, (**a**) the trend of Xingtang County and synthetic Xingtang County; (**b**) the trend of Weichang County and synthetic Weichang County.

3.1.3. The Effect of Policy Implementation on the HPCI

Whether the HPCI of Xingtang or Weichang county was used as a predictor variable, the actual value of HPCI was higher than the synthetic value after the implementation of the policy, indicating that the implementation of the policy promoted the increase of the pilot counties' share of herbivorous livestock production level in Hebei province and promoted the pilot counties' contribution to the increase of herbivorous livestock production level in Hebei province.

According to the actual trend of HPCI in the pilot counties in Figure 5a, the concentration of herbivorous livestock production in Xingtang County showed a "V-shaped" change between 2010 and 2020, reached its lowest in 2015, resumed growth after 2015, and increased at a faster rate. On the one hand, in response to the serious problem of livestock pollution, the central and local governments issued a series of pollution prevention policies, including the central government in 2011 specifying that the pollution prevention of large-scale livestock and poultry breeding should be strengthened. Shijiazhuang City is one of the most polluted areas in Hebei Province (with the highest amount of manure produced by cattle and livestock), which naturally makes it a key area for pollution control [29]. However, the high cost of farm pollution control (coupled with the fact that project support funds are used mainly for large-scale farming subjects) and the large number and wide distribution of small- and medium-scale farming subjects (who generally choose to maintain or reduce the scale of farming under the existing resource conditions to achieve effective pollution control) result in a decline in regional production levels, thus causing the share of Xingtang County, which has a good foundation for the development of the livestock industry in the whole region, to decline. The implementation of the CFFP provides an opportunity for the development of herbivorous livestock husbandry in Xingtang County to achieve the dual goals of emission reduction and transformation and upgrading of herbivorous animal husbandry, promoting the revitalization of regional herbivorous animal husbandry.

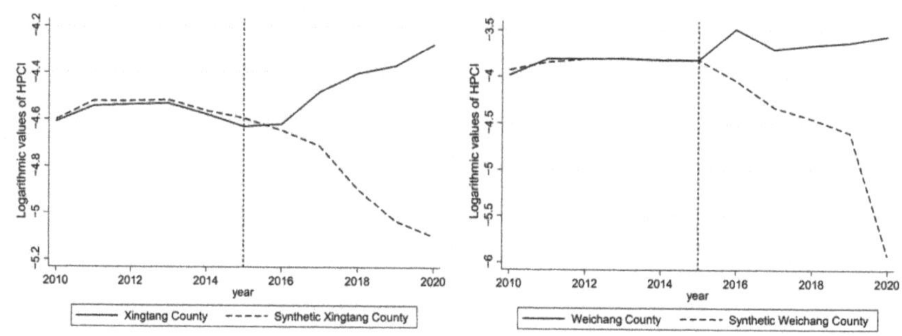

(**a**) (**b**)

Figure 5. Effect of the CFFP on the herbivorous livestock production concentration index (HPCI), (**a**) the trend of Xingtang County and synthetic Xingtang County; (**b**) the trend of Weichang County and synthetic Weichang County.

Compared with Xingtang County, the HPCI in Weichang County is on a "steady growth" trend, with a brief period of rapid growth in 2015/2016 and a rapid return to slow growth. On the one hand, in the context of the serious disconnection between agriculture and animal husbandry that brings about environmental pollution and other problems that restrict the development of animal husbandry, livestock manure in pastoral areas can be used as a resource, to a large extent, making the development of herbivorous livestock husbandry in Weichang County weakly affected by animal husbandry pollution control [30]. However, Weichang County faces problems, such as grassland ecological destruction and declining productivity, which limit the further development of herbivorous animal husbandry in Weichang County, thus preventing the contribution of the production level of herbivorous animal husbandry in Weichang County to the development of herbivorous livestock husbandry in Hebei Province from increasing in a long period of time. The implementation of the policy activates the potential of developing herbivorous livestock husbandry in the agricultural areas of the region under the condition of declining productivity in pastoral areas and gradually promoting the increase of HPCI in Weichang

County, increasing the rate of Weichang County's contribution to herbivorous livestock production in Hebei Province.

3.2. Comparison between Regions

Overall, the implementation of the policy has a significant positive effect on the development of herbivorous animal husbandry in Xingtang and Weichang counties, and the policy effect in Weichang County is significantly higher than that in Xingtang County. On the one hand, the grassland herbivorous livestock development in Weichang County, which is a semi-agricultural and semi-pastoral county, has laid the foundation for the industrial development of herbivorous livestock development in the region's agricultural areas in terms of breeding concepts, production technology, and socialization services, making Weichang County better than agricultural counties in terms of forage resource abundance and policy recognition as well as improving the possibility of policy response behavior of agricultural business entities [27]. On the other hand, in the context of grassland ecological degradation, the CFFP fits well with the needs of traditional animal husbandry transformation in the context of "conversion of grassland grazing to shed feeding" in Weichang County, and the implementation of the CFFP expands the development space of herbivorous animal husbandry in agricultural areas while stabilizing the production level of herbivorous livestock husbandry in Weichang County, which is also consistent with the findings of previous studies [31,32]. At the same time, Figure 6 shows that the policy effects of different indicators exhibit a growing trend; this indicates that the policy effects are sustainable and also proves the correctness of the policy idea that the policy is based on financial resources and drives industrial development through market mechanisms.

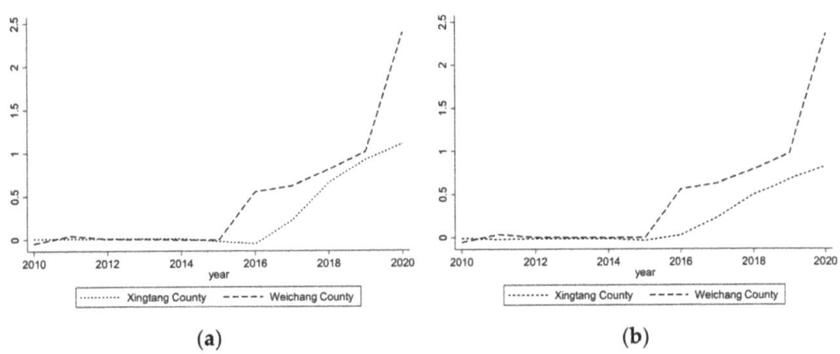

Figure 6. Effect of the CFFP on herbivorous livestock development; (**a**) effect of herbivorous livestock production levels; (**b**) effect of the herbivorous livestock production concentration index (HPCI).

In terms of the trend of the policy effect, the trends of the policy effect of the same region's herbivorous livestock production level and HPCI are similar, but there is a significant difference in the trends of the change between regions. Although the policy effect in Weichang County is higher than that in Xingtang County, the change in policy effect in Xingtang County appears to be more stable than the "fast and slow" change in Weichang County. If the policy resources are not effectively allocated between "herbivorous livestock development in agricultural areas" and "herbivorous livestock protection in pastoral areas", the policy resources will be scattered and the policy stability will be lacking, so that the production resources cannot be effectively used and the policy effect in Weichang County will be affected. The stable policy effect in Xingtang County also further confirms that the implementation of the policy has effectively tapped the potential of traditional farming areas in developing herbivorous livestock development and effectively promoted the development of a regional herbivorous livestock industry, which echoes the findings of previous studies [33].

3.3. Robustness Tests

3.3.1. Robustness Test

To test that the differences in the predictor variables in the empirical analysis are indeed due to the effects brought about by the policy rather than some other unobserved extraneous factors, a ranking test (permutation test) similar to the rank test in statistics proposed by Abadie is used here to determine how likely it is that the other control groups will appear the same as the treatment group. The idea of this test is to assume that all control groups began implementing the CFFP in 2015, to construct synthetic control subjects for the control group using the synthetic control method, to estimate the policy effect in the hypothetical case, and then to compare the policy effect actually generated in the treatment group with the policy effect generated in the urban hypothetical case in the control group. If the difference in policy effects between the two is large enough, then there is reason to believe that the policy effects are significant. The method requires synthetic control subjects to have a good fit before the policy implementation, and if a control group has a poor fit before the policy implementation, the study results will also remove the presentation of its herbivorous livestock development level difference. Figures 7–10 show the different distributions of the predictor variables.

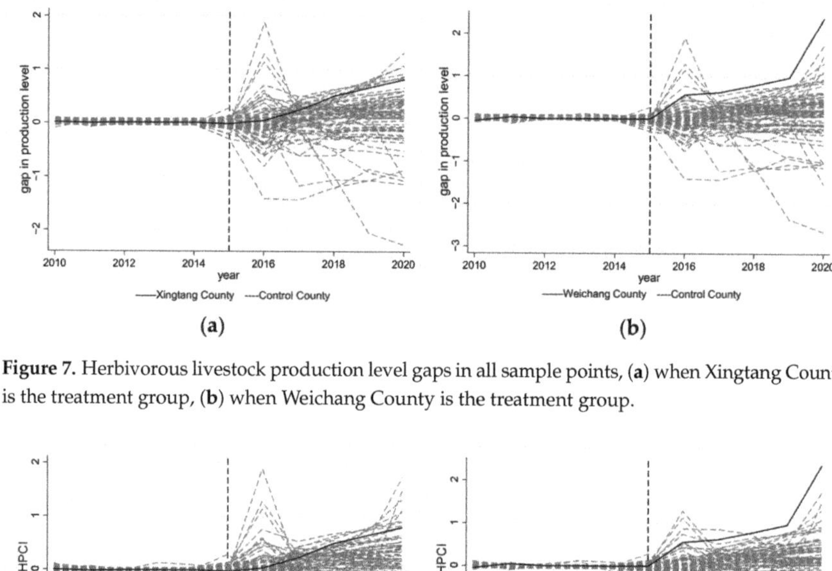

Figure 7. Herbivorous livestock production level gaps in all sample points, (**a**) when Xingtang County is the treatment group, (**b**) when Weichang County is the treatment group.

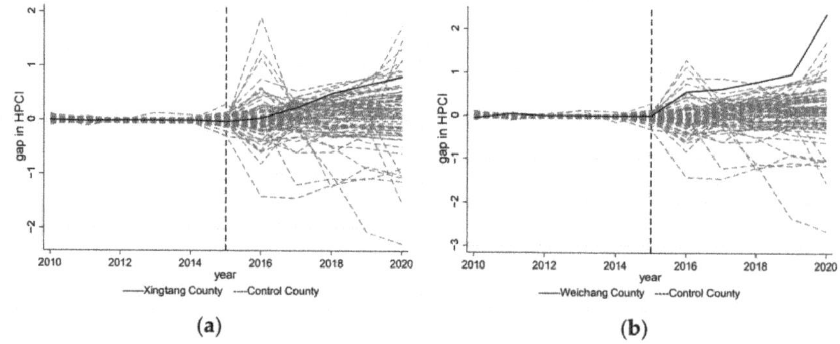

Figure 8. Herbivorous livestock industry's concentration (HPCI) gaps in all sample points, (**a**) when Xingtang County is the treatment group, (**b**) when Weichang County is the treatment group.

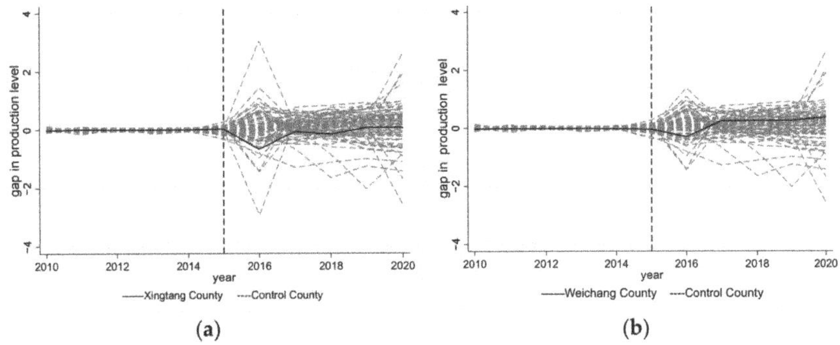

Figure 9. Sheep industry production level gaps in all sample points, (**a**) when Xingtang County is the treatment group, (**b**) when Weichang County is the treatment group.

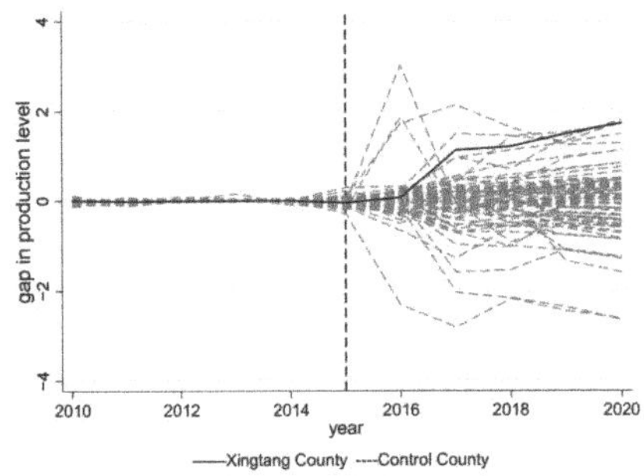

Figure 10. Cattle industry production level gaps in Xingtang County and all Control County.

Consider the level of herbivorous livestock production in Xingtang County as an example. According to Figure 7a, it can be seen that the gap between the policy effects in Xingtang County and other control group areas was not large before the implementation of the CFFP, but after the implementation of the policy, the gap between Xingtang County and other areas began to widen, and the policy effects in Xingtang County were larger than those in other areas. The likelihood of such a wide difference between the production levels of herbivorous cattle in Xingtang County and synthetic Xingtang County is 7/73 (there were 78 sample regions in the study, the remaining 73 cities were left after five areas with high RMSPE values prior to 2015 were excluded), and there is a 9.59% probability that other control groups will be similar to Xingtang County.

Similarly, according to Figures 7b, 8, 9 and 10, the probabilities that the same situation as that in treatment group will occur in other control group areas are 7/74 (9.46%), 1/71 (1.39%), 25/65 (38.46%), 0/75 (0%), 0/75 (0%), and 16/75 (24%), when the HPCI in Xingtang County, the production level of cattle livestock in Xingtang County, the level of sheep livestock in Xingtang County, the production level of herbivorous livestock in Weichang County, the HPCI in Weichang County, and the HPCI in Weichang County are used as predictor variables, respectively. This suggests that the policy effects assessed using the synthetic control method are likely to be robust.

3.3.2. A Further Examination of the Policy Impact on the Cattle Industry's Production in Weichang County—Based on the DID

The effect of the CFFP on the level of production of herbivorous livestock in Weichang County was estimated by the DID to partially compensate for the inability to find synthetic control objects and the poor fitting effect. The econometric model was set as follows:

$$Y_{it} = \beta_0 + \beta_1 treat_i * time + \beta_2 treat_i + \beta_3 time + \alpha X + \delta_i + \gamma_t + \varepsilon_{it} \quad (6)$$

where Y_{it} is the production level of cattle; $treat_i$ is the CFFP variable, with a value of '1' for the treatment group and '0' for the other control groups; *time* is the year dummy variable, with '1' after 2015 and '0' before 2015; β_1 is the net effect of the CFFP on cattle production level; X is the ensemble of control variables; δ_i is the individual fixed effect; and γ_t is the time fixed effect.

Through data analysis, it was found that the trend of cattle industry's production levels in both the treatment and control groups before policy implementation showed a slight downward trend, which is consistent with the premise of common trend with DID application. Table 1 reports the estimation results of the double difference method, and the interaction term reflects the net effect of the CFFP on the production level of the cattle industry in Weichang County. Both the least squares and fixed panel effects model results show that the interaction term coefficient is significantly positive at the 1% level, indicating that the implementation of the CFFP in the pilot counties significantly contributed to the improvement of the production level of the cattle industry in Weichang County.

Table 1. The CFFP's impact on the cattle industy's production in Weichang County (DID).

	OLS		FE	
β_1	2.51 ***	0.179 ***	0.467 ***	0.092 ***
	(0.095)	(0.078)	(0.056)	(0.029)
Constant	5.151 ***	0.211 **	5.15 ***	0.528 ***
	(0.118)	(0.263)	(0.035)	(0.526)
X	Uncontrolled	Controlled	Uncontrolled	Controlled
γ_t	Controlled	Controlled	Controlled	Controlled
δ_i	Controlled	Controlled	Controlled	Controlled
N	858	780	858	780
R2	0.06	0.934	0.036	0.899

*** $p < 0.01$ and ** $p < 0.05$.

4. Discussion

It has been discovered that effective agricultural support policies have a positive impact on industry economic growth, farm household income, and environmental protection [34,35]. On the basis of existing studies and the current situation of herbivorous livestock production in China, this study focuses on the impact of the implementation of the CFFP on the development of herbivorous livestock production in the pilot counties. The study is based on the first pilot counties of the CFFP—Xingtang and Weichang counties—and uses a synthetic control method to analyze the impact of policy implementation on the production level and HPCI in these pilot counties. The results of the study showed that the implementation of the policy as a whole was beneficial to the improvement of herbivorous livestock production levels in the pilot counties and promoted the concentration of herbivorous livestock production areas in the pilot counties in Hebei Province. Thus, the results of the study can provide theoretical support for the subsequent promotion of the policy.

At the same time, the study further analysis of the policy effects of different livestock production levels in the pilot counties, and the results are shown in Figure 11. The study found that the CFFP implementation showed a positive effect on the pilot counties' cattle industry production levels, but the positive effect on the pilot counties' sheep industry production levels was not satisfactory, especially in Xingtang County. After five years of CFFP implementation, the improvement of regional sheep industry production levels was very limited. One of the reasons is that the development of regional industries is influenced by regional industries base, the economy, and policies [36,37]. For example, the dairy and beef cattle industries have been the key industries in the development of animal husbandry in Xingtang County, which has made an important contribution to the regional economic development and farmers' income; the regional government has purposely tilted the use of policy funds toward the project areas where the dairy and beef cattle industries are concentrated, aiming to promote forage cultivation and the herbivorous livestock industry in the region. Support for the sheep industry is neglected or even squeezed due to limited policy resources.

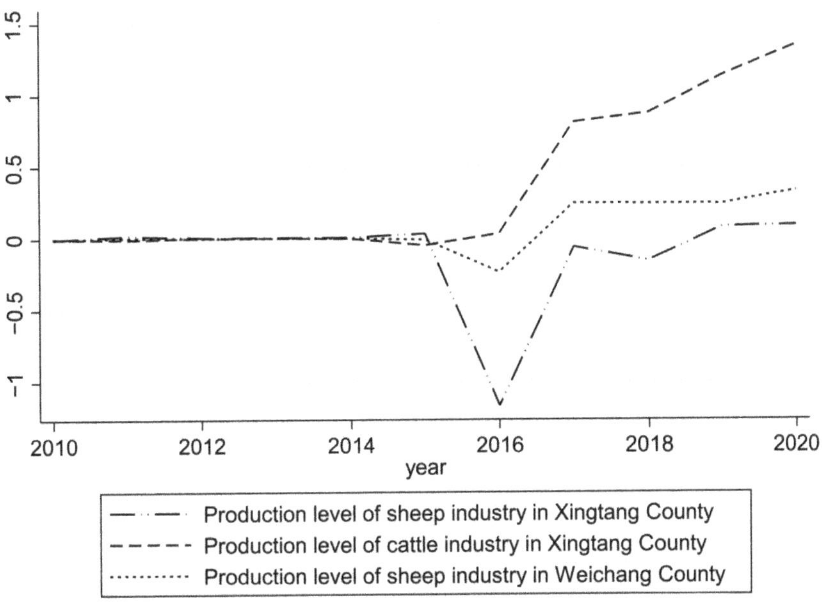

Figure 11. Effect of the CFFP on various industries' production levels.

Second, under the conditions of a market economy, the increasing consumer demand for domestic production to put forward higher requirements, including Beijing, Tianjin, and Hebei urban residents, whose per capita consumption of beef and milk per household has continued to increase over the years. The production behavior of Hebei Province, as an important supply base for livestock products in Beijing, Tianjin, and Hebei, is bound to change accordingly according to demand [38].

At the same time, combined with the policy work ideas and the actual situation, we also found that the direct beneficiaries of the policy are mainly large-scale entities (including large-scale farms, forage harvesting, and storage enterprises, etc.), while ordinary farmers, who account for a relatively large proportion of China's agricultural production and operations, do not directly benefit from the policy implementation. The policy aims to play a guiding role in financial funds, mobilize the enthusiasm of farmers in forage planting in the front end through market mechanisms, enhance the efficiency of herbivorous livestock breeding in the back end, and promote the development of herbivorous livestock breeding. However, in the market economy, given that farmers are rational economic people, in the absence of obvious interest guidance and policy inclination, there are still problems such as weak planting stability, poor forage quality, and the degree of planting–feeding combination to be improved, which will be further transferred to the breeding process [39,40]. Although the study did not specifically investigate the impact of policy implementation on general farmers, this study still has important implications for general farmers. On the one hand, the study clarified the policy implementation ideas, involving the pathways of policy implementation on forage cultivation and herbivorous livestock breeding, which is beneficial for farmers to understand the implementation content and objectives of the policy. On the other hand, the study discusses the importance of ordinary farmers to policy implementation and the neglect of policy implementation by ordinary farmers. As the direct response subject of the policy, farmers' behavioral response is the premise and foundation for the sustainable and effective implementation of the policy, which also calls for the policy to continually optimize the policy implementation content (such as broadening the scope and use of subsidies, focusing on the policy to improve the interest linkage mechanism of breeding and raising subjects, etc.), so that ordinary farmers can share the fruits of industrial development under the policy implementation.

Due to the limitation of space and focus, the study focused on the two main types of herbivorous livestock, cattle and sheep, and measured the livestock production capacity in terms of annual livestock slaughter, ignoring to a certain extent the policy effects of dairy industry development. On the other hand, the article focuses mainly on the level of herbivorous livestock production and lacks the analysis of modern livestock development indicators, such as scale, standardization, quality, and safety in the context of modern agriculture. On the basis of taking into account various levels of livestock production, the research will continue to investigate the policy effects of regional livestock production structure, scale, and production efficiency.

5. Conclusions

On the basis of CFFP implementation and herbivorous livestock development in the pilot counties of Xingtang and Weichang in Hebei Province, the study empirically analyzed the impact of CFFP implementation on herbivorous livestock development in the pilot counties using the synthetic control method. It was found that the CFFP had a significant positive effect on the improvement of herbivorous livestock production levels in the pilot counties as a whole and was conducive to enhancing the contribution of the pilot counties to the development level of the herbivorous livestock industry in Hebei Province, while differences in policy effects and change trends among different types of pilot counties and different livestock species were also found. The research provides the theoretical basis for the continued promotion of CFFP and provides direction for the subsequent optimization of the policy. The implementation of the policy should focus on the coordinated development among livestock species on the basis of regional advantages; focus on the stability of policy implementation and the rationality of project subsidies; reasonably guide farmers' policy expectations and stimulate their enthusiasm for participation; and innovate policy content and subsidy methods to allow ordinary farmers to share the policy dividends.

Author Contributions: H.Z. contributed to conceptualization, data curation, formal analysis, software, writing—original draft, and writing—review and editing; G.W. contributed to funding acquisition, supervision, and writing—editing; J.L. contributed to data curation and editing; S.H. (Shuai Hao) and S.H. (Shengnan Huang) contributed to writing—original draft preparation and supervision. All authors have read and agreed to the published version of the manuscript.

Funding: This research was funded by The Major Project of National Social Science Foundation of China (No. 21ZDA056); The National Natural Science Foundation of China (No. 41871184); and The Science and Technology Innovation Project of Chinese Academy of Agricultural Sciences (No. 10-IAED-01-2022).

Institutional Review Board Statement: Not applicable.

Data Availability Statement: The data can be found according to the corresponding data source. Scholars requesting more specific data may email the corresponding author or the first author.

Conflicts of Interest: The authors declare no conflict of interest.

References

1. Nardone, A.; Ronchi, B.; Lacetera, N.; Ranieri, M.S.; Bernabucci, U. Effects of climate changes on animal production and sustainability of livestock systems. *Livest. Sci.* **2010**, *130*, 57–69. [CrossRef]
2. Fang, J.Y.; Bai, Y.F.; Li, L.H.; Jiang, G.M.; Huang, J.H.; Huang, Z.Y.; Zhang, W.H.; Gao, S.Q. Scientific basis and practical ways for sustainable development of China's pasture regions. *Chin. Sci. Bull.* **2016**, *61*, 155–164. (In Chinese) [CrossRef]
3. Yu, X.H. Meat consumption in China and its impact on international food security: Status quo, trends, and policies. *J. Integr. Agric.* **2015**, *14*, 989–994. [CrossRef]

4. Li, S.B.; Li, X.; Ma, Q.L.; Wang, Z.Y.; Fang, F.; Zhang, D.Q. Consumer preference, behaviour and perception about lamb meat in China. *Meat Sci.* **2022**, *192*, 108878. [CrossRef] [PubMed]
5. Abbasi, T.; Abbasi, T.; Abbasi, S.A. Reducing the global environmental impact of livestock production: The minilivestock option. *J. Clean. Prod.* **2016**, *112*, 1754–1766. [CrossRef]
6. Wang, P.; Deng, X.Z.; Jiang, S.J. Diffused impact of grassland degradation over space: A case study in Qinghai province. *Phys. Chem. Earth* **2017**, *101*, 166–171. [CrossRef]
7. Bai, Y.P.; Deng, X.Z.; Zhang, Y.; Wang, C.; Liu, Y. Does climate adaptation of vulnerable households to extreme events benefit livestock production? *J. Clean. Prod.* **2019**, *210*, 358–365. [CrossRef]
8. Hou, L.Y.; Bai, W.M.; Zhang, Q.Q.; Liu, Y.H.; Sun, H.L.; Luo, Y.L.; Song, S.H.; Zhang, W.H. A new model of two-sown regime for oat forage production in an alpine region of northern China. *Environ. Sci. Pollut. Res.* **2022**, *29*, 70520–70531. [CrossRef]
9. Martin, N.P.; Russelle, M.P.; Powell, J.M.; Sniffen, C.J.; Smith, S.I.; Tricarico, J.M.; Grant, R.J. Invited review: Sustainable forage and grain crop production for the US dairy industry. *J. Dairy Sci.* **2017**, *100*, 9479–9494. [CrossRef]
10. Uddin, M.E.; Wattiaux, M.A. Effect of source and level of forage in the diet on in vitro ammonia emission from manure of Holstein and Jersey dairy cows. *JDS Commun.* **2021**, *2*, 16–20. [CrossRef]
11. Neal, J.S.; Eldridge, S.M.; Fulkerson, W.J.; Lawrie, R.; Barchia, I.M. Differences in soil carbon sequestration and soil nitrogen among forages used by the dairy industry. *Soil Biol. Biochem.* **2013**, *57*, 542–548. [CrossRef]
12. Fukase, E.; Martin, W. Who Will Feed China in the 21st Century? Income Growth and Food Demand and Supply in China. *JAE J. Agric. Econ.* **2016**, *67*, 3–23. [CrossRef]
13. Xin, L.J.; Wang, L.X.; Liu, A.M. Regional Production and Consumption Equilibrium of Feed Grain in China and Its Policy Implication. *J. Nat. Resour.* **2018**, *33*, 965–977. (In Chinese) [CrossRef]
14. Lin, Y.; Xiong, X.Z.; Samim, S.A.; Hu, Z.Q. Analysis of Water Resources and Water Environmental Carrying Capacity of Animal Husbandry in China—Based on Water Footprint Theory. *Water* **2021**, *13*, 3386. [CrossRef]
15. Shi, R.B.; Irfan, M.; Liu, G.L.; Yang, X.D.; Su, X.F. Analysis of the impact of livestock structure on carbon emissions of animal husbandry: A sustainable way to improving public health and green environment. *Front. Public Health* **2022**, *10*, 835210. [CrossRef]
16. Hou, L.L.; Fang, X.; Chen, Q.H.; Huang, J.K.; He, Y.; Rose, N.; Rozelle, S. Grassland ecological compensation policy in China improves grassland quality and increases herders' income. *Nat. Commun.* **2021**, *12*, 4683. [CrossRef]
17. Wang, H.; Hu, Y.F.; Yan, H.M.; Liang, Y.T.; Guo, X.; Ye, J.Z. Trade-off among grain production, animal husbandry production, and habitat quality based on future scenario simulations in Xilinhot. *Sci. Total Environ.* **2022**, *817*, 153015. [CrossRef]
18. Huang, H.; Hou, M.Y.; You, S.H. Urbanization and Grain Production Pattern of China: Dynamic Effect and Mediating Mechanism. *Agriculture* **2022**, *12*, 539. [CrossRef]
19. Han, C.J.; Wang, G.G.; Zhang, Y.X.; Song, L.L.; Zhu, L.Z. Analysis of the temporal and spatial evolution characteristics and influencing factors of China's herbivorous animal husbandry industry. *PLoS ONE* **2020**, *15*, 0237827. [CrossRef]
20. Abadie, A.; Diamond, A.; Hainmueller, J. Synthetic Control Methods for Comparative Case Studies: Estimating the Effect of California's Tobacco Control Program. *J. Am. Stat. Assoc.* **2010**, *105*, 493–505. [CrossRef]
21. Liu, Y.J.; Zeng, X.M. Research on the Influence of Industrial Transfer from the Property Taxes: Empirical Research from Chongqing and Shanghai. *China Ind. Econ.* **2018**, *11*, 98–116. [CrossRef]
22. Su, Z.; Hu, D. Is Inflation Targeting Effective? -New Evidence from the Synthetic Control Methods. *Econ. Res. J.* **2015**, *50*, 74–88, ISBN 0577-9154.
23. Cui, S.L.; Li, Y.J.; Jiao, X.Q.; Zhang, D. Hierarchical Linkage between the Basic Characteristics of Smallholders and Technology Awareness Determines Small-Holders' Willingness to Adopt Green Production Technology. *Agriculture* **2022**, *12*, 1275. [CrossRef]
24. Wu, G.Y.; Cheng, J.W.; Yang, F. The Influence of the Peer Effect on Farmers' Agricultural Insurance Decision: Evidence from the Survey Data of the Karst Region in China. *Sustainability* **2022**, *14*, 11922. [CrossRef]
25. Wallander, S.; Bowman, M.; Beeson, P.; Claassen, R. Farmers and Habits: The Challenge of Identifying the Sources of Persistence in Tillage Decisions. In Proceedings of the 2018 Allied Social Sciences Association (ASSA) Annual Meeting, Philadelphia, PA, USA, 5–7 January 2018; pp. 1–42.
26. Caldas, M.M.; Bergtold, J.S.; Peterson, J.M.; Graves, R.W.; Earnhart, D.; Gong, S.; Lauer, B.; Brown, J.C. Factors affecting farmers' willingness to grow alternative biofuel feedstocks across Kansas. *Biomass Bioenergy* **2014**, *66*, 223–231. [CrossRef]
27. Van Meter, D.S.; Van Horn, C.E. The Policy Implementation Process: A Conceptual Framework. *Adm. Soc.* **1975**, *6*, 445–488. [CrossRef]
28. Thornton, P.K. Livestock production: Recent trends, future prospects. *Philos. Trans. R. Soc. B Biol. Sci.* **2010**, *365*, 2853–2867. [CrossRef]
29. Meng, J.K. Differences in manure management and COD, total nitrogen and total phosphorus emission laws of different scale farms in Hebei Province. *Hebei Agric. Univ.* **2019**, *3*, 57. [CrossRef]
30. Yi, Q.; Song, K.H.; Hu, T.; Ying, T.Y. Environmental status of livestock and poultry sectors in China under current transformation stage. *Sci. Total Environ.* **2018**, *622–623*, 702–709. [CrossRef]
31. Zhao, Z.; Chen, J.C.; Bai, Y.P.; Wang, P. Assessing the sustainability of grass-based livestock husbandry in Hulun Buir, China. *Phys. Chem. Earth* **2020**, *120*, 102907. [CrossRef]
32. Yu, F.W.; Huang, X.; Wang, G.L. High-quality Development of Animal Husbandry: Theoretical Interpretation and Realization Path. *Chin. Rural. Econ.* **2021**, *4*, 85–99, ISBN 1002-8870.

33. Han, C.J.; Wang, G.G.; Yang, H.B. Agricultural Areas: A Case Study of Najitun Farm of Hulunbuir Agricultural Reclamation in China. *Land* **2022**, *11*, 691. [CrossRef]
34. Huang, J.K.; Ding, J.P. Institutional innovation and policy support to facilitate small-scale farming transformation in China. *Agric. Econ.* **2016**, *47*, 227–237. [CrossRef]
35. Cai, J.Y.; Zhang, L.G.; Tang, J.; Pan, D. Adoption of Multiple Sustainable Manure Treatment Technologies by Pig Farmers in Rural China: A Case Study of Poyang Lake Region. *Sustainability* **2019**, *11*, 6458. [CrossRef]
36. Mao, C.C.; Ma, Z.X. The Analysis of The Regional Economic Growth And The Regional Financial Industry Development Difference In China Based On The Theil Index. *Int. J. Econ. Financ. Stud.* **2021**, *13*, 128–154. [CrossRef]
37. He, G. Study on the Development of Modern Animal Husbandry in Xinjin County. Advances in Social Science, Education and Humanities Research. *Atlantis Press* **2020**, *402*, 2352–5398, ISBN 978-94-6252-904-5.
38. Mao, Y.W.; Hopkins, D.L.; Zhang, Y.M.; Luo, X. Consumption Patterns and Consumer Attitudes to Beef and Sheep Meat in China. *Am. J. Food Nutr.* **2016**, *4*, 30–39. [CrossRef]
39. Liu, S.W.; Zhang, P.Y.; Marley, B.; Liu, W.X. The Factors Affecting Farmers' Soybean Planting Behavior in Heilongjiang Province, China. *Agriculture* **2019**, *9*, 188. [CrossRef]
40. Greiner, R. Motivations and attitudes influence farmers' willingness to participate in biodiversity conservation contracts. *Agric. Syst.* **2015**, *137*, 154–165. [CrossRef]

agriculture

MDPI

Article

Scenarios for Sustainable Farming Systems for Macadamia Nuts and Mangos Using a Systems Dynamics Lens in the Vhembe District, Limpopo South Africa

Fenji Materechera * and Mary Scholes

School of Animal, Plant and Environmental Sciences (APES), University of the Witwatersrand, Johannesburg 2000, South Africa
* Correspondence: fenji.materechera@students.wits.ac.za

Abstract: Agriculture is arguably one of the most important economic sectors for South Africa's development as it is directly linked to food security. Farming systems in South Africa have been characterized by a duality where large-scale commercial farmers and small-scale farmers co-exist. The conventional approach to understanding agricultural production in the country has always viewed the two farming systems as mutually exclusive. The study argues that there are various points of interaction between the two kinds of farmers and by using a systems dynamics approach to evaluate the two farming systems this can be applied to agricultural decision making. Data were used to identify and characterise small- and large-scale farming systems of two tree crops (mangos—*Mangifera indica* L. and macadamia nuts—*Macadamia integrifolia* M&B.) in the Vhembe district of Limpopo South Africa. The interactions between the two different farmers are illustrated using Causal Loop Diagrams (CLDs) of the two farming systems under similar commodities. Results, presented as four conceptual scenarios, show that there are multiple points of interaction, such as the interdependence of farmers of macadamia nuts to meet export demands. Policy recommendations to strengthen collaboration between small-scale mango farmers and implement irrigation expansion for farmers who depend on rain-fed farming are discussed and present opportunities for the co-functioning of the two farming systems.

Keywords: interactions; scenarios; systems dynamics; agriculture; South Arica

Citation: Materechera, F.; Scholes, M. Scenarios for Sustainable Farming Systems for Macadamia Nuts and Mangos Using a Systems Dynamics Lens in the Vhembe District, Limpopo South Africa. *Agriculture* **2022**, *12*, 1724. https://doi.org/10.3390/agriculture12101724

Academic Editor: Giuseppe Timpanaro

Received: 5 September 2022
Accepted: 17 October 2022
Published: 19 October 2022

1. Introduction

Agricultural productivity plays a crucial role in South Africa's food system and sustaining the country's food security. Farming systems in South Africa are characterized by a dichotomy where large-scale commercial farmers and small-scale farmers co-exist. This is part of the legacy of the apartheid system which relegated small-scale farmers to small portions of poor-quality land in what are known as the former homeland areas or Bantustans. The result of this has been the parallel functioning of these two kinds of farmers within the context of continuous change. Large-scale commercial agriculture is regarded as the main driver of national food security in South Africa [1]. In contrast to this, economically, small-scale agriculture in South Africa enhances local economic development as it is a source of employment and keeps most of the income local as the market is predominantly localised [2]. Hendriks [3] suggests that small-scale agriculture contributes to food security at a household level as socially, especially on traditional lands, the produce is first meant to feed the household. The two farming systems are therefore indispensable. According to Dixon et al., (2001) [4] farming systems are defined as " … *a population of individual farm systems that have broadly similar resource bases, enterprise patterns, household livelihoods and constraints, and for which similar development strategies and interventions would be appropriate. Depending on the scale of the analysis, a farming system can encompass a few dozen or many millions of households.*"

High value horticultural crops are becoming increasingly significant to the South African agricultural economy as there is a demand for them on the global market [5]. Some of the most popular high value crops grown in the country that are in high demand are avocados, mangos, litchis, pecan nuts, papayas, bananas and macadamia nuts. A number of these are cultivated in the Limpopo province of South Africa due to a favourable subtropical climate. Both large- and small-scale farmers are engaged in farming high value crops in the province intended for both export and supply to local markets. Land, though not the only driver, is a key driver of agricultural production in South Africa [6]. There are numerous factors pertaining to land that impact farmers' ability to successfully produce and contribute to the country's food system which include land tenure and its associated rights, soil quality, vegetation, topography, rainfall variability and water availability amongst various others. Government policy interventions and cross-sectoral initiatives have been targeted towards addressing these land factors with increasing focus on how small-scale farmers are affected by them. Examples of government policy initiatives geared at addressing land as a driver of production include the Upgrading of Land Tenure Rights Act, Act No. 112 of 1991 and the Restitution of Land Rights act, Act No. 22 of 1994 which was later amended in 2014 [7].

The common understanding of the context of South African farming systems is that the two farmers operate farming systems that are mutually exclusive. The current study challenges this notion by suggesting that the two kinds of farmers do interact on various levels, and this can be seen in the case of agriculture in the Vhembe district of Limpopo, South Africa. According to Labadarios et al., (2011) [8] one of the characteristics of farming systems is that they are able to produce the same outcomes in different ways provided they are exposed to similar conditions. It can be inferred that the variables and processes that comprise these systems will be different based on the scale at which the farmers operate therefore necessitating management practices that are specific to the farming systems. However, farming systems that exist in the same geographical location and may be exposed to the same vagaries such as extreme weather events due to climate change in South Africa, may experience overlap in the processes and management practices that are employed.

By using a systems lens to view farming systems in the Vhembe district of Limpopo, it is possible to conceptualize the future of South African agriculture and its contribution to food security by considering the plausibility of coupling the two kinds of farming systems. This is the conceptual basis of this study. The study aims to highlight the connectivity between the two main farming systems in South Africa using systems analysis as a tool for understanding. To this end, this paper will address two objectives, namely to identify the interactions between the two farming systems using Causal Loop Diagrams (CLDs) and to develop conceptual scenarios for the co-functioning of the two farming systems under similar commodities farmed in the Vhembe District of Limpopo South Africa. In understanding the nature of the interactions within and between the two farming systems it is possible to determine the feasibility of the two systems being coupled to achieve the goal of jointly meeting the country's food security needs at all levels. Scenarios can inform future decision making, research and policy recommendations.

2. Materials and Methods

2.1. The Study Area

Limpopo is one of the largest crop producing areas in South Africa and can be regarded as a key agricultural hub. The study took place in the Vhembe district which is a district municipality located in the Northern most region of the Limpopo province of South Africa (Figure 1).

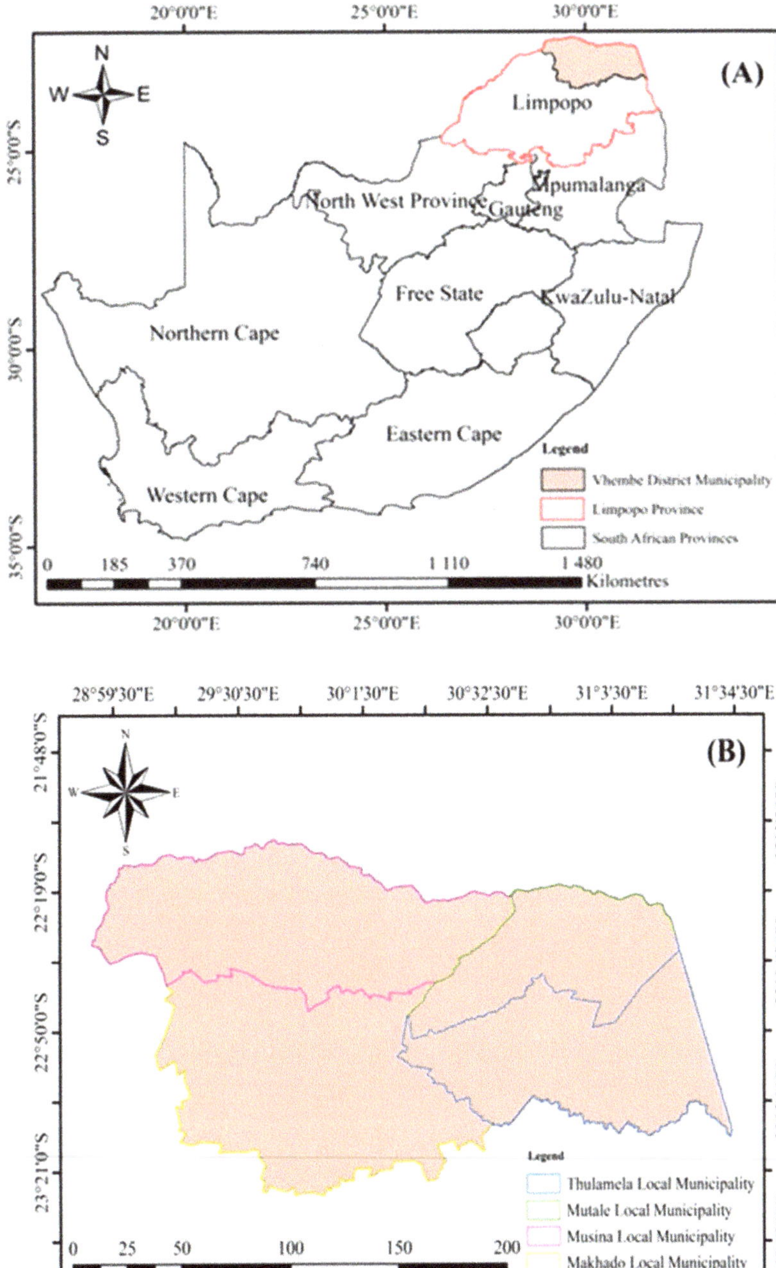

Figure 1. Map showing the location of (**A**) The Republic of South Africa' provinces and provincial boundaries, highlighting the location of the Limpopo province and the Vhembe district within the Limpopo province of South Africa and (**B**) shows the location of the four local municipalities within the Vhembe district.

The Vhembe district borders with Zimbabwe and Botswana to the north–east and Mozambique to the south–east passing through portions of the Kruger National Park [9]. The Limpopo province is comprised of five district municipalities of which the Vhembe district is one. The Vhembe district is further sub-divided into four local municipalities namely: Musina, Makhado, Mutale (renamed Collins Chabane) and Thulamela. According to the South African governance structure local municipalities are constituted by towns and their surrounding rural areas [10].

The Vhembe district has a land area of 2,140,708 ha of which only 247,757 ha is arable [11]. Agriculture is central to the livelihoods of the people in the Vhembe district and is a key contributor to employment. A reported 90% of rural communities located in the Vhembe district are dependent on agriculture to generate household income and sustain their livelihoods [12]. This is aligned with the geographical context of the Vhembe district where the district is located in an area that is predominantly rural [13]. Smallholder agriculture accounts for 70% of farming activities in the district while commercial agriculture accounts for the remaining 30% [14–16]. Numerous subtropical crops that contribute significantly to South Africa's agricultural economy particularly through exports are produced in the Vhembe district. Amongst these crops, are included commodities such as mangos, avocados, bananas, litchis, macadamia and pecan nuts. The census of commercial agriculture in 2017 recorded subtropical fruit and citrus as the biggest crop output in the district [17]. The Vhembe District Municipality's Local Economic Development Strategy in 2019 [18] reported that the Vhembe district produces 8.4% of the country's sub-tropical fruits and 6.3% of its citrus; overall amounting to 4.4% of South Africa's total agricultural output. Kom et al. [19] indicates that it is the southern side of the district i.e., the local municipalities of Thulamela and Makhado that is typically comprised of well-established white commercial horticulture farming. In contrast to this, the northern side is mostly semi-arid and is mainly utilized for livestock farming and game ranching. Horticulture in the northern region is very limited and restricted to areas where water is available.

In terms of water availability for agriculture, geographically the Vhembe district is located in a semi-arid area. Occasional droughts usually occur from May to August [13]. According to [20,21] small-scale farmers in the district mostly practice rainfed agriculture relying on seasonal rainfall which typically falls between November and March. Moeletsi et al., (2013) [20] documents that the average seasonal rainfall for the southern side of the district, identified earlier as the horticulture hub, ranges from 400 mm to 600 mm. With regard to soils, according to [14] the soils found in the southern region of the district vary significantly from one place to another; those with a higher clay and loam content tend to be found in the east and more sandy soils towards the west.

2.2. Study Design

Primary and secondary data were used to identify and characterise both small- and large-scale farming systems of three tree crops in the study area i.e., avocados, mangos and macadamia nuts. For the purpose of this paper avocados were excluded in the discussion as there was significant overlap between the interactions between large- and small-scale farmers of avocados and macadamia nuts in the study area. Mangos and macadamia nuts were selected as there were substantial differences in the interactions between the farmers of these tree crops which provided a useful means for comparison between commodities that enrich the discussion of the paper. Analysis was aimed at highlighting the connectivity of interactions within and between the two main farming systems with respect to the four drivers of production namely land, labour, capital and enterprise. Secondary data were derived from the official database of subtropical crops from the local Department of Agriculture, soil data and land type maps from the Agricultural Research Council (ARC), climate data from the Institute of Soil, Climate and Water (ISCW), related peer reviewed research papers and books. The target population was comprised of a combination of large-scale commercial and small-scale farmers of the three tree crops in the district. Initially, farms were selected based on data extracted from the subtropical database. Using a purposive sampling method [21], the criteria for site selection were determined, namely commodity, farm size, gender of the farmer and farm location (village, town and municipality). This information was available for six subtropical commodities, namely bananas—*Musa paradisiaca* L. (23); litchis—*Litchi chinensis* S. (92); avocados—*Persea americana* M. (204); mangos—*Mangifera indica* L. (528); macadamia nuts—*Macadamia integrifolia* M&B. (184); and citrus—*Citrus sps* L. (90). According to the database there are a total of 1121 documented subtropical crop farmers in the Vhembe district. The database also showed that the three commodities selected in the study were the most commonly grown commodities in the district. Mangos were selected because they formed the largest number of farms documented in the database (528 farmers). Macadamias were selected based on their significance to the South African agricultural economy as high value export crops.

Thereafter, a systematic random sampling procedure was used to select farms to ensure equal representation of farm size. This was done because the study required both farmers with smallholdings and larger holdings. Initially three size categories based on the sizes that exist in the database were selected namely, small (1–5 ha), medium, (6–13 ha) and larger (14–20 ha and above). This was later narrowed to two categories i.e., small-scale (1–10 ha) and large-scale (11 ha and above) as these provided a continuum that was context specific to the study. The classification of small-scale farmers in the South African context is complex as size is not the only factor used to determine what constitutes a small-scale farm. Other factors such as enterprise, level of mechanization and technology employed, income from farming etc. are also taken into consideration [22]. This is further reflected in the use of numerous terms to describe these kinds of farmers such as subsistence, semi-commercial, emerging etc. [16]. For this reason, the researchers used their own criterion of size to classify small-scale farmers for the specific purpose of this study. In terms of location, farms were selected that reflected equal representation of the four local municipalities located within the Vhembe district namely Musina, Makhado, Thulamela and Mutale in order to provide a comprehensive overview of farming in the district. Lastly, with regard to the criterion of gender of the farmers, a random number generation method was used to ensure equal representation of the genders across all farms. This was achieved by allocating each farmer a number using the previously mentioned criteria and placing the written numbers in a container. Numbers were then randomly picked out by the researcher to add up to a total of 12 farms. Twelve farms were selected and were made up of four samples of each of the three tree crops across the four municipalities with two small-scale and two large-scale farms and an equal distribution of male and females. After completing the site selection, a more detailed characterisation of the two farming systems based on the three commodities in relation to the four factors of production followed. Primary data were obtained by way of in-depth, on-site interviews with individual farmers. Using a snowball sampling method [23] interviews were conducted with the aim of maintaining the originally selected sample size, The result of the snowball sampling technique that was employed produced the following samples: avocados (8), macadamia nuts (7) and mangos (4). In total, 19 farmers were selected for participation in the in-depth interviews based on their willingness to participate and availability. Due to numerous challenges in accessing farms based on their extremely rural locations, data were collected at only one point in time. This influenced the exceptionally small sample size which the authors acknowledge. For the purpose of this paper the sample refers to a total of 11 farmers (7 macadamia nut farmers and 4 mango farmers).

2.2.1. Data Collection

Face-to-face farmer interviews were conducted over the duration of the two visits to the Vhembe district between October and November 2020. Ethical clearance was obtained through the University of the Witwatersrand ethics committee (protocol number: H19/09/26). Clearance was also obtained from the local Department of Agriculture through an official letter of approval. A questionnaire was used as the main data collection instrument comprised of closed and open-ended questions with the aim of collecting qualitative and quantitative data. Demographic information about the farmer was obtained through the questionnaire to obtain statistical data. Detailed information about various aspects of the four drivers of production in the context of the selected farm sites was also obtained. Open-ended questions were used to obtain more detailed responses from participants while close-ended questions were used to gather statistical information. The questionnaire was sub-divided into four key sections: land, labour, capital and enterprise.

Interviews were conducted by the researcher alongside a local who served as an interpreter due to language barriers. Interviews were mostly conducted in the local language of Vhenda. Key informant interviews were conducted with managers from processing plants for macadamia nuts and avocados. Study group meetings for the respective commodities were attended by the researcher in order to develop scenarios. These were information sharing sessions with various stakeholders (farmers, equipment suppliers, extension officers, researchers, grower association representatives, and government officials from the local Department of Agriculture) that allowed for interaction.

2.2.2. Causal Loop Diagram (CLD) Construction

The causal loop diagram (CLD) analytical tool used to represent the relationship between system variables and their dynamic feedback structures was constructed using Vensim modelling software (Ventana Systems Inc. 60 Jacob Gates Road Harvard, MA, 01451, USA, http://www.ventanasystems.com/, accessed on 4 September 2022) [24]. The overall structure of the CLD represents the links between large-scale and small-scale farmers of the two commodities (macadamia nuts and mangos) and the broader farming system. The CLDs hypothesize system behaviour and identify balancing and reinforcing feedbacks.

2.2.3. Development and Analysis of Scenarios

Scenarios were constructed using a scenario method known as 'deductive' [25] The deductive process uses multiple iterations of scenario drafts that are typically developed through stakeholder engagement facilitated through workshops [25]. Workshops allow for scenario deconstruction and revision which provide an opportunity to validate the plausibility of the scenarios. In the current study scenarios were created drawing from three iterations of scenario drafts based on (1) predominant themes on production issues arising within the farming systems of the selected commodities based on farmer interviews and secondary data (2) key informant interviews and (3) interactive study group meetings. Each narrative provided as much detail as possible with the aim of developing equally plausible futures based on a chosen time frame of 10 years (2022–2032). The 10-year time frame was preferred over a longer projected time as it provides a more realistic timeframe to imagine plausible futures based on current trends.

The process of creating scenarios involved three steps. In the first step key drivers and uncertainties surrounding production that emerged from interview responses were noted. Secondly, notes from conversations with key informants who were interviewed during field visits were used to give further detail to what these key drivers are which produced an outline of the narrative for each scenario. Key informants included technical managers for processing plants of macadamia nuts (Green Farms Nut Company and The Royal Macadamia) and representatives from the respective growers' associations (Macadamias South Africa i.e., SAMAC and the South African Avocado Growers Association i.e., SAAGA. Lastly, primary information was obtained from multistakeholder participation at local information sharing sessions termed study group meetings for the respective commodities which the researcher attended during the duration of the field work. These were used to further corroborate what the key drivers and uncertainties are and to produce the final two storylines outlining plausible scenarios for farming systems of the respective commodities in the Vhembe district by the year 2032. This iterative process is illustrated in Figure 2 below. The key driving forces of production were continuously narrowed with each iteration based on the most common themes recurring from feedback from the different participants. Themes were used as direction for what the key issues that the scenario would address should be. A 2 × 2 quadrant of four key drivers based on a scale of uncertainty vs. impact ranging from low to high (Figure 2) was used to establish what the main subject of the scenarios would be. Issues for which farmers' responses reflected a high level of uncertainty were typically used as conversation points during information sharing sessions to probe what kind of solutions could be explored to address the challenge. These aided the writing of the scenario narratives. The interactions between farmers highlighted in the CLDs were used to evaluate the scenarios.

2.2.4. Data Analysis

Descriptive statistics were used for the analysis of quantitative data [26] by calculating percentages, averages and standard errors. Chi-squared and student t-tests [27] were used to compare the means across the two farm sizes and between the three commodities. Thematic analysis was used to analyse qualitative data [28]. Participants responses to open-ended questions concerning land variables relevant to the different commodities were transcribed. Thereafter recurring responses that were mentioned were identified as major themes. Based on the themes, percentages were calculated to classify them in order of importance. Predominant themes were triangulated with quantitative data from the questionnaire and secondary data to explain phenomenon.

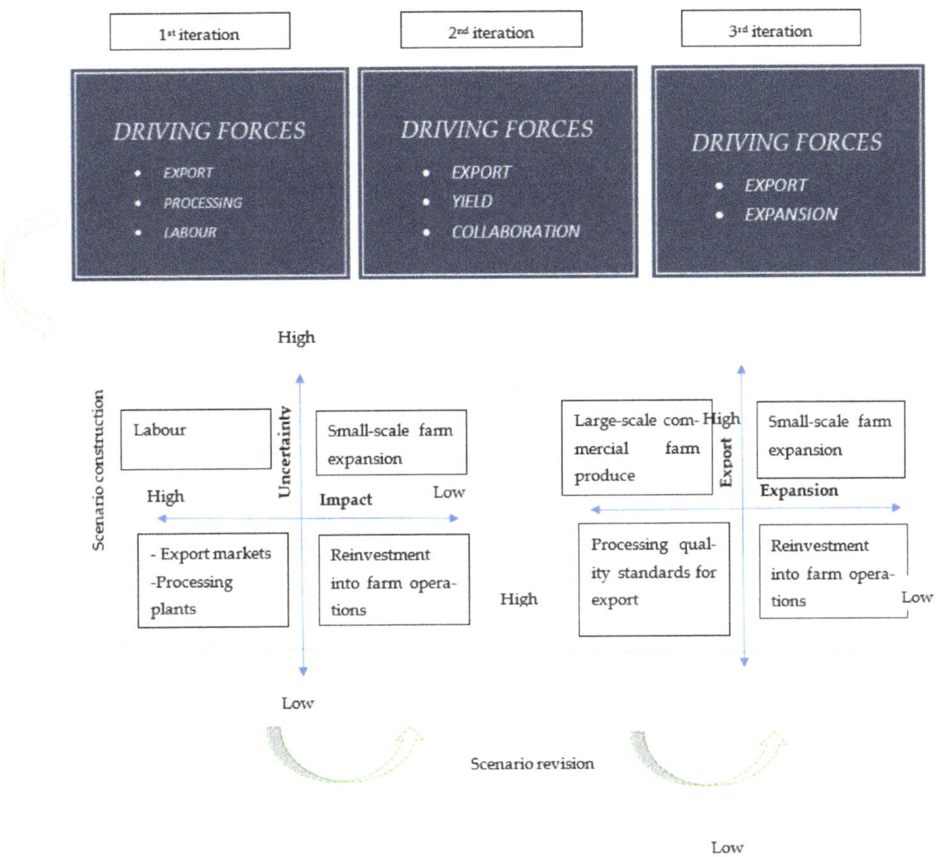

Figure 2. An example of the three iterations of the scenario development using the 2 × 2 quadrant of uncertainty vs. impact used for scenario construction for macadamia nuts based on the key drivers and uncertainties identified from interviews. Adapted with permission from Rameriz et al. [29] 2022. Rafael Ramirez.

2.3. Conceptual Framework

The two main farming systems that currently exist in South Africa can be found across the country. According to the FAO [30], both farming systems extend across the northern part of the country where the Limpopo province is located. The two farming systems are impacted by the same drivers of production i.e., land, labour, capital and enterprise [31], however respond to these drivers differently. The manner in which the two farming systems respond to the drivers of production may reveal the connectivity between the systems. A systems thinking approach is best suited to illustrate the connectivity between the farming systems and is therefore used for the study. Arnold and Wade [32] define systems thinking as *"a system of thinking about systems"*. The same authors make the assertion that systems thinking must have three components in order to be defined namely elements i.e., characteristics, interconnections i.e., the way these interactions feed back into each other or relate, and a function or purpose. System dynamics (SD) is the understanding of the relationship between integrated systems elements and how they impact each other's behaviour [33]. The integration of systems elements is done by the incorporation of concepts such as stocks, flows, feedbacks, and delays, enabling the analysis of the dynamic behaviour of the system elements over time [34]. The approach is used to describe, model, simulate, and analyse complex systems with multiple interacting elements in terms of processes, information, organisational boundaries, and strategies [35]. This conceptual understanding of systems thinking, and systems dynamics is applied in the current study as a means by which to understand the systems being analysed. The farming systems in South Africa operate against the backdrop of constantly changing economic, political, environmental and socio-economic conditions. This is the context in which the current research is positioned. Although farming systems research and farming systems analysis are well established research fields [36], little attention has been paid to how farming systems will respond to change in the future with respect to drivers of production. The study seeks to provide foresight into future farming systems in a developing country with constantly changing parameters. According to [37] when scenario analysis is used in environmental change research an important objective is exploration. Scenarios can potentially assist users to consider surprising discontinuities and developments. Scenarios are defined as *"a set of conceptual systems of equally plausible future contexts often presented as narrative descriptions typically for the purpose of providing inputs for future work"* [29]. By using scenarios derived from a systems thinking viewpoint as a tool, the study identifies four scenarios for production, for two different commodities, in farming systems in the Vhembe district of Limpopo South Africa.

3. Results and Discussion

Results are presented here in three sub-sections. Firstly, a general (for South Africa and the Vhembe district) and more detailed (by yield and income) characterization of macadamia nut and mango farming systems is presented. Secondly, CLDs are presented, and the reinforcing and balancing feedback loops are described to improve our understanding of the interconnected variables impacting production of the respective commodities in the Vhembe district. Lastly, scenario narratives derived from the key factors highlighted by the CLDs and the iterative process of scenario development are presented.

3.1. Characterization of Macadamia Nut and Mango Farming Systems

Results revealed that by 2019 South Africa was the largest macadamia producer in the world with 19,500 ha under cultivation, producing over 50,000 tonnes per year. Over 95% of South Africa's macadamia nut production is exported annually [38]. According to [39] the average yield for macadamia nuts in South Africa was 1.43 tonnes per hectare in 2019. Only 7% of macadamia nuts grown in the country are consumed by the local market. The Limpopo province is the second largest macadamia nut production area in the county after the Mpumalanga province, and the Vhembe district ranks third in order of the highest macadamia nut contributing districts in the province [12].

Results showed that mango production in South Africa has been unstable in recent years. In 2019 a total volume of 68,633 tonnes of mangos was produced in the country during that production season [40]. This may be attributed to unfavourable weather conditions. The industry makes an important contribution to direct employment in mango production and processing. In terms of the market structure, the annual crop is either sold fresh through the national fresh produce markets and as exports or processed into atchar, juice or dried mangos. The majority of mangos exported from the Limpopo province are mainly from the Mopani and Vhembe district municipalities respectively. The total export value reported by the Limpopo province was R62 million in 2019 of which R3 million was reportedly from the Vhembe district [40]. Table 1 is a summary of the characterization of the two sets of farms based on selected criteria from the farms selected in the study.

Table 1. Characterization of farm size, farm type, tonnage, yield and income by commodity for one year (2019).

Commodity	Farm Size (ha)	Farm Type		Tonnage (t)	Yield (t/ha)	Gross Annual Income (ZAR)
		Small-Scale	Large-Scale			
* Macadamia	4	√		0	0	0
Macadamia	5	√		2	0.4	10,000
Macadamia	5	√		2	0.4	150,000
Macadamia	6	√		4.2	0.7	200,000
Mean ± SD	5 ± 0.8			2.7 ± 1.3	0.5 ± 0.2	120,000
Macadamia	34		√	17	0.5	300,000
Macadamia	93		√	47	0.5	35,000,000
Macadamia	1600		√	806	0.5	40,000,000
Mean ± SD	575.7 ± 887.6			290 ± 447.1	0.5 ± 0	25,100,000
Mango	2	√		3	1.5	12,000
Mango	2	√		3	1.5	10,000
Mango	10	√		4	0.4	150,000
Mean ± SD	4.7 ± 4.6			3.3 ± 0.6	1.1 ± 0.6	57,333
Mango	15		√	4.5	0.3	20,000

* The first farmer appearing on the table was a first-time farmer who had planted trees 2 months prior to the interview and therefore did not have any yield to record.

The average gross annual income from farming amongst participants ranged between R10,000 and R40 million between the two commodities. Results revealed that macadamia farmers obtained the highest farming incomes, in both large-scale farms, average of R25,100,000, and small-scale, average of R120,000 compared to mango, R20,000 for the large-scale farmer and an average of R57,333 amongst small-scale farmers, farmers. Results of the Pearson Correlations analyses show that there is a positive statistically significant correlation between average gross annual income and farm size amongst macadamia farmers ($r = 0.763$, $p < 0.01$), and a positive significant correlation between average gross annual income and farm size amongst mango farmers ($r = 0.346$, $p < 0.01$).

Results showed that 79% of participants were male while 21% were female. The general gender profile of participants skewed towards male participants in both farm sizes and across the two commodities with only 25% of female participants who were mango farmers and no female macadamia farmers. This gender distribution is characteristic of the patriarchal context of the Limpopo province as presented in other studies conducted in the Vhembe district. This distribution is a reflection of the cultural norms and values of the Vhenda people who predominantly reside there where men generally tend to be the owners of the land. This can be viewed as a constraint, as the demographics of the broader province of Limpopo indicate that most small-scale farmers in the province are women as a result of adult males being involved in migrant labour. The small sample size obtained in the study limits detailed analysis of this aspect. The CLD for the Macadamia nut farming systems is presented in Figure 3. Followed by an explanation of the feedback loops that were identified as part of the analysis.

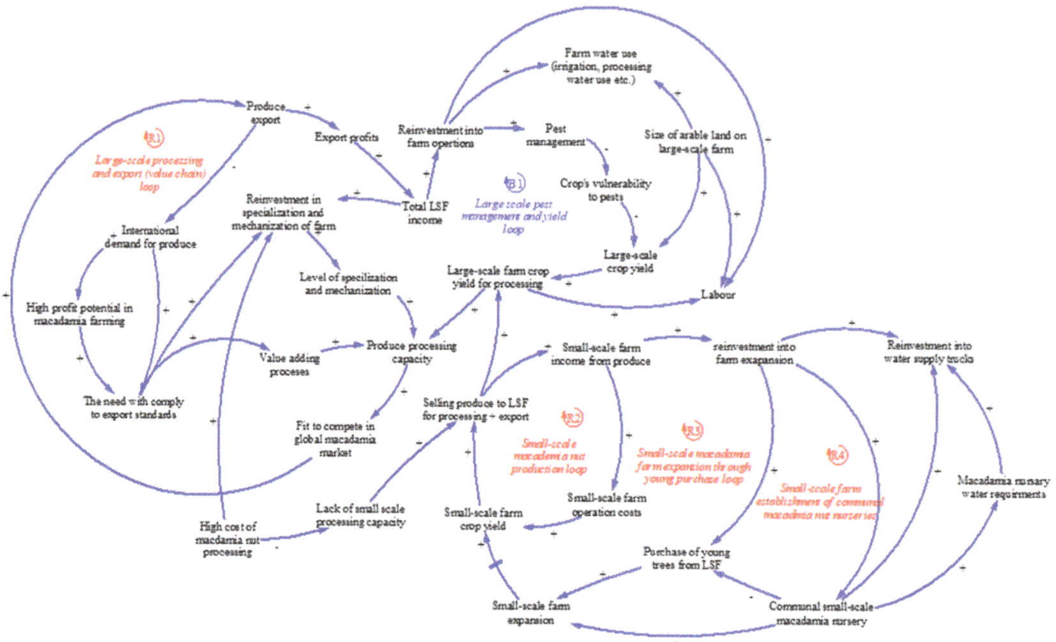

Figure 3. Causal loop diagram (CLD) showing macadamia nut farming systems in the Vhembe district, Limpopo. Arrows connect two or more variables of interest and are causal links that run in the stated direction. '+' = a positive relationship, indicating that the causality runs in the same direction (i.e., an increase in variable A will cause an increase in variable B and vice versa); '−' = a negative relationship, indicating that the causality runs in the opposite direction (i.e., an increase in variable A will cause a decrease in variable B and vice versa). The balancing feedback loops are numbered Bn and labelled in blue font. The reinforcing feedback loops are numbered Rn and labelled in red font. LSF refers to Large-scale farm/farming. Adapted with permission from Selebalo et al. [41] 2022. Itumeleng Selebalo.

3.2. *Macadamia Nut Farming Systems*

3.2.1. R1 Large-Scale Processing and Export Loop

Macadamia nuts are the fastest growing tree crop industry in the country and their production is lucrative. The demand for macadamia nuts globally is high and South Africa is currently the largest producer (in tonnes per hectare) in the world [39,42,43]. Large-scale macadamia farmers in the Vhembe district produce macadamia nuts for export and are also owners or partners in processing plants such as the Royal Macadamia located in Thohoyandou, Limpopo and Green Farms Nut Company in Levubu, Limpopo. Some profits from export sales are reinvested into farm operations of which pest management forms a component. The most common pest control strategy used by large-scale farmers is integrated pest management (IPM). Large-scale farmers contract experts to monitor their fields and thereafter recommend management interventions. This IPM approach combines techniques such as the use of resistant varieties, biological control and habitat manipulation etc., to effectively tackle pest problems. Crop vulnerability to pests such as stink bugs is decreased through investments into pest management which resultantly impacts the total annual yield positively. There is a positive causal link between the large-scale farm yields and the capacity to process the nuts for export. Large-scale farmers are able to meet processing quality standards, therefore making them fit to compete in global export markets and to make profit from export sales, thus reinforcing a cycle of export market participation.

3.2.2. R2 Small-Scale Macadamia Nut Production Loop

Small-scale macadamia farmers in the Vhembe district contribute to the macadamia nut value chain in the province reinforcing the interdependence of the two types of farmers. Nuts, produced by small-scale farmers, are transported and processed at plants owned by large-scale farmers as small-scale farms do not have the required equipment for processing and export requirements (indicated by the negative relationship "−" that is shown in the arrow between the variables high cost of macadamia nut processing and lack of small-scale processing capacity). Income made from nut sales is used to finance farm operational costs.

3.2.3. R3 Small-Scale Macadamia Farm Expansion through Young Tree Purchase Loop

Small-scale macadamia farmers in the Vhembe district use some of the profits from nut sales to reinvest in the expansion of their farms by purchasing young macadamia trees from large-scale commercial nurseries in the province (these are found in Tzaneen and are sold at a cost of R60/tree). The expansion of small-scale macadamia farms will positively impact the yield over time as trees mature; this is indicated by the delay in the CLD (the short blue line across the positive arrow between small-scale farm expansion and small-scale farm crop yield) as the causal link between farm expansion and yield is not immediate. Macadamias are long-term crops taking on average four to five years from planting before cropping commences and six to seven years before commercially viable yields are produced. This reinvestment of profits reinforces a loop of continuous farm expansion.

3.2.4. R4 Small-Scale Macadamia Farm Establishment of Communal Macadamia Nut Nurseries

Some small-scale farmers have opted to establish their own nurseries through planting trees from the yield of previous harvests and grafting. Small-scale farmers then sell young trees to other small-scale farmers within close proximity eliminating the transport costs to nurseries further afield. The establishment of macadamia nut nurseries fosters interaction and interdependence between small-scale farmers and promotes growth in the small-scale macadamia enterprise thus reinforcing a loop of continuous expansion.

3.2.5. B1 Large-Scale Pest Management and Yield Loop

Large-scale commercial macadamia nut farmers in the Vhembe district are able to reinvest income from export profits into pest management to control prevalent pests and diseases. The more farmers able to invest in integrated pest management programs, the less vulnerable the orchards become to invasion by pests. As pest vulnerability is continually reduced through pest management, yield is increased through larger numbers of trees able to produce nuts in a balancing loop in the CLD (Figure 4).

3.3. *Mango Farming Systems*

3.3.1. R1 Small-Scale Mango Farm Value Chain Loop

Small-scale mango farmers in the Vhembe district are the main suppliers of the local mango atchar (a pickled mango paste that is commonly eaten in the province forming part of a typically South African diet and sold at supermarkets) processing companies. Mango atchar processing factories (such as Gratchar located in Letaba, Mango Magic Atchar in Tzaneen and Levubu Atchar Veraardigers in Levubu) are located within the district therefore more easily accessible to the farmers. Farmers are able to make a profit from mango sales to atchar processing factories which are later reinvested into farm operations. An increased investment in farm operations results in better annual yields which ensures continued supply to processing companies creating a reinforcing loop.

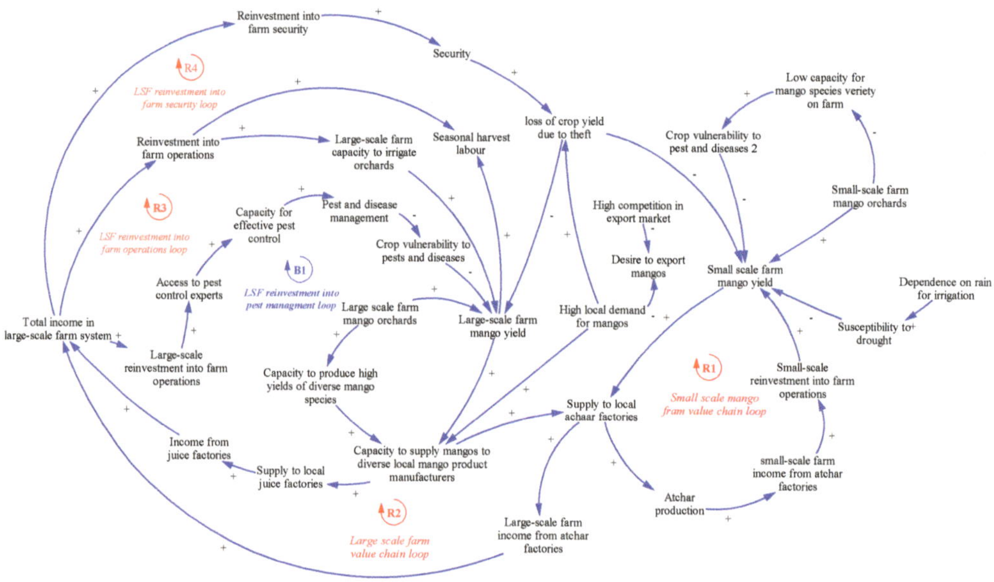

Figure 4. Causal loop diagram (CLD) showing mango farming systems in the Vhembe district, Limpopo. Arrows connect two or more variables of interest and are causal links that run in the stated direction. '+' = a positive relationship, indicating that the causality runs in the same direction (i.e., an increase in variable A will cause an increase in variable B and vice versa); '−' = a negative relationship, indicating that the causality runs in the opposite direction (i.e., an increase in variable A will cause a decrease in variable B and vice versa). The balancing feedback loops are numbered Bn and labelled in blue font. The reinforcing feedback loops are numbered Rn and labelled in red font. LSF refers to Large-scale farm/farming. Adapted with permission from Selebalo et al. [41] 2022. Itumeleng Selebalo.

3.3.2. R2 Large-Scale Mango Farm Value Chain

Large-scale mango farmers in the Vhembe district produce larger annual yields compared to small-scale farmers; these are comprised of more than one variety of mango species therefore enabling them to supply mangos to diverse markets i.e., juice manufacturers, dried fruit and mango atchar processing factories within the district, fresh produce markets, informal markets and supermarkets in other provinces. None of the farmers interviewed indicated that they supply mangos for export. The total income made from the sales to these diverse markets is used to reinvest in farm operations of which irrigation forms a part., only the large-scale farmers indicated that they irrigate while all small-scale farmers stated that they rely on rainfed agriculture. With an increased capacity to irrigate there is an increase in yield which allows farmers to supply the diverse markets creating a reinforcing loop.

3.3.3. R3 Large-Scale Mango Farm Reinvestment into Farm Operations Loop

Large-scale mango farmers in the Vhembe district are able to reinvest profits from sales into farm operations which include labour. Large-scale farmers are able to reinvest in paying seasonal labour during harvest time unlike small-scale mango farmers who rely on family members to harvest mangos.

3.3.4. R4 Large-Scale Farmer Reinvestment into Farm Security Loop

Large-scale mango farmers are able to reinvest the profits from selling produce into improving security on the farm. Theft is an ongoing challenge to the mango farmers as mangos can be sold locally by vendors in the district. The lack of adequate fencing means that surrounding communities can easily access the trees and steal large quantities of mangos (one farmer reported *"last year I was only able to harvest about a quarter of my whole farm, the rest was stolen. All that work for nothing."*) significantly impacting the quantities of mangos available for sale to markets. When farmers increase the investment in security i.e., fencing, patrol guards and watch dogs this decreases the loss of the crop due to theft and increases the overall annual yield creating a reinforcing loop.

3.3.5. B1 Large-Scale Farmer Reinvestment into Pest Management

One of the areas in which large-scale mango farmers in the Vhembe district reinvest their profits from sales is in pest management. Farmers are able to outsource pest control experts to inform their pest management activities, therefore increasing their capacity for effective pest management by implementing an integrated pest management approach that is capital intensive. Results showed that farmers made use of both spraying of pesticides and herbicides to this end. Continuous investment into effective integrated pest management decreases the vulnerability of orchards to pest invasion which creates a balancing loop that ensures good annual yields enabling continued supply of mangos to the diverse markets that large-scale farmers have access to. Due to the plethora of resource constraints experienced by small-scale farmers in the region investment in IMP is generally limited. Large-scale farmers are better equipped to make this investment.

3.4. Scenarios

3.4.1. The Macadamia Gold Rush

The global demand for macadamias continues to increase as there is an increasing public knowledge of the numerous health benefits of tree nuts and nut oils. This demand has been the key factor in market expansion. According to [44], the global macadamia nut market is expected to grow at a compound annual growth rate of 10.7% from 2021 to 2028 to reach USD 2.95 billion by 2028. South Africa remains one of the largest producers in the world and this can influence future production trends as farmers in the country aim to align with global market demands. Small-scale farmers' heavy reliance on large-scale farmers for processing in order to participate in global market supply will continue if there are no opportunities created for them to compete in terms of processing capacity. A wide range of role players need to be involved within the macadamia nut industry in order to make it competitive, efficient and dynamic. Small-scale farmers can only expand the industry if they have access to land and tenure security; results revealed that higher proportions of small-scale farmers (71%) farmed on communal land compared to large-scale farmers (29%). This speaks to the on-going land tenure reform dialogues in South Africa and the need to urgently address tenure rights of small-scale farmers in the country. In order to sustain large-scale and small-scale macadamia farmer interdependence in a manner that is mutually beneficial and equally beneficial to the country's agricultural economy, small-scale farmers need to be incorporated into the value chain in a more prominent way.

3.4.2. Exploring the Possibilities of Strengthening Small-Scale Farmer Collaboration

There is an increase in interest to farm macadamia nuts amongst small-scale farmers as the monetary gains become increasingly evident. This is well depicted in participant's contribution at a study group meeting; *"everyone is going into macadamias now because that's where the money is. If I could, I would convert my whole farm into only macadamias"*. Despite this growing interest, small-scale farmers would not be able to cope well with a major ecological or market failure in macadamia nut farming if they relied solely on the single crop. Intercropping is a highly beneficial practice for small-scale farmers as it has been established in literature on farming in South Africa that small-scale farming plays a dual role of being a source of household food security as well as generating income from sale of surplus [3]. For this reason, the practice of a monoculture cropping system, typically characteristic of large-scale commercial farmers in the country is not ideal for the small-scale farmer sustainability. The need to diversify their farming and develop a system where they are able to benefit from the yields of other food crops and supplement household income from farming with the profits from the sales of high value crops is essential. This approach should be encouraged to maintain small-scale farmers' significance as contributors to household food security as indicated in the introduction of the paper. The expansion of small-scale macadamia nut farming should therefore be supplementary to existing farming practices.

The move towards expansion of macadamia nut farming is seen in the establishment of nurseries amongst small-scale farmers from the yields of previous harvests. This is an attempt at breaking away from their dependence on commercial tree suppliers and creating a level of independence. If successful, this initiative has the potential to grow small-scale farmer's producing capacity over time. Establishing their own nurseries also presents a premise for small-scale farmer collaboration that may yield better production results. If small-scale farmers came together to increase their yields, they can continue to supply large-scale commercial processors and enter the export market at a more competitive level. The benefits of the outcomes of this interaction and interdependency between the two kinds of macadamia famers are not balanced. Small-scale farmers may not obtain profit to the full value of their produce as they only provide raw produce and are paid accordingly. Their large-scale farmer counterparts on the other hand obtain a higher profit as the produce that is sold to export markets is now value added after processing. Based on this imbalance, there is a need to explore more innovative approaches to collaboration between small-scale farmers for the purpose of enabling them to process macadamia nuts independently. Small-scale farmers could potentially band together to either rent or co-own processing facilities that they would collectively use instead of solely relying on large-scale commercial farmers for processing. There is potential to develop equipment more suited and more affordable for small-scale farmers if this is made a research and policy directive. This should serve as a model to inform government support for capacity building amongst small-scale farmers; with the aim to enable them to increase their profits from growing macadamias so that there is a balance in the benefits derived from growing macadamia nuts for both large- and small-scale growers. Lastly, although the main focus of macadamia supply in recent years has been international markets, there is potential for macadamias to become a highly sought-after commodity in local markets with changes in the South African food system leaning towards a more healthier food focus. Both large- and small-scale farmers can work together to explore how to optimize opportunities and risks.

3.4.3. Mango Supply Driven by the Demand of the Market

Mangos are highly perishable therefore necessitating careful control of packaging, transportation and distribution. This influences the South African mango value chain significantly. Unlike macadamia nuts, the market demand for mangos from farmers in the Vhembe district appears to be more localized than international. The ability to grow different cultivars based on favourable climatic conditions enables farmers in the district (both large- and small-scale) to target specific markets based on the type of mangos they produce e.g., juice making factories, processing factories (for dried fruit and mango atchar), local and provincial fresh produce markets. Currently, large-scale mango farmers from the Vhembe district supply produce to juice processing factories while small-scale famers supply atchar processing factories. This is mainly an issue of accessibility, as most atchar processing companies are located within the district much closer to where the small-scale farms are situated. They are therefore able to transport the produce to these factories faster. This becomes a more economically viable option for small-scale farmers as mangos rot easily and therefore may result in losses when they attempt to transport to further distances where the juice processing factories are found (in some cases outside of the district and province where they live). Agro-processing is the single largest market for mangos in South Africa [40]. According to the database of local subtropical fruit farmers, mango farmers make up the largest number of farmers in the district presenting an opportunity for economic gain; this however does not align with the success of the mango market distribution when compared to that of macadamia nuts and avocados. The economic profitability of processing mangos into juice is high as value is added to the raw produce once it is in the form of juice and can be preserved for longer than the mangos in their natural state. Atchar is also highly profitable as it is a popular choice as part of a low to medium income South African diet. It has a long shelf life due to the manner in which it is preserved, therefore presenting a viable economic investment. There is also potential for atchar to be sold as an export product to other countries in the region and abroad. Exploring possibilities of collaboration between farmers and agro-processors can possibly expand the value chain for the benefit of all stakeholder. Given the heterogeneity of the local mango demand, farmers in the district can invest in a more targeted approach to growing mangos, focussing on the niche markets. Mango cultivars that ripen earlier in the season are more favourable for atchar processing as opposed to cultivars that ripen mid to late season which are more suitable for the juice market, however the risks associated with ripe fruit are high e.g., theft, flies, pests etc. Solutions need to be found to minimize theft and may be associated with price control.

3.4.4. Focus on Irrigation

One of the greatest challenges for both large- and small-scale mango farmers is their reliance on rainfed agriculture as the area is semi-arid and prone to droughts. Interview responses revealed that reliance of rainfall for irrigation was the sole source of water for irrigation for mango farmers with mature orchards; localized groundwater accessed through drilled boreholes was not indicated. One of the numerous impacts of climate change is that rainfall patterns are shifting, therefore sole reliance on rainfall for cultivation is not beneficial. This is a constraint that is already recognized and is an ongoing concern for mango farmers of different scales however, small-scale farmers are particularly vulnerable to this problem and therefore need be given more attention. Irrigation is a critical factor in farmers' success and capacity to supply markets. Systems thinking is a valuable tool for finding solutions where trade-offs are involved. The mango industry needs to collaborate with water management representatives in order to maximise on production.

4. General Discussion

The value of systems thinking is illustrated in the CLDs and the scenarios for the two commodities, macadamia nuts and mangos. In understanding the interconnections between variables and the degree to which they impact each other in the present, it becomes possible to adopt a more holistic approach to decision making that informs policy and action for the future. The current study provides evidence that suggests that the coupling of large- and small-scale farmers is a viable option for agricultural development in South Africa. If both farmers can equitably contribute to the country's agricultural economy albeit through different means, it is possible to envision an economy that is supplied by the joint operation of both kinds of producers. The success of this kind of approach hinges on the implementation of a multi-pronged intervention strategy which addresses related issues simultaneously. Examples of the need for this multi-pronged intervention strategy have been highlighted in the scenarios through: (1) the need to address small-scale macadamia farmer tenure security in order for small-scale farmers to successfully expand their industry and to collaborate with one another alongside prioritizing supplying export markets; (2) the need to explore the potential of alternative sources of irrigation amongst small-scale-farmers beside their reliance on rain-fed agriculture. Sufficient irrigation could potentially improve their yields which can resultantly increase income from sales to local markets over time. Increased income from farming can enable farmers to reinvest in long-term transportation solutions that assist in accessing diverse markets. This serves as a good example of the interconnectedness of variables in the systems.

The government model for land redistribution in South Africa over the past two decades has been centred around the need to address historical inequalities in land distribution that favoured a minority of large-scale commercial farmers over the majority of small-scale farmers who were predominantly black. This is a highly contentious and politicized issue considering South Africa's history of an apartheid system. According to Materechera and Scholes, 2022 [6], the issue of overlapping use rights on communal land further complicates the challenge of a lack of tenure rights for small-scale farmers. Small-scale farmers have to contend with other community members who use the land for multiple other purposes e.g., firewood and grazing before they can consider participating in commercial activities. This presents itself a great constraint. Existing policy interventions surrounding land tenure security have mostly been targeted at land reform to improve the commercial status of previously disadvantaged farmers located in the former homeland areas. In order to be successful in this, policy directives should also include socio-economic strategies to address the issue of overlapping use rights in communal land. There is an urgent need for expansion as far as small-scale farmer irrigation systems are concerned. The potential for sustainable irrigation expansion thus becomes a factor that should inform research, decision making and policy development so that small-scale mango farmers can increase their yields and improve their market competitiveness. The application of innovation for more climate change friendly irrigation systems that are affordable and accessible to small-scale farmers becomes a necessity. The adoption of "soft-path" water harvesting for irrigation [45] is a plausible solution. This approach to bringing irrigation to rain-fed croplands involves capturing water resources in small and check dams as an alternative to the conventional centralized, capital-intensive irrigation projects that tend to be large.

Integrated pest management has the potential to be mutually beneficial for both farmers provided they are co-located. For example, a large-scale commercial farmer who implements IPM on their farm can have a positive knock-on effect on an adjacently located small-scale farmer of the same commodity. Small-scale farmers can use less capital-intensive, non-chemical approaches to IPM such as the cultivation of push and pull crops [46,47] in order to augment the activities of large-scale farmers that tend to be more capital-intensive. The implementation of this approach cancels the necessity of every area under cultivation with the same commodities to have a comprehensive IPM system in place. There may be areas without extensive IPM but benefit from being adjacent to farms that do. If farmers are willing to collaborate and make this trade-off, potentially with agreement on a certain kind of compensation for the benefits received, this interaction can foster future coupling of the two farming systems to achieve a common goal. This potential has not been explored in the current study, therefore there is no evidence of a willingness to collaborate on a regional IPM approach; this presents an opportunity for future research. Historically, the needs of the two kinds of farming systems have been addressed independently. This has contributed to the way in which they have continued to operate as separate entities. The study proposes that a systems thinking approach should inform decision making in the future.

5. Conclusions and Recommendations

Ultimately the question of whether coupling of the two farming systems in the context of meeting the country's food security needs at both national and household levels is a viable option is revisited. In understanding both the positive and negative implications of the interactions between the two groups of farmers, the scenarios derived in the study attempt to present evidence to support the conclusion of whether or not the two systems can ultimately work collaboratively in achieving food security at all levels in the future, as opposed to doing so independently as the current situation suggests. If decision making is informed by the application of systems analysis this may be an achievable goal. The study has shown that there is connectivity within and between large-scale commercial and small-scale farming systems in the Vhembe district. Applying systems analysis has shown that there are numerous points of collaboration across the two types of farmers. The use of systems analysis has also shown that the respective farmers who are co-located, respond to the drivers of production differently though farming the same commodity. This illustrates the potential for the coupling of the two farming systems and a transition from the historic dichotomous context of South African farming systems. There are research organizations such as the regional systems analysis community of the International Institute for Applied Systems Analysis (IIASA) in South Africa which is active and can help foster a systems approach to coupling the two farming systems. Collaboration between such research entities and other stakeholders in the food system value chain across different spheres of government, agribusiness and the private and public sectors can produce favourable results.

Future studies on farming in South Africa should view farming as systems incorporating the whole value chain of any commodity. This should include the social, financial and environmental values and impacts. The conceptual scenarios developed in the study could be a basis for further evaluation to determine their feasibility under various predicted changes such as those presented by the present and foreseeable impacts of climate change. The use of scenarios is recommended as a tool to inform similar studies on farming systems in South Africa e.g., Trade-offs in the adoption of IPM. Future studies must collate data from a larger sample size.

Author Contributions: Conceptualization, F.M. and M.S.; methodology, F.M.; software, F.M.; validation, F.M.; formal analysis, F.M.; investigation, F.M.; resources, M.S.; data curation, F.M.; writing—original draft preparation, F.M.; writing—review and editing, F.M. and M.S.; visualization, F.M.; supervision, M.S.; project administration, M.S. and F.M.; funding acquisition, M.S. All authors have read and agreed to the published version of the manuscript.

Funding: This research was funded by the National Research Foundation and the Department of Science and Technology, through the funding of M.C.S's SARChI Chair in Global Change and Systems Analysis (Grant number 101057).

Institutional Review Board Statement: The study was approved by the Ethics Committee of The University of the Witwatersrand (H/19/09/26 on 22 October 2019).

Informed Consent Statement: Informed consent was obtained from all subjects involved in the study.

Data Availability Statement: Data supporting results reported in this paper namely the database of sub-tropical fruits from the Department of Agriculture in Thohoyandou, Limpopo, South Africa can be obtained from the authors upon reasonable request.

Acknowledgments: The farmers from the Vhembe District are thanked for their contributions and hospitality.

Conflicts of Interest: The authors declare no conflict of interest. The funders had no role in the design of the study; in the collection, analyses, or interpretation of data; in the writing of the manuscript, or in the decision to publish the results.

References

1. PLAAS (Institute for Poverty, Land and Agrarian Studies). *Strategies to Support South African Smallholders as a Contribution to Government's Second Economy Strategy. Draft report Commissioned by the Second Economy Strategy Project*; PLAAS: Cape Town, South Africa, 2009.
2. Aliber, M.; Hall, R. Support for Smallholder Farmers in South Africa: Challenges of Scale and Strategy. *Dev. S. Afr.* **2012**, *29*, 548–562. [CrossRef]
3. Hendriks, S. Food Security in South Africa: Status Quo and Policy Imperatives. *Agrekon* **2014**, *53*, 1–24. [CrossRef]
4. Dixon, J.; Gulliver, A.; Gibbon, D. Farming Systems and Poverty: Improving Farmers' Livelihoods in a Changing World. *Exp. Agric.* **2001**, *39*, 109–110.
5. Hewett, E. High-Value Horticulture in Developing Countries: Barriers and Opportunities. *CAB Rev.* **2012**, *7*, 1–16. [CrossRef]
6. Materechera, F.; Scholes, M.C. Understanding the Drivers of Production in South African Farming Systems: A Case Study of the Vhembe District, Limpopo South Africa. *Front. Sustain. Food Syst.* **2022**, *6*, 722344. [CrossRef]
7. South African Government. Land Reform. Available online: https://www.gov.za/issues/land-reform (accessed on 24 June 2022).
8. Labadarios, D.; Mchiza, Z.J.R.; Steyn, N.P.; Gericke, G.; Maunder, E.M.W.; Davids, Y.D.; Parker, W.A. Food Security in South Africa: A Review of National Surveys. *Bull. World Health Organ.* **2011**, *89*, 891–899. [CrossRef]
9. Maponya, P. Opportunities and constraints faced by smallholder farmers in the Vhembe District, Limpopo Province in South Africa. *Circul. Econ. Sustain.* **2021**, *1*, 1387–1400. [CrossRef]
10. Independent Electoral Commission (IEC) of South Africa. More About Municipalities. 2016. Available online: https://www.elections.org.za/content/Elections/2016-Municipal-Elections/More-about-municipalities/ (accessed on 24 May 2021).

11. Setshego, M.V.; Aremu, A.O.; Mooki, O.; Otang-Mbeng, W. Nsatural resources used as folk cosmeceuticals among rural communities in Vhembe district municipality, Limpopo province, South Africa. *BMC Complement. Med. Therap.* **2020**, *20*, 1–16. [CrossRef] [PubMed]
12. Vhembe District Municipality (VDM). The Status of Agriculture in Vhembe District, Limpopo Province, Integrated Development Plan. 2014. Available online: www.vhembe.gov.za (accessed on 16 April 2021).
13. Rusere, F.; Crespo, O.; Mkuhlani, S.; Dicks, L.V. Developing pathways to improve smallholder agricultural productivity through ecological intensification technologies in semi-arid Limpopo, South Africa. *Afr. J. Sci. Technol. Innovat. Dev.* **2019**, *11*, 543–553. [CrossRef]
14. Odhiambo, J.J.; Mag, V.N. An assessment of the use of mineral and organic fertilizers by smallholder farmers in Vhembe district, Limpopo province, South Africa. *Afr. J. Agric. Res.* **2008**, *3*, 357–362. [CrossRef]
15. Oni, S.A.; Nesamvuni, A.E.; Odhiambo, J.J.O.; Dagada, M.C. *The Study of Agricultural Industry in the Limpopo Province (Executive Summary)*; School of Agriculture, Rural Development and Forestry University of Venda: Thohoyandou, South Africa, 2012.
16. Olofsson, M. Socio-economic differentiation from a class-analytic perspective: The case of smallholder tree-crop farmers in Limpopo, South Africa. *J. Agrar. Change* **2018**, *20*, 37–59. [CrossRef]
17. Statistics South Africa (StatsSA). The Extent of Food Security in South Africa | Statistics South Africa. 2017. Available online: https://www.statssa.gov.za (accessed on 20 May 2019).
18. Vhembe District Municipality. Local Economic Development Strategy. 2019. Available online: https://dergipark.org.tr/en/pub/ijefs/issue/47278/594805 (accessed on 19 June 2019).
19. Kom, Z.; Nethengwe, N.S.; Mpandeli, N.S.; Chikoore, H. Determinants of small-scale farmers' choice and adaptive strategies in response to climatic shocks in Vhembe District, South Africa. *GeoJournal* **2020**, *13*, 1–24. [CrossRef]
20. Moeletsi, M.E.; Mellaart, E.A.R.; Mpandeli, N.S.; Hamandawana, H. The use of rainfall forecasts as a decision guide for small-scale farming in Limpopo Province, South Africa. *J. Agric. Educ. Ext.* **2013**, *19*, 133–145. [CrossRef]
21. Mpandeli, S. Managing climate risks using seasonal climate forecast information in Vhembe District in Limpopo Province, South Africa. *J. Sustain. Dev.* **2014**, *7*, 68. [CrossRef]
22. Kirsten, J.; van Zyl, J. Defining small scale farmers in the South African context. *Agrekon* **1998**, *37*, 551–562. [CrossRef]
23. Etikan, I.; Alkassim, R.; Abubakar, S. Comparison of snowball sampling and sequential sampling technique. *Biometric. Biostatist. Int. J.* **2016**, *3*, 55. [CrossRef]
24. Ventana Systems, Inc. Available online: https://www.ventanasystems.com/ (accessed on 20 July 2021).
25. Van der Heijden, K. Turbulence in the Indian agricultural sector: A scenario analysis. In *Business Planning for Turbulent Times*; Routledge: England, UK, 2012; pp. 109–124.
26. Sarka, D. Descriptive statistics. In *Advanced Analytics with Transact-SQL*; Apress: Berkeley, CA, USA, 2021; pp. 3–29. [CrossRef]
27. Shen, C.; Panda, S.; Vogelstein, J. The chi-square test of distance correlation. *J. Comput. Graph. Statist.* **2021**, *16*, 1–15. [CrossRef]
28. Grodal, S.; Anteby, M.; Holm, A.L. Achieving rigor in qualitative analysis: The role of active categorization in theory building. *Acad. Manag. Rev.* **2021**, *46*, 591–612. [CrossRef]
29. Ramirez, R.; Mukherjee, M.; Vezzoli, S.; Kramer, A. Scenarios as a scholarly methodology to produce "interesting research". *Futures* **2015**, *71*, 70–87. [CrossRef]
30. Food and Agriculture Organization (FAO). Farming Systems and Poverty. Rome FAO. 2019. Available online: http://www.fao.org/farmingsystems/description_en.htm#top (accessed on 17 July 2019).
31. Dariusz, K. Changes in the relations of production factors in agriculture (the case of Poland). *Changes* **2015**, *15*, 179–188. [CrossRef]
32. Arnold, R.D.; Wade, J.P. A definition of systems thinking: A systems approach. *Procedia Comput. Sci.* **2015**, *44*, 669–678. [CrossRef]
33. Simonovic, S.P. *Managing Water Resources: Methods and Tools for a Systems Approach*; Routledge: London, UK, 2012.
34. Ford, A. *Modeling the Environment*, 2nd ed.; Island Press: Washington, DC, USA, 2009.
35. Lagnika, S.B.; Hausler, R.; Glaus, M. Modeling or Dynamic Simulation: A Tool for Environmental Management in Mining? *J. Integr. Environ. Sci.* **2017**, *14*, 1–19. [CrossRef]
36. Bawden, R.J. On the systems dimension of FSR. *J. Farming Syst. Res. Extension.* **1996**, *5*, 1–18.
37. Schweizer, V.; Kriegler, E. Improving environmental change research with systematic techniques for qualitative scenarios. *Environ. Res. Lett.* **2012**, *7*, 044011. [CrossRef]
38. Jaskiewicz, K. Macadamia nuts The New Gold of South Africa? Inclusive Value Chain Integration of Macadamia Nut Small-Scale Farmers in Limpopo, South Africa. Master's Thesis, International Development Studies, Limpopo, South Africa, 2015. Available online: https://inclusivevcc.files.wordpress.com/2015/04/macadamia-nutsthe-new-gold-of-south-africa.pdf (accessed on 19 June 2019).
39. The Macadamia Newsletter. 2020. Available online: https://themacadamia.co.za/latest-macadamia-magazine-issues (accessed on 20 October 2020).
40. DAFF. A Profile of the South African Mango Market Value Chain. 2017. Available online: https://www.nda.agric.za/doaDev/sideMenu/Marketing/Annual%20Publications/Commodity%20Profiles/field%20crops/Mango%20Market%20Value%20Chain%20Profile%202017.pdf (accessed on 19 June 2022).
41. Selebalo, I.M.; Scholes, M.C.; Clifford-Holmes, J.K. A Systemic Analysis of the Environmental Impacts of Gold Mining within the Blyde River Catchment, a Strategic Water Area of South Africa. *Water* **2021**, *13*, 301. [CrossRef]
42. Parshotam, A. Cultivating Smallholder Inclusion in Southern Africa's Macadamia Nut Value Chains. 2018. Available online: https://www.jstor.org/stable/pdf/resrep28391.pd (accessed on 15 October 2020).

43. Shabalala, M.; Toucher, M.; Clulow, A. The Macadamia Bloom—What Are the Hydrological Implications? *Sci. Hortic.* **2022**, *292*, 110628. [CrossRef]
44. Grand View Research. Available online: https://www.grandviewresearch.com/industry-analysis/macadamia-nut-market (accessed on 4 June 2022).
45. Rosa, L.; Chiarelli, D.D.; Sangiorgio, M.; Beltran-Peña, A.A.; Rulli, M.C.; D'Odorico, P.; Fung, I. Potential for sustainable irrigation expansion in a 3 C warmer climate. *Proc. Natl. Acad. Sci. USA* **2020**, *117*, 29526–29534. [CrossRef]
46. Ehler, L.E. Integrated pest management (IPM): Definition, historical development and implementation, and the other IPM. *Pest Manag. Sci.* **2006**, *62*, 787–789. [CrossRef]
47. Deguine, J.P.; Aubertot, J.N.; Flor, R.J.; Lescourret, F.; Wyckhuys, K.A.; Ratnadass, A. Integrated pest management: Good intentions, hard realities. *A review. Agron. Sustain. Dev.* **2021**, *41*, 1–35. [CrossRef]

agriculture

MDPI

Article

Improving Agricultural Green Supply Chain Management by a Novel Integrated Fuzzy-Delphi and Grey-WINGS Model

Muwen Wang [1,2,*] and Kecheng Zhang [2]

1 School of Economics and Management, Shandong Agricultural University, Taian 271018, China
2 School of Business Administration, Shandong Women's University, Jinan 250300, China
* Correspondence: 2021010113@sdau.edu.cn

Abstract: This study suggests a novel hybrid model for calculating the interrelationships between factors by integrating the Fuzzy set, Delphi, the Grey theory, and Weighted Influence Nonlinear Gauge System (WINGS) approaches in agricultural green supply chain management (AGSCM). Fuzzy Delphi helps to select 12 indicators from 19 factors by defuzzification for ambiguity associated with subjective judgment by 10 experts in data collection. Grey WINGS can illustrate the relationships, direction, and strength of factors simultaneously, which illustrates that environmental law, green consciousness, product quality, and price are the most significant factors of AGSCM. The results can help operators not only to analyze these key influencing factors, but also to understand the complex cause-and-effect relationships between these factors. This integrated model will hopefully provide a useful tool to agricultural policy makers and decision makers for sustainable development.

Keywords: AGSCM; Delphi; WINGS; Grey theory; Fuzzy set; sustainable development

Citation: Wang, M.; Zhang, K. Improving Agricultural Green Supply Chain Management by a Novel Integrated Fuzzy-Delphi and Grey-WINGS Model. *Agriculture* **2022**, *12*, 1512. https://doi.org/10.3390/agriculture12101512

Academic Editors: Moucheng Liu, Xin Chen and Yuanmei Jiao

Received: 23 August 2022
Accepted: 15 September 2022
Published: 20 September 2022

1. Introduction

Agricultural green supply chain management (AGSCM) aims to transform environmental constraints into advantages and opportunities, such as eco-brand, green consumption, and sustainable development, which is difficult to optimize influenced by complex and interactive factors intrinsic in an ever-changing complex environment, which includes global warming, the COVID-19 pandemic, and environmental pollution. In particular, agriculture is one of the largest sources of methane emissions, with the Food and Agriculture Organization (FAO) stating that the emission of greenhouse gas will increase by 30% by 2050 [1]. With the increasing concern for green and sustainable development, more and more consumers are forcing the traditional supply chain reform to become environmentally conscious with components, such as biological pesticides, renewable energy, recyclable packaging, environmentally friendly fertilizers, and so on [1,2].

As we know, the yield of agricultural products is particularly influenced by many uncertain challenges related to environmental, political, economic, social, technical, and legal dimensions, which has become a major issue affecting human beings in recent years [3]. A growing global population and a deteriorating environment have led to an increased focus on agricultural supply chains, such as resource constraints and environmental pollution [4]. With the growing environmental awareness, decision makers must take environmental factors into account in supply chain management. The implementation of environmental and social performance expands the scope of legal, social, technical, economic, and ethical properties in green supply chain management (GSCM) [5]. Furthermore, the performance of GSCM combines environmental, social, and economic dimensions, which must be considered in many interrelated operations, such as planning, production, packaging, transportation, storage, processing, distributing, publicity, and sales [6–8]. Sustainability has become a necessary obligation for enterprise development. Enterprises need to take responsibility for social and environmental issues in supply chain management [9]. However, AGSCM has become more difficult with the spread of the COVID-19 pandemic, global warming, extreme climate, and environmental pollution across the world.

Although there have been a few attempts to study agricultural green supply chain management [10–12], these studies mainly studied the factors which are independent of each other as a prerequisite assumption but ignored the interrelationships within them. This assumption may limit the development of AGSCM and the improvement of economics. However, there are many uncertain complex hierarchical factors affecting AGSCM, such as perishability, seasonality, customers' demand, and supply relationships [13]. In order to improve development of AGSCM within the restrictions of available natural resources, the decision support model must be concentrated on the real-world scenario and integrated with complicated methods to evaluate performance and the relationship of every factor [14].

Multiple-Criteria Decision-Making (MCDM) methods are designed to address complex decision-making difficulties by analyzing the structure of criteria, alternatives, and decision-makers' preference, which are suitable for assisting managers, practitioners, and developers in selecting the best options within various conflicting criteria. Saaty introduced the Analytic Hierarchy Process (AHP) as a popular MCDM approach in 1980. The hierarchical structure of AHP makes it possible to visualize the factors influencing the alternatives. Analytic Network Process (ANP) is an amplification of AHP which can take into account the intricate interdependence of decision factors in a hierarchical structure [15]. To deal with the uncertain situation, the fuzzy AHP and ANP have been used in many domains [16,17]. So, the hybrid MCDM methods have the advantage to accomplish analysis the imprecise, incomplete, or uncertain information.

In contrast to the methods mentioned above, the Decision-Making Trial and Evaluation Laboratory (DEMATEL) is an advanced and sophisticated decision-making method for addressing interdependencies by visualizing the causal interactions of indicators proposed by Gabus and Fontela [18]. DEMATEL uses mathematical tools to comprehend various specialists' perspectives on associated factors, as well as logical correlations and direct effects between these factors [19], which has been widely used in supply chain management (SCM) [20–22]. Michnik developed the Weighted Influence Nonlinear Gauge System (WINGS) approach from DEMATEL [23]. With interdependencies of factors in MCDM situations, DEMATEL simulates the direction and strength of the impact. Furthermore, WINGS simulates both the intensity and direction of the influence, in addition to the strength of the criterion, which could be utilized as a theoretical basis for AGSCM. However, classical DEMATEL and WINGS methods ignore the vagueness and uncertainty of human judgment that are so prevalent in real life. Regarding this problem, the Grey theory may successfully handle the ambiguities inherent in human subjective judgement while acquiring accurate results with a moderate data sample.

The contribution of this paper can be summarized as follows.

1. A fuzzy-Delphi and grey-WINGS approach to decision theory, which can be utilized to analyze different group choices, ambiguity, and complex interrelationships in evaluation problems, is presented in this study. The combination of a fuzzy set and grey theory can provide a more realistic representation of human judgement under ambiguous and subjective conditions.
2. The target of this study is conducive to the improvement of AGSCM by applying the current assessment approach to provide a more accurate and objective prioritization tool for AGSCM in a hazy and diverse environment. The approach is intended to assist AGSCM designers in identifying the most critical factors with the highest potential.
3. The fuzzy-Delphi and grey-WINGS method integrates four techniques, which have not been combined for illustrating mutual relationships of factors in previous studies. According to the results analysis, this research contributes significantly to improving AGSCM by providing policy and management implications.

The remainder is arranged as follows. Section 2 contains literature reviews. Section 3 consists of materials and methods. Section 4 includes research results. Section 5 is discussions. Section 6 includes conclusions.

2. Literature Review

2.1. Agriculture Green Supply Chain Management

GSM is always known as the environmental supply chain based on green manufacturing theory, which was first introduced by the Manufacturing Research Society of Michigan State University. The enterprises, merchants, and farmers within the supply chain can gain benefit from GSCM and use it as a valuable resource to improve their environmental performance [24] because it is a vital important management system involving suppliers, production plants, distributors, and customers, with the aim of minimizing the negative impacts and maximizing efficiency of resource utilization through the improvement of the whole implementation incorporating environmental concepts [25,26].

Agriculture is one of the industries that is most affected by climate. There is a clear relationship between agricultural productivity and climate fluctuations, which is especially complex and unique in developing countries [27]. Moreover, agricultural products have several specific characteristics that make agricultural supply chain management (ASCM) more complicated due to factors associated with seasonality, environment, and perishability when compared with typical supply chains [28]. In order to maintain environmental sustainability, the 'green' concept integrates environmental and ecological concerns, which has a significant impact on the environment including pollution, emissions, the health hazard to human beings, etc. [29–35]. Therefore, AGSCM has been established as an important discipline of sustainable operations management, which must be paid more attention. As more and more environmental regulations are published, AGSCM plays a proactive role in improving environmental performance and economic stability [36,37].

Environment, strategy, and logistics are the three critical components of AGSCM, involving proactive measures such as recycling, reprocessing, and monitoring of environmental standards [38,39]. In order to improve sustainable development, it is necessary that the product, package, and purchase must meet green standards. All supply chain participants must be proactive and work together to minimize negative environmental effects [40].

2.2. The Influence Factors of AGSCM

AGSCM incorporates environmental and economic elements, which are important challenges with the limitation of resources for minimizing environmental negative consequences and enhancing economic stability [36,41]. We reviewed papers with GSCM and AGSCM from 2010 to 2022. According to the operation of the supply chain, the forces driving all farmers, stakeholders, and customers should be associated with activities in AGSCM. It can be concluded that the factors influencing AGSCM include customer and stakeholder requirements and competitive advantage, both of which come from economic and social factors [42,43]. These are the motivations for successful implementation of AGSCM. On the other hand, regulation and market pressure could force companies to adopt the rules of AGSCM in the pursuit of environmental performance, such as environmental laws, competitors' pressure, suppliers' requirements, and customers' awareness [44–46]. Meanwhile, barriers are factors that hamper the implementation process of AGSCM. Some of the important barriers are cost and risk, lack of government support, financial constraints, poor supplier commitment, lack of legitimacy, technology, and resistance from the stakeholders [47–49].

2.3. Hybrid Methodology and Applications

Compared to traditional SCM, the ASCM is more difficult to measure due to the issues associated with environmental factors. Some structural methodologies have been extended in ASCM, such as AHP [50], ANP [51], and Technique for Order Preference by Similarity to an Ideal Solution (TOPSIS) [52] in Table 1. Furthermore, AGSCM is also a typical MCDM problem that requires estimation of factors based on complex objective and subjective information. In order to achieve accurate and scientific evaluation results, the fuzzy set theory might be a major tool which was initiated as a mathematical tool to handle ambiguity and fuzzy information influenced by subjective judgments. These sources of imprecision contain incomplete, nonobtainable, unquantifiable, and partial information, which exist in real life. Therefore, the fuzzy set can be employed in some decision models to analyze the factors of supply chain management [53]. In order to exploit the ambiguity and variety in articulating preferences of decisionmakers, grey theory has been utilized to assemble group fuzzy evaluations, and it can handle the preferences of various decisionmakers [54,55]. Compared with the fuzzy logit, grey theory can handle uncertainty problems with discrete values and imperfect knowledge by creating a flexible choice model with interval numbers. Its significant advantage is the capability to obtain accurate results with limited data under conditions of high variable variability [56]. By combining linguistic variables, the grey set theory can be used to evaluate uncertain conceptions related to people's subjective judgments. The implications of the grey set theory will be more significant, especially when experts make decisions based on inadequate information or when they are conscious that they lack knowledge in some scenarios. Numerous effective applications of grey system theory have been made in several fields, including business, geography, medicine, agriculture, and disaster preparedness [57,58]. So, grey system theory has been improved as an efficient approach to unresolved and ambiguous issues in recent years.

Table 1. Overview of relevant studies with MCDM methods.

Findings	Approach	Relevant Literature
Delphi and Fuzzy AHP methods are constructed to estimate the factors of green design, purchasing, production, warehousing, and logistics in supply chain practices.	Delphi, Fuzzy AHP	[50]
The research discusses the four main criteria, including product quality, production cost, customer requirements, and delivering time to select an effective supplier by Fuzzy ANP.	Fuzzy ANP	[51]
This article studied green supplier selection with the criteria of service, quality, price, and environment by using fuzzy TOPSIS.	Fuzzy TOPSIS	[52]
The case study developed a hybrid model with AHP and TOPSIS to evaluate a supply chain perspective within economic, environmental, and social dimensions.	Fuzzy AHP, TOPSIS	[59]
This work uncovered ten main factors to sustainable initiatives for ASCM, such as government pressure, stakeholder requirements, monitoring and auditing, competitive advantages, cost, and benefits, by using Fuzzy DEMATEL.	Fuzzy DEMATEL	[26]
This method proposes a model combing DEMATEL and ANP to assess indicators such as services, technology, environmental, financial, and economic dimension in sustainable supplier selection.	Fuzzy ANP, DEMATEL	[53]

DEMATEL: Decision-Making Trial and Evaluation Laboratory; TOPSIS: Technique for Order Preference by Similarity to an Ideal Solution; ANP: Analytic Network Process; AHP: Analytic Hierarchy Process.

The Delphi technique is a qualitative approach for gathering the opinions of a diverse group on a specific topic, which was proposed by the RAND Corporation. Because traditional Delphi techniques cannot deal with ambiguity, a fuzzy-Delphi method, which can handle the ambiguity and uncertainty inherent with the data, was combined by Ishikawa et al. (1993) [60]. Various applications have been employed in supply chain performance, agricultural cost, design analysis of products, healthcare, and construction [61–66]. Moreover, to analyze the complex intertwined relationships between influencing factors, scholars proposed several powerful methods including ANP [15,51], DEMATEL [56,59], and WINGS [23,67]. ANP is just a generalized version of the analytical hierarchy process proposed by Saaty, which illustrates general relations among the indicators, whereas the AHP emphasizes hierarchical relations between decision levels [68]. The ANP uses ratio scale measurements through comparisons, but unlike the AHP, it does not impose a fixed hierarchical structure. Both methods have a prerequisite assumption as no influence between criteria. ANP has been widely applied in various situations, such as location selection, project selection, and supplier selection [69–71].

DEMATEL is used to translate the interrelationships between the criteria into an understandable structural model, which was established by the Battelle Memorial Institute of the Geneva Research Center. The numbers measuring the level of influence can construct the matrices or digraphs to illustrate the interrelationship between criteria and identify the core criterion to express the performance of variables, which could also eliminate overfitting for assessment [21]. Being an update of DEMATEL, WINGS takes over the superiority of DEMATEL, including the ability to handle complex problems with various factors and the simplicity of its mathematical procedures [23]. However, it also has its own special characteristics. WINGS measures the operating factors' strength and the level of its influence, whereas DEMATEL only considers the latter. So, an improved version of WINGS can be used to evaluate the interrelationships between criteria more powerful. Especially when the criteria are distinct, it has been demonstrated that WINGS simplifies the additive agglomeration as shown in Table 2 [72].

Table 2. Comparations within different approaches.

Approach	Interdependencies of Factors	Intensity of Impact	The Strength of Factors	Group Fuzzy Assessments
FAHP	-	-	s	-
FANP	-	-	s	-
FDEMATEL	s	s	-	-
Grey–WINGS	s	s	s	s

WINGS: Weighted Influence Nonlinear Gauge System.

Based on the above analysis, the DEMATEL and WINGS techniques are superior than other traditional methods, since the input values can immediately enter the matrix, which has the advantage in calculations over the AHP/ANP approach with pairwise comparisons. However, WINGS method is superior than the classical DEMATEL method, which considers the strength of the standard., as well as the interrelationship between criteria. Unlike the previously mentioned approaches, this study combines the WINGS and DELPHI methods with fuzzy and grey theory to handle the fuzzy decision environment. Therefore, the suggested model of this paper is more accurate in describing the subjective information and more practicable in analyzing the difficult assessment problems with simple calculation. In addition, there is no instance integrating the grey theory, the fuzzy set, the WINGS, and the DELPHI approaches, which involves the ambiguity and uncertainty during the evaluation process.

3. Materials and Methods

The proposed model combining fuzzy Delphi and Grey WINGS contains two phases as in Figure 1. Firstly, identifying and finalizing the factors of AGSCM. Secondly, a cause-and-effect analysis of the components that have been selected will demonstrate how they interact.

3.1. Influencing Factors of AGSCM

Based on the status of the AGSCM and structural analysis approaches that have been applied to supply chain management, an evaluation method of the influencing elements of AGSCM has been constructed. We chose 19 factors from three dimensions including government, economy, and society, including green consciousness, competitive pressure, government subsidies, produce quality, customer demand, environmental laws, logistics, renewable material, green operation, technology, waste reduction, price of product, cost, stockholders' requirement, monitoring, social responsibilities, infrastructure, income level, and reusable packaging.

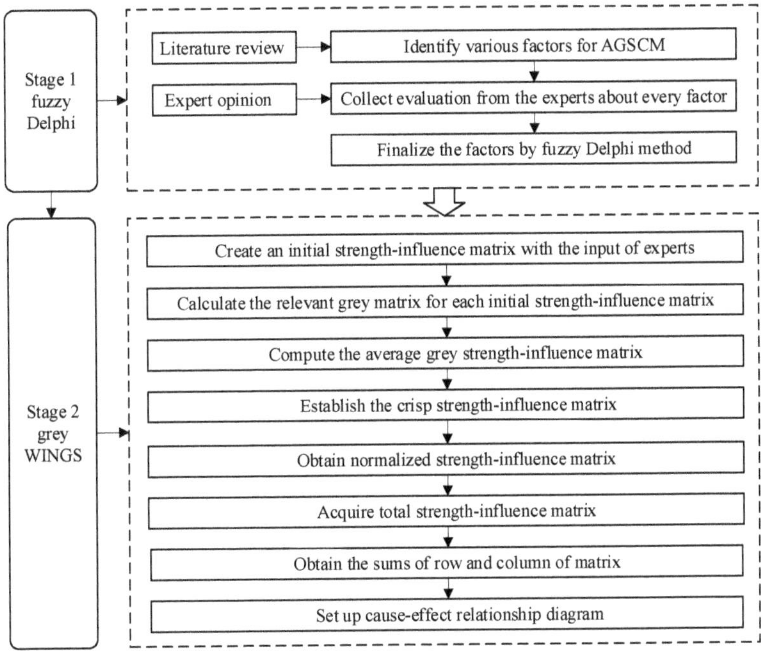

Figure 1. The framework of fuzzy Delphi and grey WINGS.

3.2. Fuzzy Delphi

The theory of fuzzy sets proposed by Zadeh to describe the ambiguity of human cognitive processes formed the basis of the fuzzy-Delphi technique. A triangular fuzzy number can be presented like $\widetilde{\lambda} = (l, o, k)$, where $l \leq o \leq k$. Then, the membership function is:

$$\theta_{\widetilde{\lambda}} = \begin{cases} \frac{x-l}{o-l}, \ x \in (l, o) \\ \frac{k-x}{k-o}, \ x \in (o, k) \\ 0, \ x \in (-\infty, l) \cup (k, \infty) \\ 1, \ x = o \end{cases} \tag{1}$$

The basic operations show as:

$$\begin{gathered} (1)\widetilde{\lambda}_1 + \widetilde{\lambda}_2 = (l_1, o_1, k_1) + (l_2, o_2, k_2) = (l_1 + l_2, o_1 + o_2, k_1 + k_2); \\ (2)\widetilde{\lambda}_1 - \widetilde{\lambda}_2 = (l_1, o_1, k_1) - (l_2, o_2, k_2) = (l_1 - l_2, o_1 - o_2, k_1 - k_2); \\ (3)\widetilde{\lambda}_1 \times \widetilde{\lambda}_2 = (l_1, o_1, k_1) \times (l_2, o_2, k_2) = (l_1 l_2, o_1 o_2, k_1 k_2); \\ (4)\widetilde{\lambda}_1 \div \widetilde{\lambda}_2 = (l_1, o_1, k_1) \div (l_2, o_2, k_2) = (l_1/k_2, o_1/o_2, k_1/l_2). \end{gathered} \tag{2}$$

where $l_1, l_2 > 0; o_1, o_2 > 0; k_1, k_2 > 0$.

The following are all fuzzy-Delphi steps:

Step 1: This process involves identifying and categorizing numerous factors that are relevant to the field under research.

Step 2: Once the criteria have been established, the experts are given the questionnaire detailing the criteria to compare by using the linguistic scale listed in Table 3. Fuzzy numbers could be transformed from experts' evaluations for each criterion. A fuzzy number referring to the cth factor suggested by the ath expert is expressed as:

$$e_{ca} = (l_{ca}, o_{ca}, k_{ca}); c = 1, 2 \ldots p; a = 1, 2 \ldots q \tag{3}$$

where p and q are the number of criteria and experts.

Table 3. Linguistic scale with triangular fuzzy number.

Linguistic Values	Numbers	Corresponding Triangular Fuzzy Number
Very unimportant	1	(0.1,0.1,0.3)
Unimportant	3	(0.1,0.3,0.5)
Normal	5	(0.3,0.5,0.7)
Important	7	(0.5,0.7,0.9)
Very important	9	(0.7,0.9,0.9)

The fuzzy number for each criterion could be estimated using triangular fuzzy numbers (E), as stated in Equation (4), which integrates the evaluations from all q experts as follows:

$$E_c = (l_{cD}, o_{cM}, k_{cH}) = \left(\min_q l^q_{cD}, \left(\prod_{a=1}^{q} o^a_{cM}\right)^{1/q}, \max_q k^q_{cH} \right) \quad (4)$$

Step 3: The fuzzy number of each assessment factor should be defuzzied using the Simple Center of Gravity (SCGM) approach to obtain the final value of each factor, which is the most prevalent approach for defuzzification [73]. This stage of SCGM involves computing the defuzzification value G using the mean approach as shown below:

$$G_c = (l_{cD} + o_{cM} + k_{cH})/3 \quad (5)$$

Step 4: A threshold value (β) must be defined to choose the most significant criteria from the expert group in order to create the list of criteria. The final step is to construct the final list of criteria based on the following threshold criteria: The criterion is chosen if $G \geq \beta$, and the criterion is omitted if $G \leq \beta$.

3.3. *Fuzzy-Delphi Grey-WINGS Model*

The main steps can be described as:

Step 1. Determine selection criteria by using the fuzzy-Delphi method.

Numerous factors relevant to AGSCM are estimated by experts. After gathering expert opinions from surveys, the triangle fuzzy numbers are utilized to determine selection criteria through the Delphi method.

Step 2. Construct an initial strength–influence matrix for all experts.

Table 4 displays the language evaluation and the related grey numbers, which could measure factor x impact over factor y using an integer scale ranging from 0 to 4, indicating "no influence", "low influence", "medium influence", "high influence", and "very high influence" between factors.

Table 4. Grey linguistic scales.

Linguistic Variables	Influence Number	Related Grey Numbers
None (N)	0	[0.0,0.0]
Low (L)	1	[0.0,0.25]
Medium (M)	2	[0.25,0.5]
High (H)	3	[0.5,0.75]
Very high (VH)	4	[0.75,1.0]

Step 3. Compute the corresponding grey matrix for the strength–influence matrix.

The ratings on the integer scale can be transformed into corresponding grey scales that give an upper range and a lower range of values. Based on the obtained grey values, the initial relation matrices are transformed into grey relation matrices, as $\otimes r^a_{xy} = \left[\underline{\otimes r}^a_{xy}, \overline{\otimes r}^a_{xy}\right]$, where x,y indicate the criterion, and a indicates the ath expert, $1 \leq a \leq q; 1 \leq x \leq p; 1 \leq y \leq p$.

Step 4. Calculate the average grey strength–influence matrix.

The average grey strength–influence matrix $\left[\otimes \widehat{r}_{xy}\right]$ can be computed by q grey relation matrices,

$$\otimes \widehat{r}_{xy} = \left(\frac{\sum_a \underline{\otimes} r^a_{xy}}{q}, \frac{\sum_a \overline{\otimes} r^a_{xy}}{q} \right) \tag{6}$$

Step 5. Obtain the crisp strength–influence matrix.

(1) Standardization of the grey number:

$$\overline{\otimes}\widetilde{r}_{xy} = \left(\overline{\otimes}\widehat{r}_{xy} - \min\overline{\otimes}\widehat{r}_{xy}\right) / \left(\max\overline{\otimes}\widehat{r}_{xy} - \min\underline{\otimes}\widehat{r}_{xy}\right) \tag{7}$$

$$\underline{\otimes}\widetilde{r}_{xy} = \left(\underline{\otimes}\widehat{r}_{xy} - \min\underline{\otimes}\widehat{r}_{xy}\right) / \left(\max\overline{\otimes}\widehat{r}_{xy} - \min\underline{\otimes}\widehat{r}_{xy}\right) \tag{8}$$

(2) Normalization of the crisp values:

$$t_{xy} = \left(\underline{\otimes}\widetilde{r}_{xy}\left(1 - \underline{\otimes}\widetilde{r}_{xy}\right) + \left(\overline{\otimes}\widetilde{r}_{xy} \times \overline{\otimes}\widetilde{r}_{xy}\right)\right) / \left(1 - \underline{\otimes}\widetilde{r}_{xy} + \overline{\otimes}\widetilde{r}_{xy}\right) \tag{9}$$

(3) Calculate the accurate total crisp values.

$$f_{xy} = \min\overline{\otimes}\widetilde{r}_{xy} + t_{xy}\left(\max\overline{\otimes}\widetilde{r}_{xy} - \min\underline{\otimes}\widetilde{r}_{xy}\right) \tag{10}$$

and $F = \left[f_{xy}\right]$

Step 6. Obtain the normalized strength–influence matrix.

$$W = \frac{1}{\sum\limits_{x=1}^{p}\sum\limits_{y=1}^{p} F_{xy}} \text{ and } B = W \times F \quad x, y \in \{1, 2, \ldots, p\} \tag{11}$$

The element of matrix B is between 0 and 1.

Step 7. Acquire the total strength–influence matrix.

The matrix Z is obtained by:

$$Z = B(I - B)^{-1} \tag{12}$$

where $Z = [z_{ca}]$, and I presents an identity matrix.

Step 8. Sum of rows and columns in matrix Z.

The sum of rows (T) and columns (L) in matrix Z can be calculated as:

$$T = [T_c] = \sum_{c=1}^{p} z_{ca}, \quad c = 1, 2, \ldots, p \tag{13}$$

$$L = [L_a] = \sum_{a=1}^{q} z_{ca} \quad , a = 1, 2, \ldots, q \tag{14}$$

T depicts the whole influence of component c as a cause affecting remaining components, while L illustrates an effect as the whole influence from other components impacting component a.

Step 9. Set up cause–effect relationship diagram.

Using the values obtained through Equations (13) and (14), a causal diagram is set up. The total impacts the given and received values by factor x, which represents the degree of prominence in the overall system.

The sum ($T + L$) presents the total effects by factor x, which represents the degree of prominence in the overall system, while ($T - L$) illustrates the net effect of factor x on the overall system. Factor x is the net cause if ($T - L$) is positive. Then, factor x is the net effect if ($T - L$) is negative.

Step 10. As shown below, a threshold value (β) is established to eliminate minor effects.

$$\beta = \frac{\sum_{x=1}^{p} \sum_{y=1}^{p} [z_{xy}]}{N} \tag{15}$$

where N is the number of factors in matrix Z.

4. Results

4.1. Data Collection and Fuzzy Delphi

The main steps can be described as:

For the purpose of gathering data, 10 experts from agricultural businesses and academics were engaged. The experts team consisted of 2 professors within the agriculture field, 2 agricultural consultants, 2 agricultural supply chain managers, 2 rural cooperative managers and 2 farmers, who all have an experience of more than 12 years. Table 5 depicts the details of these experts. The data are gathered and assessed in two stages, which are described below:

Table 5. Information of the experts.

No	Gender	Position	Work Experience
Exp 1	Male	Professor	20
Exp 2	Female	Professor	22
Exp 3	Male	Rural cooperative manager	18
Exp 4	Male	Rural cooperative manager	15
Exp 5	Male	Supply chain manager	16
Exp 6	Male	Supply chain manager	18
Exp 7	Female	Farmer	23
Exp 8	Male	Farmer	25
Exp 9	Male	Agricultural consultant	12
Exp 10	Female	Agricultural consultant	14

Exp: expert.

The fuzzy–Delphi method was used to select only those indicators significant to AGSCM that were determined through interviews and a literature review. Ten experts were given the same questionnaire based on the identified indicators, and they were asked to evaluate each factor in relation to the AGSCM by the linguistic scale shown in Table 3. Additionally, by applying the transforming procedures above, the values were converted into triangular fuzzy number to aggregate the fuzzy values of all 19 elements using Equations (3) and (5).

To select the more significant factors, the threshold defuzzification value (β) was chosen at 0.60 in this paper to determine whether to accept or reject a factor, which is larger than the normal value (0.56) for the nine-fuzzy scale [73]. Based on this threshold value of defuzzification, a total of 12 factors with values greater than 0.60 were selected, and 7 factors less than 0.60 were rejected. Table 6 lists all the selected and rejected factors.

4.2. Grey WINGS Analysis

The impact factors of AGSCM were empirically investigated using the grey-WINGS method. The significance among the 12 factors was evaluated by experts. At this stage, the grey-WINGS approach was utilized by the same 10 experts to obtain the final interrelationships and cause–effect linkages between the factors. The following sections cover the implementation of the grey-WINGS approach:

Step 1: Using the linguistic scale provided in Table 4, experts were asked to build a strength–influence matrix for factors in the AGSCM. The grey initial strength relationship matrix from the No.1 expert is displayed in Table 7.

Table 6. Finalizing factors using fuzzy Delphi.

No	Factors	Fuzzy Weight	Defuzzification	Selection	Codes
1	Green consciousness	(0.3,0.76,0.9)	0.65	√	F1
2	Competitive pressure	(0.1,0.38,0.9)	0.46	-	-
3	Government subsidies	(0.3,0.60,0.9)	0.60	√	F2
4	Produce Quality	(0.5,0.81,0.9)	0.74	√	F3
5	Customers' demand	(0.3,0.77,0.9)	0.66	√	F4
6	Environmental laws	(0.3,0.71,0.9)	0.64	√	F5
7	Logistics	(0.1,0.30,0.9)	0.43	-	-
8	Renewable material	(0.1,0.28,0.7)	0.36	-	-
9	Green operation	(0.3,0.68,0.9)	0.63	√	F6
10	Technology	(0.3,0.62,0.9)	0.61	√	F7
11	Reducing waste	(0.1,0.43,0.9)	0.48	-	-
12	Price of product	(0.3,0.7,0.9)	0.63	√	F8
13	Cost	(0.3,0.77,0.9)	0.66	√	F9
14	Stockholders' requirement	(0.3,0.64,0.9)	0.61	√	F10
15	Monitoring	(0.3,0.69,0.9)	0.63	√	F11
16	Social responsibilities	(0.1,0.48,0.9)	0.49	-	-
17	Infrastructure	(0.1,0.42,0.9)	0.47	-	-
18	Income level	(0.3,0.68,0.9)	0.63	√	F12
19	Reusable packaging	(0.1,0.28,0.7)	0.36	-	-

Codes are the abbreviations of Factors.

Table 7. The grey initial strength relationship matrix from Exp 1.

	F1	F2	F3	F4	F5	F6	F7	F8	F9	F10	F11	F12
F1	(0.75,1)	(0.75,1)	(0,0.25)	(0.5,0.75)	(0.25,0.5)	(0.25,0.5)	(0.25,0.5)	(0.25,0.5)	(0.75,1)	(0.25,0.5)	(0,0.25)	(0,0.25)
F2	(0.5,0.75)	(0.75,1)	(0,0.25)	(0,0.25)	(0,0.25)	(0.25,0.5)	(0.25,0.5)	(0,0.25)	(0,0.25)	(0,0.25)	(0,0.25)	(0.25,0.5)
F3	(0.5,0.75)	(0.5,0.75)	(0,0.25)	(0.5,0.75)	(0.5,0.75)	(0.25,0.5)	(0.25,0.5)	(0,0.25)	(0.25,0.5)	(0.25,0.5)	(0,0.25)	(0.25,0.5)
F4	(0.75,1)	(0.75,1)	(0,0.25)	(0.75,1)	(0,0.25)	(0.25,0.5)	(0.25,0.5)	(0.25,0.5)	(0.25,0.5)	(0.25,0.5)	(0,0.25)	(0.25,0.5)
F5	(0.75,1)	(0.75,1)	(0.75,1)	(0.75,1)	(0,0)	(0.5,0.75)	(0.5,0.75)	(0.5,0.75)	(0.75,1)	(0.5,0.75)	(0,0.25)	(0.25,0.5)
F6	(0.75,1)	(0.75,1)	(0.75,1)	(0.75,1)	(0.5,0.75)	(0.75,1)	(0.25,0.5)	(0.25,0.5)	(0.5,0.75)	(0.25,0.5)	(0,0.25)	(0.25,0.5)
F7	(0.75,1)	(0.75,1)	(0.25,0.5)	(0.75,1)	(0.5,0.75)	(0.5,0.75)	(0.5,0.75)	(0.5,0.75)	(0.75,1)	(0.25,0.5)	(0.25,0.5)	(0.5,0.75)
F8	(0.75,1)	(0.75,1)	(0.75,1)	(0.5,0.75)	(0,0.25)	(0,0.25)	(0.5,0.75)	(0.5,0.75)	(0.75,1)	(0.25,0.5)	(0,0.25)	(0,0.25)
F9	(0.75,1)	(0.75,1)	(0.25,0.5)	(0.5,0.75)	(0.25,0.5)	(0.5,0.75)	(0.5,0.75)	(0.25,0.5)	(0.75,1)	(0.25,0.5)	(0.25,0.5)	(0.5,0.75)
F10	(0.75,1)	(0.75,1)	(0.25,0.5)	(0.75,1)	(0.25,0.5)	(0.5,0.75)	(0.5,0.75)	(0.5,0.75)	(0.5,0.75)	(0,0.25)	(0,0.25)	(0.25,0.5)
F11	(0.75,1)	(0.75,1)	(0.25,0.5)	(0.75,1)	(0.25,0.5)	(0.25,0.5)	(0.25,0.5)	(0.5,0.75)	(0.25,0.5)	(0.25,0.5)	(0,0.25)	(0.25,0.5)
F12	(0.75,1)	(0.75,1)	(0,0.25)	(0.75,1)	(0,0.25)	(0,0.25)	(0,0.25)	(0.5,0.75)	(0.25,0.5)	(0,0.25)	(0,0.25)	(0,0.25)

Step 2: The average grey strength–influence matrix shows as Table 8. After averaging the grey initial values by Equation (6), the standardization of the grey numbers can be obtained by using Equations (7) and (8), which transform the values into the standard interval form in Table 8. Most values contain the interval [0.4,0.6]. The biggest value is [0.65,0.9], while [0.225,0.45] is the smallest interval number.

Step 3. The crisp strength–influence matrix is shown in Table 9, which is established from the average grey strength–influence matrix. The interval numbers can integrate most information, and the further analysis needs to convert the interval form to a crisp value by using Equations (9) and (10). As a result, the total crisp values are calculated as shown in Table 9, which retain four decimals for ensuring accuracy.

Table 8. Average grey matrix of expert evaluations.

	F1	F2	F3	F4	F5	F6	F7	F8	F9	F10	F11	F12
F1	[0.5,0.725]	[0.475,0.725]	[0.375,0.625]	[0.35,0.575]	[0.4,0.6]	[0.45,0.7]	[0.375,0.575]	[0.4,0.65]	[0.525,0.75]	[0.55,0.8]	[0.4,0.625]	[0.375,0.6]
F2	[0.4,0.625]	[0.6,0.825]	[0.375,0.625]	[0.275,0.525]	[0.4,0.65]	[0.425,0.675]	[0.325,0.575]	[0.3,0.5]	[0.375,0.625]	[0.45,0.7]	[0.375,0.6]	[0.4,0.65]
F3	[0.5,0.725]	[0.5,0.75]	[0.525,0.775]	[0.425,0.675]	[0.4,0.625]	[0.4,0.65]	[0.325,0.55]	[0.275,0.5]	[0.45,0.7]	[0.525,0.775]	[0.35,0.55]	[0.375,0.6]
F4	[0.475,0.725]	[0.475,0.725]	[0.425,0.675]	[0.575,0.825]	[0.35,0.6]	[0.35,0.6]	[0.25,0.5]	[0.425,0.675]	[0.3,0.55]	[0.375,0.6]	[0.35,0.575]	[0.375,0.625]
F5	[0.475,0.7]	[0.5,0.75]	[0.55,0.8]	[0.5,0.725]	[0.45,0.675]	[0.525,0.775]	[0.4,0.65]	[0.375,0.625]	[0.425,0.675]	[0.4,0.625]	[0.35,0.55]	[0.375,0.625]
F6	[0.4,0.65]	[0.35,0.575]	[0.45,0.7]	[0.35,0.6]	[0.3,0.525]	[0.5,0.75]	[0.325,0.575]	[0.35,0.6]	[0.4,0.625]	[0.375,0.625]	[0.375,0.6]	[0.225,0.45]
F7	[0.5,0.725]	[0.5,0.75]	[0.425,0.65]	[0.425,0.675]	[0.425,0.675]	[0.425,0.675]	[0.375,0.625]	[0.35,0.575]	[0.45,0.7]	[0.375,0.6]	[0.3,0.55]	[0.4,0.625]
F8	[0.65,0.9]	[0.6,0.825]	[0.525,0.775]	[0.475,0.725]	[0.3,0.55]	[0.35,0.6]	[0.45,0.7]	[0.475,0.725]	[0.45,0.675]	[0.375,0.625]	[0.25,0.5]	[0.35,0.6]
F9	[0.4,0.6]	[0.5,0.75]	[0.375,0.625]	[0.475,0.725]	[0.25,0.475]	[0.45,0.7]	[0.35,0.6]	[0.275,0.5]	[0.6,0.85]	[0.375,0.625]	[0.275,0.475]	[0.35,0.6]
F10	[0.4,0.65]	[0.4,0.625]	[0.425,0.65]	[0.45,0.675]	[0.225,0.45]	[0.225,0.475]	[0.225,0.45]	[0.35,0.6]	[0.525,0.775]	[0.475,0.725]	[0.325,0.55]	[0.275,0.525]
F11	[0.525,0.775]	[0.525,0.75]	[0.375,0.625]	[0.55,0.8]	[0.275,0.525]	[0.35,0.6]	[0.35,0.6]	[0.325,0.55]	[0.275,0.475]	[0.375,0.625]	[0.375,0.575]	[0.275,0.525]
F12	[0.45,0.675]	[0.55,0.8]	[0.375,0.625]	[0.4,0.65]	[0.25,0.45]	[0.4,0.65]	[0.325,0.55]	[0.325,0.55]	[0.4,0.65]	[0.425,0.675]	[0.4,0.65]	[0.5,0.75]

Table 9. The crisp strength–influence matrix.

	F1	F2	F3	F4	F5	F6	F7	F8	F9	F10	F11	F12
F1	0.1271	0.1468	0.0000	0.0720	0.1846	0.2231	0.1585	0.1477	0.2966	0.2061	0.1641	0.1633
F2	0.0014	0.2770	0.0000	0.0000	0.2061	0.1983	0.1177	0.0278	0.1270	0.0892	0.1367	0.2004
F3	0.1271	0.1759	0.1453	0.1488	0.1944	0.1736	0.1108	0.0000	0.2189	0.1769	0.1039	0.1633
F4	0.1036	0.1468	0.0484	0.2975	0.1477	0.1240	0.0305	0.1769	0.0352	0.0000	0.1094	0.1719
F5	0.0957	0.1759	0.1696	0.2137	0.2500	0.2975	0.2050	0.1184	0.1883	0.0277	0.1039	0.1719
F6	0.0055	0.0000	0.0727	0.0744	0.0833	0.2727	0.1177	0.0892	0.1489	0.0015	0.1367	0.0000
F7	0.1271	0.1759	0.0469	0.1488	0.2354	0.1983	0.1759	0.0833	0.2189	0.0000	0.0595	0.1905
F8	0.3327	0.2770	0.1453	0.1983	0.0892	0.1240	0.2632	0.2354	0.2079	0.0015	0.0016	0.1435
F9	0.0000	0.1759	0.0000	0.1983	0.0278	0.2231	0.1468	0.0000	0.4026	0.0015	0.0273	0.1435
F10	0.0055	0.0554	0.0469	0.1665	0.0000	0.0000	0.0000	0.0892	0.3108	0.1184	0.0820	0.0581
F11	0.1691	0.1939	0.0000	0.2727	0.0599	0.1240	0.1468	0.0556	0.0000	0.0015	0.1294	0.0581
F12	0.0643	0.2341	0.0000	0.1240	0.0278	0.1736	0.1108	0.0556	0.1577	0.0600	0.1753	0.3142

Step 4. The normalized strength–influence matrix was created in Table 10, containing the positive numbers which are less than 1 since it is necessary to control different variables within the same scale through the process of standardization as listed in Equation (11).

Table 10. The normalized crisp strength–influence matrix.

	F1	F2	F3	F4	F5	F6	F7	F8	F9	F10	F11	F12
F1	0.0070	0.0081	0.0000	0.0040	0.0102	0.0123	0.0088	0.0082	0.0164	0.0114	0.0091	0.0090
F2	0.0001	0.0153	0.0000	0.0000	0.0114	0.0110	0.0065	0.0015	0.0070	0.0049	0.0076	0.0111
F3	0.0070	0.0097	0.0080	0.0082	0.0107	0.0096	0.0061	0.0000	0.0121	0.0098	0.0057	0.0090
F4	0.0057	0.0081	0.0027	0.0164	0.0082	0.0069	0.0017	0.0098	0.0019	0.0000	0.0060	0.0095
F5	0.0053	0.0097	0.0094	0.0118	0.0138	0.0164	0.0113	0.0065	0.0104	0.0015	0.0057	0.0095
F6	0.0003	0.0000	0.0040	0.0041	0.0046	0.0151	0.0065	0.0049	0.0082	0.0001	0.0076	0.0000
F7	0.0070	0.0097	0.0026	0.0082	0.0130	0.0110	0.0097	0.0046	0.0121	0.0000	0.0033	0.0105
F8	0.0184	0.0153	0.0080	0.0110	0.0049	0.0069	0.0145	0.0130	0.0115	0.0001	0.0001	0.0079
F9	0.0000	0.0097	0.0000	0.0110	0.0015	0.0123	0.0081	0.0000	0.0223	0.0001	0.0015	0.0079
F10	0.0003	0.0031	0.0026	0.0092	0.0000	0.0000	0.0000	0.0049	0.0172	0.0065	0.0045	0.0032
F11	0.0093	0.0107	0.0000	0.0151	0.0033	0.0069	0.0081	0.0031	0.0000	0.0001	0.0072	0.0032
F12	0.0036	0.0129	0.0000	0.0069	0.0015	0.0096	0.0061	0.0031	0.0087	0.0033	0.0097	0.0174

Step 5. The total strength–influence matrix was calculated between 0.0001 and 0.0232 as shown in Table 11 after utilizing Equation (12). The diagonal line represents the strength of the factor itself, while the other positions represent the degree of influence of the factor influence on other factors.

Table 11. The total strength–influence matrix.

	F1	F2	F3	F4	F5	F6	F7	F8	F9	F10	F11	F12
F1	0.0075	0.0091	0.0003	0.0050	0.0109	0.0135	0.0096	0.0087	0.0177	0.0117	0.0097	0.0099
F2	0.0004	0.0161	0.0002	0.0006	0.0120	0.0119	0.0071	0.0018	0.0078	0.0051	0.0081	0.0118
F3	0.0074	0.0106	0.0083	0.0091	0.0114	0.0107	0.0068	0.0004	0.0133	0.0101	0.0063	0.0099
F4	0.0062	0.0089	0.0030	0.0172	0.0088	0.0078	0.0023	0.0103	0.0026	0.0002	0.0065	0.0103
F5	0.0059	0.0108	0.0098	0.0129	0.0147	0.0178	0.0122	0.0071	0.0116	0.0018	0.0064	0.0105
F6	0.0006	0.0005	0.0042	0.0047	0.0050	0.0158	0.0070	0.0052	0.0088	0.0002	0.0079	0.0004
F7	0.0074	0.0107	0.0029	0.0090	0.0138	0.0122	0.0105	0.0051	0.0132	0.0002	0.0039	0.0114
F8	0.0191	0.0166	0.0083	0.0119	0.0059	0.0082	0.0155	0.0137	0.0129	0.0005	0.0008	0.0090
F9	0.0002	0.0104	0.0001	0.0117	0.0020	0.0132	0.0086	0.0003	0.0232	0.0002	0.0019	0.0086
F10	0.0005	0.0036	0.0027	0.0098	0.0002	0.0005	0.0004	0.0051	0.0179	0.0066	0.0047	0.0037
F11	0.0097	0.0114	0.0002	0.0157	0.0039	0.0077	0.0086	0.0035	0.0006	0.0003	0.0077	0.0038
F12	0.0040	0.0138	0.0001	0.0076	0.0020	0.0105	0.0067	0.0035	0.0095	0.0035	0.0103	0.0182

Step 6. Sums of the rows T and columns L are obtained by Equations (13) and (14) in a total strength–influence matrix. Looking through the T column in Table 12, F8 has the maximum value of 0.1224, and F10 has the minimum value of 0.0558. In L column, the max value is 0.1391 for F9, and the min value is 0.0402 for F13. In addition, (T + L) values are utilized to measure the degree of prominence, and (T − L) values are computed to identify cause and effect factors in Table 12. The max value of (T + L) is F9 with 0.2196, and the min is F10 with 0.0962. On the other hand, F3 has the max value of (T − L) as 0.0641, and F6 has the min value as −0.0694. Then, a causal graph is shown in Figure 2 by placing the (T + L) data set on the horizontal axis and the (T − L) data set on the vertical axis.

Table 12. Prominence and relation of value elements.

Factors	T	L	T + L	T − L
F1	0.1136	0.0689	0.1825	0.0447
F2	0.0829	0.1225	0.2055	−0.0396
F3	0.1043	0.0402	0.1444	0.0641
F4	0.0841	0.1153	0.1994	−0.0311
F5	0.1215	0.0906	0.2121	0.0309
F6	0.0601	0.1296	0.1897	−0.0694
F7	0.1002	0.0954	0.1955	0.0048
F8	0.1224	0.0647	0.1872	0.0577
F9	0.0805	0.1391	0.2196	−0.0586
F10	0.0558	0.0404	0.0962	0.0154
F11	0.0730	0.0743	0.1473	−0.0012
F12	0.0898	0.1073	0.1971	−0.0176

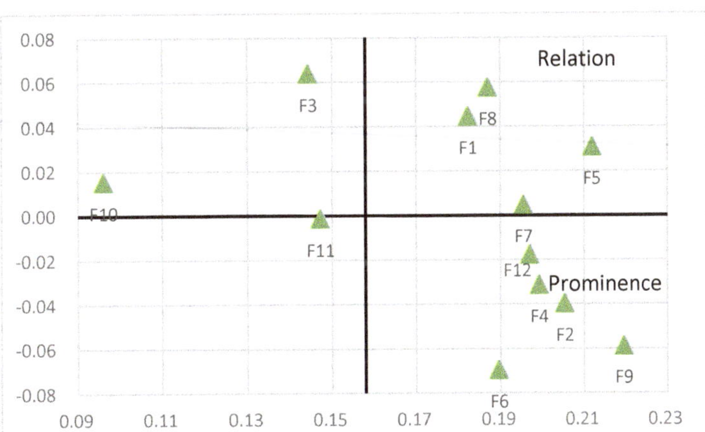

Figure 2. The cause−effect graph. Codes are the abbreviations of Factors.

Step 7. A threshold value (β) was computed using Equation (15). An interaction matrix that depicts the interrelationships between factors is created by the values greater than β as in Table 13. F6, F2, F9, F4, and F12 have more than nine interactions, respectively, whereas F1, F10, F3, and F8 have less than three interactions. Furthermore, the network diagram of interrelationships among factors can be illustrated in Figure 3.

Table 13. Interaction matrix of factors.

	F1	F2	F3	F4	F5	F6	F7	F8	F9	F10	F11	F12
F1		Δ			Δ	Δ	Δ	Δ	Δ	Δ	Δ	Δ
F2		Δ			Δ	Δ			Δ		Δ	Δ
F3		Δ	Δ	Δ	Δ	Δ			Δ	Δ		Δ
F4		Δ		Δ	Δ	Δ		Δ				Δ
F5		Δ	Δ	Δ	Δ	Δ	Δ		Δ			Δ
F6						Δ			Δ		Δ	
F7		Δ		Δ	Δ	Δ	Δ		Δ			Δ
F8	Δ	Δ	Δ	Δ		Δ	Δ	Δ	Δ			Δ
F9		Δ		Δ		Δ	Δ		Δ			Δ
F10				Δ					Δ			
F11	Δ	Δ		Δ		Δ	Δ				Δ	
F12		Δ		Δ		Δ			Δ		Δ	Δ

Δ presents the interrelationship between factors.

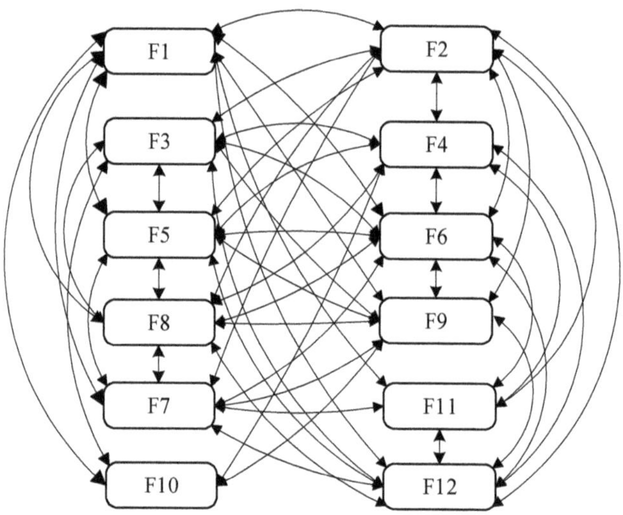

Figure 3. Network diagram of interrelationships among factors.

5. Discussion

In most cases, we encounter complex MCDM problems in which the factors are mutually influenced by each other. Due to the dependencies between various factors, it is not true that any one factor can improve the entire system. Therefore, it is necessary to identify the interrelationship of the factors in the causal group that can be improved and thus influence the entire system. Considering the above situation, this study proposes a novel combination of fuzzy-Delphi and grey-WINGS techniques to illustrate the causal relationships among the factors of AGSCM. To select the relatively more important factors, a threshold of 0.6 was set in the fuzzy-Delphi method. Furthermore, utilizing the integrated grey WINGS approach, the causal relationships between the factors can be identified by aggregating the group subjective assessment from various decisionmakers. As a result, the integrated fuzzy-DELPHI grey-WINGS methods can make a significant contribution to the MCDM employed in the AGSCM.

Based on the values of (T + L) in Table 12, the factors are prioritized as F9 > F5 > F2 > F4 > F12 > F7 > F6 > F8 > F1 > F11 > F3 > F10. Moreover, the ranking of cause–effect relationships is based on (T − L) values. Qualitative and prioritized ranking of the factors in the causal group helps to identify about how much influence each factor has. Based on positive and negative signs, the factors can be categorized into two parts as causal and effect factors in Table 12. The causal factors can be sorted as F3 > F8 > F1 > F5 > F10 > F7, and the ranking of effect factors is obtained as F11 > F12 > F4 > F2 > F9 > F6. Through Table 12 and Figure 3, produce Quality (F3) was found to be the prime causal factor with a value of 0.0641. Price of product (F8) and green consciousness (F1) followed the primary factor with values 0.0577 and 0.0447. The environmental laws (F5), stockholders' requirement (F10), and technology (F7), also can be categorized as driver factors, since the values are 0.0309, 0.0154, and 0.0048, which are greater than 0. These factors' impacts are higher than other factors, such as monitoring (F11), income level (F12), customers' demand (F4), government subsidies (F2), cost (F9), and green operation (F6). In order to demonstrate the advantage of this model, the result of DEMATEL was calculated to compare with WINGS, which is derived from DEMATEL. As shown in Table 14, most causal and effect factors are the same except for F10, which is the same factor with min T + L value between the two methods. Furthermore, F11, F7, F1, F12, F2, and F5 have a similar sequence to T + L values, but the other factors are different in both methods. The discrepancy is caused by the assumption that the WINGS considers the strength of the indicator itself, while DEMATEL omits these ingredients, which lacks a certain degree of accuracy.

Table 14. The values calculated by DEMATEL.

Factors	T	L	T + L	T − L
F1	0.0986	0.0858	0.1844	0.0129
F2	0.0887	0.0989	0.1876	−0.0102
F3	0.0942	0.0907	0.1849	0.0035
F4	0.0894	0.0905	0.1799	−0.0012
F5	0.1000	0.0866	0.1866	0.0133
F6	0.0835	0.0891	0.1726	−0.0056
F7	0.0950	0.0863	0.1813	0.0087
F8	0.0977	0.0906	0.1883	0.0071
F9	0.0858	0.0918	0.1777	−0.0060
F10	0.0815	0.0892	0.1708	−0.0077
F11	0.0867	0.0937	0.1803	−0.0070
F12	0.0887	0.0964	0.1850	−0.0077

Further analysis should be performed by categorizing all the factors into various quadrants, with factors above the X-axis being prominent as causal factors, and factors below the X-axis being effectors due to their dependence on causal factors. As illustrated in Figure 2, all the factors can be classified into four distinct clusters, where quadrant 1 is the least relevant factor or the least important factor. Monitoring (F11) lies in this group. Quadrant 2 is the causal group of factors that have a driving effect on other factors, but a weaker driving effect. Stockholders' requirement (F10) and product quality (F3) belong to this area. The shareholders generally set the goals of corporate development based on their requirements, which in turn influence various activities, including production, sales, and management operations. The next quadrant 3 is the most important and critical factor in the causal group. Green consciousness (F1), product price (F8), environmental law (F5), and technology (F7) belong to this group, thus indicating their importance to AGSCM. As discussed above, these factors have a high degree of prominence and relationship, which are priorities in AGSCM, since they can dominate other influencing factors. The fourth quadrant is for factors of high importance in the effect group, which require immediate management attention and control to improve AGSCM. Green operations (F6), cost (F9), government subsidies (F2), customer demand (F4), and income level (F12) are in this area, which integrates the activities of various parties, such as government, consumers, and companies for improving the development of AGSCM.

6. Conclusions

This study concentrates on the hierarchical evaluation structure in a complete model and proposes a novel approach using fuzzy Delphi and grey WINGS to resolve the interrelationships and incomplete information to acquire the strength and relationship between the factors of AGSCM. The practical implications and insightful conclusions of this study can be explained as follows:

With the globalization of climate change, food crisis, and the issue of the vulnerability of the agricultural supply chain, AGSCM is a complex MCDM project, which requires high priority by any organizations that are facing competition and pressure from enterprises, society, and governments. Therefore, the AGSCM needs to be improved through the optimization of influencing factors. To meet the requirements of green development, managers and policy makers strike a balance between efficiency and redundancy in the AGSCM. It is very important for the top managers to actively focus on the critical factors.

In this paper, identifying the critical factors and the corresponding causal relationships in AGSCM is the purpose. These findings suggest some preliminary guidance for the successful implementation of AGSCM. In this paper, the novel integrated method utilizes a structural modeling tool based on fuzzy Delphi and grey WINGS to evaluate the various factors of AGSCM. The fuzzy-Delphi technique is a qualitative approach for gathering opinions from various participants, which can capture the ambiguity and uncertainty in the data. By combining grey systems theory with this method, it is quite practical for integrating the preferences and views of different experts. Through the causal diagram, the factors can be divided into cause-and-effect groups. From a research perspective, this approach is valuable for assessing the relative impact and strength of the various relationships in MCDM.

The implementation of the proposed model illustrates some perspectives on the actual application and management implications of AGSCM. Some fundamental factors have been found to adjust plan and solutions. Furthermore, the cause-and-effect relationships can help to identify the factors that practitioners and researchers need to consider in AGSCM.

Product quality (F3), price of product (F8), green consciousness (F1), and environmental law (F5) are the most vulnerable causal factors of AGSCM, which need more attention. Product price (F8) and quality (F3) are the eternal concerns of consumers. Product quality (F3) is one of the main tools for marketers to position themselves in the market, which has two components: level and consistency. Agricultural product quality means the ability of an agricultural product to perform its function, including its nutrition, taste, safety, and other attributes. Price of product (F8) is the basis for establishing a diversified market mechanism, designing an efficient incentive mechanism and playing an important role in positive incentive effect, which is related to the whole process of production and marketing. Reducing the cost of green agricultural products can improve the operation of AGSCM. Environmental laws (F5) and green consciousness (F1) are the important factors for improvement of AGSCM, which refer to the activities to reduce and minimize environmental pollution of various factors. Furthermore, green consciousness (F1) improves the social image and environmental performance with new life cycle assessment, which would influence stockholders' perceptions. Environmental laws (F5) can guide agricultural production operators to scientific planting, breeding, application of pesticides, fertilizers, and other agricultural inputs. Moreover, the agricultural nonpoint pollution and other agricultural waste can also be reduced, so that AGSCM performance could be greatly developed.

Consumer demand (F4) is the number of items which consumers are able and willing to buy with any given price. The former is influenced by the level of demand for the good, the price of the good, and the price of the substitute good, while the latter is influenced by the consumer's willingness to buy and the actual income level. Thus, it can be stated that the price of the agricultural product determines the quantity of consumer demand. Stockholders' requirements (F10) are directly associated with activities of green product and process in AGSCM, as well as require incorporating green innovation for modifying product green operation, cost control, and satisfying customers' demand.

Cost(F9) is the economic value of the resources consumed to produce and sell a certain type and quantity of products measured in money. The cost of agricultural products is influenced by a variety of factors, which require focusing on. Moreover, technology (F7) is an important support to improve agricultural production capacity and competitiveness. Agricultural technology is an irreplaceable and important guarantee for the promotion of supply chain management, which is an important support to promote the development of the agricultural economy. It is necessary to strengthen government support for agricultural technology promotion, deepen the reform of the agricultural technology promotion mechanism, innovate in the agricultural technology promotion organization, and form a socialized agricultural technology service system, which is necessary to adapt to the development of AGSCM.

Government subsidies (F2) can improve the efficiency of the entire green agricultural production, thus promoting the motivation of agricultural supply chain participants to utilize green technology and supply green agricultural products. Moreover, since government subsidies can compensate some costs of green product producers, these producers can offer green products at lower prices. For the whole society, government subsidies for green agricultural products improve the willingness of consumers to pay for green consumption and increase the consumer surplus that consumers can obtain by consuming green agricultural products. Monitoring (F11) refers to the management of political, economic, and social public affairs by the relevant departments, which can supervise and manage the behavior of the subjects at all levels in the green agricultural supply chain through laws and regulations. Monitoring is not only conducive to maintaining fair development rules, but also can create a harmonious and stable social environment, thus making the green supply chain develop in a better and healthier way.

In summary, all participants of AGSCM can analyze each influencing factor and its supporting causes, or they can identify the causal links of each influencing factor through a cause–effect diagram. This can help them identify and categorize those factors and their relationships that need more attention.

This paper has some limitations. Firstly, though a sizable number of specialists took part in the investigation, there might still be some bias in the experts' assessments, and more experts can be invited to verify the statistical results of this study. Secondly, we have considered 19 factors of AGSCM, and more factors can be added at the expense of complexity. From this study, future studies could use other MCDM approaches, such as DEMATEL and ANP, and results can be compared to check the accuracy of grey WINGS. Furthermore, this proposed method could be extended to other MCDM problems in different industries, such as healthcare, the environment, pollution, transportation, etc.

Author Contributions: Conceptualization, methodology, formal analysis, writing paper, software, original draft preparation, M.W.; reviewing and editing, supervision, K.Z. All authors have read and agreed to the published version of the manuscript.

Funding: This research was funded by the National Social Science Foundation of China grant number 21BJY027.

Institutional Review Board Statement: Not applicable.

Informed Consent Statement: Not applicable.

Data Availability Statement: Not applicable.

Conflicts of Interest: The authors declare no conflict of interest.

References

1. Cui, L.; Guo, S.; Zhang, H. Coordinating a Green Agri-Food Supply Chain with Revenue-Sharing Contracts Considering Retailers' Green Marketing Efforts. *Sustainability* **2020**, *12*, 1289. [CrossRef]
2. Tomasiello, S.; Alijani, Z. Fuzzy-Based Approaches for Agri-Food Supply Chains: A Mini-Review. *Soft Comput.* **2021**, *25*, 7479–7492. [CrossRef]
3. Gardas, B.; Raut, R.; Cheikhrouhou, N.; Narkhede, B. A Hybrid Decision Support System for Analyzing Challenges of the Agricultural Supply Chain. *Sustain. Prod. Consum.* **2019**, *18*, 19–32. [CrossRef]
4. Jum'a, L.; Zimon, D.; Ikram, M. A Relationship between Supply Chain Practices, Environmental Sustainability and Financial Performance: Evidence from Manufacturing Companies in Jordan. *Sustainability* **2021**, *13*, 2152. [CrossRef]
5. Jum'a, L.; Zimon, D.; Ikram, M.; Madzik, P. Towards a Sustainability Paradigm; the Nexus between Lean Green Practices, Sustainability-Oriented Innovation and Triple Bottom Line. *Int. J. Prod. Econ.* **2022**, *245*, 108393. [CrossRef]
6. Baghizadeh, K.; Zimon, D.; Jum'a, L. Modeling and Optimization Sustainable Forest Supply Chain Considering Discount in Transportation System and Supplier Selection under Uncertainty. *Forests* **2021**, *12*, 964. [CrossRef]
7. Barman, A.; Das, R.; De, P.; Sana, S. Optimal Pricing and Greening Strategy in a Competitive Green Supply Chain: Impact of Government Subsidy and Tax Policy. *Sustainability* **2021**, *13*, 9178. [CrossRef]
8. Chiu, C.; Cheng, C.; Wu, T. Integrated Operational Model of Green Closed-Loop Supply Chain. *Sustainability* **2021**, *13*, 6041. [CrossRef]
9. Zimon, D.; Tyan, J.; Sroufe, R. Implementing Sustainable Supply Chain Management: Reactive, Cooperative, and Dynamic Models. *Sustainability* **2019**, *11*, 7227. [CrossRef]
10. Du, Y.; Zhang, D.; Zou, Y. Sustainable Supplier Evaluation and Selection of Fresh Agricultural Products Based on IFAHP-TODIM Model. *Math. Probl. Eng.* **2020**, *2020*, 4792679. [CrossRef]
11. Kumar, S.; Raut, R.; Nayal, K.; Kraus, S.; Yadav, V.; Narkhede, B. To Identify Industry 4.0 and Circular Economy Adoption Barriers in the Agriculture Supply Chain by Using ISM-ANP. *J. Clean. Prod.* **2021**, *293*, 126023. [CrossRef]
12. Swain, M.; Zimon, D.; Singh, R.; Hashmi, M.; Rashid, M.; Hakak, S. LoRa-LBO: An Experimental Analysis of LoRa Link Budget Optimization in Custom Build IoT Test Bed for Agriculture 4.0. *Agronomy* **2021**, *11*, 820. [CrossRef]
13. Park, A.; Li, H. The Effect of Blockchain Technology on Supply Chain Sustainability Performances. *Sustainability* **2021**, *13*, 1726. [CrossRef]
14. Jum'a, L.; Ikram, M.; Alkalha, Z.; Alaraj, M. Factors Affecting Managers' Intention to Adopt Green Supply Chain Management Practices: Evidence from Manufacturing Firms in Jordan. *Environ. Sci. Pollut. Res.* **2022**, *29*, 5605–5621. [CrossRef] [PubMed]
15. Saaty, T. The Modern Science of Multicriteria Decision Making and Its Practical Applications: The AHP/ANP Approach. *Oper. Res.* **2013**, *61*, 1101–1118. [CrossRef]
16. Abdullah, L.; Zulkifli, N. Integration of fuzzy AHP and interval type-2 fuzzy DEMATEL: An application to human resource management. *Expert Syst. Appl.* **2015**, *42*, 4397–4409. [CrossRef]
17. Razavitoosi, S.L.; Samani, J.M.V. Prioritizing Watersheds Using a Novel Hybrid Decision Model Based on Fuzzy DEMATEL, Fuzzy ANP and Fuzzy VIKOR. *Water Resour. Manag.* **2017**, *42*, 2853–2867. [CrossRef]
18. Lee, H.-S.; Tzeng, G.-H.; Yeih, W.; Wang, Y.-J.; Yang, S.-C. Revised DEMATEL: Resolving the Infeasibility of DEMATEL. *Appl. Math. Model.* **2013**, *37*, 6746–6757. [CrossRef]
19. Bakir, S. Exploring the Critical Determinants of Environmentally Oriented Public Procurement Using the DEMATEL Method. *J. Environ. Manag.* **2018**, *225*, 325–335. [CrossRef]
20. Patil, S.K.; Kant, R. Knowledge management adoption in supply chain: Identifying critical success factors using fuzzy DEMATEL approach. *J. Modeling Manag.* **2014**, *9*, 160–178. [CrossRef]
21. Yao, L.; Yi, Z. A DEMATEL-Based Method for Linguistic Multiple Attributes Group Decision Making Using Strict t-Norms and t-Conorms. *Systems* **2022**, *10*, 98. [CrossRef]
22. Kaur, J.; Sidhu, R.; Awasthi, A.; Chauhan, S.; Goyal, S. A DEMATEL Based Approach for Investigating Barriers in Green Supply Chain Management in Canadian Manufacturing Firms. *Int. J. Prod. Res.* **2018**, *56*, 312–332. [CrossRef]
23. Michnik, J. Weighted Influence Non-Linear Gauge System (WINGS)—An Analysis Method for the Systems of Interrelated Components. *Eur. J. Oper. Res.* **2013**, *228*, 536–544. [CrossRef]
24. Zimon, D.; Madzik, P.; Sroufe, R. The Influence of ISO 9001 & ISO 14001 on Sustainable Supply Chain Management in the Textile Industry. *Sustainability* **2020**, *12*, 4282. [CrossRef]
25. Gong, R.; Xue, J.; Zhao, L.; Zolotova, O.; Ji, X.; Xu, Y. A Bibliometric Analysis of Green Supply Chain Management Based on the Web of Science (WOS) Platform. *Sustainability* **2019**, *11*, 3459. [CrossRef]
26. Mangla, S.; Luthra, S.; Rich, N.; Kumar, D.; Rana, N.; Dwivedi, Y. Enablers to Implement Sustainable Initiatives in Agri-Food Supply Chains. *Int. J. Prod. Econ.* **2018**, *203*, 379–393. [CrossRef]
27. Fu, H.; Li, J.; Li, Y.; Huang, S.; Sun, X. Risk Transfer Mechanism for Agricultural Products Supply Chain Based on Weather Index Insurance. *Complex. Constr. Mega Infrastruct. Proj.* **2018**, *2018*, 2369423. [CrossRef]
28. Deng, L.; Xu, W.; Luo, J. Optimal Loan Pricing for Agricultural Supply Chains from a Green Credit Perspective. *Sustainability* **2021**, *13*, 2365. [CrossRef]
29. Tseng, M.; Islam, M.; Karia, N.; Fauzi, F.; Afrin, S. A Literature Review on Green Supply Chain Management: Trends and Future Challenges. *Resour. Conserv. Recycl.* **2019**, *141*, 145–162. [CrossRef]

30. Yang, C.; Lien, S. Governance Mechanisms for Green Supply Chain Partnership. *Sustainability* **2018**, *10*, 2681. [CrossRef]
31. Herrmann, F.; Barbosa-Povoa, A.; Butturi, M.; Marinelli, S.; Sellitto, M. Green Supply Chain Management: Conceptual Framework and Models for Analysis. *Sustainability* **2021**, *13*, 8127. [CrossRef]
32. Tarigan, Z.; Siagian, H.; Jie, F. Impact of Enhanced Enterprise Resource Planning (ERP) on Firm Performance through Green Supply Chain Management. *Sustainability* **2021**, *13*, 4358. [CrossRef]
33. Lintukangas, K.; Kahkonen, A.; Ritala, P. Supply Risks as Drivers of Green Supply Management Adoption. *J. Clean. Prod.* **2016**, *112*, 1901–1909. [CrossRef]
34. Zaid, A.; Jaaron, A.; Bon, A. The Impact of Green Human Resource Management and Green Supply Chain Management Practices on Sustainable Performance: An Empirical Study. *J. Clean. Prod.* **2018**, *204*, 965–979. [CrossRef]
35. Yu, Y.; Zhang, M.; Huo, B. The Impact of Relational Capital on Green Supply Chain Management and Financial Performance. *Prod. Plan. Control* **2021**, *32*, 861–874. [CrossRef]
36. Govindan, K.; Khodaverdi, R.; Vafadarnikjoo, A. Intuitionistic Fuzzy Based DEMATEL Method for Developing Green Practices and Performances in a Green Supply Chain. *Expert Syst. Appl.* **2015**, *42*, 7207–7220. [CrossRef]
37. Hu, Q.; Xu, Q.; Xu, B. Introducing of Online Channel and Management Strategy for Green Agri-Food Supply Chain Based on Pick-Your-Own Operations. *Int. J. Environ. Res. Public Health* **2019**, *16*, 1990. [CrossRef]
38. Long, Q.; Tao, X.; Shi, Y.; Zhang, S. Evolutionary Game Analysis Among Three Green-Sensitive Parties in Green Supply Chains. *IEEE Trans. Evol. Comput.* **2021**, *25*, 508–523. [CrossRef]
39. Sharma, V.; Chandna, P.; Bhardwaj, A. Green Supply Chain Management Related Performance Indicators in Agro Industry: A Review. *J. Clean. Prod.* **2017**, *141*, 1194–1208. [CrossRef]
40. Han, Z.; Huo, B. The Impact of Green Supply Chain Integration on Sustainable Performance. *Ind. Manag. Data Syst.* **2020**, *120*, 657–674. [CrossRef]
41. Jeng, D.J.-F. Generating a Causal Model of Supply Chain Collaboration Using the Fuzzy DEMATEL Technique. *Comput. Ind. Eng.* **2015**, *87*, 283–295. [CrossRef]
42. Govindan, K.; Azevedo, S.; Carvalho, H.; Cruz-Machado, V. Lean, Green and Resilient Practices Influence on Supply Chain Performance: Interpretive Structural Modeling Approach. *Int. J. Environ. Sci. Technol.* **2015**, *12*, 15–34. [CrossRef]
43. Chang, C. Selection or Influence? The Position-Based Method to Analyzing Behavioral Similarity in Adolescent Social Networks. *Int. J. Adolesc. Youth* **2022**, *27*, 149–165. [CrossRef]
44. Singh, M.; Jawalkar, C.; Kant, S. Analysis of Drivers for Green Supply Chain Management Adaptation in a Fertilizer Industry of Punjab (India). *Int. J. Environ. Sci. Technol.* **2019**, *16*, 2915–2926. [CrossRef]
45. Bimpikis, K.; Fearing, D.; Tahbaz-Salehi, A. Multisourcing and Miscoordination in Supply Chain Networks. *Oper. Res.* **2018**, *66*, 1023–1039. [CrossRef]
46. Mohseni, S.; Baghizadeh, K.; Pahl, J. Evaluating Barriers and Drivers to Sustainable Food Supply Chains. *Math. Probl. Eng.* **2022**, *2022*, 4486132. [CrossRef]
47. Rejeb, A.; Rejeb, K.; Keogh, J.; Zailani, S. Barriers to Blockchain Adoption in the Circular Economy: A Fuzzy Delphi and Best-Worst Approach. *Sustainability* **2022**, *14*, 3611. [CrossRef]
48. Kumar, A.; Dixit, G. An Analysis of Barriers Affecting the Implementation of E-Waste Management Practices in India: A Novel ISM-DEMATEL Approach. *Sustain. Prod. Consum.* **2018**, *14*, 36–52. [CrossRef]
49. Govindan, K.; Nasr, A.; Karimi, F.; Mina, H. Circular Economy Adoption Barriers: An Extended Fuzzy Best-Worst Method Using Fuzzy DEMATEL and Supermatrix Structure. *Bus. Strategy Environ.* **2022**, *31*, 1566–1586. [CrossRef]
50. Zhou, Y.; Xu, L.; Shaikh, G. Evaluating and Prioritizing the Green Supply Chain Management Practices in Pakistan: Based on Delphi and Fuzzy AHP Approach. *Symmetry* **2019**, *11*, 1346. [CrossRef]
51. Wang, C.-N.; Nguyen, V.T.; Duong, D.H.; Do, H.T. A Hybrid Fuzzy Analytic Network Process (FANP) and Data Envelopment Analysis (DEA) Approach for Supplier Evaluation and Selection in the Rice Supply Chain. *Symmetry* **2018**, *10*, 221. [CrossRef]
52. Banaeian, N.; Mobli, H.; Fahimnia, B.; Nielsen, I.E.; Omid, M. Green Supplier Selection Using Fuzzy Group Decision Making Methods: A Case Study from the Agri-Food Industry. *Comput. Oper. Res.* **2018**, *89*, 337–347. [CrossRef]
53. Nasri, S.A.; Ehsani, B.; Hosseininezhad, S.J.; Safaie, N. A Sustainable Supplier Selection Method Using Integrated Fuzzy DEMA TEL-ANP-DEA Approach (Case Study: Petroleum Industry). *Environ. Dev. Sustain.* **2022**. [CrossRef]
54. Alkharabsheh, A.; Moslem, S.; Oubahman, L.; Duleba, S. An Integrated Approach of Multi-Criteria Decision-Making and Grey Theory for Evaluating Urban Public Transportation Systems. *Sustainability* **2021**, *13*, 2740. [CrossRef]
55. Nguyen, N.; Wang, C.; Dang, L.; Dang, L.; Dang, T. Selection of Cold Chain Logistics Service Providers Based on a Grey AHP and Grey COPRAS Framework: A Case Study in Vietnam. *Axioms* **2022**, *11*, 154. [CrossRef]
56. Sun, H.; Mao, W.; Dang, Y.; Xu, Y. Optimum Path for Overcoming Barriers of Green Construction Supply Chain Management: A Grey Possibility DEMATEL-NK Approach. *Comput. Ind. Eng.* **2022**, *164*, 107833. [CrossRef]
57. Chen, X.; Ding, Y.; Cory, C.; Hu, Y.; Wu, K.; Feng, X. A Decision Support Model for Subcontractor Selection Using a Hybrid Approach of QFD and AHP-Improved Grey Correlation Analysis. *Eng. Constr. Archit. Manag.* **2021**, *28*, 1780–1806. [CrossRef]
58. Kumar, A.; Anbanandam, R. Analyzing Interrelationships and Prioritising the Factors Influencing Sustainable Intermodal Freight Transport System: A Grey-DANP Approach. *J. Clean. Prod.* **2020**, *252*, 119769. [CrossRef]
59. Xu, J.; Jiang, X.; Wu, Z. A Sustainable Performance Assessment Framework for Plastic Film Supply Chain Management from a Chinese Perspective. *Sustainability* **2016**, *8*, 1042. [CrossRef]

60. Ishikawa, A.; Amagasa, M.; Shiga, T.; Tomizawa, G.; Tatsuta, R.; Mieno, H. The Max-Min Delphi Method and Fuzzy Delphi Method via Fuzzy Integration. *Fuzzy Sets Syst.* **1993**, *55*, 241–253. [CrossRef]
61. Cascella, M.; Miceli, L.; Cutugno, F.; Di Lorenzo, G.; Morabito, A.; Oriente, A.; Massazza, G.; Magni, A.; Marinangeli, F.; Cuomo, A.; et al. A Delphi Consensus Approach for the Management of Chronic Pain during and after the COVID-19 Era. *Int. J. Environ. Res. Public Health* **2021**, *18*, 13372. [CrossRef] [PubMed]
62. Markou, M.; Michailidis, A.; Loizou, E.; Nastis, S.A.; Lazaridou, D.; Kountios, G.; Allahyari, M.S.; Stylianou, A.; Papadavid, G.; Mattas, K. Applying a Delphi-Type Approach to Estimate the Adaptation Cost on Agriculture to Climate Change in Cyprus. *Atmosphere* **2020**, *11*, 536. [CrossRef]
63. van der Schans, M.; Yu, J.; Martin, G. Digital Luminaire Design Using LED Digital Twins—Accuracy and Reduced Computation Time: A Delphi4LED Methodology. *Energies* **2020**, *13*, 4979. [CrossRef]
64. Liu, S.; Li, Y.; Fu, S.; Liu, X.; Liu, T.; Fan, H.; Cao, C. Establishing a Multidisciplinary Framework for an Emergency Food Supply System Using a Modified Delphi Approach. *Foods* **2022**, *11*, 1054. [CrossRef] [PubMed]
65. Mei, W.-B.; Hsu, C.-Y.; Ou, S.-J. Research on the Evaluation Index System of the Construction of Communities Suitable for Aging by the Fuzzy Delphi Method. *Environments* **2020**, *7*, 92. [CrossRef]
66. Feng, Y.; Zhang, Z.; Tian, G.; Fathollahi-Fard, A.M.; Hao, N.; Li, Z.; Wang, W.; Tan, J. A Novel Hybrid Fuzzy Grey TOPSIS Method: Supplier Evaluation of a Collaborative Manufacturing Enterprise. *Appl. Sci.* **2019**, *9*, 3770. [CrossRef]
67. Wang, W.; Tian, Z.; Xi, W.; Tan, Y.R.; Deng, Y. The Influencing Factors of China's Green Building Development: An Analysis Using RBF-WINGS Method. *Build. Environ.* **2021**, *188*, 107425. [CrossRef]
68. Neira-Rodado, D.; Ortíz-Barrios, M.; De la Hoz-Escorcia, S.; Paggetti, C.; Noffrini, L.; Fratea, N. Smart Product Design Process through the Implementation of a Fuzzy Kano-AHP-DEMATEL-QFD Approach. *Appl. Sci.* **2020**, *10*, 1792. [CrossRef]
69. Pourmehdi, M.; Paydar, M.; Asadi-Gangraj, E. Reaching Sustainability through Collection Center Selection Considering Risk: Using the Integration of Fuzzy ANP-TOPSIS and FMEA. *Soft Comput.* **2021**, *25*, 10885–10899. [CrossRef]
70. Uygun, Ö.; Kaçamak, H.; Kahraman, Ü.A. An Integrated DEMATEL and Fuzzy ANP Techniques for Evaluation and Selection of Outsourcing Provider for a Telecommunication Company. *Comput. Ind. Eng.* **2015**, *86*, 137–146. [CrossRef]
71. Hatefi, S.M.; Tamošaitienė, J. An integrated fuzzy DEMATEL-fuzzy ANP model for evaluating construction projects by considering interrelationships among risk factors. *J. Civ. Eng. Manag.* **2019**, *25*, 114–131. [CrossRef]
72. Wang, M.; Tian, Y.; Zhang, K. The Fuzzy Weighted Influence Nonlinear Gauge System Method Extended with D Numbers and MICMAC. *Complex Intell. Syst.* **2022**. [CrossRef]
73. Padilla-Rivera, A.; Merveille, N. Social Circular Economy Indicators: Selection through Fuzzy Delphi Method. *Sustain. Prod. Consum.* **2021**, *26*, 101–110. [CrossRef]

agriculture

MDPI

Article

The Impact of Digital Technology on Land Rent-Out Behavior: Information Sharing or Exclusion?

Xiaofan Zuo [1] and Zhisheng Hong [2,*]

1 College of Humanities and Development Studies, China Agricultural University, Beijing 100193, China; zuoxiaofan@cau.edu.cn
2 Institute of Science and Development, Chinese Academy of Sciences, Beijing 100190, China
* Correspondence: hongzhisheng@casisd.cn

Abstract: In the digital age, it is critical to understand the nexus between digital technology (DT) and land rent-out behavior (LRB). It has implications for reducing the rate of land abandonment to achieve sustainable agricultural development. A large dataset (n = 5233) dating from 2016 and coming from the China Family Panel Studies (CFPS) is used to explore the impact of DT on LRB by applying several econometric models, also including the "Recursive Bivariate Probit (RBP) model" and "Chain Multiple Mediation effect (CMM) model". We provide empirical evidence that the DT's information sharing effect positively impacted LRB, while an opposite effect is observed by the "digital divide (DT_GAP)" i.e., information exclusion that negatively impacted LRB. We further test the effect of two other variables, namely "digital information dependence" and "non-farm jobs" supposed as mediating factors of DT and DT_GAP in influencing LRB, respectively in a positive and negative way. In particular, the variable "nonfarm jobs" plays a mediating role conditional on the variable "digital information dependence" as a mediating variable at the first level. In addition, statistical tests reveal that the impact of DT and the DT_GAP on LRB is not significant in terms of regional preferences but is significant in terms of age of householder and household income level.

Keywords: digital technology; land rent-out behavior; digital divide; China; RBP model; CMM model; CFPS

Citation: Zuo, X.; Hong, Z. The Impact of Digital Technology on Land Rent-Out Behavior: Information Sharing or Exclusion? *Agriculture* **2022**, *12*, 1046. https://doi.org/10.3390/agriculture12071046

Academic Editor: Massimo Monteleone

Received: 20 April 2022
Accepted: 12 July 2022
Published: 19 July 2022

1. Introduction

Measuring Digital Development: Facts and Figures 2021, published by International Telecommunication Union (ITU), shows that global Internet penetration is 59.5% as of 2020 and measures that it will reach 63% in 2021 [1]. As the largest developing country in the world, China has an Internet penetration rate of 73%, with 78.3% in urban areas and 59.3% in rural areas [2]. The information dividend released by the development of the Internet has contributed to the economic and social development of the world. The land is an important resource in agricultural production. Promoting the land production factor mobility is a key link to achieving the improvement of agricultural production efficiency. As a largely agricultural country, highly fragmented land and smallholder are the basic characteristics of China's agriculture. However, low level of agricultural mechanization, high degree of land fragmentation, and small-scale family farms are also the characteristics of the agricultural development constraints faced by most developing countries or regions. Therefore, promoting the land production factor mobility and integrating finely fragmented land are inevitable requirements for improving agricultural production efficiency and achieving sustainable development of the agricultural activity. Digital technology (DT) breaks through the limitation of time and space and brings about a change in information transmission. There are advantages to optimizing the allocation of land resources and promoting the mobility of land resources. However, it is undeniable that there is a digital divide (DT_GAP) caused by unevenness and inadequacy in the development of DT. It can produce information exclusion and weaken the positive impact brought by DT. China has realized the digital management of national land use status in 2014 [3]. In recent years, the Chinese government has actively promoted the reform of rural land digitization. It has pushed forward the digital management process of registration, transfer, and distribution of rural land. Taking China as an example, we reveal the impact of DT and the DT_GAP on land rent-out behavior (LRB) and how this impact can be interpreted. It is meaningful for developing countries to reduce the rate of land abandonment, improve agricultural efficiency, and achieve sustainable farmer livelihoods.

The nexus between DT and income levels has received extensive academic attention and has been thoroughly researched. Existing studies have strongly confirmed that the development of DT plays a positive role in global economic growth and poverty alleviation [4,5]. The impact of DT on agriculture development has also been extensively researched. Agricultural information, which is effectively supplied by DT, controls damage to crops by adverse factors (such as natural disasters) and achieves increased agricultural production [6]. At the same time, the distribution of production factors and the structure of cultivation are optimized by information access from DT, thus increasing agricultural productivity [7]. Agricultural productivity and efficiency are improved by artificial intelligence, which is an important application of DT, while the problem of labor shortages and sustainable agricultural development are addressed effectively [8]. For developing countries, the information problems that prevented smallholders from accessing markets are solved by the application of DT in agricultural production [9]. It is specifically practiced in China where DT is embedded in agricultural production. Agricultural cell phone SMS services had appeared in the Chinese agricultural market in the early 2000s. Farmers' price search costs before the market launch of agricultural products are reduced by SMS services, which improves farmers' position in the market. Therefore, farmers use agricultural information technology to obtain more information and increase the selling price of their agricultural products [10]. With the rise of e-commerce, e-commerce clustered villages (e.g., Taobao villages) promote e-commerce down to the rural market. The cluster development of rural e-commerce has broadened the channels for agricultural product sales [11,12].

The land is one of the key elements of agricultural production and has been focused on by agricultural economics. Good resource allocation can effectively improve productivity. Some studies have shown that the effective allocation of resources and the improvement of agricultural productivity are promoted by land transfer (i.e., an active land buying and selling market). When land transfer promotes large-scale operation, agricultural productivity is effectively improved and farmers' agricultural income is increased [13,14]. Philippines land reform, which included government land allocation and prohibition of alienation, reduces average farm size by 34% and agricultural productivity by 17%, which is a negative example [15]. In China's land reform, the Chinese central government has proposed the "Three Rights Separation" ("Three Rights Separation" refers to the separation of ownership right, contracting right (disposal right) and operation right of land. In China, the transfer of agricultural land refers to the transfer of operation rights). It encourages the transfer of operation rights to professional farmers to increase farm income by increasing the operation scale as far as possible [16,17]. In 2019, the scale of transferred land in China accounts for 28.94% of the total land [18]. In terms of land rental characteristics, land rentals from smallholders to other operators are very limited. Such characteristics highlight the long-term nature of smallholder agricultural production in China and the obstacles to expanding agricultural production on a larger scale [19].

LRB is driven by many factors, including economic factors, cultural factors, and individual characteristics [20]. For example, people who have experienced famine are more reluctant to rent out their land [21]. Household labor migration also has an effect on LRB, and this effect varies by the size of labor migration and region difference [22]. At the same time, numerous studies have shown that there is a deviation between land rental willingness and behavior [23–25]. In practice, the deviation arise mainly from the imperfection of the land rental market and related systems, and the lack of property income for farmers in process of land rental [24]. At this stage, the nexus between DT, more precisely Internet technology, and land use is also more fully justified. DT, represented by Internet technology, can enhance the accessibility of modern technologies in agriculture (e.g., agricultural machinery) and improve land use efficiency [26]. Meanwhile, DT can significantly improve information asymmetry in agricultural markets, while reducing cropland abandonment. An empirical study based on a sample of 8031 farming households showed that Internet use can reduce the abandonment of cropland by 43.20% [27].

It is clear that the mobility of land production factor is essential to improving land utilization [28]. However, the nexus between DT and land rental has been explored only preliminarily and is to a very limited extent. Related research has concluded that farmers' land rental behavior (including rent out and rent in) was significantly facilitated by access to agricultural information through the Internet [29]. Among them, the information-seeking ability is an important impact path of the Internet on land rental [30]. The negative impact of DT is also not negligible. DT_GAP contributes to the widening of the household wealth gap [31], the further polarization of the educational divide [32], exacerbating inequalities in healthcare accessibility [33], and exclusion of the aging population [34]. Meanwhile, DT also has negative effects on individuals' behaviors and perceptions, such as DT can exacerbate people's pessimism [35,36]. However, existing studies have not focused on the nexus between the DT_GAP and land rental.

In summary, there is a consensus in the existing literature on the positive role of DT, represented by the Internet, in promoting economic development and poverty alleviation. The positive impact of DT in promoting agricultural production efficiency and land utilization with information empowerment is also widely discussed. At the present stage, although the amount of literature on the impact of DT on land rental is limited, the positive effect of DT agricultural land rental has been initially affirmed. Undeniably, the existing studies still have the following shortcomings. On the one hand, existing studies have not paid attention to the impact of the DT_GAP generated by the uneven development of DT on LRB. On the other hand, in terms of the available literature, exploring the nexus between DT and LRB is still insufficient, and the mechanism of DT's impact on LRB has not been interpreted in depth.

Based on the existing literature, we will analyze the information sharing effect of DT's impact on LRB and how this impact can be interpreted. Meanwhile, we will also analyze the information exclusion effect of DT_GAP's impact on LRB and how this impact can be interpreted. Our research will enrich the studies on the nexus between DT and LRB, and fill the gap in the studies of DT_GAP's impact on LRB.

2. Theoretical Analysis and Research Hypotheses

2.1. Information Sharing and Exclusion of DT

In market economic activities, the theory of information asymmetry assumes that different people have different knowledge of information. Those who have more adequate information tend to be in a more advantageous position, while those who are poorly informed tend to be in a more disadvantageous position [37]. The imbalance caused by information asymmetry impacts the efficiency of market allocation. For example, the information-advantaged party always captures the surplus generated by information asymmetry. In the era of mobile Internet, as a representative of DT, the Internet not only shortens the time distance of information transmission but also improves the timeliness of the information and crosses the geographical limitation. The universal and shared nature of the Internet has reduced information asymmetry. It has improved access to information for the individuals who are in information disadvantaged. Meanwhile, the ICTs revolution has the potential to create new means of social exclusion [38], which is mainly reflected in the information and knowledge inequality brought about by DT_GAP.

In areas of abundantly DT application scenarios, the information sharing effect of DT has fully alleviated information dislocation. Information grabbing will be alleviated or eliminated with the widespread use of DT. In reality, however, a fully covered scenario for the application of DT does not exist. The primary DT_GAP is generated by hardware exclusion, which is mainly reflected in the lack of broadband access for the population in developmentally disadvantaged areas. The secondary DT_GAP is generated by use exclusion, which is mainly reflected in information exclusion of specific populations, including racial exclusion, aging exclusion, and economic or social development exclusion [39–41]. It can be considered that both the primary and secondary DT_GAP form a potential information exclusion. The DT_GAP exacerbates information grabbing by worsening the original information asymmetry. Therefore, while DT exerts its information-sharing effect, the DT_GAP generated by the uneven development of DT brings about an information exclusion effect.

2.2. Impact of DT on LRB: Theoretical Analysis and Research Hypothesis

The framework diagram of the theoretical analysis of DT's impact on LRB was reported in Figure 1.

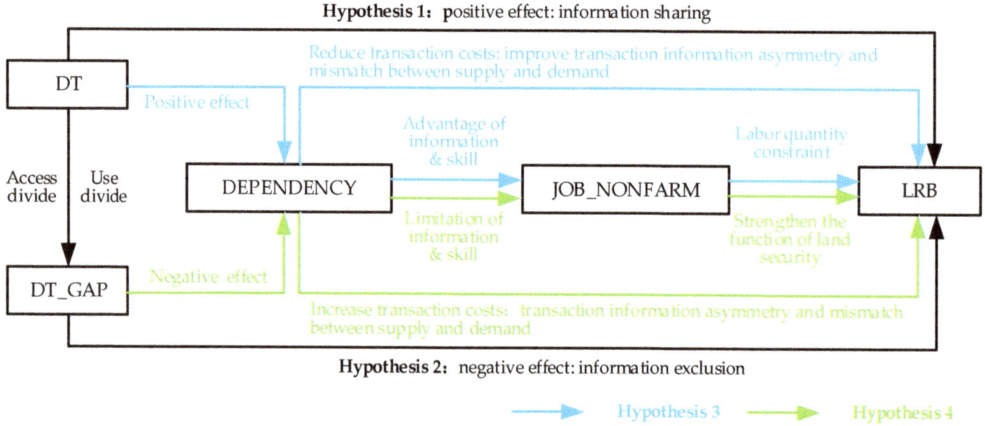

Figure 1. Theoretical analysis framework diagram: impact of DT on LRB.

Information mismatch and information asymmetry are the specific forms of transaction costs in the land rental process, while the search cost of matching supply and demand is reduced by DT. For the side of land rent-out, DT improves the bargaining power and increases the benefits of land rental [42]. Meanwhile, DTs are better than traditional information technologies in terms of timeliness and convenience of information. Therefore, the information advantage of DT increases the farmers' dependence on digital information channels (DEPENDENCY). With the increase in farmers' DEPENDENCY, the transaction costs caused by information asymmetry and supply-demand mismatch in land transactions are reduced, further facilitating the formation of LRB decisions for farmers' households. In the Internet era, the DEPENDENCY is critical for farmers to access nonfarm jobs (JOB_NONFARM). The DEPENDENCY gives farmers more advantage of information. Job seekers, who use DT, get better quality jobs than those who use traditional media [43]. Rural mobile workers with DT skills and DEPENDENCY have access to higher quality income. This is because their skills advantage and information advantage realize the substitution of low-skilled labor groups.

It is undeniable that there is a widespread real dilemma of inadequate and uneven development of DT, namely the DT_GAP problem, which is reflected in the access divide and the use divide. Individuals who use the Internet will further develop Internet knowledge, widening the gap between them and those who do not use the Internet [44], ultimately, there is an information exclusion effect on individuals who do not have access to digital or the Internet. DT_GAP deprives or excludes some individuals from accessing digital information channels, which is attributed to Internet access restrictions or lack of skills to use the Internet. Both of these restrictions reduce or prevent some groups from DEPENDENCY. This will further reduce the likelihood that farmers have access to JOB_NONFARM opportunities. The increase of JOB_NONFARM opportunities will weaken the social security function of land to some extent [45,46]. In other words, the decrease of JOB_NONFARM opportunities will strengthen the social security function of land, then reduce the probability of LRB.

Since the theory of New Economics of Labor Migration was proposed, the nexus between labor migration and factor markets in the place of emigration has been concerned [47]. Agricultural laborers engage in non-agricultural production, leading to a reduction in the number of laborers engaged in agricultural production, which changes the ratio of land to labor factors, resulting in a mismatch between the number of laborers and the scale of the existing agricultural industry. In short, the number of laborers constrains the scale of agricultural production. It can be expected that after the transfer of labor originally involved in agricultural production to the non-agricultural production sector, farm households will reconfigure the ratio of land to labor factors through the land rental market [48].

Based on the above analysis, we propose the following research hypothesis to be tested:

Hypothesis 1 (H1). *The information sharing effect of DT has a significant, direct, and positive effect on farmers' LRB.*

Hypothesis 2 (H2). *The information exclusion effect of DT (DT_GAP) has a significant negative effect on the LRB of farmers.*

Hypothesis 3 (H3). *DEPENDENCY and JOB_NONFARM are indirect factors, i.e., mediators of DTU, positively influencing LRB in sequence with each other.*

Hypothesis 4 (H4). *Similar to H3, DEPENDENCY and JOB_NONFARM are indirect factors, i.e., mediators of DT_GAP, negatively influencing LRB in sequence with each other.*

3. Data Sources, Variables, and Empirical Methods

3.1. Data Sources

China Family Panel Studies (CFPS) data were used in this article. This dataset was provided by the Institute of Social Science Survey (ISSS) of Peking University. CFPS focuses on the economic and non-economic welfare of Chinese residents, and many research topics, including economic activities, educational achievements, family relations and family dynamics, population migration, health, etc. It is a national, large-scale, multidisciplinary social survey project, which uses computer-assisted survey technology to conduct interviews [49]. CFPS program follows the relevant laws and policies of the People's Republic of China regarding the protection of personal information.

CFPS uses the implicit stratification method to draw multi-stage probability samples, and the samples of each sub-sample frame are obtained by three stages of drawing. The first-stage sample is county-level administrative units, the second-stage sample is village-level administrative units, and the third-stage sample is households. In the third stage, the sampling frame is constructed using the map address method, and the sample households are drawn using circular equiprobable sampling with random starting points. Through data cleaning, we obtained a sample of 5233 rural households, distributed across 455 communities in 25 provincial administrations in a new cross-sectional dataset.

3.2. Variable Settings and Basic Descriptive Statistics

The variables, definitions, and descriptive statistics are reported in Table 1.

Table 1. Variables, definitions, and descriptive statistics.

Variables (n = 5233)	Definition	Mean	Std. Dev. [1]
LRB	Whether the Interviewed household has LRB; 1 = yes, 0 = no	0.155	0.362
DT	1 = using the Internet; 0 = not using the Internet	0.219	0.414
AGE	Age of the householder (years)	51.975	13.547
GENDER	Gender of householder, 1 = male, 0 = female	0.563	0.496
HEALTH	Self-reported health of householder: from 1 = very healthy to 5 = very unhealthy	3.230	1.261
EDUCATION	Years of education of householder	6.127	4.213
PARTY	Householder's political identity as a member of the Communist Party of China, 1 = yes, 0 = no	0.077	0.267
AGE_F	Average age of household members (years)	48.567	12.488
HEALTH_F	Average self-reported health of family members: from 1 = very healthy to 5 = very unhealthy	3.129	0.966
EDUCATION_F	Average education of family members	6.078	3.445
MARRY	1 = married; 0 = other	0.869	0.338
FAMILYSIZE	Number of family members (living together)	4.101	1.999
PINCOME_F	Family net income per capita (logarithmic processing, yuan [2])	8.641	1.181
JOB_NONFARM	Household members engage in non-farm jobs, 1 = yes, 0 = no	0.720	0.449
DEPENDENCY	Dependence on Internet information channels: from 1 = unimportant to 5 = very important	1.829	1.353
DT_LEARNING	Frequency of using the Internet for learning, from 0 = infrequently to 7 = always	0.589	1.513
DT_WORKING	Frequency of using the Internet for working, from 0 = infrequently to 7 = always	0.471	1.340
DT_SOCIAL	Frequency of using the Internet for social interaction, from 0 = infrequently to 7 = always	1.166	2.443
DT_ENTERTAINMENT	Frequency of using the Internet for entertainment, from 0 = infrequently to 7 = always	1.035	2.234
DT_TRADE	Frequency of using the Internet for business activities, from 0 = infrequently to 7 = always	0.429	1.095
EASTERN [3]	1 = interviewed household is located in eastern China region, 0 = otherwise	0.245	0.430
CENTRAL	1 = interviewed household is located in central China region, 0 = otherwise	0.265	0.441
WESTERN	1 = interviewed household is located in the western China region, 0 = otherwise	0.367	0.482

[1] Std. Dev. refers to standard deviation. [2] Yuan is the Chinese currency: 1 USD = 6.49 Yuan (31 December 2015 onshore CNY closing price from Forex Capital Markets. New York). [3] According to the classification method of the National Bureau of Statistics of China, the provincial administrative regions of mainland China are divided into eastern, central, western, and northeastern regions. The eastern part includes Beijing, Tianjin, Hebei, Shanghai, Jiangsu, Zhejiang, Fujian, Shandong, Guangdong, and Hainan; The central part includes Shanxi, Anhui, Jiangxi, Henan, Hubei, and Hunan; The western part includes Inner Mongolia, Guangxi, Chongqing, Sichuan, Guizhou, Yunnan, Tibet, Shaanxi, Gansu, Qinghai, Ningxia, and Xinjiang; The northeast part includes Liaoning, Jilin, and Heilongjiang.

Explained variables. LRB indicates whether the respondent farm household had land rent-out behavior in 2015. The mean value of LRB shows that 15.5% of the overall sample had LRB.

Core explanatory variables. DT indicates whether or not household members used the Internet in 2015. The value of 1 is assigned if any member of the household uses the Internet, and 0 is assigned otherwise. The mean value of DT indicates that 21.9% of rural households have been able to access and use the Internet.

Control variables. Refer to the existing literature on the behavioral decisions of rural residents or households [35,50]. Eleven control variables covering both individual characteristics and household characteristics were selected in our research. For aspects of individual characteristics, we selected individual characteristic variables such as AGE, GENDER, HEALTH, EDUCATION, and PARTY of the householder (Since the CFPS questionnaire does not respond to specific information about the head of household. We have chosen to substitute information on the head of household for the household financial respondent (decision maker) by referring to the substitution guidelines commonly used in microdata studies). For aspects of family characteristics, we selected AGE_F, HEALTH_F, EDUCATION_F, MARRY, FAMILYSIZE, and PINCOME_F as control variables for household characteristics. The specific definitions and basic descriptive statistics for all control variables were shown in Table 1. The descriptions were not repeated here.

Auxiliary variables. In accordance with the research design, relevant auxiliary variables were introduced to discuss the impact of the information sharing effect and exclusion effect of DT on LRB. The variable JOB_NONFARM indicates whether or not a household member was engaged in nonfarm jobs during 2015. The mean value shows that 72.0% of rural households were engaged in nonfarm jobs, indicating that nonfarm jobs have become the main employment option for rural households in China. The variable DEPENDENCY indicates the level of dependence on Internet information channels. Both of these variables are mediating variables for discussing the impact of DTU on LRB. We also introduced the frequency of Internet usage scenarios, including the frequency of use in five scenarios: learning, work, social, entertainment, and trade activities, to measure DT_GAP. In terms of frequency of use in different scenarios, social interaction is one of the most frequent scenarios.

Regional variables. To control regional differences and counter the impact of possible unmeasured omitted variables on the model estimation results. Three regional variables were set by using the northeastern region of China as the reference region, namely EASTERN, CENTRAL, and WESTERN.

3.3. *Empirical Methods*

3.3.1. Probit Model

Based on the characteristics of the data distribution of our chosen explanatory variable LRB (dichotomous variables), the Probit model was selected to test the effect of DT on LRB. The specific numerical derivation process of Probit was not shown anymore, and the equation of the model was set in the form shown in Equation (1). In Equation (1), LRB_i denotes the LRB of the i-th sample household, $i = 1, 2, \ldots, 5233$. DTU_i denotes the DT of the i-th sample household, CV_{ir} denotes the r-th control variable of the i-th sample household, $r = 1, 2, \ldots, 11$. RV_{ik} denotes the k-th regional control variable for the i-th sample household, $k = 1, 2, 3$. β_0, β_1, β_{2r}, β_{3k} denotes the coefficient to be estimated, respectively. ε_i denotes the random error term.

$$LRB_i = \beta_0 + \beta_1 DTU_i + \beta_{2r} CV_{ir} + \beta_{3k} RV_{ik} + \varepsilon_i \quad (1)$$

3.3.2. Recursive Bivariate Probit (RBP) Model

There may be endogeneity between DT and LRB arising from omitted variables. That is, there may be some important explanatory variables that are omitted due to database limitations or subjective preference of the researcher, and these explanatory variables may be correlated with the model's disturbance term, i.e., the omitted explanatory variables are correlated with the existing explanatory variables, resulting in biased model estimates in Probit model. For this reason, RBP model with instrumental variables is constructed to predict the effect of endogenous dichotomous explanatory variables on dichotomous explanatory variables, and the model equation is set to the form shown in Equation (2). In Equation (2), IV_i is the instrumental variable selected for the endogenous variable DT. In this article, the mean value of DT in the same community (excluding the sample itself) is selected (DT_Mean). It is clear that DT_Mean cannot have a direct impact on LRB. Meanwhile, the behavior and cognition of groups within a community can have an effect on the behavior and cognition of individuals, which we call the neighborhood effect or endogenous interaction effect [50]. Therefore, DT_Mean satisfies the principle of bounded exclusion for the selection of instrumental variables. α_0, α_1, α_{2r}, α_{3k}, μ_0, μ_1, μ_{2r}, and μ_{3k} denote the coefficients to be estimated in the two equations respectively. ξ_{1i} and ξ_{2i} denote the random error term in the two equations, respectively. The other parameters have the same meaning as in Equation (1).

The use of the RBP model requires the existence of correlation requirement for the two perturbation terms of the two equations in Equation (2). *athrho* is the parameter that tests whether the perturbation terms are correlated. If *athrho* passes the significance test, i.e., the original hypothesis that the two perturbation terms are not correlated is rejected, indicating that the use of the RBP model is necessary, and the results of the Probit model are biased.

$$\begin{cases} DT_i = \alpha_0 + \alpha_1 IV_i + \alpha_{2r} CV_{ir} + \alpha_{3k} RV_{ik} + \xi_{1i} \\ LRB_i = \mu_0 + \mu_1 DT_i + \mu_{2r} CV_{ir} + \mu_{3k} RV_{ik} + \xi_{2i} \end{cases} \quad (2)$$

3.3.3. Principal Component Analysis (PCA): Measurement of DT_GAP

To provide a more comprehensive measure of the DT_GAP, we have introduced the frequency of Internet use for learning, work, social, entertainment, and business activities as a comprehensive measure of DT_GAP. PCA method is used to reduce the dimensionality and extract the principal components (DT_PCA) as the main source of data for the calculation of DT_GAP. PCA method is a way of replacing the original variables with a new set of mutually uncorrelated composite variables by regrouping them. The extraction of principal components by the PCA method is a process of dimensionality reduction while retaining more data efficiency [51].

We need to perform a correlation test on the different components selected before moving on to PCA. At the 1% statistical level, the bartlett test passes the significance test. It indicates that the original hypothesis that all variables are uncorrelated with each other is rejected. The test parameter KMO = 0.898 indicates that the sum of squares of the simple correlation coefficients of the different components is much greater than the sum of squares of the partial correlation coefficients. In other words, there is a strong correlation between the different components. Based on all results of the statistical test above, PCA is allowed to be continued.

Table 2 reports the relevant test values for the PCA process. The rules for PCA selection require that the eigenvalue of the selected principal component factor needs to be greater than 1. However, we also have to meet another requirement is that the cumulative variance contribution rate of the principal component be 0.85 or higher. For this reason, we selected both factor1 and factor2 as the components in PCA. We further calculated the uniqueness of the variables, all of which are less than 0.6, indicating that the variance explained by the common factors is large. Therefore, all the factors selected in this article satisfy the uniqueness requirement of the PCA method.

Table 2. Relevant test values for the PCA process.

Factors	Eigenvalue	Proportion	Cumulative
Factor1	4.689	0.782	0.782
Factor2	0.522	0.087	0.868
Factor3	0.298	0.050	0.918
Factor4	0.245	0.041	0.959
Factor5	0.155	0.026	0.985
Factor6	0.091	0.015	1.000
Variables	**Factor1**	**Factor2**	**Uniqueness**
DT	0.930	−0.177	0.103
DT_LEARNING	0.814	0.217	0.291
DT_WORKING	0.770	0.280	0.329
DT_SOCIAL	0.911	−0.183	0.136
DT_ENTERTAINMENT	0.891	−0.170	0.177
DT_TRADE	0.834	0.109	0.292

Further, the extracted factors were summed up using proportion as the weight and calculated as shown in Equation (3). In Equation (3), DT_PCA_i denotes the calculated principal component and DT_GAP_i denotes the digital divide. f_{1i} indicates the proportion of the factor1, $value_1$ indicates the eigenvalue corresponding to the factor1. j indicates the number of factors, according to the above calculation, two factors should be extracted. So, the value of j here is 2. Max() denotes that extracting the maximum value of a variable. Assigned a value to DT_GAP_i by calculating the specific difference between the maximum value of DT_PCA_i and the DT_PCA_i of each sample. The other parameters have the same meaning as in Equations (1) and (2).

$$\begin{cases} DTU\_PCA_i = \frac{f_{1i}*value_1+\ldots+f_{ji}*value_j}{value_1+\ldots+value_j} \\ DTU\_GAP_i = \text{Max}(DTU\_PCA_i) - DTU\_PCA_i \end{cases} \quad (3)$$

3.3.4. Chain Multiple Mediating Effects (CMM) Model

The CMM model is suitable for testing mediating effects that contain two or more mediating variables, and there mediating variables are related to each other [52]. Compared with the general mediating effects model, the advantage of the CMM model is that it takes full account of the relationship between the mediating variables. Therefore, the CMM model can be effective in reducing errors in model estimation.

Theoretical analysis has shown that the mediating variable DEPENDENCY impacts LRB through another mediating variable JOB_NONFARM. This means that the two mediating variables are related to each other and apply to the CMM model. A prerequisite for constructing CMM model is that DT has a significant effect on LRB. Obviously, this prerequisite was confirmed in the benchmark regression model and robustness tests above. According to the interpretation of the path of DT's impact on LRB in the theoretical analysis, we need to construct four equations to test all the impact path of DT on LRB (as shown in Equations (4)–(7)).

$$LRB_i = \beta_0 + \beta_1 DT_i + \beta_{2r} CV_{ir} + \beta_{3k} RV_{ik} + \varepsilon_i \quad (4)$$

$$DEPENDENCY_i = \lambda_0 + \lambda_1 DT_i + \lambda_{2r} CV_{ir} + \lambda_{3k} RV_{ik} + \zeta_{1i} \quad (5)$$

$$JOB\_NONFARM_i = \eta_0 + \eta_1 DEPENDENCY_i + \eta_2 DT_i + \eta_{3r} CV_{ir} + \eta_{4k} RV_{ik} + \zeta_{2i} \quad (6)$$

$$LRB_i = \sigma_0 + \sigma_1 DT_i + \sigma_2 DEPENDENCY_i + \sigma_3 JOB\_NONFARM_i + \sigma_{4r} CV_{ir} + \sigma_{5k} RV_{ik} + \zeta_{3i} \quad (7)$$

Equation (4) is the same as Equation (1). In Equation (5), $DEPENDENCY_i$ is the explained variable and DT_i is the explanatory variable. λ_0, λ_1, λ_{2r}, and λ_{3k} are the coefficients to be estimated and ζ_{1i} is the random error term. In Equation (6), $JOB\_NONFARM_i$ is the explained variable, $DEPENDENCY_i$ and DT_i are the explanatory variables. η_0, η_1, η_2, η_{3r}, and η_{4k} are all indicators coefficients to be estimated and ζ_{2i} is the random error term. In Equation (7), LRB_i is the explained variable.DT_i, $DEPENDENCY_i$, and $JOB\_NONFARM_i$ are the explanatory variables. σ_0, σ_1, σ_2, σ_3, σ_{4r} and σ_{5k} are the coefficients to be estimated and ζ_{3i} is the random error term. All other variables, which are not explained above in Equations (5)–(7), have the same meaning as the variables in Equations (1)–(4).

In order to show and illustrate the mainly coefficients that need to be examined in the CMM model and the path of DT's impact on the LRB more clearly, we drew a schematic diagram of the CMM model (as shown in Figure 2).

To verify the "DT- DEPENDENCY -LRB" impact mechanism, we need to examine whether the λ_1 and σ_2 coefficients both pass the significance test. To verify the "DT-JOB_NONFARM -LRB" impact mechanism, we need to examine whether the η_2 and σ_3 coefficients both pass the significance test. To verify the "DT- DEPENDENCY-JOB_NONFARM -LRB" impact mechanism, we need to examine whether the λ_1, η_1 and σ_3 coefficients all pass the significance test. The above method also applies when we examine the chain mediating effect of DEPENDENCY and JOB_NONFARM in the impact of DT_GAP on LRB.

Further, It needs to be specifically stated that if the coefficient of DT's impact on the mediating variable (e.g., λ_1) and the coefficient of mediating variable's impact on the LRB (e.g., σ_2) do not all pass the significance test, but only at least one coefficient (e.g., λ_1 or σ_2) passes the significance test. At this point, we need to conduct the Sobel test [53]. If the coefficients pass the Sobel test, we can consider that the mediating effect still holds.

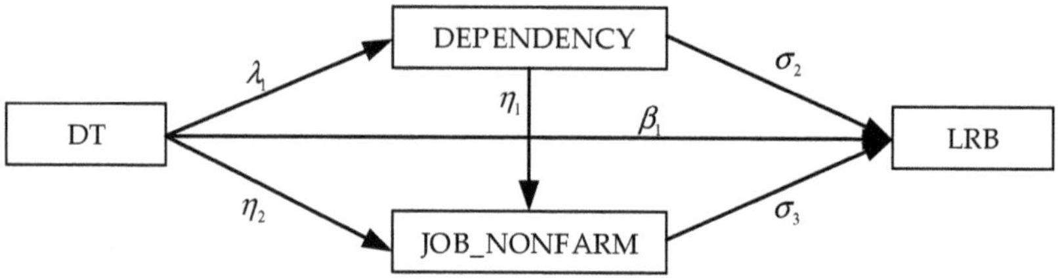

Figure 2. Schematic diagram of the CMM model.

4. Analysis of Results

4.1. Analysis of DT's Impact Paths on LRB

The results of benchmark model test for the impact of DT on LRB are reported in Table 3. The Probit model is applied in columns (1)–(3), and we put into the control variables and regional variables sequentially for regression. Column (4) reports the results of the marginal effects test of column (3). The test results show that DT exerts a significant positive effect on LRB at the 1% or 5% statistical level, regardless of whether control variables and regional variables are included in the model. It shows that the information sharing effect of DT has a positive impact on LRB, i.e., DT can significantly enhance the formation of LRB. From the results reported in column (4), DT increases probability of LRB by 6.5%. At this point, Hypothesis 1 is initially verified.

The results of the impact of control variables on LRB are also reported in Table 3. The AGE has a positive effect on LRB at the 1% significance level. The probability of LRB increase by 0.2% for each 1-year increase in AGE. The results of the GENDER's impact on LRB show that gender have a negative effect on LRB at the 1% significance level, and the marginal effect result indicates that female has a greater probability (3.2%) of conducting LRB than male. The higher the AGE_F, the higher the probability of LRB, which is similarity with the results of the AGE's impact on LRB. As the HEALTH_F continues to deteriorate, labor resource may be inadequate or rapidly shift to other industries with higher labor compensation rates (secondary and tertiary industries), further impacting LRB. At the 5% level of significance, the probability of LRB is elevated by 2.1% for each 1-unit declined in HEALTH_F. At the 1% level of significance, the probability of LRB is 7.8% lower for households in married status compared to those otherwise. To some extent, it means that MARRY promote the household to carry out agricultural production, due to those who are in married status have a lower probability of LRB. The impact of FAMILYSIZE on LRB is not robust but has a negative effect on LRB at the 10% significance level. The marginal effect results show that the probability of LRB decreased by 0.5% for each-1 person increased in FAMILYSIZE. This result further illustrates the importance of labor in the process of engaging in agricultural production. PINCOME_F is an important indicator of household livelihood status [54]. The better the economic status of the households, the higher the probability of LRB. At the 1% significance level, the probability of LRB increased by 2.4% for each 1-unit increased in PINCOME_F. It illustrates that agricultural production has become a non-preferred choice for Chinese farm households to maintain their livelihood. As the income level increases, the willingness of farm households to engage in agricultural production decreases, and the probability of LRB increases. So that means, it is limited that the positive effect of agricultural production on the improvement of household economic status. The results of the regional variables' impact on LRB indicate that the differences exist in LRB across regions (compared with the northeast region). Detailed interpretation is not performed here.

Table 3. The impact of DT on LRB: benchmark model test results.

Variables	Benchmark Model: Probit			
	(1)	(2)	(3)	(4)
DT	0.119 **	0.304 ***	0.289 ***	0.065 ***
	(0.050)	(0.064)	(0.064)	(0.014)
AGE		0.011 ***	0.010 ***	0.002 ***
		(0.003)	(0.003)	(0.001)
GENDER		−0.143 ***	−0.140 ***	−0.032 ***
		(0.047)	(0.048)	(0.011)
HEALTH		−0.009	−0.010	−0.002
		(0.027)	(0.028)	(0.006)
EDUCATION		0.006	0.006	0.001
		(0.009)	(0.010)	(0.002)
PARTY		0.053	0.051	0.011
		(0.081)	(0.082)	(0.018)
AGE_F		0.008 ***	0.008 **	0.002 **
		(0.003)	(0.003)	(0.001)
HEALTH_F		0.091 **	0.094 **	0.021 **
		(0.036)	(0.036)	(0.008)
EDUCATION_F		0.013	0.009	0.002
		(0.012)	(0.012)	(0.003)
MARRY		−0.333 ***	−0.344 ***	−0.078 ***
		(0.060)	(0.060)	(0.014)
FAMILYSIZE		−0.019	−0.024*	−0.005 *
		(0.012)	(0.012)	(0.003)
PINCOME_F		0.113 ***	0.104 ***	0.024 ***
		(0.020)	(0.020)	(0.005)
EASTERN			0.229 ***	0.052 ***
			(0.076)	(0.017)
CENTRAL			0.292 ***	0.066 ***
			(0.076)	(0.017)
WESTERN			0.041	0.009
			(0.078)	(0.018)
N	5233	5233	5233	5233

Standard errors in parentheses. * $p < 0.1$, ** $p < 0.05$, *** $p < 0.01$.

To further test the robustness of the results in the benchmark model, we re-examined the impact of DT on LRB in two approaches. Table 4 reports robustness test results of the impact of DT on LRB.

In the first approach, the RBP model is used to address omitted variables. The test results in column (1) of Table 4 show that DT has a significant positive effect on LRB in the RBP model, the marginal effect test results for the RBP model are reported in column (2), DT increase the probability of LRB by 3.1%. In contrast, the increasing effect of DT on LRB in the benchmark model was 6.5%. The difference between the two results indicates that the positive effect of DT on LRB without considering endogeneity was overestimated. The parameter *athrho* passes the significance test, indicating that the RBP model we constructed is reasonable and valid.

In the second approach, we further tested the robustness of DT's impact on LRB by replacing proxy variables. We have obtained the variable DT_PCA in process of measuring DT_GAP. Column (3) reports the results of impact of DT_PCA on LRB. At the 1% statistical significance level, the results show that DT_PCA has a positive impact on LRB. The results of the IV-Probit model test are reported in column (4). In a similar method to the selection of IV for DT, the mean value of DT_PCA in the community (excluding the sample itself) is used as a IV for DT_PCA. The test results report in column (4) are consistent with column (3). The marginal effect results of column (4) is reported in column (5), which show that the probability of LRB increase by 19.3% for every 1 unit increase in DT_PCA.

Table 4. Robustness test results for DT's impact on LRB.

Variables	RBP Model		Probit Model	IV-Probit Model	
	(1)	(2)	(3)	(4)	(5)
DT		0.031 ***			
		(0.005)			
DT_Mean	0.728 ***				
	(0.126)				
DT_PCA			0.065 ***	0.189 ***	0.193 ***
			(0.014)	(0.036)	(0.038)
Control variables	Yes	Yes	Yes	Yes	Yes
Regional control	Yes	Yes	Yes	Yes	Yes
Parameter: athrho	−0.328 ***		/	/	/
	(0.084)		/	/	/
Wald test of exogeneity	/		/	12.95 ***	/
Wald F Statistics	/		/	289.93	/
N	5233	5233	5233	5233	5233

Standard errors in parentheses. *** $p < 0.01$. The IV-Probit model test results are obtained by a two-stage estimation method, column (3) reports the result of second stage. The marginal effect results of column (4) are reported in column (5). "Wald test of exogeneity" passes the significance test, indicating that the model rejects the original hypothesis that the explanatory variables are exogenous, meaning that the IV has strong explanatory power [55]. "Wald F" test value greater than 10, indicates that the IV is not weak [56].

Based on the results of the empirical tests above, the positive effect of DT on LRB has been verified by replacing the estimation method and replacing the proxy variables. The robustness of the benchmark model is verified. Therefore, hypothesis 1 is further verified.

Further, we interpreted how DT impacts LRB. Table 5 reports the results of DT's impact on LRB in CMM model. The test results in columns (1)–(4) correspond to Equations (4)–(7).

Table 5. Results of CMM model of DT's impact on LRB.

Variables	LRB	DEPENDENCY	JOB_NONFARM	LRB
	(1)	(2)	(3)	(4)
DT	0.289 ***	1.393 ***	0.073	0.208 ***
	(0.064)	(0.049)	(0.084)	(0.076)
DEPENDENCY			0.042 *	0.045 **
			(0.024)	(0.022)
JOB_NONFARM				0.273 ***
				(0.066)
Control variables	YES	YES	YES	YES
Regional control	YES	YES	YES	YES
N	5233	5233	5233	5233

Standard errors in parentheses. * $p < 0.1$, ** $p < 0.05$, *** $p < 0.01$. Based on the data distribution characteristics of the explanatory variables, columns (1), (2), and (4) are estimated using the Probit model, and column (3) is estimated using the Ordered Probit model.

Firstly, the test results in column (1) show that the significant positive effect of DT on LRB, which becomes the premise of CMM model to test the impact path of DT on LRB. Secondly, at the 1% statistical level, the results in column (2) show that DT positive impact DEPENDENCY. And the test results in column (3) show that DT does not play a direct effect on JOB_NONFARM, indicating that JOB_NONFARM does not play a mediating effect in DT's impact on LRB independently. Thirdly, the positive effect of DEPENDENCY on JOB_NONFARM is confirmed at the 10% significance level. In column (4), DT, DEPENDENCY, and JOB_NONFARM have a significant positive effect on LRB at the 1%, 5%, and 1% statistical levels respectively.

Therefore, all test results of the CMM model show that "DT- DEPENDENCY -LRB" and "DT-DEPENDENCY-JOB_NONFARM-LRB" impact path pass the significance test. We further conducted the Sobel test on the impact path of "DT-JOB_NONFARM-LRB", and the statistical results do not pass the Sobel test. So, the impact path of "DT-JOB_NONFARM-LRB" is not statistically valid. Therefore, we can consider that JOB_NONFARM is not able to independently play a mediating effect in the process of DT's impact on LRB, but JOB_NONFARM can play a significant mediating effect after the first transmission through DEPENDENCY. Up to this point, hypothesis 3 is verified.

4.2. Analysis of DT_GAP's Impact Paths on LRB

The DT_GAP is the major manifestation of the information exclusion effect, which emerged during the development of DT.

Table 6 reports test results of DT_GAP's impact on LRB. Similarly, the strategy of sequentially placement of control variables and regional variables are also used to test the impact of DT_GAP on LRB. At the 1% significance level, DT_GAP has a significant negative effect on LRB is reported in columns (1)–(3). Further, the mean value of DT_GAP within the community (excluding the sample itself) as IV is used to construct the IV-Probit model, and the parameters associated with the selected IV passed the test. The test results of the IV-Probit model are reported in column (4), which are consistent with the results reported in columns (1)–(3). In summary, from the test results of benchmark model and IV model, the significant negative effect of DT_GAP on LRB is confirmed. To this point, hypothesis 2 is verified.

Table 6. Results of the DT_GAP's impact on LRB.

Variables	Probit Model			IV-Probit Model
	(1)	(2)	(3)	(4)
DT_GAP	−0.033 ***	−0.069 ***	−0.065 ***	−0.189 ***
	(0.010)	(0.014)	(0.014)	(0.036)
Control variables	NO	Yes	Yes	Yes
Regional control	NO	NO	Yes	Yes
Wald test of exogeneity				13.07 ***
Wald F				289.93
N	5233	5233	5233	5233

Standard errors in parentheses. *** $p < 0.01$. "Wald test of exogeneity" passes the significance test, indicating that the model rejects the original hypothesis that the explanatory variables are exogenous, meaning that the IV has strong explanatory power [55]. "Wald F" test value greater than 10, indicates that the IV is not weak [56].

Refer to the CMM model used in impact path of DT on LRB. Similarly, CMM model for the impact of DT_GAP on LRB is constructed. Table 7 reports the results of DT_GAP's impact on LRB in CMM model. From the test results in column (1), DT_GAP has a significant negative effect on LRB, which is consistent with the test results above. The test result of the significant negative effect of DT_GAP on DEPENDENCY is reported in column (2). At the 5% significance level, the test results in column (3) show that DT_GAP has no significant effect on JOB_NONFARM, and DEPENDENCY has a significant positive effect on JOB_NONFARM. In the test results in column (4), DT_GAP still has a significant negative effect on LRB, both of DEPENDENCY and JOB_NONFARM exert positive effect on LRB at the 10% and 1% significance levels. Based on all test results in Table 7, DT_GAP reduce the probability of LRB by weakening DEPENDENCY is confirmed.

Meanwhile, DT_GAP decrease DEPENDENCY, then DEPENDENCY decrease probability of JOB_NONFARM, ultimately JOB_NONFARM decrease the probability of LRB. Further, the Sobel test reveals that JOB_NONFARM cannot play an independent mediating effect in the process of DT_GAP' impact on LRB, the mediating effect of JOB_NONFARM must rely on DEPENDENCY to be realized. To this point, hypothesis 4 is verified.

Table 7. Results of CMM model of DT_GAP's impact on LRB.

Variables	LRB	DEPENDENCY	JOB_NONFARM	LRB
	(1)	(2)	(3)	(4)
DT_GAP	−0.065 ***	−0.313 ***	0.005	−0.051 ***
	(0.014)	(0.011)	(0.019)	(0.016)
DEPENDENCY			0.056 **	0.040 *
			(0.024)	(0.022)
JOB_NONFARM				0.280 ***
				(0.066)
Control variables	YES	YES	YES	YES
Regional control	YES	YES	YES	YES
N	5233	5233	5233	5233

Standard errors in parentheses. * $p < 0.1$, ** $p < 0.05$, *** $p < 0.01$. Based on the data distribution characteristics of the explanatory variables, columns (1), (2), and (4) are estimated using the Probit model, and column (3) is estimated using the Ordered Probit model.

4.3. Heterogeneity Analysis: Impact of DT and DT_GAP on LRB

Based on the empirical analysis above, we have interpreted and verified how the information sharing effect of DT exerts a positive impact on LRB and how the information exclusion effect of DT_GAP exerts a negative impact on LRB. Further, for a more comprehensive understanding of the impact of DT and DT_GAP on LRB. We explored the effects of DT and DT_GAP on LRB from the perspective of heterogeneity in regional, age of householder, and household income levels.

Firstly, we examined the impact of DT and DT_GAP on LRB from the perspective of regional heterogeneity. The results of the impact of DT and DT_GAP on LRB from regional heterogeneity perspective are reported in Table 8. The test results show that DT and DT_GAP exert significant effects on LRB in the eastern, central, and western regions, with DT exerting a positive effect and DT_GAP exerting a negative effect. In contrast, in the northeast region, both DT and DT_GAP do not pass the significance test on LRB. From the group regression results, the impact of DT on LRB and DT_GAP on LRB differ between regions at the significance level and extent. However, the differences test does not pass the significance test. Therefore, we can consider that the impact of DT on LRB and DT_GAP on LRB is not significantly different between regions. However, the test of regional grouped regression is not useless. It still illustrates the robustness of the positive effect of DT on LRB and the negative impact of DT_GAP on LRB.

Secondly, we examined the impact of DT and DT_GAP on LRB from the perspective of householder's age heterogeneity. Table 9 reports the test results of the impact of DT and DT_GAP on LRB from the perspective of householder's age heterogeneity. We divided the age of householder in all samples into four groups: under-30 years old, 30 to 50 years old, 50 to 70 years old, and over-70 years old. Neither DT nor DT_GAP exert a significant effect on LRB in the regressions for the under-30 and over-70 age groups.

From the results of DT's impact on LRB reported in columns (1)–(3). DT does not exert a significant effect on LRB in the subgroup under-30 years old. The marginal effect of the positive effect of DT on LRB reaches 0.043 in the subgroup regression of 30 to 50 years old. In the subgroup regression of 50–70 years old, the marginal effect of the positive effect of DT on LRB reaches 0.045. Therefore, we conclude that the positive effect of DT on LRB progressively decreases as the age of householder increases in the sample of 30–70 years old.

From the results of impact of DT_GAP on LRB reported in columns (5)–(7). The test results show that DT_GAP does not play a significant effect on LRB in the grouping of under-30. In the subgroup regression of 30 to 50 years old, the marginal effect of the negative effect of DT_GAP on LRB reaches 0.009. In the subgroup regression of 50–70 years old, the marginal effect of DT_GAP on LRB reaches 0.012. Therefore, we conclude that in the sample below 70 years old, as the age of householder increases, the negative effect of the impact of DT_GAP on LRB gradually elevated. Meanwhile, the both of differences tests pass the significance test at the 5% level, indicating that the changes in the effects of DT and DT_GAP on LRB are statistically significant in different householder's age groups.

Table 8. Regional heterogeneity: test results of the impact of DT & DT_GAP on LRB.

Variables	Eastern	Central	Western	Northeast
	(1)	(2)	(3)	(4)
DT	0.407 ***	0.320 ***	0.259 **	0.017
	(0.123)	(0.118)	(0.115)	(0.196)
Control variables	Yes	Yes	Yes	Yes
Regional control	No	No	No	No
Differences test		3.08		
N	1282	1385	1918	648
Variables	**Eastern**	**Central**	**Western**	**Northeast**
	(5)	**(6)**	**(7)**	**(8)**
DT_GAP	−0.066 ***	−0.090 ***	−0.050 **	−0.024
	(0.026)	(0.025)	(0.025)	(0.045)
Control variables	Yes	Yes	Yes	Yes
Regional control	No	No	No	No
Differences test		2.04		
N	1282	1385	1918	648

Standard errors in parentheses. ** $p < 0.05$, *** $p < 0.01$. The Probit model is applied to columns (1)–(8). "Differences test" uses method of seemingly unrelated regression (SUR), see Greene (2003) for the details of the method [57].

Table 9. Age of householder heterogeneity: test results of the impact of DT & DT_GAP on LRB.

Variables	Age < 30	30 ≤ Age < 50	50 ≤ Age < 70	Age ≥ 70
	(1)	(2)	(3)	(4)
DT	0.174	0.223 **	0.205 *	0.000
	(0.257)	(0.090)	(0.117)	(.)
Control variables	Yes	Yes	Yes	Yes
Regional control	Yes	Yes	Yes	Yes
Differences test		10.05 **		
N	361	1841	2523	505
Variables	**age < 30**	**30 ≤ age < 50**	**50 ≤ age < 70**	**age ≥ 70**
	(5)	**(6)**	**(7)**	**(8)**
DT_GAP	−0.028	−0.045 **	−0.056 **	0.000
	(0.041)	(0.020)	(0.029)	(.)
Control variables	Yes	Yes	Yes	Yes
Regional control	Yes	Yes	Yes	Yes
Differences test		9.67 **		
N	361	1841	2523	505

Standard errors in parentheses. * $p < 0.1$, ** $p < 0.05$. The marginal effects of DT on LRB in columns (2) and (3) are 0.043 ** (0.018) and 0.045 * (0.026), respectively. The marginal effects of DT_GAP on LRB in columns (2) and (3) are −0.009 ** (0.004) and −0.012 ** (0.006), respectively. The Probit model is applied to columns (1)–(8). "Differences test" uses method of seemingly unrelated regression (SUR), see Greene (2003) for the details of the method [57].

Thirdly, we examined the impact of DT and DT_GAP on LRB from the perspective of household income level heterogeneity. Table 10 reports the test results of DT's and DT_GAP's impact on LRB from the perspective of household income level heterogeneity. According to the data distribution of the PINCOME_F of all samples, we defined income of households below the 25% quantile as low-income households and income of households above the 75% quantile as high-income households.

Columns (1) and (2) report the impact of DT on LRB with different income level, and the test results show that DT has a more positive effect on LRB of low-income level households compared to high-income. Columns (3) and (4) report the impact of DT_GAP on LRB of households with different income level, and the test results show that DT_GAP has a more negative effect on LRB of low-income level households compared to high-income level households. These results pass the difference test at the 1% significance level. Therefore, we conclude that the information sharing effect of DT is significantly pro-poor, but the information exclusion effect of DT_GAP on low-income households also has a significant preference.

Table 10. Household income level heterogeneity: test results of the impact of DT & DT_GAP on LRB.

Variables	Low-Income	High-Income	Low-Income	High-Income
	(1)	(2)	(3)	(4)
DT	0.513 ***	0.139		
	(0.166)	(0.110)		
DT_GAP			−0.128 ***	−0.013
			(0.039)	(0.022)
Control variables	Yes	Yes	Yes	Yes
Regional control	Yes	Yes	Yes	Yes
Differences test	3.64 *		6.89 ***	
N	1308	1308	1308	1308

Standard errors in parentheses. * $p < 0.1$, ** $p < 0.05$, *** $p < 0.01$. The marginal effects of DT on LRB in columns (1) and (2) are 0.095 ** (0.031) and 0.037(0.029), respectively. The marginal effects of DT_GAP on LRB in columns (3) and (4) are −0.024 *** (0.007) and −0.004 (0.006), respectively. The Probit model is applied to columns (1)–(8). "Differences test" uses method of seemingly unrelated regression (SUR), see Greene (2003) for the details of the method [57].

5. Discussion and Conclusions

5.1. Discussion

Digital technologies (e.g., internet, blockchain, etc.) can provide a positive role in facilitating transactions in land, real estate, etc. [58,59]. However, the application of DT in land rental transaction market is still limited in developing countries or regions. Our research, based on a large dataset in China, finds that DT can significantly increase the probability of LRB for farmers (6.5%). It provides new empirical evidence that the application of DT can play the positive role in the process of land rental.

Existing studies have confirmed that Internet use can facilitate famer's land rental behavior, but there are shortcomings of small dataset and insufficient interpretation of the impact paths. A very important finding in our research is that JOB_NONFARM and DEPENDENCY are mediating variables for the impact of DT on LRB, and JOB_NONFARM needs to rely on DEPENDENCY to exert the mediating effect but cannot exert independently, i.e., path of "DT-DEPENDENCY-JOB_NONFARM-LRB" is feasible, but path of "DT -JOB_NONFARM-LRB" is not. In other words, the conclusion of previous studies that DT can directly impact land rental behavior through JOB_NONFARM is inaccurate or biased [29]. So, our research is based on a large dataset ($n = 5233$) and fully interprets how DT impacts LRB, improving on the shortcomings of existing studies.

In addition, we focus on the negative effect brought by DT, or namely the information exclusion effect brought by DT_GAP. Our empirical results confirm that DT_GAP has a negative effect on LRB, which means that DT_GAP produces information exclusion and is detrimental to the formation of an efficient land rental market. It compensates for the shortcoming that existing studies have not focused on DT_GAP's impact on land rental behavior.

The results of the heterogeneity analysis showed that youngers are able to promote LRB more effectively with DT (compared to elders), and information exclusion with DT_GAP appeared to be more effective in elders. The results of such a test fully demonstrate that DT has produced an information exclusion effect on the elderly. It reflects the fact that the aging DT_GAP has become an important manifestation of the DT_GAP [34,60]. Although DT has a positive impact on LRB of low-income groups, it is interesting to note that DT_GAP also has more negative impact on LRB of low-income groups (compared to high-income groups). Such results suggest that DT does mitigate the position of low-income groups in the information market, but it is undeniable that more low-income groups may be informationally deprived due to information asymmetry [61].

The negative impact of DT's information exclusion effect on the elderly and low-income groups are only parts of many negative effects. As DT_GAP continues to expand, the phenomenon of new social exclusions may be derived [38,62].

Our research findings have implications for policy formulation. On the one hand, the government should promote the digitization of the land rental market to facilitate the efficient allocation of land resources and reduce the rate of land abandonment. On the other hand, the government should improve internet quality (e.g., broadband access rates, etc.), promote internet coverage, especially expand mobile internet coverage in remote rural areas (e.g., 4G and 5G communication base stations, etc.), and optimize the adaptation of digital applications between different groups, with particular attention to the digital divide of the ageing.

However, there are still certain shortcomings in our research. DT measurement variables are limited by data availability, and the measurement variables of DT and DT_GAP are highly homogeneous, which makes it difficult to interpret the net effect of DT's and DT_GAP's impact on LRB. In addition, we only use DEPENDENCY and JOB_NONFARM as mediating variables to interpret the impact path of DT and DT_GAP on LRB, it still needs to be strengthened. To this end, further exploring the net effect of DT and DT_GAP on LRB, and the more comprehensive impact path of DT and DT_GAP on land rental behavior (including rent out and rent in) are the next research that needs to focus on.

5.2. Conclusions

Our empirical results validate Hypotheses 1–4, which we propose based on our theoretical analysis. Overall, the findings of our study can be summarized in three points.

First, we found that the information sharing effect of DT exerts a significant positive impact on LRB, while the information exclusion effect of DT_GAP exerts a significant negative effect on LRB.

Second, another important finding is that JOB_NONFARM and DEPENDENCY are mediating variables in process of DT's and DT_GAP's impact on LRB, but JOB_NONFARM needs to rely on the transmission of DEPENDENCY to exert a mediating effect and does not exert independently.

Third, the impact of DT on LRB has a clear preference for lower age groups (30–70 age range) as well as a preference for lower income. However, the effect of DT_GAP on LRB has a clear preference for higher age groups (lower-70 age range) as well as a preference for lower income.

Author Contributions: X.Z. contributed to conceptualization; data curation; formal analysis; software; writing—original draft; writing—review & editing. Z.H. contributed to Funding acquisition; Supervision; Validation; Writing—review & editing. All authors have read and agreed to the published version of the manuscript.

Funding: The authors would like to thank the funding support from President's Youth Fund of Institute of Science and Development, Chinese Academy of Sciences (grant no. E0X3751Q01). We are grateful for the data provided by the ISSS of Peking University.

Institutional Review Board Statement: Not applicable.

Informed Consent Statement: Not applicable.

Data Availability Statement: The CFPS data set used in this manuscript can be downloaded from the following website, http://www.isss.pku.edu.cn/cfps/ (accessed on 7 April 2022).

Conflicts of Interest: The authors declare no conflict of interest.

References

1. Data Source: ITU. Available online: https://www.itu.int/en/ITU-D/Statistics/Pages/facts/default.aspx (accessed on 7 April 2022).
2. Data Source: China Internet Network Information Center. Available online: http://www.cnnic.com.cn/IDR/ReportDownloads/ (accessed on 7 April 2022).
3. Data Source: Ministry of Natural Resources of the People's Republic of China. Available online: http://www.mnr.gov.cn/dt/zb/2017/tdly/beijingziliao/201401/t20140108_2127423.html/ (accessed on 7 April 2022).
4. Paunov, C.; Rollo, V. Has the internet fostered inclusive innovation in the developing world? *World Dev.* **2016**, *78*, 587–609. [CrossRef]
5. Relich, M. The impact of ICT on labor productivity in the EU. *Inf. Technol. Dev.* **2017**, *23*, 706–722. [CrossRef]
6. Subramanian, A. Harnessing digital technology to improve agricultural productivity? *PLoS ONE* **2021**, *16*, e0253377. [CrossRef] [PubMed]
7. Nakasone, E.; Torero, M.; Minten, B. The Power of Information: The ICT Revolution in Agricultural Development. *Annu. Rev. Res. Econ.* **2014**, *6*, 533–550. [CrossRef]
8. Lakshmi, V.; Corbett, J. How Artificial Intelligence Improves Agricultural Productivity and Sustainability: A Global Thematic Analysis. In Proceedings of the 53rd Hawaii International Conference on System Sciences, Maui, HI, USA, 1 July 2020.
9. Deichmann, U.; Goyal, A.; Mishra, D. Will digital technologies transform agriculture in developing countries? *Agri. Econ.* **2016**, *47*, 21–33. [CrossRef]
10. Xu, Z.; Zheng, F.; Chen, J. Digital divided or digital provided? The effective supply of information and the farm-gate price: An empirical study from micro-level. *China Econ. Quart.* **2013**, *12*, 1513–1536.
11. Changyu, L.; Jiale, L.; Jing, L. Rural E-commerce and New Model of Rural Development in China: A Comparative Study of "Taobao Village" in Jiangsu Province. *Asian Agri. Resn.* **2015**, *7*, 35.
12. Yi, S. E-commerce strategy for agricultural product transaction market based on information asymmetry. *Agr. Food Ind. Hi-Tech* **2016**, *27*, 138–143.
13. Cheni, B.; Mai, N.; Wangi, D. Land Circulation, Agricultural Productivity and Rural Household Income. *J. World Econ.* **2020**, *43*, 97–120.
14. Zuo, X.; Lu, J. Effects of agricultural land transfer on rural poverty reduction from the perspective of poverty vulnerability. *Res. Sci.* **2020**, *42*, 274–285. [CrossRef]
15. Adamopoulos, T.; Restuccia, D. Land Reform and Productivity: A Quantitative Analysis with Micro Data. *Am. Econ. J-Macroecon* **2020**, *12*, 1–39. [CrossRef]
16. Wang, Q.X.; Zhang, X.L. Three rights separation: China's proposed rural land rights reform and four types of local trials. *Land Use Policy* **2017**, *63*, 111–121. [CrossRef]
17. Yan, J.M.; Yang, Y.M.; Xia, F.Z. Subjective land ownership and the endowment effect in land markets: A case study of the farmland "three rights separation" reform in China. *Land Use Policy* **2021**, *101*, 105137. [CrossRef]
18. Data source: Department of Policy and Reform, Ministry of Agriculture and Rural Affairs. In *2019 Annual Report on China's Rural Policy and Reform Statistics*; China Agricultural Press: Beijing, China, 2020.
19. Rogers, S.; Wilmsen, B.; Han, X.; Wang, Z.J.-H.; Duan, Y.; He, J.; Li, J.; Lin, W.; Wong, C. Scaling up agriculture? The dynamics of land transfer in inland China. *World Dev.* **2021**, *146*, 105563. [CrossRef]
20. Wang, J.; Xu, Y.; Zou, L.; Wang, Y. Does Culture Affect Farmer Willingness to Transfer Rural Land? Evidence from Southern Fujian, China. *Land* **2021**, *10*, 594. [CrossRef]
21. Deng, X.; Xu, D.; Zeng, M.; Qi, Y. Does early-life famine experience impact rural land transfer? Evidence from China. *Land Use Policy* **2019**, *81*, 58–67. [CrossRef]
22. Gao, J.; Song, G.; Sun, X. Does labor migration affect rural land transfer? Evidence from China. *Land Use Policy* **2020**, *99*, 105096. [CrossRef]
23. Zhang, L.; Cao, Y.; Bai, Y. The impact of the land certificated program on the farmland rental market in rural China. *J. Rural Stud.* **2019**, *93*, 165–175. [CrossRef]
24. Zhang, Y.; Halder, P.; Zhang, X.; Qu, M. Analyzing the deviation between farmers' Land transfer intention and behavior in China's impoverished mountainous Area: A Logistic-ISM model approach. *Land Use Policy* **2020**, *94*, 104534. [CrossRef]
25. Echegaray, F.; Hansstein, F.V. Assessing the intention-behavior gap in electronic waste recycling: The case of Brazil. *J. Clean. Prod.* **2017**, *142*, 180–190. [CrossRef]
26. Goyal, A. Information, direct access to farmers, and rural market performance in central India. *Amer. Econ. J. App. Econ.* **2010**, *2*, 22–45. [CrossRef]
27. Deng, X.; Xu, D.; Zeng, M.; Qi, Y. Does Internet use help reduce rural cropland abandonment? Evidence from China. *Land Use Policy* **2019**, *89*, 104243. [CrossRef]
28. Fei, R.; Lin, Z.; Chunga, J. How land transfer affects agricultural land use efficiency: Evidence from China's agricultural sector. *Land Use Policy* **2021**, *103*, 105300. [CrossRef]
29. Liu, Z.; Xin, X.; Lv, Z. Does farmers' access to agricultural information on the internet promote the land transfer. *J. Agrotech. Econ.* **2021**, *2*, 100–111.

30. Weng, F.; Zhang, Q.; Huo, X. The Impact of Internet Use on Farmland Transfer of Professional Apple Growers: An Analysis of the Mediation Eect of Information Search, Social Capital and Credit Acquisition. *China Land Sci.* **2021**, *35*, 63–71.
31. Wang, M.; Yin, Z.; Pang, S.; Li, Z. Does Internet development affect urban-rural income gap in China? An empirical investigation at provincial level. *Inf. Dev.* **2021**, 02666669211035484. [CrossRef]
32. Iske, S. Differences in Internet usage-social inequality and informal education. *Soc. Work Soc.* **2005**, *3*, 215–223.
33. Gann, B. Combating digital health inequality in the time of coronavirus. *J. Consum. Health Int.* **2020**, *24*, 278–284. [CrossRef]
34. Hargittai, E.; Piper, A.M.; Morris, M.R. From internet access to internet skills: Digital inequality among older adults. *Univers. Access Inf. Soc.* **2019**, *18*, 881–890. [CrossRef]
35. Zuo, X.F.; Hong, Z.S. The Impact of Internet Use on Perception of the Poor-Rich Gap: Empirical Evidence from China. *Sustainability* **2022**, *14*, 3488. [CrossRef]
36. Kraut, R.; Patterson, M.; Lundmark, V.; Kiesler, S.; Mukophadhyay, T.; Scherlis, W. Internet paradox: A social technology that reduces social involvement and psychological well-being? *Am. Psychol.* **1998**, *53*, 1017. [CrossRef] [PubMed]
37. Akerlof, G.A. The market for "lemons": Quality uncertainty and the market mechanism. In *Uncertainty in Economics*; Elsevier: Amsterdam, The Netherlands, 1978; pp. 235–251.
38. Tewathia, N.; Kamath, A.; Ilavarasan, P.V. Social inequalities, fundamental inequities, and recurring of the digital divide: Insights from India. *Technol. Soc.* **2020**, *61*, 101251. [CrossRef]
39. Jackson, L.A.; Zhao, Y.; Kolenic, A.; Fitzgerald, H.E.; Harold, R.; Von Eye, A. Race, gender, and information technology use: The new digital divide. *Cyberpsychology Behav.* **2008**, *11*, 437–442. [CrossRef] [PubMed]
40. Abbey, R.; Hyde, S. No country for older people? Age and the digital divide. *J. Inf. Commun. Ethics Soc.* **2009**, *7*, 225–242. [CrossRef]
41. Ogbo, E.; Brown, T.; Gant, J.; Sicker, D. When Being Connected is not Enough: An Analysis of the Second and Third Levels of the Digital Divide in a Developing Country. *J. Inf. Policy* **2021**, *11*, 104–146. [CrossRef]
42. Aker, J.C.; Mbiti, I.M. Mobile Phones and Economic Development in Africa. *J. Econ. Perspect.* **2010**, *24*, 207–232. [CrossRef]
43. Feldman, D.C.; Klaas, B.S. Internet job hunting: A field study of applicant experiences with on-line recruiting. *Hum. Res. Manag.* **2002**, *41*, 175–192. [CrossRef]
44. David, P.A.; Foray, D. Economic fundamentals of the knowledge society. *Policy Futur. Edu.* **2003**, *1*, 20–49. [CrossRef]
45. Xu, Q.; Lu, Y. Off-farm employment, social security function of land, and land transfer. *Chin. J. Popul. Sci* **2018**, *5*, 30–41.
46. Feng, S.; Heerink, N. Are farm households' land renting and migration decisions inter-related in rural China? *NJAS-Wagening. J. Life Sci.* **2008**, *55*, 345–362. [CrossRef]
47. Stark, O.; Taylor, J.E. Migration incentives, migration types: The role of relative deprivation. *Econ. J.* **1991**, *101*, 1163–1178. [CrossRef]
48. Rahman, S. Determinants of agricultural land rental market transactions in Bangladesh. *Land Use Policy* **2010**, *27*, 957–964. [CrossRef]
49. From the Introduction of the CFPS on ISSS. Available online: http://www.isss.pku.edu.cn/cfps/ (accessed on 7 April 2022).
50. Zuo, X.; Kang, M.; Lu, J. The impact of social interaction and Internet use on rural residents' willingness to sort domestic waste. *Res. Sci.* **2022**, *44*, 47–58. [CrossRef]
51. Abdi, H.; Williams, L. Principal component analysis. *Wiley Interdiscip. Rev. Comput. Stat.* **2010**, *2*, 433–459. [CrossRef]
52. Hayes, A.F. *Introduction to Mediation, Moderation, and Conditional Process Analysis: A Regression-Based Approach*; Guilford publications: New York, NY, USA, 2017.
53. Sobel, M.E. Asymptotic confidence intervals for indirect effects in structural equation models. *Sociol. Meth.* **1982**, *13*, 290–312. [CrossRef]
54. Serrat, O. The sustainable livelihoods approach. In *Knowledge Solutions*; Springer: Berlin, Germany, 2017; pp. 21–26.
55. Agresti, A. Categorical Data Analysis. John Wiley & Sons: Hoboken, NJ, USA, 2003.
56. Staiger, D.; Stock, J.H.; Watson, M.W. The NAIRU, unemployment and monetary policy. *J. Econ. Perspect.* **1997**, *11*, 33–49. [CrossRef]
57. Greene, W.H. *Econometric Analysis*; Pearson Education: New Delhi, India, 2003.
58. Saull, A.; Baum, A.; Braesemann, F. Can digital technologies speed up real estate transactions? *J. Prop. Investig. Financ.* **2020**, *38*, 349–361. [CrossRef]
59. Lemieux, V.L. Evaluating the use of blockchain in land transactions: An archival science perspective. *Eur. Prop. Law J.* **2017**, *6*, 392–440. [CrossRef]
60. Gilleard, C.; Higgs, P. Internet use and the digital divide in the English longitudinal study of ageing. *Eur. J. Ageing* **2008**, *5*, 233–239. [CrossRef]
61. Naudé, W.; Vinuesa, R. Data deprivations, data gaps and digital divides: Lessons from the COVID-19 pandemic. *Big Data Soc.* **2021**, *8*, 20539517211025545. [CrossRef]
62. Selwyn, N.; Gorard, S.; Williams, S. Digital divide or digital opportunity? The role of technology in overcoming social exclusion in US education. *Edu. Polic.* **2001**, *15*, 258–277. [CrossRef]

agriculture

MDPI

Article

Genetic Diversity of Fish in Aquaculture and of Common Carp (*Cyprinus carpio*) in Traditional Rice–Fish Coculture

Yingying Ye [1,2], Weizheng Ren [3], Shixiang Zhang [4], Lufeng Zhao [1], Jianjun Tang [1], Liangliang Hu [1,5,*] and Xin Chen [1]

[1] College of Life Sciences, Zhejiang University, Hangzhou 310058, China; yeyy@zjou.edu.cn (Y.Y.); zhaolf1995@zju.edu.cn (L.Z.); chandt@zju.edu.cn (J.T.); chen-tang@zju.edu.cn (X.C.)
[2] National Engineering Research Center for Marine Aquaculture, Zhejiang Ocean University, Zhoushan 316022, China
[3] College of Forestry, Henan Agriculture University, Zhengzhou 450002, China; renwz@henau.edu.cn
[4] Agricultural and Rural Bureau of Yongjia County, Yongjia 325100, China; 22107036@zju.edu.cn
[5] College of Fisheries and Life Science, Shanghai Ocean University, Shanghai 201306, China
* Correspondence: huliangliang@zju.edu.cn

Abstract: The genetic diversity of cultured species (e.g., plants and fish) has decreased as intensive agriculture and aquaculture have increased in recent decades. Maintaining genetic diversity in agriculture is a significant concern. To test whether aquaculture affects the genetic diversity of aquatic animals and whether traditional agriculture could help maintain genetic diversity, we conducted a meta-analysis to quantify the genetic diversity of cultured and wild populations. We also examined the genetic diversity and population genetic structure of common carp (*Cyprinus carpio*) in the traditional rice–fish coculture in the south of Zhejiang Province, China, using 20 microsatellite loci. The results of the meta-analysis showed a negative overall effect size of all cultured aquatic animals that were tested both when weighted by population replicate and when weighted by the inverse of variance. Aquaculture has caused a general decline in the genetic diversity of many cultured aquatic animals. The results from the survey of a traditional rice–fish coculture system in the south of Zhejiang Province of China showed high levels of genetic diversity in all 10 sampled populations (mean Na = 7.40, mean Ne = 4.57, mean I = 1.61, mean He = 0.71, and mean Ho = 0.73). Both the conventional analysis and a model-based analysis revealed a high and significant genetic divergence among the 10 sampled populations all over the three counties (F_{ST} value ranged from 0.00 to 0.13, and *Nei*'s genetic distance ranged from 0.07 to 0.62). Populations within Yongjia and Jingning counties were also genetically differentiated, respectively. Furthermore, molecular variance (AMOVA), membership coefficients estimated by STRUCTURE, PCoA, and migration network analysis supported the findings from pairwise F_{ST} values. Our results suggest that the traditional rice–fish coculture plays an important role in maintaining the genetic diversity of carp cocultured in rice paddies and future policies should favor the conservation of the rice–fish system and raise the awareness of farmers on methods to maintain carp genetic diversity.

Keywords: traditional agriculture; rice–fish system; aquatic animals; meta-analysis; genetic diversity; microsatellite analysis

Citation: Ye, Y.; Ren, W.; Zhang, S.; Zhao, L.; Tang, J.; Hu, L.; Chen, X. Genetic Diversity of Fish in Aquaculture and of Common Carp (*Cyprinus carpio*) in Traditional Rice–Fish Coculture. *Agriculture* **2022**, *12*, 997. https://doi.org/10.3390/agriculture12070997

Academic Editor: Mark O. Winfield

Received: 14 May 2022
Accepted: 8 July 2022
Published: 11 July 2022

1. Introduction

The rapid development of modern intensive agriculture has contributed to improving global food output and ensuring food security [1]. However, with the intensive development of agriculture, the use of high chemical inputs and high-yield varieties causes the reduction in genetic diversity in agriculture [2,3]. There has been much concern over how to conserve and manage genetic diversity [4–6]. In contrast to modern intensive agriculture, traditional agriculture developed by local farmers using indigenous natural and social resources often nourishes rich genetic diversity [7], which is critical in providing germplasm and maintaining ecosystem services and would help improve local food security for an uncertain future [8–10].

Aquaculture is a kind of agriculture that has experienced rapid growth in the past decades as a reliable source of protein for the human diet [11–13]. Although the aquaculture industry is increasingly important, it has brought great environmental and ecological risks, including water pollution and degradation of germplasm resources of aquatic species [14]. For example, there are concerns about the release of non-native species that may escape from aquaculture and cause negative genetic impacts on wild species; further work in understanding and mitigating those risks is justified. [15]. However, compared with staple crops and livestock, the development and conservation of aquatic animals have not received as much attention [16]. Due to the high fertility of many aquatic animals, farmers usually use a small number of brood stock, which leads to inbreeding, genetic drift, and, thus, the reduction in genetic diversity [17–19]. As genetic diversity is the primary resource in the successful artificial propagation of any aquatic animals, understanding the effect of aquaculture on the genetic diversity of aquatic animals is essential.

Studies have shown that traditional rice–fish systems have maintained several types of common carp [20–22]. The coculture of rice and fish is an integrated agri-aquaculture system (IAAS) that combines rice cultivation with aquaculture, which is a typical traditional farming system in southern China [20]. In the rice–fish coculture system, common carp (*C. carpio*) is the major aquatic animal raised in the paddy field, where the environment is characterized by shallow water [21].

In the present study, the effect sizes of genetic diversity (i.e., *Na* and *He*) in cultured and wild populations of a variety of aquatic animals, including mollusk, arthropod, echinoderm, carp, perch, flounder, salmon, catfish, puffer, and herring, were assessed by a meta-analysis based on 117 studies. We also catalogued the genetic diversity and genetic variation of common carp cocultured in paddies (*C. carpio*) in three counties (i.e., Jingning, Qingtian, and Yongjia) of Zhejiang Province, China, using 20 polymorphic microsatellite loci. Our objectives were to evaluate the impact of aquaculture activities on the genetic diversity of aquatic animals and characterize the genetic diversity of carp cocultured in paddies in the southern Zhejiang Province of China.

2. Materials and Methods

2.1. Meta-Analysis

A systematic search of the literature was conducted across two databases: the Web of Science (1900–2021) and CNKI (1970–2021) in March 2022. No restrictions were considered either on the language or on the publication date. A combination of search terms used to search for the topic was as follows: "genetic diversity" OR "genetic variability" AND nature* OR wild AND farmed OR cultured OR hatchery OR artificial OR cultivated AND fish. The pre-specified eligibility criteria for research to be selected in the meta-analysis database were that (1) the studies used microsatellite markers, (2) the studies included cultured and wild populations of the same aquatic animals, and (3) cultured and wild populations were isolated from each other and had no gene exchange.

The species of aquatic animals, number of cultured and wild populations, mean of the number of alleles per locus (*Na*), and mean of the expected heterozygosity (*He*) were extracted from data reported in each piece of literature. The natural log (ln)-transformation of the response ratio *R* was used to calculate effect sizes [23]:

$$lnR = \ln\frac{\overline{Y_1}}{\overline{Y_2}} = \ln\overline{Y_1} - \ln\overline{Y_2}$$

The variance of *lnR* was calculated as:

$$V_{\ln R} = \frac{S_1^2}{n_1\overline{Y}_1^2} + \frac{S_2^2}{n_2\overline{Y}_2^2}$$

where $\overline{Y_1}$ and $\overline{Y_2}$ represent the means of genetic diversity of cultured and wild populations, respectively, S_1^2 and S_2^2 represent the variance of genetic diversity of the cultured and wild populations, respectively, and n_1 and n_2 represent the numbers of cultured and wild populations, respectively.

The weight of the effect sizes is calculated in two ways: (1) weighting by the inverse of variance ($\frac{1}{V_{lnR}}$) and (2) weighting by the population replicate:

$$W = Np = \frac{n_1 n_2}{n_1 + n_2}$$

Because the calculation of S_1^2 or S_2^2 is not allowed when the number of cultured or wild populations was 1 (i.e., $n_1 = 1$ or $n_2 = 1$), we excluded those items of research that had only one population of cultured or wild aquatic animals when we weighted the effect sizes by the inverse of variance. The meta-analysis was performed in Metawin v2.1 with 95% confidence intervals (*CIs*) [23].

2.2. Sample Collection and DNA Extraction

The traditional rice–fish coculture system located in southern Zhejiang Province of China has a long history, of more than 1200 years, and is listed as a Globally Important Agriculture Heritage System (GIAHS) [20,24]. The fish populations with breeding introduction on purpose or by chance were excluded from the sample collection. Those isolated local populations were sampled in this study to avoid the influences of genetic exchange with modern varieties. A total of 166 carp cocultured in rice paddies were collected from 10 locations across three counties (i.e., Jingning, Qingtian, and Yongjia in the south of Zhejiang Province, China) (Figure 1 and Table 1). All of these locations have a long history of rice–fish coculture. The total genomic DNA extraction was obtained from the tail fin of each individual using a commercial DNA extraction kit (Sangon Biotech Co., Ltd. Shanghai, China). After the quality of DNA was examined through the 1% agarose gel electrophoresis, the extracted DNA was stored at −20 °C before further polymerase chain reactions (PCRs).

Table 1. Collection details for *C. carpio* cocultured in paddies in the south of Zhejiang Province, China.

County	Village	Abbreviation	Sample Size	Geographic Locations
Jingning	Hexi	HX	12	119.69° E 27.93° N
	Chengzhao	CZ	8	119.61° E 27.96° N
	Luci	LC	11	119.40° E 27.87° N
Qingtian	Jizhai	JZ	10	120.18° E 28.46° N
	Wenxi	WX	14	120.39° E 28.18° N
	Wukeng	WK	18	120.41° E 28.24° N
	Xiaozhoushan	XZS	31	120.39° E 28.20° N
Yongjia	Bilian	BL	18	120.56° E 28.32° N
	Daruoyan	DRY	9	120.61° E 28.27° N
	Minao	MA	35	120.51° E 28.30° N

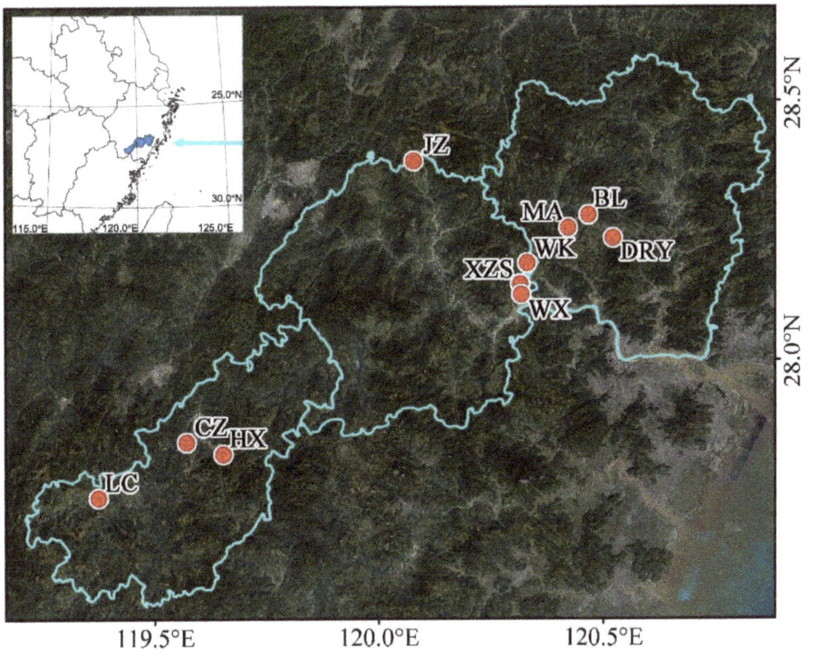

Figure 1. The sampling locations of common carp (*C. carpio*) in the traditional rice–fish coculture system in the south of Zhejiang Province, China.

2.3. Microsatellite Analysis

We selected 20 microsatellite loci for *C. carpio* from the literature (Table S1) [25–29]. The forward primers were labeled with a fluorescent dye (-FAM, TAMRA, or HEX) at the 5′ end. Microsatellite polymorphism of each DNA sample was analyzed by PCR, which was performed in a final volume of 15 μL reaction containing 50 ng of DNA, 1.5 pmol of each forward and reverse primer, and 7.5 μL *Taq* MasterMix (Cwbiotech. Co. Ltd., Beijing, China). Cycling conditions for all assays included initial denaturation at 94 °C for 3 min followed by 30 cycles at 94 °C for 30 s, 50–60 °C for 30 s, and 72 °C for 1 min and final elongation at 72 °C for 7 min. Sequencing was performed on the ABI3730xl platform by Sangon Biotech Co. Ltd. (Shanghai, China).

2.4. Genetic Data Analysis

2.4.1. Genetic Diversity

Micro-Checker v2.2.3 software was used to double-check the effect of null alleles and allele scoring errors before data analysis [30]. For each microsatellite locus, we assessed the number of alleles per locus (*Na*), the effective number of alleles per locus (*Ne*), Shannon's diversity index (*I*), expected heterozygosity (*He*), observed heterozygosity (*Ho*), and the fixation index (*Fis*) using GenAlEx v6.5 [31].

The linkage disequilibrium method (LD) was used to estimate the effective population size for each carp population by NeEstimator v2 [32]; the lowest allele frequency used was 0.01 and the confidence interval was 95%. The two-phased model (TPM) with 90% single-step mutations and 10% multiple-step mutations with 1000 replications and the mode-shift test [33] based on an L-shaped distribution of allele frequency under mutation–drift equilibrium were used to assess whether populations of the sampled carp had experienced recent bottlenecks by using Bottleneck v1.2.02 software [34]. Statistical significance at each locus was evaluated by a one-tailed Wilcoxon sign-rank test [35].

2.4.2. Genetic Variation

To estimate the level of genetic variation among population pairs, pairwise F_{ST} values and the exact test p values were calculated using Arlequin v3.5 [36]. The *Nei*'s genetic distance was assessed by GenAlEx v6.5 [31]. The molecular variance (AMOVA) was also assessed by Arlequin v3.5 [36]. The software Structure v2.3.4 was used for the clustering analysis based on the Bayesian method (admixture model, K set 1 to 7, 20 runs, MCMC = 1,000,000, burn-in = 25,000) [37]. The results were submitted to an online tool, Structure Harvester v0.6094 [38], to obtain the best K value. Principal coordinates analysis (PCoA) of the correlation matrix was used to further investigate the relationships between individuals using GenAlEx v6.5 [31].

The directional relative migration patterns among populations were estimated by the web-based software divMigrateOnline using the F_{ST} statistic as a measure of genetic differentiation [39]. The significance of asymmetrical migration patterns among populations was tested using 1000 bootstrap iterations. Additionally, the mantel test (10,000 repetitions) for isolation by distance (IBD) was performed between genetic distance and geographical distance (i.e., Euclidean distance based on latitude and longitude) via R software with ggplot2, diveRsity, and reshape packages [40].

3. Results

3.1. Meta-Analysis

The meta-analysis data set was derived from 117 articles for which we weighted the data by population replicate and a further 77 articles for which we weighted the data by the inverse of the variance (Figure 2, Tables S2 and S3). According to the taxonomic status of species in publications, species were divided into 10 groups, including mollusk, arthropod, echinoderm, and seven groups of bony fish in chordate (i.e., carp, perch, flounder, salmon, catfish, puffer, and herring). Echinoderm and puffer were only used when weighted by population replicate.

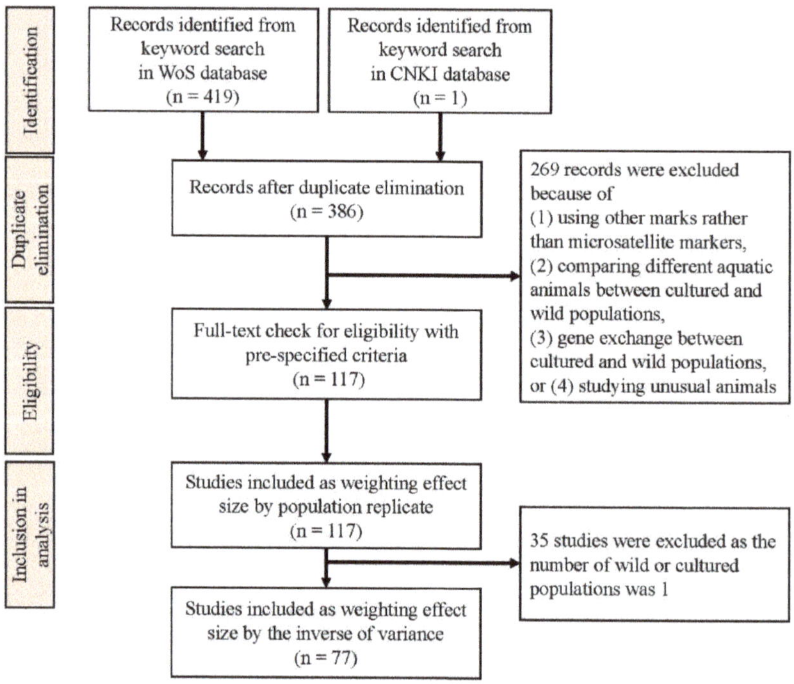

Figure 2. Selection of literature to be included in the meta-analysis data set.

Results from the meta-analysis showed the negative effect size of all cultured aquatic animals that were tested both when weighted by population replicate and weighted by the inverse of variance. The levels of genetic diversity decrease were different in different aquatic animals (Table 2). In the results when weighted by the inverse of variance, the highest effect sizes of *Na* and *He* were in flounder and salmon, respectively, and the genetic diversities of cultured populations decreased by 38.44% and 10.73% from wild populations, respectively. In the results when weighted by population replicate, the highest effect sizes of *Na* and *He* were in echinoderm and carp, respectively, and the genetic diversities of cultured populations decreased by 31% and 10% from wild populations, respectively. The overall effect size was −0.23 (CI: −0.32 to −0.16) for *Na* and −0.08 (CI: −0.13 to −0.04) for *He*, respectively, when weighted by the inverse of variance. Similarly, the overall effect sizes were −0.24 (CI: −0.33 to −0.15) for *Na* and −0.05 (CI: −0.07 to −0.03) for *He*, respectively, when weighted by population replicate. For carp, the reduction in genetic diversity of cultured populations was 12% or 24% by *Na* and 5% or 10% by *He* when weighted by the inverse of variance or when weighted by population replicate, respectively.

Table 2. Meta-analysis of the effect of aquaculture on different aquatic animals.

Class	Studies	Weight	Genetic Diversity Indices	Effect Size	CI-l	CI-u	Decrease (%)
Salmon	21	*1/Var*	*Na*	−0.30	−0.18	−0.43	26
Flounder	3	*1/Var*	*Na*	−0.49	−0.24	−0.65	38
Perch	10	*1/Var*	*Na*	−0.14	0.01	−0.32	13
Arthropod	3	*1/Var*	*Na*	−0.05	0.01	−0.62	5
Mollusk	16	*1/Var*	*Na*	−0.22	−0.07	−0.69	20
Carp	14	*1/Var*	*Na*	−0.13	−0.03	−0.23	12
Herring	2	*1/Var*	*Na*	−0.43	−0.13	−0.45	35
Catfish	3	*1/Var*	*Na*	−0.07	0.03	−0.28	7
Arapaima	2	*1/Var*	*Na*	−0.07	−0.04	−0.24	6
Overall	74	*1/Var*	*Na*	−0.23	−0.16	−0.32	20
Salmon	21	*1/Var*	*He*	−0.11	−0.05	−0.23	11
Flounder	3	*1/Var*	*He*	−0.03	−0.01	−0.12	3
Perch	9	*1/Var*	*He*	−0.04	−0.03	−0.07	4
Arthropod	3	*1/Var*	*He*	−0.01	0.00	−0.10	1
Mollusk	16	*1/Var*	*He*	−0.05	−0.02	−0.13	5
Carp	14	*1/Var*	*He*	−0.05	−0.02	−0.09	5
Herring	2	*1/Var*	*He*	−0.07	−0.02	−0.12	7
Catfish	3	*1/Var*	*He*	0.00	0.05	−0.06	0
Arapaima	2	*1/Var*	*He*	−0.09	−0.09	−0.19	9
Overall	73	*1/Var*	*He*	−0.08	−0.04	−0.13	8
Salmon	25	*Np*	*Na*	−0.13	0.06	−0.27	12
Flounder	7	*Np*	*Na*	−0.24	0.12	−0.50	21
Perch	20	*Np*	*Na*	−0.27	−0.09	−0.46	24
Arthropod	3	*Np*	*Na*	−0.23	0.01	−0.62	21
Mollusk	28	*Np*	*Na*	−0.31	−0.11	−0.62	27
Carp	16	*Np*	*Na*	−0.28	−0.14	−0.55	24
Echinoderm	2	*Np*	*Na*	−0.36	0.09	−0.52	31
Herring	3	*Np*	*Na*	−0.16	−0.13	−0.25	15
Catfish	4	*Np*	*Na*	−0.36	0.24	−0.81	30
Puffer	3	*Np*	*Na*	−0.25	−0.07	−0.54	22
Overall	111	*Np*	*Na*	−0.24	−0.15	−0.33	21

Table 2. *Cont.*

Class	Studies	Weight	Genetic Diversity Indices	Effect Size	CI-l	CI-u	Decrease (%)
Salmon	25	*Np*	*He*	−0.02	0.07	−0.09	2
Flounder	7	*Np*	*He*	−0.08	−0.05	−0.10	7
Perch	19	*Np*	*He*	−0.06	−0.01	−0.11	6
Arthropod	3	*Np*	*He*	−0.03	0.00	−0.07	3
Mollusk	28	*Np*	*He*	−0.06	−0.01	−0.12	6
Carp	16	*Np*	*He*	−0.11	−0.05	−0.14	10
Echinoderm	2	*Np*	*He*	0.00	0.01	−0.03	0
Herring	2	*Np*	*He*	−0.01	0.00	−0.02	1
Catfish	4	*Np*	*He*	−0.04	0.04	−0.09	4
Puffer	3	*Np*	*He*	−0.05	−0.01	−0.12	4
Overall	109	*Np*	*He*	−0.05	−0.03	−0.07	5

1/Var is the inverse of effect size variance; *Np* is the number of populations; *Na* is allele number; *He* is expected heterozygosity; CI-l and CI-u are the lower and upper limits of bootstrap confidence intervals, respectively. Decrease (%) is the reduction in genetic diversity of cultured populations compared with their corresponding wild populations.

3.2. Genetic Diversity within Carp Populations in Rice–Fish Coculture

All 20 microsatellite loci were polymorphic in the sampled carp (Table S3). The average numbers of alleles (*Na*) ranged from 5.80 (HX) to 10.40 (MA), the effective numbers of alleles (*Ne*) ranged from 3.86 (LC) to 5.70 (BL), Shannon's diversity indices (*I*) ranged from 1.42 (WX) to 1.85(BL), the expected heterozygosity (*He*) values ranged from 0.68 (LC) to 0.76 (CZ and BL), the observed heterozygosity (*Ho*) values ranged from 0.68 (WX) to 0.76 (MA), and the fixation indices (*Fis*) ranged from −0.02 (CZ) to 0.08 (JZ) (Table 3). Mean *Na* = 7.40, mean *Ne* = 4.57, mean *I* = 1.61, mean *He* = 0.71, and mean *Ho* = 0.73 (Table 3).

Table 3. The genetic characteristics of the 10 carp populations based on 20 microsatellite loci.

Pop.	*Na*	*Ne*	*I*	*Ho*	*He*	*Fis*
HX	5.80	4.35	1.53	0.75	0.74	0.00
CZ	7.00	4.70	1.63	0.76	0.73	−0.02
LC	5.85	3.86	1.44	0.68	0.69	0.01
JZ	7.25	4.52	1.64	0.68	0.74	0.08
WX	5.85	3.92	1.42	0.68	0.68	0.00
WK	7.50	4.42	1.59	0.71	0.72	0.03
XZS	7.75	4.07	1.54	0.70	0.71	0.01
BL	9.65	5.70	1.85	0.76	0.77	0.03
DRY	6.90	4.81	1.62	0.70	0.73	0.07
MA	10.40	5.29	1.83	0.72	0.76	0.06
Mean	7.40	4.57	1.61	0.71	0.73	0.03

The effective population size estimates of the 10 populations ranged from 8.6 (CZ, CI = 5.7–13.6) to infinity (JZ, CI = 148.9–infinity and DRY, CI = 58.3–infinity) (Table 4). The results from the bottleneck tests showed that no heterozygote excess was significant in any of the populations, indicating that there was no recent genetic bottleneck in the sampled carp populations in the south of Zhejiang Province, China (Table 4). In addition, a normal L-shaped distribution pattern of the allele frequency from the mode-shift test also suggested the lack of bottleneck events in the recent history of carp coculture in rice paddies in the south of Zhejiang Province, China.

Table 4. Effective population size estimates with 95% confidence intervals and results from the bottleneck analysis for the 10 *C. carpio* populations using 20 microsatellite loci.

Pop.	Effective Population Size Estimate	95% Confidence Intervals		Bottleneck Test
		Lower Bound	Upper Bound	TPM (p-Value)
HX	9.9	8.0	12.5	0.63575
CZ	8.6	5.7	13.6	0.06155
LC	61.5	27.6	Inf	0.83501
JZ	Inf	148.9	Inf	0.99585
WX	124.8	46.4	Inf	0.47816
WK	61.8	41.4	114.2	0.98802
XZS	54.2	44.0	69.3	0.99928
BL	209.6	98.4	Inf	0.99884
DRY	Inf	58.3	Inf	0.87726
MN	38.0	34.1	42.6	0.99940
Total	63.4	60.2	66.9	0.99985

3.3. Genetic Differentiation among Populations

The pairwise F_{ST} values ranged from 0.00 (WX-WK) to 0.13 (WX-CZ), and *Nei*'s genetic distances ranged from 0.070 (WX-WK) to 0.620 (WX-CZ) (Figure 3). Among all 45 F_{ST} values, 37 values were statistically significant ($p < 0.001$; p-value after adjusting for multiple comparisons = 0.05/45), revealing remarkable differentiation of carp cocultured in paddies in the south of Zhejiang Province, China. Pairwise F_{ST} analyses also indicated that populations within Yongjia and Jingning counties were genetically different. Weak genetic differentiation was found within Qingtian County except that JZ was significantly different from WX and XZS. AMOVA revealed that the genetic variations among populations and within populations contributed 5% ($p < 0.01$) and 95% ($p < 0.01$) to the total genetic variation, respectively (Table 5).

Table 5. The AMOVA of carp cocultured in paddies in three counties based on 20 microsatellite loci.

Source of Variation	d.f.	Sum of Square	Variance Component	% of Variation	p Value
Among populations	9	173.541	0.37651 Va	5	0.001
Within populations	165	2311.173	7.17756 Vb	95	0.001
Total	174	2484.714	7.55407		

The Structure Harvester analysis identified K = 2 as the most probable cluster number of the 10 populations, and the second identified K value was K = 3 (Figure 4A). The Structure clustering analysis revealed two major genetic clusters (the red cluster and the green cluster, Figure 4B). The carp from Qingtian County were mainly assigned into the red cluster, while the carp from Jingning County were mainly assigned into the blue cluster. The two clusters equally made up Yongjia County. In the case of K = 3, another cluster (indicated in blue) was mainly separated from Jingning County. The result of the principal coordinates analysis (PCoA) based on *Nei*'s genetic distance is presented in Figure 5. The first and second axes explained 45% and 25% of the total variance, respectively. No obvious clustering was found among the three counties. The samples from Yongjia County were located in the center, while the samples from Jingning County were relatively discrete.

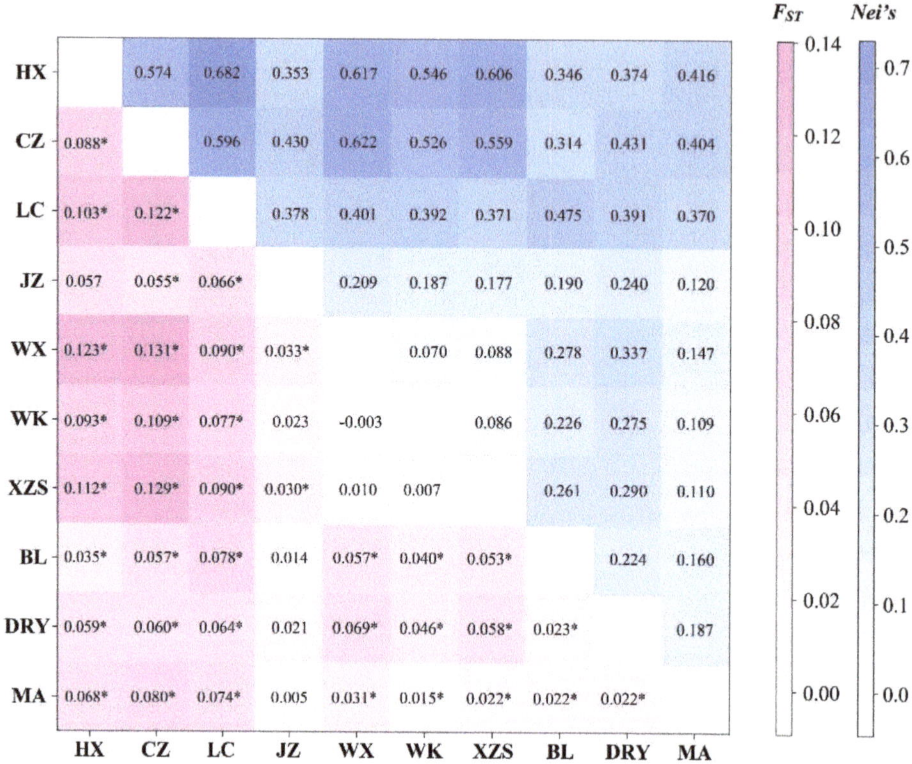

Figure 3. Pairwise differentiation estimates (F_{ST}) (below the diagonal) and *Nei*'s genetic distances (above the diagonal) of the 10 *C. carpio* populations based on 20 microsatellite loci. Values with * indicate statistical significance ($p < 0.001$; p-value after adjusting for multiple comparisons = 0.05/45).

The directional relative migration network for the studied carp populations indicated that WX, WK, XZS, JZ, and MA were core populations that had a high level of genetic exchange with other populations (i.e., migration in directional relative migration networks), the first four of which belong to Qingtian County, whereas HX, LC, CZ, DRY, and BL were peripheral populations with a low level of genetic exchange (Figure 6), the first three of which belong to Jingning County. No significant asymmetric migration pattern was detected. The test of isolation by distance (IBD) proved that there was a significantly positive correlation between the genetic distances and the geographical distances of the 10 carp populations ($R^2 = 0.579$ and $p = 0.005$, Figure 7). This indicated that the genetic differentiation between the sampled carp populations in southern Zhejiang Province of China was mainly affected by geographical distance.

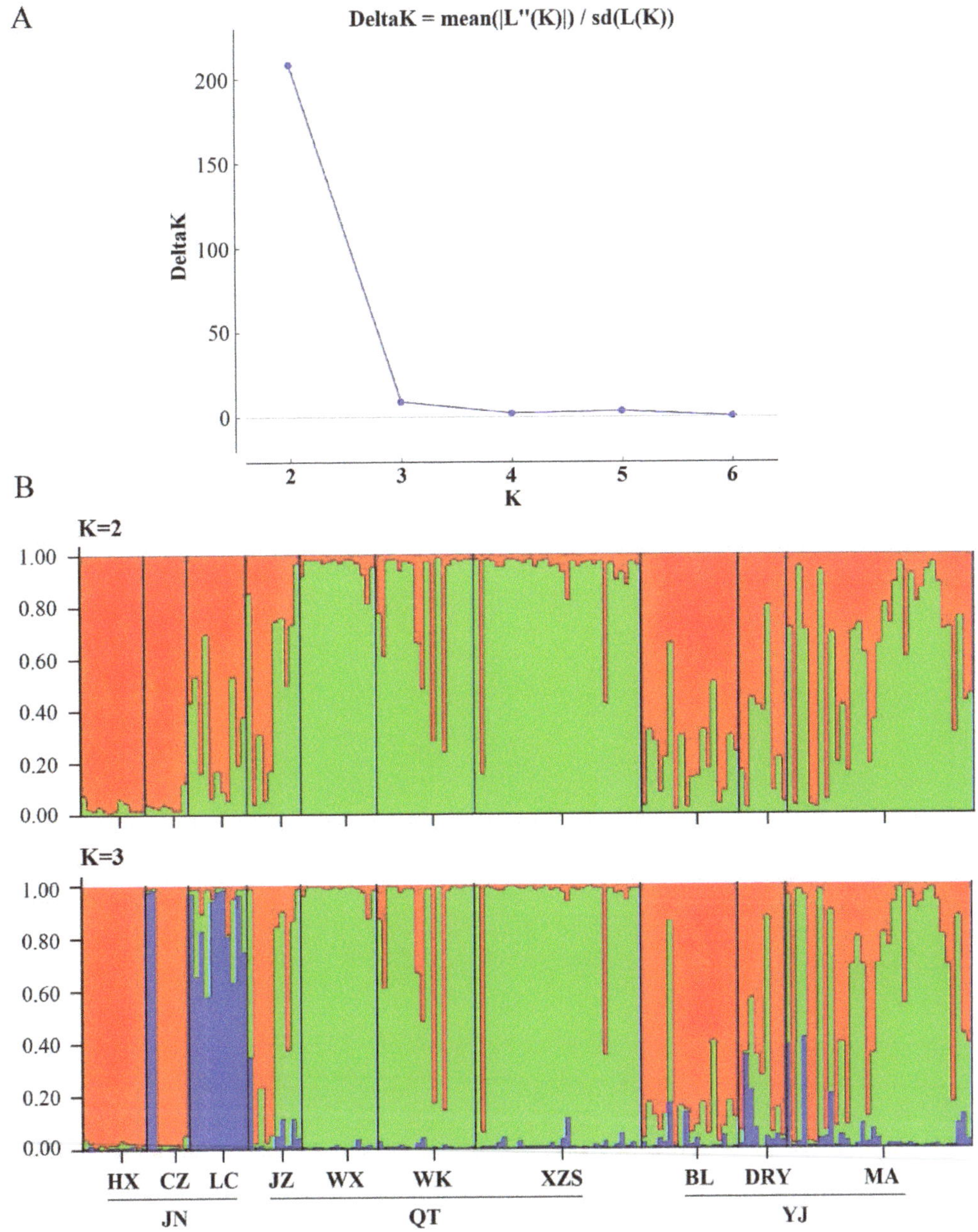

Figure 4. (**A**) Selection of K value in the structure analysis. (**B**) Genetic structure of the 10 carp populations based on 20 microsatellite loci in the case of K = 2 or 3. Colors represent the membership of each individual to the different clusters.

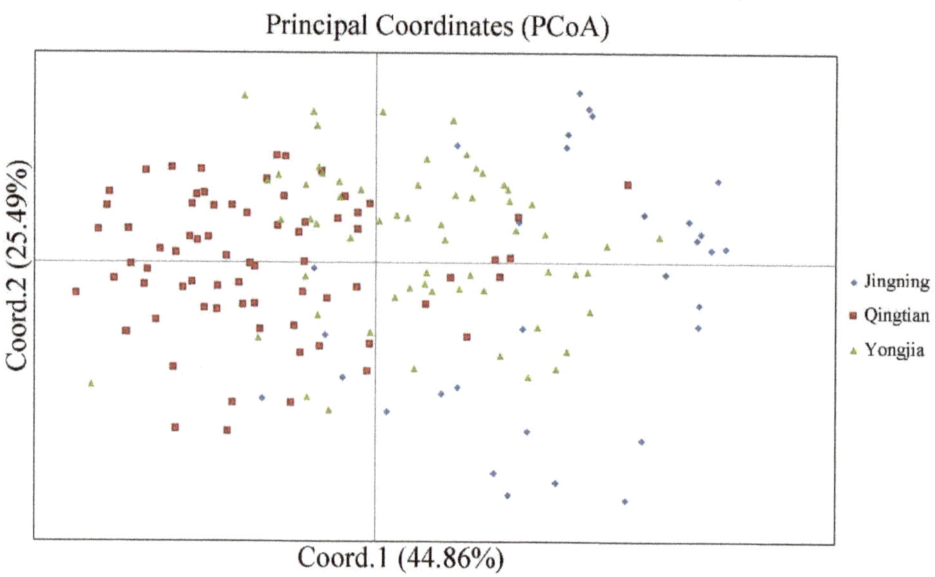

Figure 5. Principal coordinates analysis of the sampled carp based on *Fst* values.

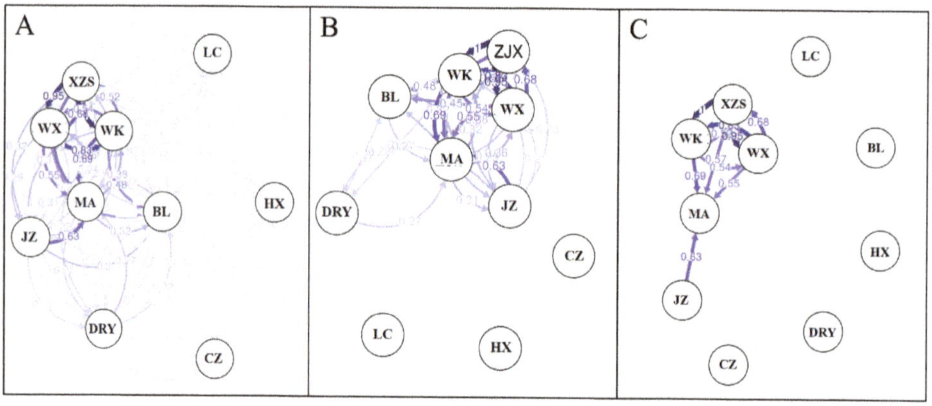

Figure 6. Directional relative migration networks of the studied carp populations constructed with divMigrate. Numbers on the arrows represent the relative migration coefficients derived from F_{ST} statistics. Line shading and thickness increase with the relative strength of gene flows. When larger than 0.05 (**A**), 0.2 (**B**), or 0.5 (**C**), the coefficients are displayed.

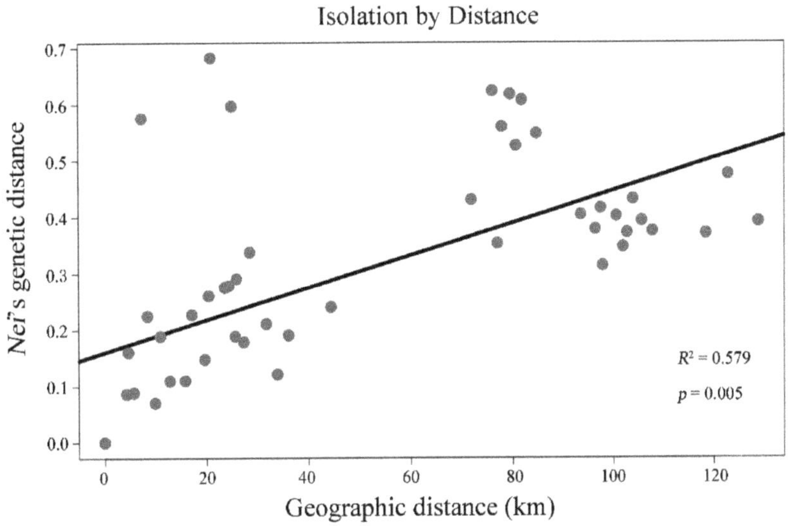

Figure 7. Relationship between the genetic distance and geographic distance among the studied carp populations.

4. Discussion

Our meta-analysis showed that the genetic diversity of cultured populations of most tested aquatic animals was obviously lower than that of wild populations and that the change in genetic diversity differed among different types of aquatic animals. Those results indicate that aquaculture could generally reduce the genetic diversity of many cultured aquatic animals. An additional concern is the reduction in some wild populations and, hence, the reduction in those pools of genetic diversity. Actually, some tested species in this study may not have a statistically significant reduction in genetic diversity (e.g., catfish, arthropods). New techniques are being used to maximize genetic diversity in cultured species [41–43]. The genetic resources of aquatic animals were not significantly paid attention to until the 1990s [44,45]. A small effective population size and poor breeding management were considered to be the main causes of decline in genetic diversity in cultured populations [46]. For example, Machado-Schiaffino et al. [47] found the imbalance in the sex ratio in breeding causes a decline in genetic diversity of the fish *Salmo salar*. Fazzi-Gomes et al. [48] found a loss of genetic diversity and high inbreeding rates in farmed populations of the fish *Arapaima gigas* throughout the Amazon basin due to genetic bottlenecks caused by the domestication process and the founding effect. The decrease in the genetic diversity of aquatic animals will impair the adaptation and fitness of aquatic animals (e.g., productivity and disease resistance) [49–51]. As the proportion of aquaculture is increasing in the supply of aquatic products, it is urgent to protect the genetic diversity of aquatic animals. To ensure the long-term sustainability of aquatic stocks, the breeding program should be taken seriously. The selective breeding program can provide farmers a high rate of economic return by creating wide variations and improving hereditary traits [52,53]. As the breeding program progresses, it is important to collect as much allelic variability as possible, which can maintain the level of genetic variability of aquatic animals [54]. Genetic diversity should be paid more attention to in the future as a guide for choosing brood stock to form the base population for selective breeding programs and for ongoing monitoring of the levels of inbreeding and genetic drift [55,56].

The genetic diversity of *C. carpio* cocultured in paddies in southern Zhejiang Province was at a high level throughout our study. Ren [57] conducted a literature review to evaluate the genetic diversity indices of wild carp populations (mean Na = 7.71, He = 0.71) and cultured carp populations (mean Na = 5.37, He = 0.62) from 55 relevant published papers.

The results of our study (mean $Na = 7.40$, $He = 0.71$) were similar to the genetic diversity of wild carp populations and higher than cultured carp populations. This suggests that traditional agricultural systems play a role in in situ conservation of the genetic diversity of carp cocultured in paddies in southern Zhejiang Province. The *C. carpio* preserved in the traditional rice–fish coculture system is a landrace that has well adapted to the paddy environment, with strong resistance to pests and diseases and adaptation to the fierce habitat changes during rice cultivation [58,59]. In the traditional agricultural system, the conservation of carp depends on the recognition, collection, and introduction of new strains by local smallholders, resulting in the diversity accumulation of genotypes and alleles in landraces [60,61]. Farmers often exchanged germplasm (e.g., selection and exchange of brood stock or seed) of carp cocultured in paddies in the traditional agricultural system. Germplasm exchange has been found to be an important factor in maintaining genetic diversity of crops and livestock in many traditional agricultural systems [9,62–64]. In our study, the carp cocultured in paddies in Qingtian County were rich in body color. The practice of having a mixed culture of carp with diverse color types is in favor of the diversified use of natural food resources by fish in paddy fields, which promotes fish productivity [21]. Therefore, the demand for diverse carp colors may be the reason for maintaining germplasm exchange. In addition, in Yongjia County, the exchange of carp with different colors is also related to marriage customs; this custom is still preserved in some areas [57]. Accordingly, the traditional techniques and culture noted in Zhejiang are an excellent source of maintaining genetic diversity. It is somewhat encouraging that, with proper management, such as those learned from traditional agriculture, genetic losses in agriculture could be minimized or avoided [9,63].

Pairwise F_{ST} value in our study revealed significant genetic differentiation of carp populations among the three counties. This genetic division may be due to geographic isolation and the differences between farmers from different counties in the selection preference of carp. The results also suggested a significant genetic differentiation within Yongjia County and within Jingning County, while a weak genetic differentiation was found within Qingtian County, except for JZ. This could be due to the fact that the sampling sites of Yongjia and Jingning were located in remote and isolated mountainous areas, while Qingtian County, as a GIAHS site, still maintains a large area of the traditional rice–fish farming system: more than 90% of rice paddies are stocked with fish [57]. Traditional farmers have created a unique sharing system in which farmers interdependently select parental carp and produce and exchange fry in Qingtian County [21]. Larger relative migration values from the migration network estimated by divMigrateOnline of the XZS population to other populations indicated that XZS was most likely the source population, whereas the populations from Jingning County (i.e., LC, HX, and CZ) might be the sink populations. The results strongly suggested that JZ was genetically distinct from the other Qingtian populations. This population exhibited the lowest F_{ST} values with populations from Yongjia County compared to the other three populations from Qingtian County. Membership coefficients estimated by Structure and the migration network strongly suggested that the JZ population was genetically distinct from other Qingtian populations. The JZ population and populations from Yongjia County had close relationships. This could be due to the transfer of seed stock/brood stock from JZ to the Yongjia area. A genetic structure analysis in the present study showed that the genetic structure of carp populations cocultured in paddies in southern Zhejiang Province was mainly determined by germplasm exchange caused by breeding introduction.

The rice–fish coculture system can be a sustainable agricultural model that improves farm productivity, provides an opportunity to improve the economic benefits of farmers, and improves the utilization of paddy and water resources [65]. However, the development of modern agriculture reduces the exchange of germplasm resources and activities of the rice–carp coculture in paddies among farmers. To ensure the long-term sustainability of germplasm resources of carp cocultured in paddies, a scientific and feasible monitoring program of genetic diversity in the existing populations should be formulated and proper measures of development and conservation should be strengthened. Firstly, we suggest the government should formulate policies on promoting the practice of rice–fish coculture, on training farmers to maintain carp diversity by continuing to exchange brood stock, and on monitoring every 5 to 10 years. Secondly, in the breeding process, to avoid the reduction in genetic diversity caused by inbreeding, the number of parents and effective population size should be increased to at least the minimum level of heterozygosity required. Thirdly, seed stock or brood stock selection should be performed locally to avoid genetic pollution of the local population. In addition, it is necessary to reduce environmental pollution and protect the paddy habitats and resources for traditional agriculture and local carp populations, thus ensuring the sustainable development of the rice–fish coculture system in southern Zhejiang Province of China.

5. Conclusions

The meta-analysis results, both when weighted by population replicate and when weighted by the inverse of variance, showed an overall negative effect size in genetic diversity of all tested aquatic animals. This indicates that aquaculture activities have caused a general decline in the genetic diversity of aquatic animals, although at different levels for different types of aquatic animals. We detected high levels of genetic variation in all 10 populations of carp cocultured in paddies (*C. carpio*) in the traditional rice–fish coculture system in southern Zhejiang Province of China. Low levels of an effective population size were detected in most of the *C. carpio* populations. No bottleneck events have recently occurred in these populations. Both conventional and model-based population genetic analyses suggested significant genetic divergence among the three counties. Pairwise F_{ST} values suggested genetic differentiation within Yongjia County and Jingning County, while no obvious genetic difference between sampled populations was found in Qingtian County with the exception of the JZ population. We suggest formulating policies, training farmers, and monitoring regularly for maintaining the rice–fish coculture system and using a sufficient number of parents in the breeding process to avoid inbreeding and genetic erosion of local carp cocultured in paddies.

Supplementary Materials: The following supporting information can be downloaded at https://www.mdpi.com/article/10.3390/agriculture12070997/s1. Table S1: Information of 20 polymorphic microsatellite loci for *C. carpio*. Table S2: Source data in meta-analysis of the effect of aquaculture on aquatic animals (weighted by the inverse of variance). Table S3: Source data in meta-analysis of the effect of aquaculture on aquatic animals (weighted by population replicates).

Author Contributions: Conceptualization, Y.Y. and W.R.; methodology, L.H. and X.C.; software, L.Z.; validation, S.Z., J.T., L.H. and X.C.; formal analysis, Y.Y. and W.R.; investigation, L.Z.; resources, S.Z., J.T. and X.C.; data curation, L.H.; writing–original draft preparation, Y.Y. and W.R.; writing–review and editing, L.H. and X.C.; visualization, J.T.; supervision, L.H.; project administration, L.H. and X.C.; funding acquisition, L.H. and X.C. All authors have read and agreed to the published version of the manuscript.

Funding: The study was supported by the National Natural Science Foundation of China (31770481, 31661143001), Zhejiang Provincial Key Research and Development Project (2022C02058, LGN22C030002), and Guangdong Provincial Key Research and Development Project (2021B0202030002).

Institutional Review Board Statement: Not applicable.

Informed Consent Statement: Not applicable.

Data Availability Statement: The data presented in this study are available upon request from the corresponding author.

Conflicts of Interest: The authors declare no conflict of interest.

References

1. Pingali, P.L. Green Revolution: Impacts, limits, and the path ahead. *Proc. Natl. Acad. Sci. USA* **2012**, *109*, 12302–12308. [CrossRef] [PubMed]
2. Cheng, W.; D'Amato, A.; Pallante, G. Benefit sharing mechanisms for agricultural genetic diversity use and on-farm conservation. *Econ. Politica* **2020**, *37*, 337–355. [CrossRef]
3. Castilla, F.; Montilla-Bascón, G.; Bekele, W.A.; Howarth, C.J.; Prats, E. Population genomics of mediterranean oat (*A. sativa*) reveals high genetic diversity and three loci for heading date. *Theor. Appl. Genet.* **2021**, *134*, 2063–2077.
4. van de Wouw, M.; Kik, C.; Hintum, T.; van Treuren, R.; Visser, B. Genetic erosion in crops: Concept, research results and challenges. *Plant Genet. Resour.* **2010**, *8*, 1–15. [CrossRef]
5. Boettcher, P.J.; Hoffmann, I. Protecting indigenous livestock diversity. *Science* **2011**, *334*, 1058. [CrossRef]
6. Aerts, R.; Berecha, G.; Honnay, O. Protecting coffee from intensification. *Science* **2015**, *347*, 139. [CrossRef]
7. Rajpurohit, D.; Jhang, T. In situ and ex situ conservation of plant genetic resources and traditional knowledge. In *Plant Genetic Resources and Traditional Knowledge for Food Security*; Springer: Singapore, 2015; pp. 137–162.
8. Priyanka, V.; Kumar, R.; Dhaliwal, I.; Kaushik, P. Germplasm conservation: Instrumental in agricultural biodiversity—A review. *Sustainability* **2021**, *13*, 6743. [CrossRef]
9. Jarvis, D.I.; Brown, A.H.D.; Cuong, P.H.; Collado-Panduro, L.; Latournerie-Moreno, L.; Gyawali, S.; Tanto, T.; Sawadogo, M.; Mar, I.; Sadiki, M.; et al. A global perspective of the richness and evenness of traditional crop-variety diversity maintained by farming communities. *Proc. Natl. Acad. Sci. USA* **2008**, *105*, 8160. [CrossRef]
10. Achtak, H.; Ater, M.; Oukabli, A.; Santoni, S.; Kjellberg, F.; Khadari, B. Traditional agroecosystems as conservatories and incubators of cultivated plant varietal diversity: The case of fig (*Ficus carica* L.) in Morocco. *BMC Plant Biol.* **2010**, *10*, 28. [CrossRef]
11. Garlock, T.; Asche, F.; Anderson, J.; Bjørndal, T.; Kumar, G.; Lorenzen, K.; Tveteras, R. A global blue revolution: Aquaculture growth across regions, species, and countries. *Rev. Fish Sci. Aquac.* **2020**, *28*, 107–116. [CrossRef]
12. Ahmad, A.; Abdullah, S.R.S.; Hasan, H.A.; Othman, A.R.; Ismail, N.I. Aquaculture industry: Supply and demand, best practices, effluent and its current issues and treatment technology. *J. Environ. Manag.* **2021**, *287*, 112271. [CrossRef] [PubMed]
13. Napier, J.A.; Haslam, R.P.; Olsen, R.E.; Tocher, D.R.; Betancor, M.B. Agriculture can help aquaculture become greener. *Nat. Food* **2020**, *1*, 680–683. [CrossRef]
14. Wang, C.; Li, Z.; Wang, T.; Xu, X.; Zhang, X.; Li, D. Intelligent fish farm—The future of aquaculture. *Aquacult. Int.* **2021**, *29*, 2681–2711. [CrossRef] [PubMed]
15. Ju, R.T.; Li, X.; Jiang, J.J.; Wu, J.; Liu, J.; Strong, D.R.; Li, B. Emerging risks of non-native species escapes from aquaculture: Call for policy improvements in China and other developing countries. *J. Appl. Ecol.* **2020**, *57*, 85–90. [CrossRef]
16. Tan, M.P.; Wong, L.L.; Razali, S.A.; Afiqah-Aleng, N.; Mohd Nor, S.A.; Sung, Y.Y.; de Peer, Y.V.; Sorgeloos, P.; Danish-Daniel, M. Applications of next-generation sequencing technologies and computational tools in molecular evolution and aquatic animals conservation studies: A short review. *Evol. Bioinform.* **2019**, *15*, 1–5. [CrossRef]
17. Aguiar, J.D.P.; Gomes, P.F.F.; Hamoy, I.G.; Santos, S.E.B.D.; Schneider, H.; Sampaio, I. Loss of genetic variability in the captive stocks of tambaqui, *Colossoma macropomum* (Cuvier, 1818), at breeding centres in Brazil, and their divergence from wild populations. *Aquac. Res.* **2018**, *49*, 1914–1925. [CrossRef]
18. Lind, C.E.; Evans, B.S.; Knauer, J.; Taylor, J.J.U.; Jerry, D.R. Decreased genetic diversity and a reduced effective population size in cultured silver-lipped pearl oysters (*Pinctada maxima*). *Aquaculture* **2009**, *286*, 12–19. [CrossRef]
19. Loukovitis, D.; Ioannidi, B.; Chatziplis, D.; Kotoulas, G.; Magoulas, A.; Tsigenopoulos, C.S. Loss of genetic variation in Greek hatchery populations of the European sea bass (*Dicentrarchus labrax* L.) as revealed by microsatellite DNA analysis. *Mediterr. Mar. Sci.* **2015**, *16*, 197–200. [CrossRef]
20. Xie, J.; Wu, X.; Tang, J.J.; Zhang, J.E.; Luo, S.M.; Chen, X. Conservation of traditional rice varieties in a Globally Important Agricultural Heritage System (GIAHS): Rice-fish coculture. *Agric. Sci. China.* **2011**, *10*, 101–105. Available online: http://www.fao.org/giahs/giahs-home/en/ (accessed on 10 March 2022). [CrossRef]
21. Ren, W.; Hu, L.; Guo, L.; Zhang, J.; Tang, L.; Zhang, E.; Zhang, J.; Luo, S.; Tang, J.; Chen, X. Preservation of the genetic diversity of a local common carp in the agricultural heritage rice–fish system. *Proc. Natl. Acad. Sci. USA* **2018**, *115*, E546–E554. [CrossRef]
22. Xie, J.; Hu, L.; Tang, J.; Wu, X.; Li, N.; Yuan, Y.; Yang, H.; Zhang, J.; Luo, S.; Chen, X. Ecological mechanisms underlying the sustainability of the agricultural heritage rice–fish coculture system. *Proc. Natl. Acad. Sci. USA* **2011**, *108*, E1381–E1387. [CrossRef] [PubMed]
23. Hedges, L.V.; Gurevitch, J.; Curtis, P.S. The meta-analysis of response ratios in experimental ecology. *Ecology* **1999**, *80*, 1150–1156. [CrossRef]
24. Li, Q.; Ling, Q.F.; Wang, L.; Wang, D.P.; Tong, G.J. Study on the morphological differences of four common carp populations. *J. Anhui Agric. Sci.* **2011**, *39*, 15404–15405.
25. Crooijmans, R.P.M.A.; Bierbooms, V.A.F.; Komen, J.; VanderPoel, J.J.; Groenen, M.A.M. Microsatellite markers in common carp (*Cyprinus carpio* L.). *Anim. Genet.* **1997**, *28*, 129–134. [CrossRef]
26. David, L.; Rajasekaran, P.; Fang, J.; Hillel, J.; Lavi, U. Polymorphism in ornamental and common carp strains (*Cyprinus carpio* L.) as revealed by AFLP analysis and a new set of microsatellite markers. *Mol. Genet. Genom.* **2001**, *266*, 353–362. [CrossRef]
27. Wei, D.W.; Lou, Y.D.; Sun, X.Y.; Shen, J.B. Screening of microsatellite markers in common carp. *Zool. Res.* **2001**, *223*, 238–241.

28. Wang, D.; Liao, X.L.; Cheng, L.; Yu, X.M.; Tong, J.G. Development of novel EST-SSR markers in common carp by data mining from public EST sequences. *Aquaculture* **2007**, *271*, 558–574. [CrossRef]
29. Yue, G.H.; Ho, M.Y.; Orban, L.; Komen, J. Microsatellites within genes and ESTs of common carp and their applicability in silver crucian carp. *Aquaculture* **2004**, *234*, 85–98. [CrossRef]
30. Oosterhout, C.V.; Hutchinson, W.F.; Wills, D.P.; Shipley, P. MICRO-CHECKER: Software for identifying and correcting genotyping errors in microsatellite data. *Mol. Ecol.* **2004**, *4*, 535–538. [CrossRef]
31. Peakall, R.; Smouse, P.E. GenAlEx 65: Genetic analysis in excel population genetic software for teaching and research an update. *Bioinformatics* **2012**, *28*, 2537–2539. [CrossRef]
32. Do, C.; Waples, R.S.; Peel, D.; Macbeth, G.; Tillett, B.J.; Ovenden, J.R. NeEstimator v2: Re-implementation of software for the estimation of contemporary effective population size (Ne) from genetic data. *Mol. Ecol. Resour.* **2014**, *14*, 209–214. [CrossRef] [PubMed]
33. Luikart, G.; Allendorf, F.; Cornuet, J.; Sherwin, W. Distortion of allele frequency distributions provides a test for recent population bottlenecks. *J. Hered.* **1998**, *89*, 238–247. [CrossRef] [PubMed]
34. Piry, S.; Luikart, G.; Cornuet, J.M. Bottleneck: A computer program for detecting recent reductions in the effective population size using allele frequency data. *J. Hered.* **1999**, *90*, 502–503. [CrossRef]
35. Luikart, G.; Cornuet, J.M. Empirical evaluation of a test for identifying recently bottlenecked populations from allele frequency data. *Conserv. Biol.* **1998**, *12*, 228–237. [CrossRef]
36. Excoffier, L.; Lischer, H.E. Arlequin suite ver 35: A new series of programs to perform population genetics analyses under Linux and Windows. *Mol. Ecol. Resour.* **2010**, *10*, 564–567. [CrossRef]
37. Pritchard, J.K.; Stephens, M.; Donnelly, P. Inference of population structure using multilocus genotype data. *Genetics* **2000**, *155*, 945–959. [CrossRef]
38. Earl, D.A.; vonHoldt, B.M. STRUCTURE HARVESTER: A website and program for visualizing STRUCTURE output and implementing the Evanno method. *Conserv. Genet. Resour.* **2012**, *4*, 359–361. Available online: http://taylor0.biology.ucla.edu/structureHarvester/ (accessed on 27 March 2022). [CrossRef]
39. Sundqvist, L.; Keenan, K.; Zackrisson, M.; Prodöhl, P.; Kleinhans, D. Directional genetic differentiation and relative migration. *Ecol. Evol.* **2016**, *6*, 3461–3475. Available online: https://popgen.shinyapps.io/divMigrate-online/ (accessed on 29 March 2022). [CrossRef]
40. Oksanen, J.; Kindt, R.; O'Hara, B. Vegan: R functions for vegetation ecologists. *Date Access* **2005**, *15*, 2014.
41. Abdelrahman, H.; ElHady, M.; Alcivar-Warren, A.; Allen, S.; Al-Tobasei, R.; Bao, L.; Beck, B.; Blackburn, H.; Bosworth, B.; Buchanan, J.; et al. Aquaculture genomics, genetics and breeding in the United States: Current status, challenges, and priorities for future research. *BMC Genom.* **2012**, *18*, 1–23.
42. Reading, B.J.; McGinty, A.S.; Clark, R.W.; Hopper, M.S.; Woods III, L.C.; Baltzegar, D.A. Genomic enablement of temperate bass aquaculture (family Moronidae). In *Breeding and Culture of Perch and Bass*; Science China Press (Chinese Academy of Sciences): Beijing, China, 2018.
43. Andersen, L.K.; Clark, R.W.; Hopper, M.S.; Hodson, R.G.; Schilling, J.; Daniels, H.V.; Woods, L.C., III; Kovach, A.I.; Berlinsky, D.L.; Kenter, L.W.; et al. Methods of domestic striped bass (*Morone saxatilis*) spawning that do not require the use of any hormone induction. *Aquaculture* **2021**, *533*, 736025. [CrossRef]
44. Gibson, J.; Gamage, S.; Hanotte, O.; Iñiguez, L.; Maillard, J.; Rischkowsky, B.; Semambo, D.; Toll, J.; Gibson, J. *Options and Strategies for the Conservation of Farm Animal Genetic Resources: Report of an International Workshop, AGROPOLIS, Montpellier, France, 7–10 November 2005*; CGIAR System-Wide Genetic Resources Programme (SGRP)/Bioversity International: Rome, Italy, 2006; Volume 5.
45. Harvey, B. *Blue Genes: Sharing and Conserving the World's Aquatic Biodiversity*; Routledge: London, UK, 2013.
46. Norris, A.T.; Bradley, D.G.; Cunningham, E.P. Microsatellite genetic variation between and within farmed and wild Atlantic salmon (*Salmo salar*) populations. *Aquaculture* **1999**, *180*, 247–264. [CrossRef]
47. Machado-Schiaffino, G.; Dopico, E.; Garcia-Vazquez, E. Genetic variation losses in Atlantic salmon stocks created for supportive breeding. *Aquaculture* **2007**, *264*, 59–65. [CrossRef]
48. Fazzi-Gomes, P.F.; Aguiar, J.D.P.; Marques, D.; Fonseca Cabral, G.; Moreira, F.C.; Rodrigues, M.D.N.; Silva, C.S.; Hamoy, I.; Santos, S. Novel microsatellite markers used for determining genetic diversity and tracing of wild and farmed populations of the Amazonian giant fish *Arapaima gigas*. *Genes* **2021**, *12*, 1324. [CrossRef]
49. Danzmann, R.G.; Ferguson, M.M.; Allendorf, F.W. Genetic variability and components of fitness in hatchery strains of rainbow trout. *J. Fish Biol.* **1989**, *35*, 313–319. [CrossRef]
50. Koehn, R.K.; Diehl, W.J.; Scott, T.M. The different contribution by individual enzymes of glycolysis and protein catabolism to the relationship between heterozygosity and growth rate in the coot clam Milinia lateralis. *Genetics* **1988**, *118*, 121–130. [CrossRef]
51. Tiira, K.; Laurila, A.; Peuhkuri, N.; Piironen, J.; Ranta, E.; Primmer, C.R. Aggressiveness is associated with genetic diversity in landlocked salmon (*Salmo salar*). *Mol. Ecol.* **2003**, *12*, 2399–2407. [CrossRef]
52. Ponzoni, R.W.; Nguyen, N.H.; Khaw, H.L.; Ninh, N.H. Accounting for genotype by environment interaction in economic appraisal of genetic improvement programs in common carp *Cyprinus carpio*. *Aquaculture* **2008**, *285*, 47–55. [CrossRef]

53. Ninh, N.H.; Ponzoni, R.W.; Nguyen, N.H.; Woolliams, J.A.; Taggart, J.B.; McAndrew, B.J.; Penman, D.J. A comparison of communal and separate rearing of families in selective breeding of common carp (*Cyprinus carpio*): Responses to selection. *Aquaculture* **2013**, *408*, 152–159. [CrossRef]
54. Gjerde, B. Design of breeding programs. In *Selection and Breeding Programs in Aquaculture*; Gjedrem, T., Ed.; Springer: Dordrecht, The Netherlands, 2005; pp. 173–195.
55. Loughnan, S.R.; Smith-Keune, C.; Jerry, D.R.; Beheregaray, L.B.; Robinson, N.A. Genetic diversity and relatedness estimates for captive barramundi (*Lates calcarifer*, Bloch) broodstock informs efforts to form a base population for selective breeding. *Aquac. Res.* **2016**, *47*, 3570–3584. [CrossRef]
56. Abebe, A.T.; Kolawole, A.O.; Unachukwu, N.; Chigeza, G.; Gedil, M. Assessment of diversity in tropical soybean (*Glycine max* (L.) Merr.) varieties and elite breeding lines using single nucleotide polymorphism markers. *Plant Genet. Resour.* **2021**, *19*, 20–28. [CrossRef]
57. Ren, W.Z. The Genetic Diversity in Traditional Rice-Fish System. Ph.D. Thesis, Zhejiang University, Hangzhou, China, 2016.
58. Rahman, M.L.; Shahjahan, M.; Ahmed, N. Tilapia farming in Bangladesh: Adaptation to climate change. *Sustainability* **2021**, *13*, 7657. [CrossRef]
59. Ji, D.; Su, X.; Yao, J.; Zhang, W.; Wang, R.; Zhang, S. Genetic Diversity and Genetic Differentiation of Populations of Golden-Backed Carp (*Cyprinus carpio* var. *Jinbei*) in Traditional Rice Fields in Guizhou, China. *Animals* **2022**, *12*, 1377. [CrossRef] [PubMed]
60. Pusadee, T.; Jamjod, S.; Chiang, Y.C.; Rerkasem, B.; Schaal, B.A. Genetic structure and isolation by distance in a landrace of Thai rice. *Proc. Natl. Acad. Sci. USA* **2009**, *106*, 13880–13885. [CrossRef] [PubMed]
61. Alvarez, N.; Garine, E.; Khasah, C.; Dounias, E.; Hossaert-McKey, M.; McKey, D. Farmers' practices, metapopulation dynamics, and conservation of agricultural biodiversity on-farm: A case study of sorghum among the Duupa in sub-sahelian Cameroon. *Biol. Conserv.* **2005**, *121*, 533–543. [CrossRef]
62. Leclerc, C.; D'Eeckenbrugge, G.C. Social organization of crop genetic diversity. The G × E × S Interaction Model. *Diversity* **2011**, *4*, 1–32. [CrossRef]
63. Labeyrie, V.; Thomas, M.; Muthamia, Z.K.; Leclerc, C. Seed exchange networks, ethnicity, and sorghum diversity. *Proc. Natl. Acad. Sci. USA* **2016**, *113*, 98–103. [CrossRef]
64. Delêtre, M.; McKey, D.B.; Hodkinson, T.R. Marriage exchanges, seed exchanges, and the dynamics of manioc diversity. *Proc. Natl. Acad. Sci. USA* **2011**, *108*, 18249–18254. [CrossRef]
65. Bashir, M.A.; Liu, J.; Geng, Y.; Wang, H.; Pan, J.; Zhang, D.; Rehim, A.; Aon, M.; Liu, H. Co-culture of rice and aquatic animals: An integrated system to achieve production and environmental sustainability. *J. Clean. Prod.* **2020**, *249*, 119310. [CrossRef]

agriculture

MDPI

Communication

Improved Production of Mashua (*Tropaeolum tuberosum*) Microtubers MAC-3 Morphotype in Liquid Medium Using Temporary Immersion System (TIS-RITA®)

Gilmar Peña-Rojas [1], Roxana Carhuaz-Condori [1], Vidalina Andía-Ayme [2], Victor A. Leon [3] and Oscar Herrera-Calderon [4,*]

1 Laboratory of Cellular and Molecular Biology, Biological Sciences Faculty, Universidad Nacional de San Cristóbal de Huamanga, Portal Independencia 57, Ayacucho 05003, Peru; gilmar.pena@unsch.edu.pe (G.P.-R.); roxana.carhuaz@unsch.edu.pe (R.C.-C.)
2 Laboratory of Food Microbiology, Biological Sciences Faculty, Universidad Nacional de San Cristóbal de Huamanga, Portal Independencia 57, Ayacucho 05003, Peru; vidalina.andia@unsch.edu.pe
3 Department of Biology, New York University, New York, NY 10003, USA; victorleon@nyu.edu
4 Department of Pharmacology, Bromatology and Toxicology, Pharmacy and Biochemistry Faculty, Universidad Nacional Mayor de San Marcos, Lima 15001, Peru
* Correspondence: oherreraca@unmsm.edu.pe; Tel.: +51-956-550-510

Abstract: Essential molecules are embedded within the millenary crop *Tropaeolum tuberosum* (mashua); these compounds are critical for the Andean people's traditional diet and extensively utilized by the pharmaceutical industry in Peru. In the Andean region, conventional cropping techniques generate microtubers susceptible to a viral infection, which substantially endangers mashua's production. Therefore, we developed an innovative in vitro technique condition for enhancing the agriculture process for micro tubers production. The temporary immersion system (TIS) permits the production of high-quality microtubers in a reduced space, a lower amount of time, and in large quantities compared with tubers grown under traditional conditions. To obtain *T. tuberosum*'s microtubers via TIS, we propagated seedlings, utilizing TIS-RITA® vessels. A set of immersion frequency times were evaluated. Interestingly, results showed that immersion at 2 min every 3 h was more beneficial compared with 2 min every 5 h based on microtubers produced after 10 weeks from the treatments, revealing an efficient frequency setting which outputted improved microtubers quality and production.

Keywords: microtubers; temporary immersion system; *Tropaeolum tuberosum*; mashua

Citation: Peña-Rojas, G.; Carhuaz-Condori, R.; Andía-Ayme, V.; Leon, V.A.; Herrera-Calderon, O. Improved Production of Mashua (*Tropaeolum tuberosum*) Microtubers MAC-3 Morphotype in Liquid Medium Using Temporary Immersion System (TIS-RITA®). *Agriculture* **2022**, *12*, 943. https://doi.org/10.3390/agriculture12070943

Academic Editors: Moucheng Liu, Xin Chen and Yuanmei Jiao

Received: 17 May 2022
Accepted: 27 June 2022
Published: 29 June 2022

1. Introduction

The mashua is a millenary crop that contains substantial nutritional and medical properties [1–3]. This crop originated from the Andean region [4], and is considered the fourth most crucial Andean root among other tubers such as potatoes, oca, and olluco [5]. The mashua is a propagation crop cultivated over the latest centuries across the Andean mountains in Peru, Bolivia, Ecuador, Venezuela, and Colombia [5,6]. This formidable tuber has managed to grow under nutrient-deprived soil conditions and at high altitudes without fertilizers or pesticides, outstanding for its resistance against harsh conditions in contrast to other contemporary crops [7,8].

Traditionally, the Andean mashua is propagated for production purposes as other tubers within the Andean crop fields [9]. Additionally, when the mashua is cultivated under field conditions, it necessitates between 6 and 8 months to properly attain its vegetative cycle stages, and in several cases, the tubers become virally infected, dramatically impacting the crop's production [6]. Therefore, rural and local mashua production within the Andean region does not ensure suitable seed quality and necessitates alternative cropping techniques to improve biological features such as growth and vigor. In vitro techniques are utilized to improve crop production and decrease time constraints; therefore, sculpting innovative in vitro techniques for tuber cropping is needed.

One in vitro technique to harness the growth of seedlings is the modulation of the frequency and duration of immersion times [10] during the tuber early development. The TIS (temporary immersion system) strategy enables the rapid and efficient propagation of several plants with keen agricultural interest. The TIS enhances the growing speed and ensures the optimal quality of the plant tissue generated in vitro [11]. The TIS permits the production of high-quality pathogen-free seedlings and microtubers in vitro at any time throughout the year [12]. Moreover, it reduces large-scale crop production costs [13], automatizes the cropping process, and permits proper propagation by utilizing liquid media to ensure seedlings vigor [11].

The TIS enhances the growing speed and ensures the optimal quality of the plant tissue generated in vitro [11], enabling rapid and efficient propagation of several plants with strong agricultural interest. The TIS technique initiates by inducing air pressure flow through an air compressor; this de novo pressure elevates the liquid media permitting contact with the explants localized inside the chamber intermittently. As the air injection subsides within the system, and the media descends by gravity, the atmosphere remodels within the system, facilitating a robust growth and substantial improvement for the seedlings' development [14]. Some critical factors defining the TIS technique are the following: the optimization of the total volume inside the vessels, the supplement in the media, the vitrification, the ethylene accumulation, and the carbon dioxide. Seedlings development and growth can be harnessed by modifying the frequency and duration of the immersion time [10].

Likewise, TIS-RITA® includes a structured container divided into two vessels: a superior vessel hosting the plants and an inferior vessel containing the media. The applied overpressure to the inferior vessel propels the media towards the superior compartment generating bubbles grazing the plant tissues. At this stage, seedlings temporarily submerge as overpressure is delivered. During the immersion period, the media falls by gravity. One result is the altered atmosphere inside the container [12]. The other critical parameter is the immersion time, involved in efficient sprout micropropagation, microtuberization, and somatic embryogenesis [12]. The TIS is regularly utilized for increasing in vitro propagation coefficients compared with field conditions. For example, these methods have been in used in other crop species such as bananas [15], anthurium [16], sugar cane [17], and potato microtubers [18].

In this study, we concentrated on the production of mashua MAC-3 morphotype via a novel in vitro procedure. MAC-3 morphotype (purple mashua) is known for its high antioxidant activity, total phenolic, tannins, total flavonoids, and total anthocyanins [19]. Recently, we had faithfully propagated *Solanum tuberosum*, *Oxalis tuberosa*, and *Ullucus tuberosum* in vitro utilizing TIS-RITA [20]. In a previous report, we found that a specific frequency condition in TIS-RITA substantially enhances the mashua's microtuber propagation [21]. Therefore, we explored vital settings to generate high-quality seeds, utilizing a set of immersion frequencies to obtain improved seeds of the *T. tuberosum* MAC-3 morphotype that would positively impact the crop production for the Andean community in South America.

2. Materials and Methods

2.1. Micropropagation Study

Mashua seedlings (*T. tuberosum* Ruiz & Pav.) derived from a MAC-3 morphotype from the germplasm bank of the Cellular and Molecular Biology Laboratory (UNSCH, Ayacucho, Peru) were propagated in vitro (Figure 1). Seedlings were maintained using Murashige and Skoog 1962 (MS) solid medium. After 30 days of culture, the seedlings grown in solid medium were transferred to flasks containing 100 mL of MS liquid medium supplemented with 3% sucrose at a pH of 5.6 to obtain seedling vigor; the flasks were kept under constant agitation on an orbital shaker. Culture conditions were 19 ± 2 °C; 16 h of light and 8 h of darkness with a relative humidity between 60% and 70% during the multiplication phase.

Figure 1. *T. tuberosum* Ruiz and Pav. "mashua" MAC-3 morphotype used in the TIS to obtain microtubers.

2.2. Production of Microtubers via TIS-RITA

TIS-RITA vessels were used according to Etienne et al. [12]. Murashige and Skoog (MS) liquid medium were prepared inside the RITA vessels, supplemented with 2 ppm BAP and 8% sucrose, at a pH 5.6; the vessels were sterilized at 121 °C for 15 min. This study evaluated one immersion time point and two frequencies: 2 min every 3 h and 2 min every 5 h. Next, TIS-RITA vessels were incubated under the constant temperature of 19 °C ± 2 °C, for a total of 10 weeks in total darkness. Produced microtubers were harvested off the culture vessels, and samples were rinsed off with tap water to remove any excess media. These microtubers were placed on trays covered with filter paper to remove humidity. Finally, the microtubers' fresh weights (g) were evaluated using an analytical scale and a vernier ruler to measure the size (cm).

2.3. Data Analysis

The output data were statistically analyzed by using a random design format with double replicates. The variation analysis was performed to compare the size and weight of microtubers in a non-parametric U-Mann–Whitney test.

3. Results and Discussion

Using the RITA® temporary immersion system, it was possible to obtain mashua microtubers from MAC-3 morphotype using the two immersion frequencies. However, there was a significant difference between the two immersion frequencies in obtaining the size of the microtubers; a size of 1.09 cm was achieved at an immersion frequency of every three hours for two minutes, compared with 0.86 cm at a frequency of every five hours for two minutes respectively (Figure 2A), these findings being statistically significant ($p = 0.0017$; U-Mann Whitney test). According to Akita et al. [22], potato microtubers (*Solanum tuberosum* L.) were obtained using a laboratory-scale fermenter with a weight of more than 0.2 g. Montoya et al. [23] achieved the greatest number and size of *Solanum tuberosum* shoots using TIS with an immersion frequency of three hours. Escalona et al. [24] in the cultivation of *Ananas commosus* achieved a higher multiplication rate using an immersion time of two minutes and a frequency of three hours. Likewise, Cabrera et al. [25] obtained a greater number and size of *Dioscorea alata* microtubers with significant differences in relation to other immersion times using an immersion time of 15 min and after 18 weeks of culture. Etienne et al. [12]

stated that *Solanum tuberosum* microtubers and *Coffea arabica* somatic embryos produced in temporary immersion bioreactors developed satisfactorily after planting.

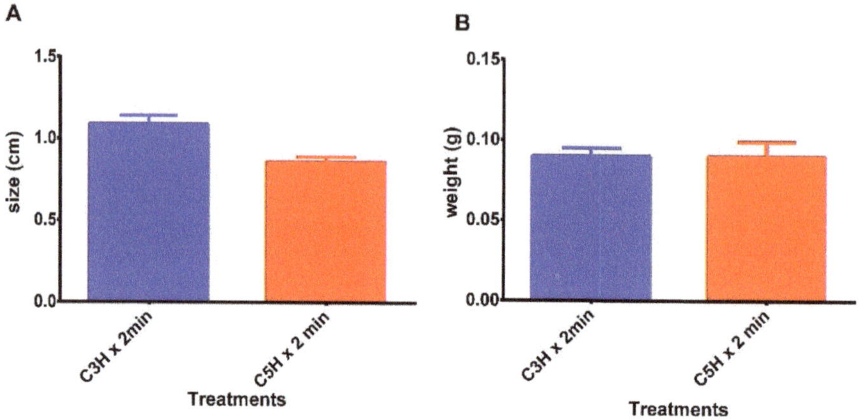

Figure 2. (**A**) Mean comparison of the microtuber size and (**B**) weight from *T. tuberosum* "mashua" obtained in two TIS treatments: 2 min every 3 h (C3H × 2 min); 2 min every 5 h (C5H × 2 min).

On the other hand, using a TIS, 52 mashua microtubers from a MAC-3 morphotype were obtained at a frequency of every three hours of immersion for two minutes compared with a frequency of five hours for two minutes in which 39 microtubers were obtained. Moreover, Igarza et al. [26] acquired an average of between five and seven potato microtubers of the "Andinita" variety using a TIS. Montoya et al. [23] achieved greater efficiency in the in vitro tuberization of *Solanum tuberosum* variety Diacol Capiro when used in temporary immersion bioreactors and in MS medium supplemented with 1 ppm of 6-Benzylaminopurine (BAP) and 8% sucrose; in addition, the microtubers obtained in a TIS allowed the formation of tubers under field conditions. Gopal et al. [27]) concluded that microtubers produced in media without abscisic acid (ABA) during and containing high concentrations of sucrose and BAP can be stored for 12 months.

Regarding the fresh weight of the mashua, an average of 0.09 g was achieved in both immersion frequency treatments; therefore, there was no statistically significant difference in the results obtained between both treatments (Figure 2B). Igarza et al. [25] achieved an average fresh weight that did not exceed 3.5 g using an immersion system to obtain potato microtubers cv. "Andinita".

The immersion system allows obtaining microtubers in two and a half months (Figure 3), significantly reducing the production time compared with the production of mashua tubers in the field, which generally requires between six to nine months. In addition, under this production system it is possible to obtain high-quality seeds, allowing ex situ conservation in a germplasm bank, and fundamentally for use in seed management and improvement programs.

Figure 3. (**a**) Conservation in germplasm bank. (**b**) Micropropagation in solid medium. (**c**) Propagation in liquid medium. (**d**) Obtaining microtubers in a temporary immersion from *T. tuberosum* "mashua" MAC-3 after a 10-week culture. (**e**) The frequency: (Left): three hours for two minutes, (right): five hours for two minutes.

4. Conclusions

It was possible to obtain MAC-3 microtubers in the TIS RITA® using the Murashige and Skoog medium supplemented with 8% sucrose, 2 ppm BAP, and with an immersion frequency of every 3 h for 2 min. The TIS RITA® is an efficient alternative for the production of high-quality seeds. Furthermore, it would lead to obtaining virus-free microtubers as well as reducing the harvest time compared with traditional production techniques.

Author Contributions: Conceptualization, G.P.-R. and R.C.-C.; methodology, G.P.-R.; formal analysis, V.A.-A.; investigation, G.P.-R., R.C.-C., V.A.-A. and O.H.-C.; writing—original draft preparation, V.A.L.; writing—review and editing, V.A.L.; visualization, O.H.-C.; project administration, G.P.-R.; funding acquisition, G.P.-R. All authors have read and agreed to the published version of the manuscript.

Funding: This research was funded by the CONCYTEC, Ministerio de Educacion, Perú. (MINEDU-CONCYTEC) project 199-2015-FONDECYT—UNSCH.

Institutional Review Board Statement: Not applicable.

Data Availability Statement: Not applicable.

Acknowledgments: Authors would like to thank FONDECYT and the Department of Biology of the New York University.

Conflicts of Interest: The authors declare no conflict of interest.

References

1. Apaza Ticona, L.N.; Tena Pérez, V.; Bermejo Benito, P. Local/Traditional Uses, Secondary Metabolites and Biological Activities of Mashua (*Tropaeolum tuberosum* Ruíz & Pavón). *J. Ethnopharmacol.* **2020**, *247*, 112152. [CrossRef]
2. Apaza Ticona, L.; Arnanz Sebastián, J.; Serban, A.M.; Rumbero Sánchez, Á. Alkaloids Isolated from *Tropaeolum tuberosum* with Cytotoxic Activity and Apoptotic Capacity in Tumour Cell Lines. *Phytochemistry* **2020**, *177*, 112435. [CrossRef]
3. Ticona, L.A.; Sánchez, Á.R.; Estrada, C.T.; Palomino, O.M. Identification of TRPV1 Ion Channels Agonists of *Tropaeolum tuberosum* in Human Skin Keratinocytes. *Planta Med.* **2021**, *87*, 383–394. [CrossRef]
4. Campos, D.; Noratto, G.; Chirinos, R.; Arbizu, C.; Roca, W.; Cisneros-Zevallos, L. Antioxidant Capacity and Secondary Metabolites in Four Species of Andean Tuber Crops: Native Potato (*Solanum* Sp.), Mashua (*Tropaeolum tuberosum* Ruiz & Pavón), Oca (*Oxalis Tuberosa* Molina) and Ulluco (*Ullucus Tuberosus* Caldas). *J. Sci. Food Agric.* **2006**, *86*, 1481–1488. [CrossRef]
5. Pissard, A.; Arbizu, C.; Ghislain, M.; Bertin, P. Influence of Geographical Provenance on the Genetic Structure and Diversity of the Vegetatively Propagated Andean Tuber Crop, Mashua (*Tropaeolum tuberosum*), Highlighted by Intersimple Sequence Repeat Markers and Multivariate Analysis Methods. *Int. J. Plant Sci.* **2008**, *169*, 1248–1260. [CrossRef]
6. Grau, A.; Dueñas, R.O.; Cabrera, C.N.; Hermann, M. *Mashua Tropaeolum Tuberosum Ruíz & Pav*; Promoting the Conservation and Use of Underutilized and Neglected Crops 25; International Potato Center: Lima, Peru, 2003; Volume 52, p. 427.
7. Ortega, O.R.; Duran, E.; Arbizu, C.; Ortega, R.; Roca, W.; Potter, D.; Quiros, C.F. Pattern of Genetic Diversity of Cultivated and Non-Cultivated Mashua, *Tropaeolum tuberosum*, in the Cusco Region of Perú. *Genet. Resour. Crop. Evol.* **2007**, *54*, 807–821. [CrossRef]
8. Arbizu, C.; García, E.R. *Catálogo de Los Recursos Fitogenéticos de Raíces y Tubérculos Andinos*; Programa de Investigación de Cultivos andinos, Facultad de Ciencias Agrarias, Universidad Nacional San Cristóbal de Huamanga: Ayacucho, Peru, 1986.
9. Lim, T.K. *Edible Medicinal and Non-Medicinal Plants: Volume 12, Modified Stems, Roots, Bulbs*; Universiteitsbibliotheek Gent: Ghent, Belgium, 2016; pp. 1–690. [CrossRef]
10. Jäger, A.K.; Schottländer, B.; Smitt, U.W.; Nyman, U. Somatic Embryogenesis in Cell Cultures of *Thapsia garganica*: Correlation between the State of Differentiation and the Content of Thapsigargins. *Plant Cell Rep.* **1993**, *12*, 517–520. [CrossRef]
11. Alvarenga Venutolo, S. Micropropagación Masiva de *Stevia rebaudiana* Bertoni en Sistemas de Inmersión Temporal. *Cultiv. Trop.* **2015**, *36*, 50–57.
12. Etienne, H.; Berthouly, M. Temporary Immersion Systems in Plant Micropropagation. *Plant Cell Tissue Organ Cult.* **2002**, *69*, 215–231. [CrossRef]
13. So Young, P.; Murthy, H.N.; Kee Yoeup, P. Mass Multiplication of Protocorm-like Bodies Using Bioreactor System and Subsequent Plant Regeneration in Phalaenopsis. *Plant Cell Tissue Organ Cult.* **2000**, *63*, 67–72. [CrossRef]
14. Maldonado, E.R.; de Francisco, L.E.R.; Gómez, O.A.; Cerda, M.E.C. Diseño y Construcción de Un Sistema de Inmersión Temporal. *Cent. Agríc.* **2003**, *30*, 69–72.
15. Pérez, M.B.; Pérez, M.B.; Vega, V.M.; Gálvez, E.O.; Delgado, M.T.; Torres, J.L.; Jova, M.C.; Pino, A.S.; Cabrera, A.R.; Toledo, M.B.; et al. Empleo de Sistemas de Inmersión Temporal Como Alternativa Para La Propagación in Vitro Del Cultivar de Plátano Vianda INIVITPV06-30 (*Musa* AAB). *Biotecnol. Veg.* **2012**, *12*, 53–57.

16. Alamilla Magaña, J.C.; Caamal Velazquez, J.H.; Criollo Chan, M.A.; Vera Lopez, J.E.; Reyes Montero, J.A. Biofábricas y Biorreactores de Inmersión Temporal: Propagación in Vitro de *Anthurium andreanum* L., y Su Viabilidad Económica. *Agro Product.* **2019**, *12*, 23–29. [CrossRef]
17. Villegas, A.B.; Villegas, A.B.; Aguila, Z.O.; Vázquez, M.J.; Fernández, O.R.; García-Aguila, L.; Feria, M. de Empleo de Los Sistemas de Inmersión Temporal Para La Producción de Vitroplantas de Caña de Azúcar. *Biotecnol. Veg.* **2002**, *2*, 201–206.
18. Gautam, S.; Solis-Gracia, N.; Teale, M.K.; Mandadi, K.; da Silva, J.A.; Vales, M.I. Development of an in Vitro Microtuberization and Temporary Immersion Bioreactor System to Evaluate Heat Stress Tolerance in Potatoes (*Solanum tuberosum* L.). *Front. Plant Sci.* **2021**, *12*, 1659. [CrossRef]
19. Rojas, G.P.; Rojas, G.P.; Sanchez, H.; Barahona, I.R.; Ayme, V.A.; Segura-Turkowsky, M.; Jimenez, R.E. Alternative Inputs for Micropropagation of *Solanum tuberosum*, *Ullucus tuberosus* and *Oxalis tuberosa* in Semisolid and Liquid Medium and Temporary Immersion System. *Trop. Subtrop. Agroecosyst.* **2020**, *23*, 41.
20. Peña, G.; Peña, G.; Carhuaz, R.; Davalos, J.; Ayme, V.A. Use of Rita® Temporary Immersion System to Obtain Microtubers of Several Mashua (*Tropaeolum tuberosum* Ruiz & Pavón) Morphotypes. *Trop. Subtrop. Agroecosyst.* **2020**, *23*, 84.
21. Akita, M.; Takayama, S. Stimulation of Potato (*Solanum tuberosum* L.) Tuberization by Semicontinuous Liquid Medium Surface Level Control. *Plant Cell Rep.* **1994**, *13*, 184–187. [CrossRef]
22. Montoya, N.; Castro, D.; Díaz, J.; Ríos, D. Tuberización in Vitro de Papa (*Solanum tuberosum* L), Variedad Diacol Capiro, En Biorreactores de Inmersión Temporal y Evaluación de Su Comportamiento En Campo. *Rev. Cienc.* **2008**, *16*, 288–295.
23. Escalona, M.; Lorenzo, J.C.; González, B.; Daquinta, M.; González, J.L.; Desjardins, Y.; Borroto, C.G. Pineapple (*Ananas comosus* L. Merr) Micropropagation in Temporary Immersion Systems. *Plant Cell Rep.* **1999**, *18*, 743–748. [CrossRef]
24. Cabrera, M.; Gómez, R.; Espinosa, E.; López, J.; Medero, V.; Basail, M.; Santos, A. Yam (*Dioscorea alata* L.) Microtuber Formation in Temporary Immersion System as Planting Material. *Biotecnol. Apl.* **2011**, *28*, 268–271.
25. Igarza Castro, J.; Agramonte, D.; de Feria, M.; Jaime, J.; Pérez, M.; San Román, M. Obtención de Microtubérculos de Papa Cv. 'Andinita' En Sistemas de Inmersión Temporal. *Biotecnol. Veg.* **2011**, *11*, 59–62.
26. Gopal, J.; Chamail, A.; Sarkar, D. In Vitro Production of Microtubers for Conservation of Potato Germplasm: Effect of Genotype, Abscisic Acid, and Sucrose. *Vitr. Cell. Dev. Biol.-Plant* **2004**, *40*, 485–490. [CrossRef]
27. Moreno, M.; Oropeza, M. Efecto de Las Hormonas Vegetales y El Fotoperiodo En La Producción de Microtubérculos de Papa (*Solanum tuberosum* L.). *Rev. Colomb. Biotecnol.* **2017**, *19*, 29–38. [CrossRef]

agriculture

MDPI

Article

Can Agricultural Machinery Harvesting Services Reduce Cropland Abandonment? Evidence from Rural China

Ping Xue [1], Xinru Han [1], Yongchun Wang [2] and Xiudong Wang [1,*]

[1] Institute of Agricultural Economics and Development, Chinese Academy of Agricultural Sciences, Beijing 100081, China; 82101201276@caas.cn (P.X.); hanxinru@caas.cn (X.H.)

[2] Agricultural Information Institute, Chinese Academy of Agricultural Sciences, Beijing 100081, China; wangyongchun@caas.cn

* Correspondence: wangxiudong@caas.cn; Tel.: +86-10-8210-6163

Abstract: Ending hunger, achieving food security, and promoting sustainable agriculture are the main targets of sustainable development goals. It is well known that cropland resources are the most essential factor in achieving sustainable development goals. However, China has been facing the problem of a continuous reduction in cropland resources. Reducing the abandonment of cropland has become an important way to curb the reduction in cropland resources. Can agricultural machinery harvesting services reduce cropland abandonment in rural China? To answer this scientific question, this study employs the Survey for Agriculture and Village Economy data from 8345 samples of 12 provinces in rural China. The extended regression models (i.e., the extended probit regression model and the extended interval regression model) are used to empirically analyze the relationship between agricultural machinery harvesting services accessed by farmers and cropland abandonment. The results are as follows. Agricultural machinery harvesting services accessed by farmers significantly reduced the probability of cropland abandonment and the proportion of the area of abandoned cropland in farmers' contracted cropland area decreased by 18.5% and 20.3%, respectively. Moreover, the heterogeneity analysis results showed that farmers' access to agricultural machinery harvesting services significantly reduced cropland abandonment in small-scale groups, without elderly households, with nonagricultural income groups, and in the eastern region. This study also provides some policy implications for policymakers to reduce cropland abandonment in rural China.

Keywords: cropland abandonment; agricultural machinery harvesting services; extended regression models; rural China

Citation: Xue, P.; Han, X.; Wang, Y.; Wang, X. Can Agricultural Machinery Harvesting Services Reduce Cropland Abandonment? Evidence from Rural China. *Agriculture* **2022**, *12*, 901. https://doi.org/10.3390/agriculture12070901

Academic Editors: Moucheng Liu, Xin Chen and Yuanmei Jiao

Received: 30 May 2022
Accepted: 20 June 2022
Published: 21 June 2022

1. Introduction

Ending hunger, achieving food security, and promoting sustainable agriculture are the main targets of sustainable development goals worldwide [1]. In this context, how to use limited cropland resources to feed a large population has become a key issue for the present and future in China, even though China has fed 20% of the world's population with approximately 7% of the world's cropland [2].

Cropland resources are the most essential factor for ensuring food security and promoting the development of sustainable agriculture [3,4]. According to the statistics of FAO in 2019, China's cropland area ranks third in the world, which is only lower than that of India and the United States. In terms of the per capita cropland area, however, China is only 0.09 hectares, lower than the 0.12 hectares in India and 0.48 hectares in the United States. This evidence shows that China still faces the problem of insufficient cropland resources. In addition, with the acceleration of urbanization and industrialization, the continuous reduction in cropland resources has further exacerbated this problem [3,5]. Previous studies mainly put forward two ways to curb the reduction of cropland resources: improving the existing cropland use efficiency, and reducing cropland abandonment [6–8]. Admittedly, improving the use efficiency of cropland alone is not enough to deal with cropland resource reduction, which should also reduce cropland abandonment.

It is well known that cropland abandonment has already become a common phenomenon in the world, which includes both developed countries (e.g., the United States, Australia, and Japan) and developing countries (e.g., China, Chile, Latin America, and Southeast Asia) [9–12]. It has become an increasingly important issue in China since 2000. For example, China's government issued the "Urgent Notice on Resuming the Production of Abandoned Cropland as soon as possible" in 2004, indicating that cropland has been abandoned to varying degrees in some areas. Moreover, large-scale cropland abandonment has occurred in China since 2005, especially in the mountainous counties [11]. This has posed challenges and threats to China's food security and sustainable agriculture [5,13,14]. It has also caused a series of environmental issues, such as the loss of agro-biodiversity and species richness, soil erosion, shallow landslides, and desertification [15–18]. In this context, the Chinese government has paid great attention to this issue and promulgated a series of policies. For example, the "Guiding Opinions on the Overall Utilization of Abandoned Cropland to Promote the Development of Agricultural Production" emphasized the importance and urgency of curbing the abandonment of cropland. In addition, it pointed out that one means of alleviating the abandonment of cropland is by cultivating agricultural professional service organizations to provide services for migrant workers and farmers with weak labor ability [19]. This also provides some inspiration for our study.

Much research has been done on the reasons for cropland abandonment. On the one hand, some studies indicate that rural laborers' migration to cities is the main factor that leads to cropland abandonment, such as Xu et al. (2018) [10] and Gao et al. (2020) [20]. In particular, with the arrival of the Lewis turning point in rural China in 2003, the era of the unlimited supply of rural labor force has passed [21]. China's unique household contract responsibility system, that is, that farmers have only the right to use the land but not the right to sell, has restricted farmers from selling cropland. In this case, the migrant workers can only transfer out of their cropland, therefore, the part of the cropland that cannot be transferred out of, will be abandoned. On the other hand, high agricultural production costs, such as the high investment costs of agricultural machinery, are also the main factor leading to cropland abandonment [22]. In addition, the croplands that are located far away from the villages and towns may be abandoned [23–25]. To sum up, the low agricultural production capacity, due to the lack of an agricultural labor force and operation equipment, is the main factor leading to the abandonment of cropland.

In terms of the driving mechanism for reducing cropland abandonment, previous research mainly explored land transfer [7,26], population aging [27], agricultural cooperatives [28], internet use [29], etc. Few studies have focused on the critical factors (i.e., agricultural production capacity) in the reduction of cropland abandonment. Agricultural mechanization services may be an effective way of improving agricultural production capacity, which is a special form of helping farmers to achieve the mechanized operation of part or all of the agricultural production links in rural China. It can alleviate rural labor shortages, reduce agricultural production costs, and improve agricultural mechanization levels [30–33]. This may effectively reduce cropland abandonment. In particular, harvesting is the most time-consuming and labor-intensive step in agricultural production (i.e., the "heaviest" of the agricultural production links) [30,34], which is more likely to lead farmers to abandon cropland. Therefore, this paper mainly focuses on agricultural machinery harvesting services (AMHSs).

In summary, the main aim of this paper was to explore whether AMHSs accessed by farmers can reduce cropland abandonment. To achieve this aim, we used the data of the Survey for Agriculture and Village Economy (SAVE) in 2019 and 2020 and employed the extended regression models (ERMs). Precisely, the main questions answered in this study are as follows: Can AMHSs reduce cropland abandonment in rural China? What is the heterogeneity in the impact of AMHSs on cropland abandonment?

Compared with the previous studies, there are mainly three marginal contributions of our study. First, different to previous quantitative studies, which mainly focus on the reasons for cropland abandonment, such as Deng et al. (2018) [27], Xu et al. (2018) [10], etc., the main aim of our study was to qualitatively explore the factors for reducing cropland abandonment. Second, we employed the extended regression models (i.e., the extended probit regression model and the extended interval regression model) to circumvent the endogenous problems caused by the reverse causality between AMHSs access and cropland abandonment and the problem of self-selection. Moreover, compared with IV-Probit or IV-Tobit, this model is suitable for binary endogenous explanatory variables. Third, this is the first study, to the best of our knowledge, to analyze whether access to AMHSs can reduce cropland abandonment in rural China.

2. Methods and Data

2.1. Theoretical Framework

In this section, we constructed a theoretical framework to analyze the relationship between access to AMHSs and cropland abandonment. As rational economic men, farmers maximize their income mainly through the rational allocation of labor and land resources [35]. With the growth of off-farm wages, farmers tend to allocate more labor resources to the nonagricultural sectors, which will lead to the reduction of labor input in agricultural production [24,36]. This phenomenon has induced serious cropland abandonment in rural China, that is, most of their cropland has been abandoned due to the lack of a sufficient labor force to manage it [10,13]. Although land transfers can alleviate the cropland abandonment to a certain extent, the rural land transfer market is still imperfect and the land transfer degree is still low [37]. Moreover, most of the migrant workers cannot get social security in cities, and most of the farmers still have a "land complex", so they would rather abandon the cropland than transfer it out [11,38,39]. In addition, the expensive input of agricultural machinery also prevents farmers from investing in machinery to replace labor input for agricultural production [40].

The agricultural mechanization services may provide a feasible approach to reducing cropland abandonment caused by the above dilemmas, especially in AMHSs. Figure 1 shows the theoretical framework for the impact of agricultural mechanization services access on cropland abandonment.

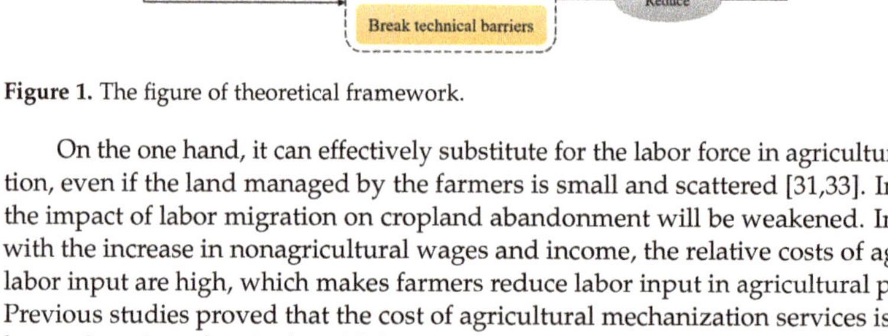

Figure 1. The figure of theoretical framework.

On the one hand, it can effectively substitute for the labor force in agricultural production, even if the land managed by the farmers is small and scattered [31,33]. In this case, the impact of labor migration on cropland abandonment will be weakened. In addition, with the increase in nonagricultural wages and income, the relative costs of agricultural labor input are high, which makes farmers reduce labor input in agricultural production. Previous studies proved that the cost of agricultural mechanization services is relatively lower than that of agricultural labor input, especially in the labor-intensive production links (such as the harvesting links) [41,42]. This will prevent farmers from abandoning their cropland due to high labor input costs in agricultural production.

On the other hand, purchasing agricultural machinery is expensive for most farmers, especially for the low-income level groups [43], which leads farmers to give up their cropland and leave agricultural production, due to a lack of agricultural machinery. Moreover, there are high technical barriers for most farmers to use agricultural machinery [44]. Therefore, farmers' access to agricultural mechanization services may be a better way to carry out agricultural production, which can alleviate the impact of the low level of agricultural mechanization on cropland abandonment. To sum up, agricultural mechanization services (mainly AMHSs in this study) may reduce cropland abandonment, which still needs to be tested by subsequent empirical analysis.

Accordingly, we mainly propose the following two hypotheses:

Hypothesis 1. *AMHSs accessed by farmers can reduce cropland abandonment in rural China.*

Hypothesis 2. *AMHSs can effectively alleviate the impact of labor migration on cropland abandonment.*

2.2. Study Methods

The extended regression models (ERMs) were employed in this study to evaluate the impact of AMHSs accessed by farmers on cropland abandonment. On the one hand, one dependent variable is binary (i.e., whether farmers abandon cropland), and another dependent variable (the proportion of the area of abandoned cropland in farmers' contracted cropland area) is a truncated variable in this study. On the other hand, the key variable, whether farmers access AMHSs, is a binary that may have a reverse causality with the dependent variables. In addition, the AMHSs accessed by farmers is a self-selection process, which will produce selection bias due to unobserved factors of farmers (such as agricultural management ability and the ability to accept new things). This can lead to endogeneity problems that make the estimated results biased. In this context, previous studies mainly adopted the IV-Probit and IV-Tobit model for a binary dependent variable and a truncated dependent variable, respectively. The above models only fit continuous endogenous covariables [45], while the key endogenous variable (i.e., whether farmers access AMHSs) is binary in our study. Therefore, we adopted the extended regression models (ERMs), which can fit the binary endogenous covariables. In particular, we adopted an extended probit regression for the binary dependent variable and an extended interval regression for the truncated dependent variable. These two benchmark models are given as:

$$LA_i = \alpha_0 + \alpha_1 AMHS_i + \alpha_2 X_i + \alpha_3 Year + \alpha_4 Region + \mu_i \quad (1)$$

$$LA_{ip} = \beta_0 + \beta_1 AMHS_i + \beta_2 X_i + \beta_3 Year + \beta_4 Region + \mu_i \quad (2)$$

where LA_i and LA_{ip} represent whether farmers i abandon cropland and the proportion of the area of abandoned cropland in the contracted cropland area of farmers i, respectively; $AMHS_i$ represents whether farmers i access *AMHSs*; X_i are the vectors of other control variables; *Year* and *Region* represent dummy variables of year and provinces, respectively; α_0–α_4 and β_0–β_4 are the vectors of the parameters; and μ_i and μ_i are the error terms.

This study also introduced an instrumental variable to circumvent the endogeneity problem. Following Kung (2002) [46] and Deng et al. (2018) [7] etc., this study selected the percentage of other farmers in the same village who access AMHSs as an instrumental variable. On the one hand, this instrumental variable, related to endogenous covariables, is satisfied, that is, the percentage of the other farmers in the same village who access AMHSs directly affects the probability that the focal farmer accesses AMHSs. On the other hand, it should not be related to the dependent variables, i.e., the percentage of other farmers in the same village who access AMHSs does not directly affect the focal farmer's abandoned cropland. Thus, this instrumental variable is reasonable for our study. The instrumental variable is calculated as follows:

$$IV\_AHMS_{ni_l} = (\sum_{i \neq i_l}^{j} AHMS_i)/(j-1) \tag{3}$$

where $IV\_AHMS_{ni_l}$ represents the probability of access to AMHSs by other farmers in the village n except for farmers i_l; j represents the number of samples surveyed in village n.

2.3. Data Source

This study used micro-level data from the 2019 and 2020 Survey for Agriculture and Village Economy (SAVE) [47,48]. This survey is launched and conducted annually by the Institute of Agricultural Economics and Development (IAED) at the Chinese Academy of Agricultural Sciences (CAAS). It covers 37 counties, 65 towns, and 292 villages in 12 provinces of Hebei, Jilin, Heilongjiang, Zhejiang, Anhui, Fujian, Henan, Hunan, Sichuan, Yunnan, Shaanxi, and Xinjiang (Figure 2). Moreover, it also includes surveys of rural households and villages, which provide abundant information about the rural households, household income, land use, villages, etc.

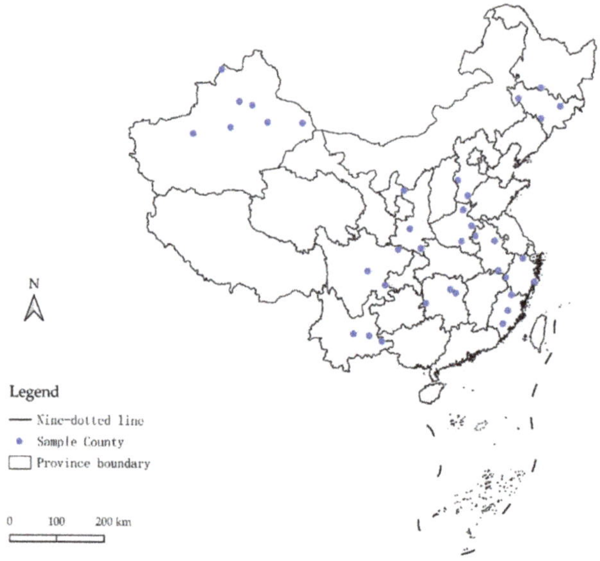

Figure 2. The geographical location of the study areas.

To accurately analyze the relationship between AMHSs' access and cropland abandonment, this study processed the data as follows: (1) The national fixed base Consumer Price Index (CPI, 2012 = 100) was used to process income-related variables to eliminate the impact of inflation; (2) This study mainly focused on the impact of the AMHSs accessed by the cropland actual operators on the abandonment of cropland, so the samples with zero actual cropland area are deleted. Finally, 8345 samples were used in this study, which includes 4518 samples from 2019 and 3827 samples from 2020. As mentioned above, unbalanced panel data were used in this study.

2.4. Definition of the Model Variables

This study focuses on the impacts of AMHSs accessed by farmers on cropland abandonment in rural China. To achieve this goal, we defined the dependent variables, key variables, and other control variables, as follows.

Following Xu et al. (2018) [10], Deng et al. (2018) [27], this study defined cropland abandonment through the behavior and degree of abandoned cropland as dependent variables. Namely, the behavior of abandoned cropland refers to whether farmers abandoned cropland. The degree of abandoned cropland refers to the proportion of the area of abandoned cropland in farmers' contracted cropland area. The key variable was defined by the AMHSs accessed by farmers, i.e., whether farmers accessed AMHSs in agricultural production. The theoretical analysis above shows that labor migration is the main reason for cropland abandonment, so this study also defined the proportion of nonagricultural income in total income as a key variable. In addition, this study also defined other control variables that may affect the dependent variables. According to the previous research related to the driving mechanisms of cropland abandonment (e.g., Wang et al. (2022) [26], Ma et al. (2020) [28]), this study defined the control variables as the characteristics of the household head (e.g., gender, age, years of education, village cadre status, multiple occupations, and internet access with mobile phone), the household (i.e., the proportion of children, the proportion of seniors, and access to credit), the agricultural production (i.e., the area of cropland, the number of land blocks, agricultural machinery ownership, transfers of land out, purchase of agricultural production insurance, and participation in a cooperative), and the village characteristics (i.e., the disputes relating to contracted land, location). In addition, dummy variables for the year and provinces were also included in this study.

3. Results and Analysis

3.1. Descriptive Statistics Analysis

The descriptive statistics analysis results are given in Table 1. For cropland abandonment, a total of 14.58 percent of farmers chose to abandon their cropland, and the average proportion of the area of abandoned cropland in farmers' contracted cropland area was 8.46 percent. These results were in accordance with China's actual situation, that most farmers do not abandon their cropland [7,26]. The average proportion of farmers who accessed AMHSs was 29.81 percent. For the characteristics of the household head, most of the household heads were male, the average age was 54.10, and the average years of education were 7.58. The proportion of household heads who were village cadre was 12.13 percent, who engaged in multiple occupations was 35.18 percent, and who accessed the internet with a mobile phone was 42.62 percent. For the characteristics of the household, the average proportion of children and seniors was 11.91 percent and 13.69 percent, respectively. A total of 13.11 percent of households had accessed credit. The proportion of nonagricultural income in the total household income was 60.26 percent, which also suggested that nonagricultural income has become an important part of farmers' income, with the rapid urbanization and industrialization. For the characteristics of agricultural production, the average area of cropland was 1.14 hectares and the average number of land blocks was 4.59. A total of 43.12 percent of households owned agricultural machinery, 47.69 percent of households transferred land out, 30.90 percent of households had purchased agricultural production insurance, and 11.37 percent of households participated in a cooperative. For the characteristics of villages, 36.83 percent of the villages had disputes relating to contracted land, and 13.16 percent of the villages were now located in the town.

Table 1. Descriptive statistics analysis results.

Variables	Description	Mean	SD
Cropland abandonment	Whether farmers abandoned cropland (1 = Yes; 0 = No)	0.15	0.35
The proportion of cropland abandonment	The proportion of the area of abandoned cropland in farmers' contracted cropland area (%)	8.46	25.41
AMHSs access	Whether household accesses AMHSs (1 = Yes; 0 = No)	0.30	0.46
Gender	Gender of household head (1 = Male; 0 = Female)	0.94	0.24
Age	Age of household head (Years)	54.10	10.31
Education	Years of education of household head (Years)	7.58	3.03
Village cadre status	Whether the household head is a village cadre (1 = Yes; 0 = No)	0.12	0.33
Multiple occupations	Whether household head engaged in multiple occupations (1 = Yes; 0 = No)	0.35	0.48
Internet access	Whether household head accesses the internet with mobile phone (1 = Yes; 0 = No)	0.43	0.49
Proportion of children	The proportion of children under the age of 14 (%)	11.91	16.13
Proportion of seniors	The proportion of seniors over the age of 65 (%)	13.69	25.82
Access to credit	Whether household has access to credit (1 = Yes; 0 = No)	0.13	0.34
Proportion of nonagricultural income	The proportion of nonagricultural income in the total household income (%)	60.26	36.12
Area of cropland	Area of cropland of household management (ha)	1.14	1.92
Land blocks	Number of land blocks	4.59	4.29
Agricultural machinery ownership	Whether household owns agricultural machinery (1 = Yes; 0 = No)	0.43	0.50
Transfers of land out	Whether household transfers land out (1 = Yes; 0 = No)	0.48	0.50
Agricultural production insurance	Whether household purchases agricultural production insurance (1 = Yes; 0 = No)	0.31	0.46
Cooperative participation	Whether household participates in a cooperative (1 = Yes; 0 = No)	0.11	0.32
Contracted land dispute	Whether contracted land disputes occur in the village (1 = Yes; 0 = No)	0.37	0.48
Village location	Whether the village is located in the town (1 = Yes; 0 = No)	0.13	0.34
Observations		8345	

3.2. The Impacts of AMHSs Access on Cropland Abandonment

We employed an extended probit regression and an extended interval regression to empirically analyze the impacts of AMHSs access on the behavior and degree of cropland abandonment. The identification strategy of adding control variables step by step was used in these models, where the first regression only controlled the dummy variables of year and provinces, and the second regression added other control variables based on the first regression. The results are shown in Table 2; Model 1 and Model 2 show the extended probit regression model estimation results for whether farmers abandoned their cropland; and Model 3 and Model 4 show extended interval regression model estimation results for the proportion of cropland abandoned by farmers. According to these models, the results of the endogenous test (i.e., H0: endogenous variables are independent of the dependent variables) are all significant at the level of 5%, which indicates that the endogenous variables are related to the dependent variables, and it is appropriate to add the instrumental variable to these models. The results of AMHSs access significantly reduce the cropland abandonment, and Hypothesis 1 was verified. This is in accordance with Deng et al. (2018) [7], who indicated that agricultural mechanization services can alleviate the abandonment of cropland. Based on this, the interpretation of the model results is mainly based on Model 2 and Model 4.

Table 2. The estimation results of cropland abandonment.

Variables	Cropland Abandonment		The Proportion of Cropland Abandonment	
	Model 1	Model 2	Model 3	Model 4
AMHSs access	−0.202 *** (0.068)	−0.185 ** (0.073)	−0.244 ** (0.109)	−0.203 * (0.115)
Gender		0.103 (0.077)		0.138 (0.119)
Age		0.002 (0.002)		0.001 (0.003)
Education		−0.017 *** (0.007)		−0.022 ** (0.010)
Village cadre status		−0.070 (0.057)		−0.103 (0.088)
Multiple occupations		0.015 (0.040)		−0.028 (0.062)
Internet access		0.161 *** (0.039)		0.262 *** (0.061)
Proportion of children		0.132 (0.117)		0.186 (0.180)
Proportion of seniors		−0.136 (0.084)		−0.255 * (0.130)
Access to credit		0.021 (0.057)		0.007 (0.089)
Proportion of nonagricultural income		0.111 * (0.063)		0.231 ** (0.097)
Area of cropland		−0.054 *** (0.016)		−0.114 *** (0.027)
Land blocks		0.026 *** (0.005)		0.039 *** (0.007)
Agricultural machinery ownership		0.025 (0.043)		−0.022 (0.065)
Transfers of land out		0.195 *** (0.038)		0.088 (0.059)
Agricultural production insurance		−0.019 (0.043)		−0.034 (0.067)
Cooperative participation		−0.014 (0.060)		−0.060 (0.093)
Contracted land dispute		0.011 (0.040)		0.108 * (0.061)
Village location		0.219 *** (0.053)		0.497 *** (0.082)
Year dummy	Yes	Yes	Yes	Yes
Province dummies	Yes	Yes	Yes	Yes
Constant	−1.829 *** (0.096)	−2.063 *** (0.193)	−2.963 *** (0.183)	−3.078 *** (0.316)
Instrumental variable	Yes	Yes	Yes	Yes
Endogenous test	0.143 *** (0.047)	0.143 *** (0.050)	0.107 ** (0.047)	0.104 ** (0.050)
Wald $\chi 2$	523.02 ***	613.00 ***	363.88 ***	405.97 ***
Observations	8345	8345	8345	8345

*** $p < 0.01$, ** $p < 0.05$ and * $p < 0.1$.

As shown in Model 2, the AMHSs accessed by farmers reduced the cropland abandonment at the 5% statistical significance level and the probability of farmers reducing cropland abandonment was 18.5%. In terms of the impact of other control variables on farmers' cropland abandonment, the years of education of household heads significantly reduced the abandonment of cropland. The household heads' access to the internet with a mobile phone can significantly increase cropland abandonment, because it can promote farmers participation in off-farm work through convenient access to employment information [20,49,50]. Moreover, the proportion of nonagricultural income in the total household income also has a significant and positive impact on cropland abandonment, which also proved that off-farm employment is a major factor leading to cropland abandonment [10]. In addition, the land status, such as the area of cropland, the number of land blocks, and the land transfers, is also an important determinant of cropland abandonment. Specially, the larger the area of cropland managed by farmers, the lower the probability of abandoning their cropland. This is consistent with Yan et al. (2016) [51], who suggested that expanding the scale of cropland management is an effective way of reducing the abandonment of cropland. However, the number of the land blocks has significantly increased the cropland abandonment. Zhang et al. (2014) [25] also proved that much-fragmented cropland has been abandoned in rural China. Farmers' transfers of land out also significantly increased the probability of cropland abandonment. In terms of the characteristics of the village, farmers tended to abandon their cropland when their village was now located in the town, which also proved that the process of urbanization has accelerated the abandonment of cropland [52].

As shown in Model 4, the AMHSs accessed by farmers significantly reduced the proportion of the area of abandoned cropland in farmers' contracted cropland area, and farmers' access to AMHSs can reduce the proportion of cropland abandonment by 20.3%. This study will not detail all of the regression results here to save space. It is worth noting that the proportion of seniors has a significant and negative impact on cropland abandonment. This is in line with Deng et al. (2018) [27], who indicated that elderly farmers help curb cropland abandonment. Many migrant workers do not want to give up their rural land use rights to maintain social security and benefits [38,53], so they may give the cropland to the elderly farmers in the household for management. In addition, the villages with disputes relating to contracted land significantly increased the proportion of abandoned cropland, which also suggested the importance of stable use rights of cropland.

3.3. Robustness Check

In this section, we tested the robustness of the estimation results. First, following Xu et al. (2018) [10], we used a Probit model and a Tobit model to test the robustness of the results estimated by an extended probit regression and an extended interval regression, respectively. As shown in Table 3, AMHSs access reduced cropland abandonment in both the Probit and Tobit models but the results were not significant. This indicates that estimation results are biased in the above two models, due to ignoring the endogenous problems. Second, we also compared the results estimated by the IV-Probit model and the IV-Tobit model. The results are given in Table 3; the exogenous Wald test values are both significantly non-zero at the 5% statistical level, rejecting the hypothesis that all of the explanatory variables are exogenous. Moreover, the access to AMHSs has significantly reduced the cropland abandonment in the above two models, while the coefficients estimated by the IV-Probit model were bigger than the extended probit regression model and estimated by the IV-Tobit model were smaller than the extended interval regression model.

In addition, we also compared the results estimated by different key variables to test the robustness [29]. Our study selected the variable of whether farmers' access to the machinery plowing, sowing, and harvesting services (i.e., comprehensive mechanized services (CMSs)) to approximately replace the original key variable. In particular, machinery plowing, sowing, and harvesting are the main links of agricultural production. The farmers' access to the above three links of services can represent their agricultural production capacity. Moreover, AMHSs are an important part of CMSs. As shown in Table 3, the results of the endogenous test are both significant at the level of 1%, which also proved that the endogenous variables are related to the dependent variables. CMSs' access significantly reduced the probability of cropland abandonment and the proportion of the area of abandoned cropland in farmers' contracted cropland area.

Table 3. Robustness check models estimation results.

Variables	Cropland Abandonment			The Proportion of Cropland Abandonment		
	Probit Model	IV-Probit Model	Different Key Variable [a]	Tobit Model	IV-Tobit Model	Different Key Variable [b]
AMHSs access	−0.014 (0.043)	−0.199 ** (0.078)	- -	−0.005 (0.040)	−0.132 * (0.072)	- -
CMSs access	- -	- -	−0.381 *** (0.083)	- -	- -	−0.423 *** (0.132)
Control variables	Yes	Yes	Yes	Yes	Yes	Yes
Year dummy	Yes	Yes	Yes	Yes	Yes	Yes
Province dummies	Yes	Yes	Yes	Yes	Yes	Yes
Constant	−2.099 *** (0.193)	−2.060 *** (0.193)	−1.995 *** (0.193)	−1.878 *** (0.185)	−1.858 *** (0.185)	−3.015 *** (0.316)
Instrumental variable	No	Yes	Yes	No	Yes	Yes
Wald test of exogeneity	-	8.14 ***	-	-	4.48 **	-
Endogenous test	-	-	0.274 ***	-	-	0.210 ***
Observations	8345	8345	8345	8345	8345	8345

*** $p < 0.01$, ** $p < 0.05$ and * $p < 0.1$; [a] the results estimated by the extended probit regression model; [b] the results estimated by the extended interval regression model.

In summary, the above results confirmed that the results estimated by the extended probit regression model and the extended interval regression model are robust. In addition, this also proved that the IV-Probit model and IV-Tobit model cannot fit the binary endogenous covariables well.

3.4. Heterogeneity Analysis

According to the results of Table 2, the control variables of the area of cropland under household management, the proportion of seniors, and the proportion of nonagricultural income all have a significant impact on cropland abandonment. Thus, this study further analyzed the heterogeneity of the impact of AMHSs access on cropland abandonment across different scales of cropland managed by farmers, the household composition (i.e., whether this was a household with seniors), and nonagricultural income (i.e., with and without nonagricultural income). Furthermore, we also analyzed the heterogeneity across different regions (i.e., the eastern region includes the provinces of Hebei, Zhejiang, and Fujian, the central region includes the provinces of Jilin, Heilongjiang, Anhui, Henan, and Hunan, and the western region includes the provinces of Sichuan, Yunan, Shaanxi, and Xinjiang). To save space, this study only listed the results of the impact of AMHSs access on whether farmers abandon their cropland. The results are shown in Table 4.

For the different cropland scales, AMHSs accessed by farmers significantly reduced cropland abandonment only in small-scale groups, and the impacts were higher than the full sample. This may be because the ability of small-scale farmers to manage cropland is weaker and the relative costs of cropland abandonment are smaller than those of medium- and large-scale farmers. For the household composition groups, a group of households without seniors gaining access to AMHSs significantly reduced the cropland abandonment, and the impacts were higher than the full sample. However, there were no significant impacts in the group of households with seniors. The above results also proved that those households with seniors were less likely to abandon their cropland, as the elderly in the households can manage the cropland [54]. For nonagricultural income, the AMHSs accessed by farmers significantly reduced cropland abandonment only in those households with nonagricultural income. These results verified Hypothesis 2, that is, that AMHSs can effectively alleviate the impact of labor migration on cropland abandonment. In addition, for different regions, the AMHSs accessed by farmers significantly reduced cropland abandonment only in the eastern region, with a higher level of economic development and more nonagricultural employment opportunities. This is consistent with Deng et al. (2018) [7], that is, the regions with a higher nonagricultural employment rate have more abandoned cropland.

Table 4. Heterogeneity analysis results.

Variables	Different Cropland Scale Groups			Household Composition		Nonagricultural Income		Different Region		
	Small-Scale	Medium-Scale	Large-Scale	With Seniors	Without Seniors	With Nonagricultural Income	Without Nonagricultural Income	Eastern	Central	Western
AMHSs access	−0.670 ***	−0.170	0.121	0.009	−0.264 ***	−0.192 **	−0.091	−0.525 ***	−0.058	−0.227
	(0.135)	(0.132)	(0.147)	(0.149)	(0.086)	(0.077)	(0.287)	(0.174)	(0.101)	(0.168)
Control variables	Yes	Yes	Yes	Yes	Yes	Yes	Yes	Yes	Yes	Yes
Year dummy	Yes	Yes	Yes	Yes	Yes	Yes	Yes	Yes	Yes	Yes
Province dummies	Yes	Yes	Yes	Yes	Yes	Yes	Yes	Yes	Yes	Yes
Constant	−5.995	−3.115 ***	−1.452 ***	−1.629 ***	−2.232 ***	−2.086 ***	−1.695 **	−1.869 ***	−1.907 ***	−1.878 ***
	(296.596)	(0.446)	(0.350)	(0.378)	(0.238)	(0.204)	(0.738)	(0.496)	(0.304)	(0.279)
Instrumental variable	Yes	Yes	Yes	Yes	Yes	Yes	Yes	Yes	Yes	Yes
Endogenous test	0.380 ***	0.201 **	−0.043	0.022	0.190 ***	0.137 ***	0.148	0.264 **	0.083	0.165
	(0.084)	(0.094)	(0.092)	(0.101)	(0.058)	(0.053)	(0.185)	(0.114)	(0.070)	(0.105)
Wald χ2	323.30 ***	245.40 ***	206.50 ***	178.03 ***	473.53 ***	547.13 ***	52.87 ***	72.84 ***	312.43 ***	341.16 ***
Observations	2946	2702	2697	2628	5717	7345	1000	1496	3630	3219

*** $p < 0.01$, ** $p < 0.05$.

In addition, this study further analyzed the heterogeneity of household composition and nonagricultural income (i.e., with groups with nonagricultural income and containing seniors, with groups with nonagricultural income but without seniors). The results are shown in Table A1 (Appendix A), the AMHSs accessed by farmers significantly reduced cropland abandonment only in those groups of households with nonagricultural income but without seniors. This also proved that the rural elderly labor force can curb the impact of labor migration on cropland abandonment to a certain extent.

4. Discussion

Prior studies have proved that labor migration and high agricultural production costs are the main factors causing the abandonment of cropland [10,22]. Based on these factors and the actual situation of China's agricultural production, this study mainly focuses on the problem of whether AMHSs accessed by farmers can reduce cropland abandonment. We used the data from 8345 samples collected by the Survey for Agriculture and Village Economy in 2019 and 2020, and employed the extended probit regression model and the extended interval regression model to empirically analyze the relationship between AMHSs access and cropland abandonment. Our results revealed that AMHSs accessed by farmers can significantly reduce cropland abandonment in rural China. The research results can provide a theoretical reference for the government to promote agricultural mechanization services and reduce cropland abandonment.

Interestingly, the heterogeneity analysis results of our study showed that AMHSs accessed by farmers significantly reduced cropland abandonment in small-scale groups, groups of households without seniors, groups of households with nonagricultural income, those located in the eastern region, and groups of households with nonagricultural income but without seniors. On the one hand, household management by smallholders (smallholder refers to those who operate cropland area that is less than 3.33 hectares) is still the main form of agricultural management in China, and accounts for more than 98.00% of the total number of farmers [55]. They may not purchase agricultural machinery for agricultural production due to the small management scale, fragmented cropland, and high fixed costs of the agricultural machinery. In this case, the low level of agricultural mechanization, imperfect land transfer market, and high nonagricultural wages will make them more inclined to abandon their cropland, while accessing AMHSs can effectively reduce cropland abandonment. This finding is consistent with the Chinese government's policy aimed at promoting modern agricultural practices to small farmers through developing the agricultural mechanization services market. On the other hand, population aging is a social phenomenon faced by all countries in the world, including China with 13.50 percent elderly people [56]. However, the surplus elderly labor force can still manage the cropland when the young and middle-aged labor force participate in non-agricultural work [26,54]. This undoubtedly proves that the elderly labor force may be an important resource to manage the cropland, which will reduce the cropland abandonment caused by labor migration. However, if the elderly labor force lacks farm successors, future land use issues should be a concern for scholars and governments, which may threaten food security and sustainable agriculture [57]. There are regional differences in the impacts of AMHSs access on cropland abandonment, which mainly has a significant impact on the eastern region. In addition, this study also proved that AMHSs access can alleviate the impact of labor migration on cropland abandonment.

This study provides some policy implications to reduce cropland abandonment. Our results show that AMHSs accessed by farmers can reduce cropland abandonment, which implies that policymakers should actively promote the development of agricultural mechanization services and build a perfect services market, especially for labor-intensive services (e.g., AMHSs). In addition, heterogeneity analysis showed that AMHSs have a more significant impact on reducing the abandonment of cropland by small-scale farmers, which also implies they may be the main group engaged in cropland abandonment in rural China. Thus, policymakers should strengthen the agricultural production support policies for small-scale farmers, such as subsidies for the use of agricultural mechanization services. This will help to realize the organic connection between small farmers and modern agriculture practice. Although elderly farmers can alleviate the cropland abandonment to some extent, the government should focus on farm successors in the future and continuously optimize the mode of agricultural mechanization services to better help farmers manage cropland.

This study mainly has two limitations as follows:

(1) This study only analyzed the impacts of AMHSs on cropland abandonment, while the impact of other services (such as agricultural machinery plowing, sowing, and irrigation services) was not analyzed. Although the AMHSs represent one of the "heaviest" agricultural production links. Thus, future research is required to explore the impact of other services on cropland abandonment, so as to provide a more comprehensive reference for the developing agricultural mechanization services and reducing cropland abandonment;
(2) Given the limitations of the paper length and questionnaire design, the potential channels (e.g., land transfer) of the impacts of AMHSs access on cropland abandonment have not been explored. Although we found that AMHSs can effectively alleviate the impact of labor migration on cropland abandonment, we still need to explore other channels in future research. In this case, we can provide more evidence on how to reduce cropland abandonment in rural China.

5. Conclusions

Based on the above analysis, the major conclusions are as follows:

(1) AMHSs accessed by farmers significantly reduced the probability of cropland abandonment by 18.5%;
(2) AMHSs accessed by farmers significantly reduced the proportion of the area of abandoned cropland in farmers' contracted cropland area by 20.3%;
(3) Heterogeneity analysis results showed that farmers' access to AMHSs significantly reduces cropland abandonment in small-scale groups, groups without elderly households, with nonagricultural income groups, in the eastern region, and in groups with nonagricultural income but without seniors.

In conclusion, our study confirms that AMHSs accessed by farmers can reduce cropland abandonment in rural China, which is also beneficial to ending hunger, achieving food security, and promoting sustainable agriculture.

Author Contributions: Conceptualization, P.X. and X.W.; methodology, P.X., Y.W. and X.H.; software, P.X.; validation, P.X., Y.W. and X.H.; formal analysis, P.X., X.H. and Y.W.; investigation, P.X., X.H. and X.W.; resources, P.X. and X.H.; data curation, P.X.; writing—original draft preparation, P.X. and X.W.; writing—review and editing, P.X., X.H., Y.W. and X.W.; visualization, P.X. and X.W.; supervision, X.W.; project administration, P.X. and X.W.; funding acquisition, X.W. All authors have read and agreed to the published version of the manuscript.

Funding: This research was funded by The Agricultural Science and Technology Innovation Program, grant number 10-IAED-08-2022, 10-IAED-RC-04-2022; and The National Key Research and Development Project, grant number 2020YFD1001205-1.

Institutional Review Board Statement: Not applicable.

Informed Consent Statement: Not applicable.

Data Availability Statement: The data presented in this study are available within the article.

Conflicts of Interest: The authors declare no conflict of interest.

Appendix A

Table A1. Heterogeneity analysis results.

Variables	Household Composition and Nonagricultural Income	
	With Nonagricultural Income and Seniors	**With Nonagricultural Income but without Seniors**
AMHSs access	−0.060 (0.158)	−0.250 *** (0.091)
Control variables	Yes	Yes
Year dummy	Yes	Yes
Province dummies	Yes	Yes
Constant	−1.956 *** (0.481)	−2.318 *** (0.249)
Instrumental variable	Yes	Yes
Endogenous test	0.033 (0.109)	0.177 *** (0.062)
Wald $\chi2$	172.70 ***	421.71 ***
Observations	2345	5000

*** $p < 0.01$.

References

1. United Nations. *The Sustainable Development Goals Report 2021*; United Nations Publications: New York, NY, USA, 2021.
2. Zhang, J. China's success in increasing per capita food production. *J. Exp. Bot.* **2011**, *62*, 3707. [CrossRef] [PubMed]
3. Fei, R.; Lin, Z.; Chunga, J. How land transfer affects agricultural land use efficiency: Evidence from China's agricultural sector. *Land Use Policy* **2021**, *103*, 105300. [CrossRef]
4. Restuccia, D.; Santaeulalia-Llopis, R. *Land Misallocation and Productivity*; FAO: Rome, Italy, 2015.
5. Deng, X.; Huang, J.; Rozelle, S.; Zhang, J.; Li, Z. Impact of urbanization on cultivated land changes in China. *Land Use Policy* **2015**, *45*, 1–7. [CrossRef]
6. Lin, H.-C.; Hülsbergen, K.-J. A new method for analyzing agricultural land-use efficiency, and its application in organic and conventional farming systems in southern Germany. *Eur. J. Agron.* **2017**, *83*, 15–27. [CrossRef]
7. Deng, X.; Xu, D.; Qi, Y.; Zeng, M. Labor Off-Farm Employment and Cropland Abandonment in Rural China: Spatial Distribution and Empirical Analysis. *Int. J. Environ. Res. Public Health* **2018**, *15*, 1808. [CrossRef] [PubMed]
8. Prishchepov, A.V.; Müller, D.; Dubinin, M.; Baumann, M.; Radeloff, V.C. Determinants of agricultural land abandonment in post-Soviet European Russia. *Land Use Policy* **2013**, *30*, 873–884. [CrossRef]
9. Baumann, M.; Kuemmerle, T.; Elbakidze, M.; Ozdogan, M.; Radeloff, V.C.; Keuler, N.S.; Prishchepov, A.V.; Kruhlov, I.; Hostert, P. Patterns and drivers of post-socialist farmland abandonment in Western Ukraine. *Land Use Policy* **2011**, *28*, 552–562. [CrossRef]
10. Xu, D.; Deng, X.; Guo, S.; Liu, S. Labor migration and farmland abandonment in rural China: Empirical results and policy implications. *J. Environ. Manag.* **2019**, *232*, 738–750. [CrossRef]
11. Huang, Y.; Li, F.; Xie, H. A Scientometrics Review on Farmland Abandonment Research. *Land* **2020**, *9*, 263. [CrossRef]
12. Díaz, G.I.; Nahuelhual, L.; Echeverría, C.; Marín, S. Drivers of land abandonment in Southern Chile and implications for landscape planning. *Landsc. Urban Plan.* **2011**, *99*, 207–217. [CrossRef]
13. Long, H. Land use policy in China: Introduction. *Land Use Policy* **2014**, *40*, 1–5. [CrossRef]

14. Han, X.; Chen, Y.; Wang, X. Impacts of China's bioethanol policy on the global maize market: A partial equilibrium analysis to 2030. *Food Secur.* **2022**, *14*, 147–163. [CrossRef]
15. Van der Zanden, E.H.; Verburg, P.H.; Schulp, C.J.E.; Verkerk, P.J. Trade-offs of European agricultural abandonment. *Land Use Policy* **2017**, *62*, 290–301. [CrossRef]
16. Agnoletti, M. Rural landscape, nature conservation and culture: Some notes on research trends and management approaches from a (southern) European perspective. *Landsc. Urban Plan.* **2014**, *126*, 66–73. [CrossRef]
17. Gariano, S.L.; Petrucci, O.; Rianna, G.; Santini, M.; Guzzetti, F. Impacts of past and future land changes on landslides in southern Italy. *Reg. Environ. Chang.* **2017**, *18*, 437–449. [CrossRef]
18. García-Ruiz, J.M.; Noemí, L.-R. Hydrological and erosive consequences of farmland abandonment in Europe, with special reference to the Mediterranean region—A review. *Agric. Ecosyst. Environ.* **2011**, *140*, 317–338. [CrossRef]
19. Minstry of Agriculture and Rural Affairs of the People's Repubulic of China. *Guiding Opinions on the Overall Utilization of Abandoned Cropland to Promote the Development of Agricultural Production*; Minstry of Agriculture and Rural Affairs of the People's Repubulic of China: Beijing, China, 2021.
20. Gao, J.; Song, G.; Sun, X. Does labor migration affect rural land transfer? Evidence from China. *Land Use Policy* **2020**, *99*, 105096. [CrossRef]
21. Zhang, X.; Jin, Y.; Wang, S. China has reached the Lewis turning point. *China Econ. Rev.* **2011**, *22*, 542–554. [CrossRef]
22. Lieskovský, J.; Bezák, P.; Špulerová, J.; Lieskovský, T.; Koleda, P.; Dobrovodská, M.; Bürgi, M.; Gimmi, U. The abandonment of traditional agricultural landscape in Slovakia—Analysis of extent and driving forces. *J. Rural Stud.* **2015**, *37*, 75–84. [CrossRef]
23. MacDonald, D.; Crabtree, J.R.; Wiesinger, G.; Dax, T.; Stamou, N.; Fleury, P.; Gutierrez Lazpita, J.; Gibon, A. Agricultural abandonment in mountain areas of Europe: Environmental consequences and policy response. *J. Environ. Manag.* **2000**, *59*, 47–69. [CrossRef]
24. Xie, H.; Wang, P.; Yao, G. Exploring the Dynamic Mechanisms of Farmland Abandonment Based on a Spatially Explicit Economic Model for Environmental Sustainability: A Case Study in Jiangxi Province, China. *Sustainability* **2014**, *6*, 1260–1282. [CrossRef]
25. Zhang, Y.; Li, X.; Song, W. Determinants of cropland abandonment at the parcel, household and village levels in mountain areas of China: A multi-level analysis. *Land Use Policy* **2014**, *41*, 186–192. [CrossRef]
26. Wang, J.; Cao, Y.; Fang, X.; Li, G.; Cao, Y. Does land tenure fragmentation aggravate farmland abandonment? Evidence from big survey data in rural China. *J. Rural Stud.* **2022**, *91*, 126–135. [CrossRef]
27. Deng, X.; Xu, D.; Zeng, M.; Qi, Y. Landslides and Cropland Abandonment in China's Mountainous Areas: Spatial Distribution, Empirical Analysis and Policy Implications. *Sustainability* **2018**, *10*, 3909. [CrossRef]
28. Ma, W.; Zhu, Z. A Note: Reducing Cropland Abandonment in China: Do Agricultural Cooperatives Play a Role. *J. Agric. Econ.* **2020**, *71*, 929–935. [CrossRef]
29. Deng, X.; Xu, D.; Zeng, M.; Qi, Y. Does Internet use help reduce rural cropland abandonment? Evidence from China. *Land Use Policy* **2019**, *89*, 104243. [CrossRef]
30. Yang, J.; Huang, Z.; Zhang, X.; Reardon, T. The Rapid Rise of Cross-Regional Agricultural Mechanization Services in China. *Am. J. Agric. Econ.* **2013**, *95*, 1245–1251. [CrossRef]
31. Zhang, X.; Yang, J.; Thomas, R. Mechanization outsourcing clusters and division of labor in Chinese agriculture. *China Econ. Rev.* **2017**, *43*, 184–195. [CrossRef]
32. Qing, Y.; Chen, M.; Sheng, Y.; Huang, J. Mechanization services, farm productivity and institutional innovation in China. *China Agric. Econ. Rev.* **2019**, *11*, 536–554. [CrossRef]
33. Qian, L.; Lu, H.; Gao, Q.; Lu, H. Household-owned farm machinery vs. outsourced machinery services: The impact of agricultural mechanization on the land leasing behavior of relatively large-scale farmers in China. *Land Use Policy* **2022**, *115*, 106008. [CrossRef]
34. Qiu, T.; Luo, B. Do small farms prefer agricultural mechanization services? Evidence from wheat production in China. *Appl. Econ.* **2021**, *53*, 2962–2973. [CrossRef]
35. Schultz, T.W. *Transforming Traditional Agriculture*; The Commercial Press: Beijing, China, 1964.
36. Ji, X.; Qian, Z.; Zhang, L.; Zhang, T. Rural Labor Migration and Households' Land Rental Behavior: Evidence from China. *China World Econ.* **2018**, *26*, 66–85. [CrossRef]
37. Ding, C. Policy and praxis of land acquisition in China. *Land Use Policy* **2007**, *24*, 1–13. [CrossRef]
38. Whalley, J.; Zhang, S. A numerical simulation analysis of (Hukou) labour mobility restrictions in China. *J. Dev. Econ.* **2007**, *83*, 392–410. [CrossRef]
39. Rupelle, M.; Deng, Q.; Shi, L.; Vendryes, T. Land Rights Insecurity and Temporary Migration in Rural China. *Social Science Electronic Publishing*. 2009. Available online: https://d-nb.info/999277979/34 (accessed on 29 May 2022).
40. Wang, X.; Yamauchi, F.; Otsuka, K.; Huang, J. Wage Growth, Landholding, and Mechanization in Chinese Agriculture. *World Dev.* **2016**, *86*, 30–45. [CrossRef]
41. Qiu, T.; Boris Choy, S.T.; Li, S.; He, Q.; Luo, B. Does land renting-in reduce grain production? Evidence from rural China. *Land Use Policy* **2020**, *90*, 104311. [CrossRef]
42. Yi, Q. Adoption of Agricultural Mechanization Services among Maize Farmers in China: Impacts of Population Aging and Off-farm Employment. In Proceedings of the 30th International Conference of Agricultural Economists, Vancouver, BC, USA, 28 July–2 August 2018.

43. Sims, B.; Hilmi, M. *Agricultural Mechanization A Key Input for Sub-Saharan African Smallholders*; Integrated Crop Management (FAO): Rome, Italy, 2016; Volume 23.
44. Sims, B.; Heney, J. Promoting Smallholder Adoption of Conservation Agriculture through Mechanization Services. *Agriculture* **2017**, *7*, 64. [CrossRef]
45. Newey, W.K. Efficient estimation of limited dependent variable models with endogenous explanatory variables. *J. Econom.* **1987**, *36*, 231–250. [CrossRef]
46. Kung, J.K.-S. Off-Farm Labor Markets and the Emergence of Land Rental Markets in Rural China. *J. Comp. Econ.* **2002**, *30*, 395–414. [CrossRef]
47. Han, X.; Xue, P.; Zhang, N. Impact of Grain Subsidy Reform on the Land Use of Smallholder Farms: Evidence from Huang-Huai-Hai Plain in China. *Land* **2021**, *10*, 929. [CrossRef]
48. Xue, P.; Han, X.; Elahi, E.; Zhao, Y.; Wang, X. Internet Access and Nutritional Intake: Evidence from Rural China. *Nutrients* **2021**, *13*, 2015. [CrossRef]
49. Lu, Y.; Xie, H.; Xu, L.C. Telecommunication externality on migration: Evidence from Chinese villages. *China Econ. Rev.* **2016**, *39*, 77–90. [CrossRef]
50. Aker, J.C. Dial "A" for agriculture: A review of information and communication technologies for agricultural extension in developing countries. *Agric. Econ.* **2011**, *42*, 631–647. [CrossRef]
51. Yan, J.; Yang, Z.; Li, Z.; Li, X.; Xin, L.; Sun, L. Drivers of cropland abandonment in mountainous areas: A household decision model on farming scale in Southwest China. *Land Use Policy* **2016**, *57*, 459–469. [CrossRef]
52. Long, H.; Tu, S.; Ge, D.; Li, T.; Liu, Y. The allocation and management of critical resources in rural China under restructuring: Problems and prospects. *J. Rural Stud.* **2016**, *47*, 392–412. [CrossRef]
53. Ma, X.; Heerink, N.; Ierland, E.; Shi, X. Land tenure insecurity and rural-urban migration in rural China. *Pap. Reg. Sci.* **2014**, *95*, 383–406. [CrossRef]
54. Lu, H.; Xie, H.; Yao, G. Impact of land fragmentation on marginal productivity of agricultural labor and non-agricultural labor supply: A case study of Jiangsu, China. *Habitat Int.* **2019**, *83*, 65–72. [CrossRef]
55. NBSC. *Communique of the Third National Agricultural Census (No. 2)*; National Bureau of Statistics: Beijing, China, 2017.
56. NBSC. *Communique of the Seventh National Population Census (No. 5)*; National Bureau of Statistics: Beijing, China, 2020.
57. Zou, B.; Mishra, A.K.; Luo, B. Aging population, farm succession, and farmland usage: Evidence from rural China. *Land Use Policy* **2018**, *77*, 437–445. [CrossRef]

agriculture

MDPI

Article

Yield Responses of Grain Sorghum and Cowpea in Binary and Sole Cultures under No-Tillage Conditions in Limpopo Province

Tlou E. Mogale [1,2,*], Kingsley K. Ayisi [1,2], Lawrence Munjonji [1,2] and Yehenew G. Kifle [3]

1 Risk and Vulnerability Science Centre (RSVC), University of Limpopo, Polokwane 0727, South Africa; kwabena.ayisi@ul.ac.za (K.K.A.); lawrence.munjonji@ul.ac.za (L.M.)
2 Department of Plant Production, Soil Science and Agricultural Engineering, University of Limpopo, Polokwane 0727, South Africa
3 Department of Math and Statistics, University of Maryland Baltimore County, Baltimore, MD 21250, USA; yehenew@umbc.edu
* Correspondence: queenmogale78@gmail.com

Abstract: Climate change is severely disrupting ecosystem services and crop productivity, resulting in lower crop growth and yields. Studies have emphasized the importance of assessing conservation practices through crop modelling to improve cropland productivity. There is a lack of accurate information in the performance of conservation practices as well as data for improved crop modelling. No-tillage sorghum–cowpea intercrop experiments were established to assess the productivity of four sorghum cultivars and cowpea at two densities of 37,037 and 74,074 per plants and generate data for improved crop modelling. The leaf area index (LAI) varied in sorghum cultivars and cowpea densities during the two growing seasons. Cultivars Enforcer and NS5511 produced the highest grain yields of 4338 kg per ha and 2120 kg per ha, respectively, at Syferkuil. Ofcolaco's Enforcer and Avenger were the highest yielding cultivars at Ofcolaco, with mean yields of 2625 kg per ha and 1191 kg per ha, respectively. At Syferkuil, cowpea yield was 93% and 77% more in sole compared to binary cultures during the growing seasons at Syferkuil. At Ofcolaco, sole yielded approximately 96% more grain than binary. The findings confirm that for the sorghum–cowpea intercrop to improve overall system productivity, cowpea density should be increased.

Keywords: climate-smart agriculture; grain yield; yield components; intercropping system; land equivalent ratio

Citation: Mogale, T.E.; Ayisi, K.K.; Munjonji, L.; Kifle, Y.G. Yield Responses of Grain Sorghum and Cowpea in Binary and Sole Cultures under No-Tillage Conditions in Limpopo Province. *Agriculture* **2022**, *12*, 733. https://doi.org/10.3390/agriculture12050733

Academic Editors: Moucheng Liu, Xin Chen and Yuanmei Jiao

Received: 14 April 2022
Accepted: 17 May 2022
Published: 23 May 2022

1. Introduction

Grain sorghum and cowpea are two of the most important grain crops grown in South Africa, particularly in Limpopo Province, where they are staple foods for many subsistence farmers [1]. When conditions are favourable, smallholder farmers can produce up to 20,000 tons per ha of grain sorghum [2]. Cowpeas are also grown in the province for domestic consumption, with the excess sold at the local market to generate revenue. Temperature extremes and precipitation fluctuations have long hampered grain sorghum production in Southern Africa [3,4]. Furthermore, anthropogenic activities such as conventional agriculture, overuse of chemical fertilizers, and continuous cultivation of the same crop on the same plot of land have contributed approximately 12% of the greenhouse gases emitted into the atmosphere globally [5,6]. These practices' negative impact has also contributed to severe land degradation [7,8].

Agriculture must become more productive and diverse to cope with climate change and increased natural resource constraints [9]. Producing more food with fewer resources while preserving and improving farmers' livelihoods is a global challenge. Adopting climate-smart agricultural practices such as intercropping, and conservation tillage can boost crop productivity and alleviate food insecurity in many Limpopo province areas [10]. Intercropping is defined as the simultaneous cultivation of two crops on the same plot of land [11], whereas a no-tillage system is the practice of preparing the soil with minimal soil disturbance [12]. The two systems are widely used around the world due to their efficient use of resources such as land and water, as well as their ability to improve soil fertility and crop intensification. Intercropping system combined with no-tillage system have proved to improve the crop productivity through soil moisture conservation [13].

The most common system used in South Africa is maize-legume intercropping. However, with average maize production threatened by climate change, sorghum has been projected to be one of the most viable substitute crops due to its ability to withstand the harsh conditions in South Africa. As a result, sustainable grain sorghum management and crop use as a maize substitute can secure food for the general populace while mitigating climate change scenarios [14]. Intercropping grain sorghum with cowpea improves soil fertility due to nitrogen fixation by the legume crop. Crop models can be used to assess the productivity of traditional agronomic practices such as intercropping systems in a changing climate. However, in South Africa, the availability of data required to run crop model simulations remains a challenge [15]. The main goal was to evaluate the productivity of four sorghum cultivars (Avenger, Enforcer, Titan, and NS5511) intercropped with cowpea (betch witch) under two cowpea densities and to generate data that can aid in climate-smart practices and crop model analysis.

2. Materials and Methods

2.1. Experimental Sites

A field experiment was carried out in two distinct agro-ecological regions of Limpopo province during the 2018–19 and 2020–21 cropping seasons. The first location was the University of Limpopo Experimental Farm in Syferkuil, which was located at 23°50′02.7″ S and 29°41′25.5″ E. The area receives 350 to 500 mm of rainfall per year, with average maximum and minimum temperatures of 15 °C and 30 °C, respectively. The second location was Itemeleng Ba-Makhutjwa Primary Cooperative at Farmers Field at Ofcolaco, which was located at 24°06′38.3″ S and 30°23′11.8″ E near Tzaneen. Ofcolaco receives approximately 650 to 700 mm of rainfall per year, with an average maximum and minimum temperatures of 18 °C and 35 °C, respectively. The two locations also have different soil types: sandy-clay at Syferkuil and clay-loam at Ofcolaco [16]. The experimental sites were both previously used to plant soybeans, followed by two years of fallow under no-till dryland conditions.

2.2. Weather Conditions

Two automatic weather stations near or at the experimental sites were used to provide daily weather data. At the University of Limpopo experimental farm (Syferkuil), the weather station was located at the farm whereas, at Ofcolaco, a rain gauge placed at the site and an automatic weather station situated 27.9 km from the experimental site were used to access daily weather data during the period of experimentation.

2.3. Soil Samples

Soil samples were collected before planting at the depth of 0–30 cm and 30–60 cm using a random sampling method at the two experimental sites. A total of four composite samples per sampling depth from each location, representing the experimental blocks was collected and analysed in the laboratory for chemical and physical properties (Table 1). The samples were sieved to pass through a 2 mm sieve and analysed for chemical properties. Phosphorus (P), potassium (K), calcium (Ca), magnesium (Mg), zinc (Zn), manganese (Mn) and copper (Cu) were following the procedure of Mehlich-III multi-nutrient extraction method. Soil pH was determined in potassium chloride (KCl) [17], soil bulk density using a metal ring at each soil depth following the procedure of [18]. Available mineral nitrogen (N) was determined using the colorimetric method for ammonium and nitrate. The bray method was used to determine available phosphorus (P), cation exchange capacity (CEC) following the procedure of [19]. Walkley and Black method were used to determine organic carbon (org. C). Soil particle size was determined using the hydrometre method [20]. Before planting, Syferkuil soil had higher K, Ca and Mg macronutrients and low Phosphorus P compared to the soil from Ofcolaco. However, Ofcolaco soil had higher micronutrients Zn, Mn and Cu compared to Syferkuil soil. The results further indicated that soil from Ofcolaco has high organic carbon of 1.38% compared to Syferkuil which had about 0.6% organic carbon.

Table 1. Pre-planting soil chemical and physical properties from Syferkuil and Ofcolaco in the two seasons.

Soil Properties	Syferkuil		Ofcolaco	
	2018/19	2020/21	2018/19	2020/21
P (mg/kg)	22.00	26.89	53.75	29.3
K (mg/kg)	433.00	276.36	234.00	158.99
Ca (mg/kg)	1119.75	1059.61	917.25	742.73
Mg (mg/kg)	558.50	592.455	152.25	156.54
Exch. Acidity (cmol/kg)	0.03	0.02	0.04	0.03
Total cations (cmol/kg)	11.32	14.35	6.47	6.65
Acid sat. (%)	0.00	0.00	0.75	0.66
pH (KCL)	6.35	-	6.06	-
Zn (mg/kg)	1.48	2.77	5.48	7.75
Mn (mg/kg)	17.50	13.64	48.25	37.98
Cu (mg/kg)	4.08	2.89	5.13	4.48
org. C (%)	0.60	0.63	1.38	1.37
N (%)	0.05	0.07	0.05	0.06
Clay (%)	30.00	-	23.25	-
Fine silt (%)	7.50	-	8.25	-
Coarse silt and sand (%)	65.50	-	72.25	-
Texture class	Sandy clay loam	-	Clay loam	-

2.4. Experimental Design and Management

Prior to planting, the land at both locations was prepared by first reducing the size of weeds using a motorised slasher, followed by the application of Roundup, a non-selective, systematic, broad-spectrum glyphosate-based post-emergence herbicide one month after slashing. A 250 mL volume of Roundup was used in 10 L of water. The trial was planted 10 days after herbicide application as randomised complete block design (RCBD) in a factorial arrangement with four blocks (replications) under a no-tillage condition. The experimental treatments comprised four grain sorghum cultivars namely Avenger, Enforcer, Titan and NS5511 and two cowpea (var. Betch Witch) densities. Sorghum and cowpea were planted in both sole and binary cultures. Grain sorghum density was maintained at 37,037 plants per ha for each cultivar. Each experimental unit was 3.0 m × 3.6 m consisting of four rows of sorghum and four rows of cowpea in the intercropped treatment. The net plot size was 604.8 square metres at each experimental site. For grain sorghum, seeds were planted at inter- and intra-row spacings of 0.9 m and 0.3 m, respectively. Cowpea was planted at an inter-row spacing of 0.9 m and intra-row spacings of 0.3 and 0.15 m to obtain treatment densities of 37,037 and 74,074 plants per ha, respectively. The spacing between sorghum and cowpea in the intercropped treatment was thus 0.45 m. The trials were planted on the 17 January 2019 and 20 November 2020 at Syferkuil, whereas at Ofcolaco, the planting dates were 23 March 2019 and 21 November 2020. Each experimental unit received phosphorus in a form of superphosphate (10.5% P) at 20 kg P per ha, based on preplant soil fertility analysis. Nitrogen was applied as Limestone Ammonium Nitrate (LAN) (28% N) at a rate of 100 kg N per ha in a split application of 50 kg N per ha each at planting and knee height of grain sorghum. All fertilisers were banded along the row. Standard crop management practices including thinning, weeding, and pest control for both crops were monitored and addressed when necessary throughout the cropping season. Aphids and stalk borer infestation in cowpea and grain sorghum were controlled using Cypermethrin 200 cm. Hundred and twenty (120) mL of Cypermethrin was diluted with 64 L of water. The damage due to bird attack on sorghum grains from flowering to physiological maturity was prevented by covering sorghum heads using a protective translucent nylon mesh net at the onset of the milk stage.

2.5. Data Collection

Leaf Area Index (LAI) data was collected from two weeks after emergence per experimental unit and continued every two weeks until physiological maturity. The data were collected using AccuPAR LAI Ceptometre LP-80 (Decagon Devices, Inc., Pullman, WA, USA) on middle rows of binary and sole cultures of grain sorghum and cowpea between 10:00 a.m. and 1:00 p.m. LAI on individual fully expanded flag leaves of three plants within an experimental unit was measured at 3 min interval. In the 2020/21 cropping season, cowpea at Ofcolaco failed to produce grain. Hence, only the grain yield of the 2018/19 cropping season from Ofcolaco is presented in this paper. At harvesting, 10 plants with their heads were sampled from two middle rows within an area of 2.7 square metres to determine biomass and grain yield. All cowpea plants from a 2.7-square-metre area were harvested with pods to determine grain yield and biomass. Cowpea leaves that dropped to the ground were retrieved on a continuous basis after flowing to add to the final biomass at harvest. Biomass was oven-dried in the laboratory at 65 °C for 72 h and weighed using a weighing balance to get the weight of dry matter. Grains collected from a 2.7-square-metres area were taken to the laboratory to determine grain yield and yield components. Grain yield was determined by weight of grains per plot and converted to kg per ha. Three grain sorghum from the harvested heads were sampled from 10 heads harvested to determine head weight and head length. The 3 plants were threshed separately to determine seed weight per head as well as shelled head weight. We determined 1000 seed weight by counting and weighing 1000 grain sorghum seeds. Cowpea pod weight was obtained by weight pods collected per plot in 2.7 per square metres and 100 seed weight was determined by counting as well as weighing 100 cowpea seeds. Harvest index (HI) and land equivalent ratio (LER) for each crop were calculated using the following formulas:

$$\text{HI (\%)} = (\text{Grain yield})/(\text{stover yield}+\text{grain yield}) * 100 \quad (1)$$

$$\text{LER} = \text{YSbinary}/\text{YSsole} + \text{YCbinary}/\text{YCsole} \quad (2)$$

where YSbinary is yield of sorghum in intercropping, YSsole is yield of sorghum in sole culture, YCbinary is yield of cowpea in intercropping, and YCsole is yield of cowpea in sole culture.

2.6. Data Analysis

After checking the relevant model assumptions including normality, independence, and constant variance, we used a multivariate analysis of variance (ANOVA) model to fit each response variable using the Statistical Analysis System (21 SAS version 9.4). In grain sorghum, the four cultivars were regarded as factor 1 and the cropping system as factor 2. In the case of cowpea, the cropping system was factor 1 while density was factor 2. For LAI, days after planting (time), cultivars and cropping system were tested for interaction for grain sorghum. The LAI interaction for cowpea was tested among days after planting, cropping system, and density. The interaction of yield and yield components, as well as the harvest index of grain sorghum, was tested between cultivars and cropping system. In cowpea, the interaction was tested between cropping systems and density. Mean separation was performed where the means were different using the least significant difference (LSD) at probability levels of $p \leq 0.05$. Land equivalent ratio (LER) was used to assess the productivity and effectiveness of the intercropping system.

3. Results

3.1. Weather Conditions during Growing Seasons

Syferkuil had daily average minimum and maximum temperatures of 12 °C and 27 °C, respectively, with a total rainfall of 349 mm in 2018/19 and 292 mm in the 2020/21 growing period (Figure 1). Rains of about 156.49 mm and 10 mm were received throughout the planting period at Syferkuil in the 2018/19 and 2020/21 cropping seasons.

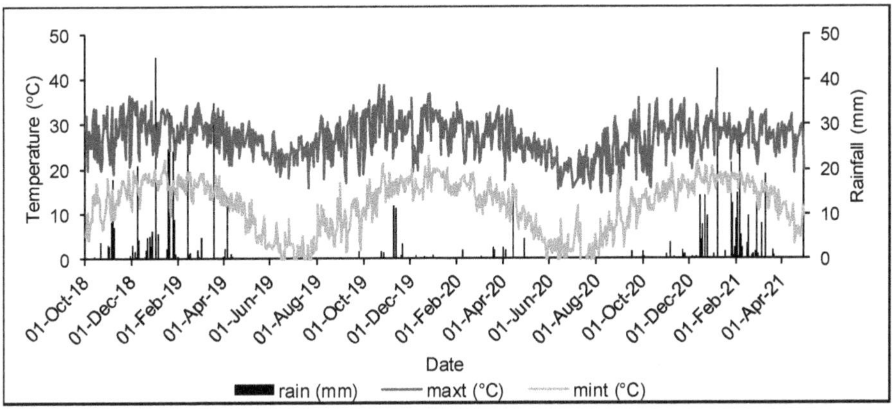

Figure 1. Syferkuil daily rainfall, maximum and minimum temperature during 2018/19 and 2020/21 cropping seasons.

At Ofcolaco, the maximum and minimum temperatures across the two seasons were 31 °C and 18 °C, respectively, with a total rainfall of 261 mm in 2018/19 and 608 mm in 2020/21. During planting months Ofcolaco received rainfall of 5 mm in 2018/19 and 38 mm in 2020/21. The highest rainfall (about 130 mm) in 2018/19 was received in December, when minimum and maximum temperatures were 22 °C and 35 °C, respectively. These were higher compared to the other months. However, in 2020/21, the highest rainfall was received in December, when temperatures were lower compared to other months (Figure 2).

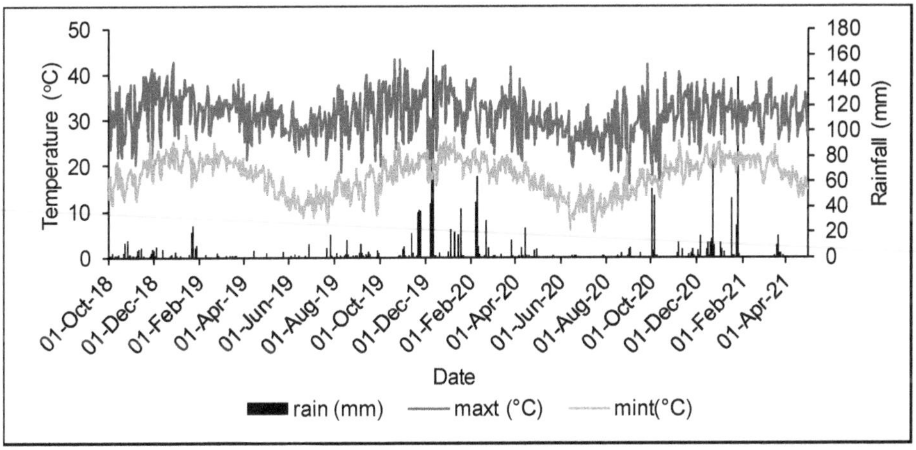

Figure 2. Ofcolaco daily rainfall, maximum and minimum temperature during 2018/19 and 2020/21 cropping seasons.

3.2. *Grain Yield and Yield Components of Sorghum and Cowpea*

The cropping system and density of the companion cowpea crop had no effect on grain yield of sorghum cultivars at the test sites over two seasons. Grain sorghum cultivars, on the other hand, showed a significant variation ($p \leq 0.05$) in grain yield over the two cropping seasons at Syferkuil and Ofcolaco (Figures 3 and 4). The results from Syferkuil revealed that cultivars Enforcer and NS5511 outperformed Avenger and Titan, with an average grain yield of 4153 kg per ha during the 2018/19 cropping season, while Avenger and Titan produced an average yield of 2607 kg per ha. According to the results, 85.86 kg per ha more grain yield was harvested in 2018/19 at this location than in 2020/21. The cultivar NS5511 with yield of 2120 kg per ha outperformed the cultivars Enforcer, Avenger, and Titan, which had mean yields of 1942 kg per ha, 1652 kg per ha, and 1561 kg per ha, respectively (Figure 3).

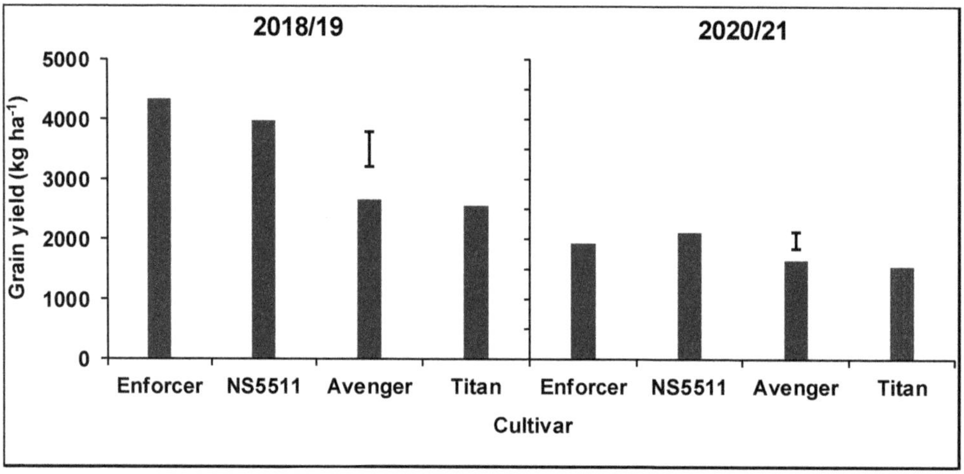

Figure 3. Grain yield of four sorghum cultivars evaluated at Syferkuil during 2018/19 and 2020/21 cropping seasons. Vertical bars represent LSD value ($p \leq 0.05$) for mean separation.

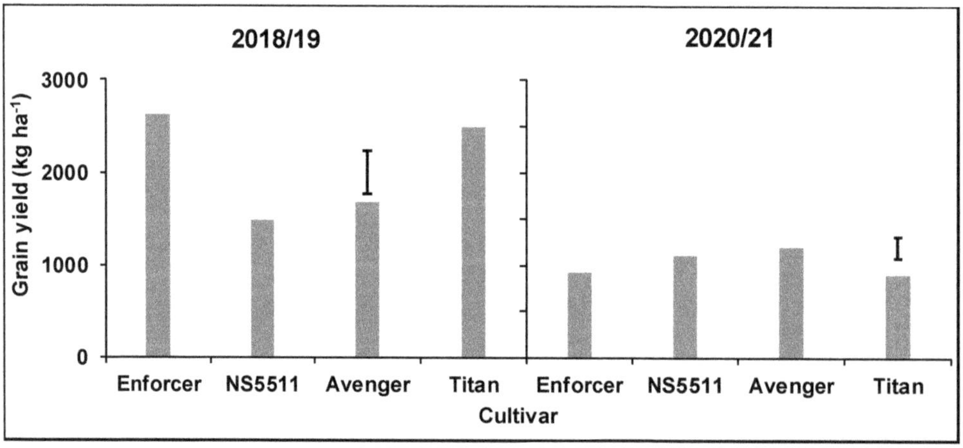

Figure 4. Grain yield (GY) of four grain sorghum cultivars evaluated at Ofcolaco during 2018/19 and 2020/21 cropping seasons. Vertical bars represent LSD value ($p \leq 0.05$) for mean separation.

The grain yield of the sorghum cultivars at Ofcolaco was inconsistent across seasons (Figure 4). Enforcer and Titan, for example, produced higher grain yields than NS5511 and

Avenger in the 2018/19 cropping seasons, averaging 2562 kg per ha and 1584 kg per ha, respectively. However, in 2020/21, NS5511, Avenger, and Enforcer outperformed Titan, which produced a yield of 910 kg per ha.

Harvest index (HI) based on grain production differed significantly ($p \leq 0.05$) between grain sorghum cultivars at the two locations and cropping seasons. Across the two cropping seasons and two locations, Enforcer consistently had the highest harvest index compared to the other cultivars (Figure 5). NS5511 had the second highest harvest index at Syferkuil compared to Avenger and Titan during the 2018/19 cropping season, but the HI were similar in the other seasons and locations.

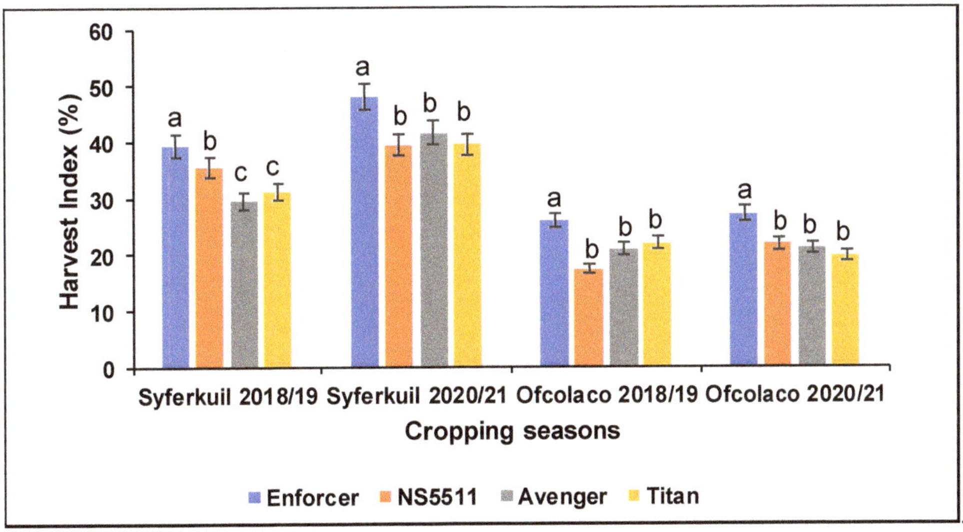

Figure 5. Harvest index of four grain sorghum cultivars in the two agro-ecological regions across different cropping seasons. Different letters indicate that the means were different at $p \leq 0.05$.

Regarding grain sorghum yield components, a significant variation ($p \leq 0.05$) was observed among the grain sorghum cultivars at Syferkuil during the two cropping seasons except for 1000 seed weight and seed weight per head, which did not differ during the 2020/21 cropping season (Table 2). The cultivar Enforcer was generally superior in most of the yield components compared to the other cultivars during the 2018/19 cropping season at this location, except for shelled head weight. The cultivar NS5511 had a relatively higher 1000-seed weight and seed weight per head compared to Avenger and Titan. The cultivar Avenger had a lower seed weight per head and harvest index compared to the cultivar Titan, regardless of having a longer head length, shelled head weight, and head weight compared to the other cultivars in the 2018/19 cropping season. In the 2020/21 cropping season, all the cultivars had a high head length and harvest index compared to cultivar NS5511 (Table 2). The results further revealed that cultivar Avenger produced fewer seeds per head compared to all other cultivars but had a relatively higher head length and shelled head weight. The mean head length and shelled head weight were 29.09 cm and 18.82 g, respectively.

At Ofcolaco, the results indicated that all yield components significantly differed among the grain sorghum cultivars during the two cropping seasons, except head length, which did not vary in 2020/21 (Table 3). The cultivar Avenger was superior in many of the yield components measured compared to all other cultivars except 1000 seed weight and harvest index during the 2018/19 cropping season. Furthermore, the seed weight per head of Avenger and NS5511 (48.15 g per head and 40.10 g per head) was higher than the grand mean of 30.47 g per head. However, the two cultivars (Avenger and NS5511) had lower HI compared to the grand mean. The results further indicated that Enforcer and Titan obtained a higher average HI of 23.94% compared to Avenger and NS5511, with an average of 19.13%. However, the two cultivars (Avenger and NS5511) obtained about 63.79% more seed weight head per head compared to Enforcer and Titan. In the 2020/21 cropping season, the results showed that although there was no statistical variation among the cultivars, the cultivar Avenger had the tendency to produce a higher head length. Although there was no statistically significant difference between cultivars Avenger and NS5511, Avenger had higher head weight and seed weight per head. The cultivar (Avenger) also had a high shelled head weight of 14.26 g per head and a higher 1000 seed weight of 6.29 g compared to all the other cultivars.

Table 2. Yield components of four grain sorghum cultivars evaluated at Syferkuil during 2018/19 and 2020/21 cropping seasons.

Cultivars	Head Length (cm)	Head Weight (g Head^{-1})	Shelled Head Weight (g Head^{-1})	1000-Seed Weight (g)	Seed Weight Head (g Head^{-1})
			Syferkuil 2018/19		
Enforcer	27.54 ^{a}	109.13 ^{a}	47.01 ^{ab}	28.17 ^{a}	61.21 ^{a}
NS5511	25.07 ^{b}	92.39 ^{b}	43.06 ^{ab}	23.88 ^{b}	49.03 ^{b}
Avenger	26.08 ^{ab}	77.19 ^{bc}	49.65 ^{a}	21.76 ^{c}	27.49 ^{c}
Tittan	25.34 ^{b}	71.76 ^{c}	39.93 ^{b}	27.82 ^{a}	31.80 ^{c}
$p \leq 0.05$	*	*	*	*	*
Grand mean	26	87.62	44.91	25.41	42.38
LSD value	1.79	16.09	8.93	1.51	156.3
			Syferkuil 2020/21		
Enforcer	28.59 ^{a}	108.97 ^{ab}	14.47 ^{b}	39.41	90.13
NS5511	26.54 ^{b}	112.15 ^{a}	16.55 ^{ab}	43.02	90.83
Avenger	29.09 ^{a}	98.35 ^{b}	18.82 ^{a}	38.61	82.39
Tittan	28.67 ^{a}	99.31 ^{b}	17.45 ^{a}	41.03	81.95
$p \leq 0.05$	*	*	*	ns	ns
Grand mean	28.22	104.7	16.82	40.52	86.33
LSD value	1.22	11.6	2.93	6.39	12.92

Means followed by the same letter are not significantly different based on LSD ($p \leq 0.05$). * = Significantly different at $p \leq 0.05$; ns = not significantly different at $p \leq 0.05$.

Table 3. Yield components of four grain sorghum cultivars evaluated at Ofcolaco during 2018/19 and 2020/21 cropping season.

Cultivars	Head Length (cm)	Head Weight (g Head^{-1})	Shelled Head Weight (g Head^{-1})	1000-Seed Weight (g)	Seed Weight Head (g Head^{-1})
			Ofcolaco 2018/19		
Enforcer	25.61 ^{ab}	28.33 ^{c}	7.43 ^{b}	35.69 ^{c}	17.71 ^{b}
NS5511	21.91 ^{b}	50.04 ^{b}	6.68 ^{b}	45.76 ^{a}	40.10 ^{a}
Avenger	30.95 ^{a}	70.03 ^{a}	12.91 ^{a}	43.59 ^{ab}	48.15 ^{a}
Tittan	29.59 ^{a}	24.69 ^{c}	8.08 ^{b}	39.98 ^{bc}	15.91 ^{b}
$p \leq 0.05$	*	*	*	*	*
Grand mean	27.02	43.27	8.78	41.26	30.47
LSD value	8.81	11.99	2.23	5.68	9.82
			Ofcolaco 2020/21		
Enforcer	30.1	28.37 ^{b}	7.40 ^{b}	4.09 ^{b}	24.29 ^{b}
NS5511	30.34	40.36 ^{a}	9.87 ^{b}	4.55 ^{b}	35.80 ^{a}
Avenger	30.98	44.53 ^{a}	14.26 ^{a}	6.29 ^{a}	38.24 ^{a}
Tittan	30.91	32.37 ^{b}	9.88 ^{b}	4.47 ^{b}	27.91 ^{b}
$p \leq 0.05$	ns	*	*	*	*
Grand mean	30.58	36.41	10.35	4.85	30.81
LSD value	2.02	7.63	1.31	30.2	6.62

Means followed by the same letter are not significantly different based on LSD ($p \leq 0.05$). * = Significantly different at $p \leq 0.05$; ns = not significantly different at $p \leq 0.05$.

During the 2018/19 cropping season, cowpea grain yield was 63 percent higher under high density versus low density at Syferkuil (Figure 6). However, grain yield was 32% higher under high density compared to low density in the 2020/21 cropping season.

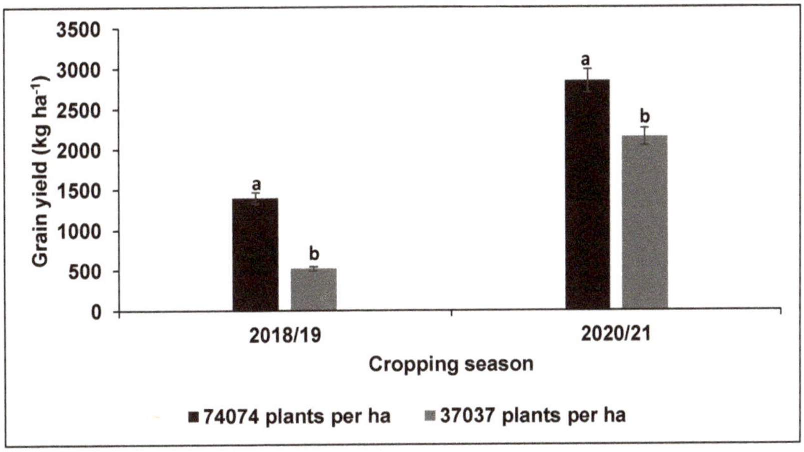

Figure 6. Grain yield of cowpea under two densities of cowpea grown at Syferkuil during contrasting seasons. Different letters indicate that the means were different at $p \leq 0.05$.

In sole compared to binary culture, cowpea produced a higher grain yield in sole with a mean of 1534 kg per ha and 992 kg per ha in high and low density, respectively, during the 2018/19 cropping season (Figure 7). Although in binary cultures there was no statistical difference between treatments, the grain yield of cowpea was higher when intercropped with Titan, followed by NS5511, with a grain yield of 852 kg per ha and 718 kg per ha, respectively. In the 2020/21 cropping season, grain yield was significantly affected by the cropping system. Similar to the 2018/19 cropping season, the results indicated that cowpea attained a higher grain yield when grown in sole compared to binary culture, with a mean of 5045 kg per ha in high density sole and 3411 kg per ha in low density sole (Figure 7).

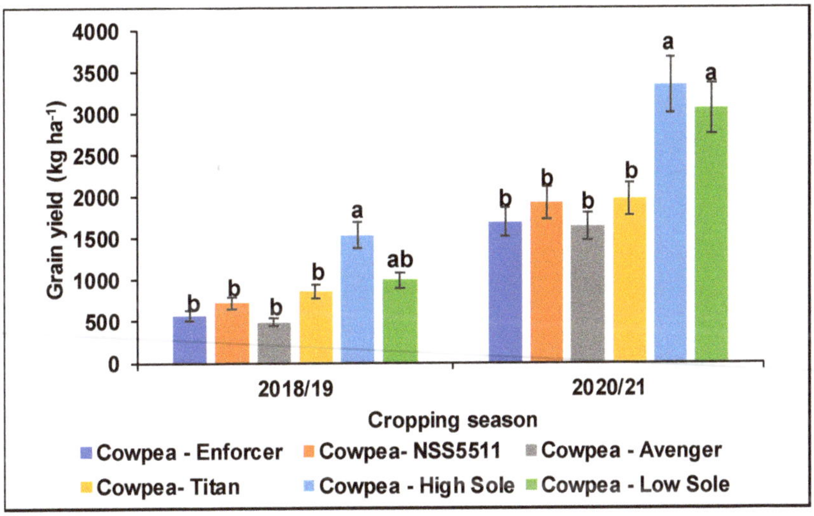

Figure 7. Grain yield of cowpea under in binary and sole cultures grown at Syferkuil during two contrasting seasons. Different letters indicate that the means were different at $p \leq 0.05$.

Grain yield among cowpea treatments was higher under high cowpea density compared to lower density, with means of 3175 kg per ha and 1233 kg per ha, respectively, at Ofcolaco during the 2018/19 cropping season (Figure 8).

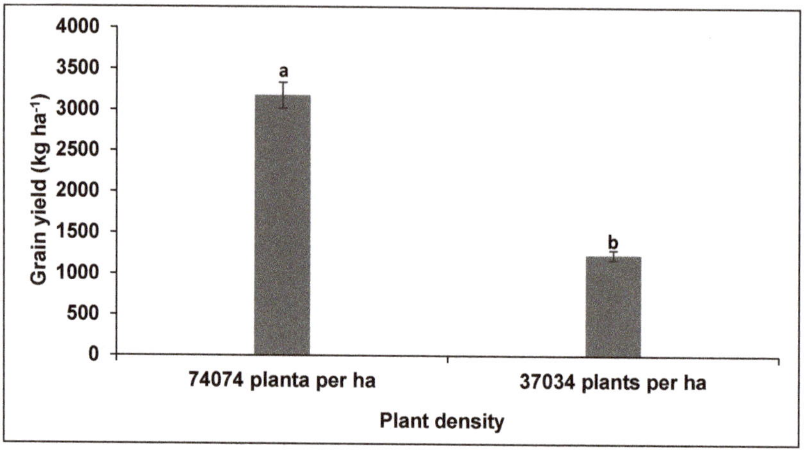

Figure 8. Grain yield of cowpea under two densities of cowpea grown at Ofcolaco during the 2018/19 cropping season. Different letters indicate that the means were different at $p \leq 0.05$.

The results from Ofcolaco revealed that, in binary cultures, cowpea attained the highest yield of 1701 kg per ha when intercropped with Avenger followed by when intercropped with Titan, which produced 1508 kg per ha (Figure 9). Although intercropping with Enforcer attained the lowest grain yield compared to all treatments in binary and sole cultures, a higher harvest index was obtained by this treatment compared to binary cultures.

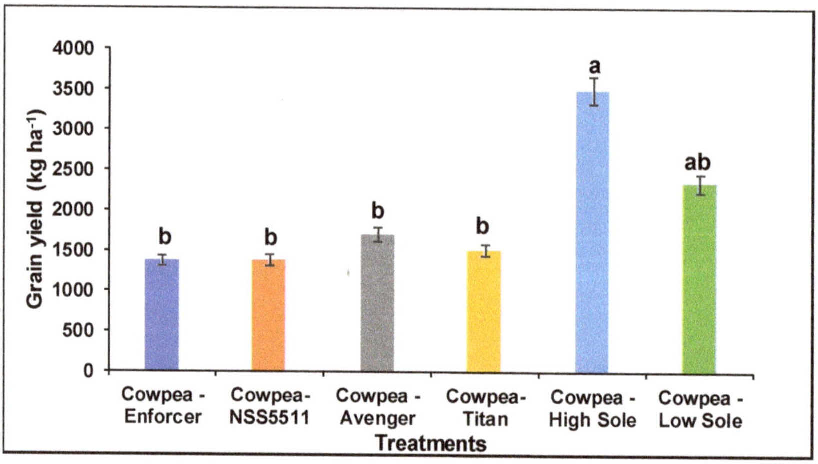

Figure 9. Grain yield of cowpea under in binary and sole cultures grown at Ofcolaco during 2018/19 cropping season. Different letters indicate that the means were different at $p \leq 0.05$.

There was no significant variation ($p \leq 0.05$) in the cowpea harvest index according to the cropping system at Syferkuil and Ofcolaco during the two cropping seasons. Sole cowpea under high density had a higher harvest index compared to the other cowpea treatments during the 2018/19 and 2020/21 cropping seasons at the two locations (Figure 10). The cowpea intercrop with Avenger had the lowest harvest index during the 2018/19 cropping season at Syferkuil. Furthermore, cowpea intercrop with Enforcer and Titan had a higher harvest index compared to sole cowpea in low density culture during the same season. At Ofcolaco and Syferkuil during the 2018/19 and 2020/21 cropping seasons, respectively, binary cultures were not statistically different (Figure 10).

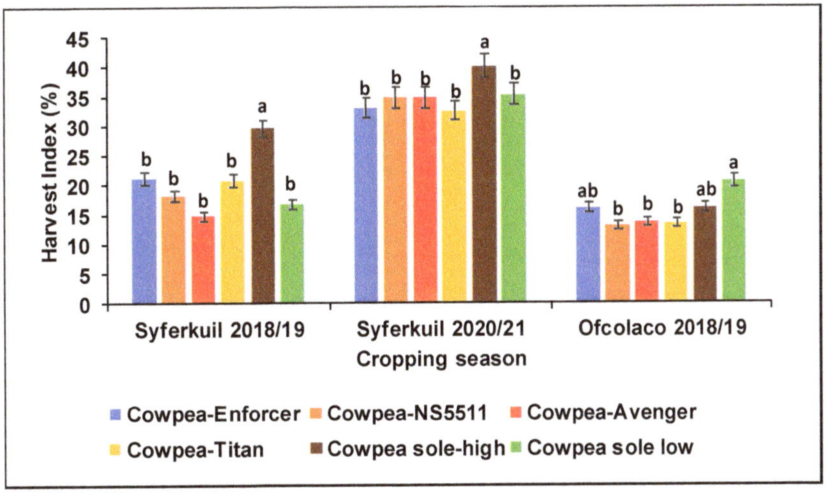

Figure 10. Harvest index of cowpea in binary and sole cultures at Syferkuil and Ofcolaco during 2018/19 and 2020/21 cropping seasons. Different letters indicate that the means were different at $p \leq 0.05$.

Assessing the yield components, the weight of 100 seeds was not significantly different between binary and sole cultures of cowpea at Syferkuil during the 2018/19 cropping season. However, significant variation ($p \leq 0.05$) was found for this yield component in the 2020/21 cropping season. Pod weight per plot was influenced by the cropping system in both seasons at this location (Table 4). The weight of 100 seeds was not significantly affected by the cropping system at Ofcolaco among cowpea treatments in binary and sole cultures during the 2018/19 cropping season. However, pod weight per plot was significantly ($p \leq 0.05$) affected by the intercropping system for cowpea treatments. The cowpea sole under high density resulted in a high pod weight per plot compared to all other treatments (Table 4).

Table 4. Yield components of cowpea in binary and sole cultures evaluated at Syferkuil and Ofcolaco during the 2018/19 and 2020/21 cropping season.

Treatments	Syferkuil 2018/19		Syferkuil 2020/21		Ofcolaco 2018/19	
	100-seed weight	pod weight per plot	100-seed weight	pod weight per plot	100-seed weight	pod weight per plot
Cowpea–Enforcer	16.17	139.73 [c]	15.54 [b]	336.56 [b]	14.71	364.10 [b]
Cowpea–NSS5511	16.24	167.23 [c]	14.51 [c]	384.06 [b]	14.68	355.97 [b]
Cowpea–Avenger	16.17	114.72 [c]	14.65 [c]	321.87 [b]	14.29	440.97 [b]
Cowpea–Titan	16.78	199.10 [bc]	15.53 [b]	383.44 [b]	14.88	307.22 [b]
Cowpea–High Sole	16.22	325.51 [a]	15.54 [b]	681.02 [a]	14.96	778.94 [a]
Cowpea–Low Sole	16.39	285.19 [ab]	16.61 [a]	398.36 [b]	14.74	715.51 [a]
$p \leq 0.05$	ns	*	*	*	ns	*
Grand mean	16.33	205.25	15.39	417.55	14.71	493.79
LSD value	0.86	79.31	0.69	99.31	1.05	170.16

Means followed by the same letter are not significantly different based on LSD ($p \leq 0.05$). * = Significantly different at $p \leq 0.05$; ns = not significantly different at $p \leq 0.05$.

3.3. Partial and Total Land Equivalent Ratio (LER) of Sorghum and Cowpea

The partial land equivalent ratio of cowpea ranged from 0.4 to 0.7 at Syferkuil during the 2018/19 and 2020/21 cropping seasons, respectively. The partial of grain sorghum at Syferkuil was between 0.7 and 1.3 in the 2018/19 and 2020/21 cropping seasons. At Ofcolaco, the partial land equivalent ratio was between 0.4 and 0.6 for cowpea and 0.8–1.4 for grain sorghum in the 2018/19 and 2020/21 cropping seasons. The total LER was above

1.0 in all grain sorghum and cowpea intercrop treatments (Table 5). At Syferkuil, Enforcer had a higher LER when intercropped with low cowpea density compared to high cowpea density, with means of 1.8 and 1.3, respectively, during the 2018/19 season. Avenger had a total LER of 1.6 and 1.7 under low and high density, respectively. However, Titan obtained 1.5 and 1.6 total LER under low and high density, respectively. The results also indicated that Avenger and NS5511 intercropped with cowpea high density had a total LER of 1.7, whereas NS5511 and Titan intercropped with low density had a total LER of 1.6 in the 2018/19 cropping season. In the 2020/21 cropping season, Titan intercropped with cowpea under low and high density had a total LER of 1.8 and 1.9, respectively. Enforcer intercropped with cowpea low density had the lowest total LER of 1.3 compared to all treatments. At Ofcolaco, total LER ranged from 1.4 to 1.9, with the highest observed in NS5511 intercropped with cowpea high density (Table 5).

Table 5. Total land equivalent ratio of grain sorghum and cowpea at Syferkuil and Ofcolaco during 2018/19 and 2020/21 cropping seasons.

Treatments	Syferkuil 2018/19	Syferkuil 2020/21	Ofcolaco 2018/19
Enforcer + Cowpea-low	1.7	1.3	1.7
Enforcer + Cowpea-high	1.3	1.4	1.2
NSS5511 + Cowpea-low	1.5	1.6	1.9
NSS5511 + Cowpea-high	1.7	1.6	1.3
Avenger + Cowpea-low	1.6	1.5	1.6
Avenger + Cowpea-high	1.7	1.7	1.5
Titan + Cowpea-low	1.6	1.8	1.5
Titan + Cowpea-high	1.5	1.8	1.6

Low = 37,037 plants per ha, high = 74,074 plants per ha.

3.4. *Leaf Area Index of Sorghum and Cowpea in Binary and Sole Cultures*

At Syferkuil, leaf area index (LAI) was significantly different ($p \leq 0.05$) among grain sorghum cultivars at Syferkuil during the 2018/19 and 2020/21 cropping seasons (Figure 11). NS5511 had a higher LAI compared to the other sorghum cultivars, followed by Enforcer during the 2018/19 cropping season. However, Enforcer was superior compared to the other cultivars in the 2020/21 growing season.

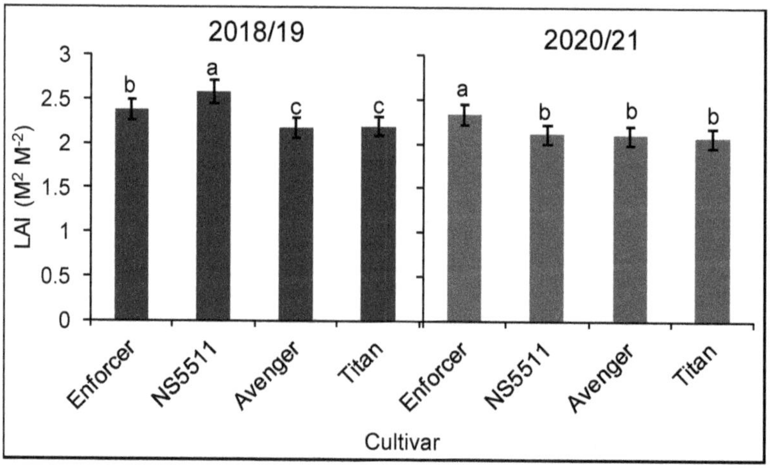

Figure 11. Leaf area index of four grain sorghum cultivars evaluated at Syferkuil during 2018/19 and 2020/21 cropping seasons. Different letters indicate that the means were different at $p \leq 0.05$.

There was no variation among grain sorghum cultivars for LAI at Ofcolaco during the 2018/19 cropping season. However, in 2020/21 there was a significant variation in LAI among the cultivars (Figure 12). The results revealed that NS5511 and Avenger were higher than Enforcer and Titan during the 2020/21 cropping season.

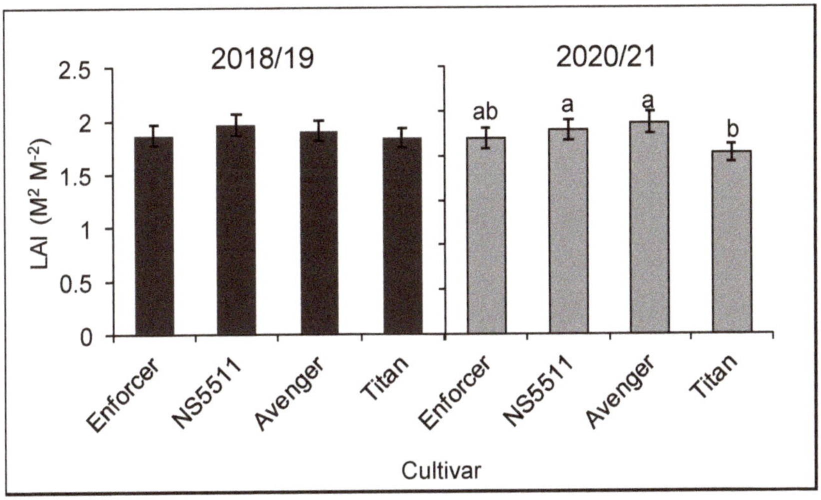

Figure 12. Leaf area index of four grain sorghum cultivars at Ofcolaco during 2018/19 and 2020/21 cropping seasons. Different letters indicate that the means were different at $p \leq 0.05$.

There was a significant interaction effect between the cropping system and days after planting of cowpea at Syferkuil during the 2018/19 and 2020/21 cropping seasons. The results indicated that, in the 2018/19 cropping season, cowpea treatments had higher LAI at 63DAP, excluding cowpea sole under high density. Cowpea sole high density started at a higher rate and remained steady until 83DAP (Figure 13). During the 2020-21 cropping season, cowpea treatments started at a low rate and increased until 67DAP, then decreased until 104DAP (Figure 13).

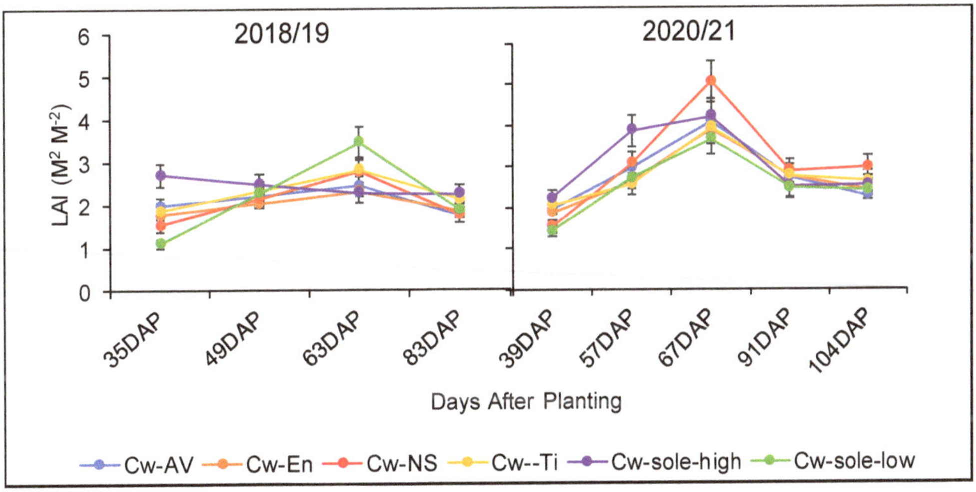

Figure 13. Leaf area index of cowpea treatments in binary and sole cultures at Syferkuil during 2018/19 and 2020/21 cropping seasons.

The results from Ofcolaco were similar to those at Syferkuil, with significant interaction occurring between the cropping system and days after planting during the two cropping seasons. Cowpea treatments had a similar trend during the 2018/19 growing season, with the highest LAI being between 49DAP and 83DAP (Figure 14). However, in the 2020–21 cropping season, cowpea treatments had fluctuating LAI across the days after planting.

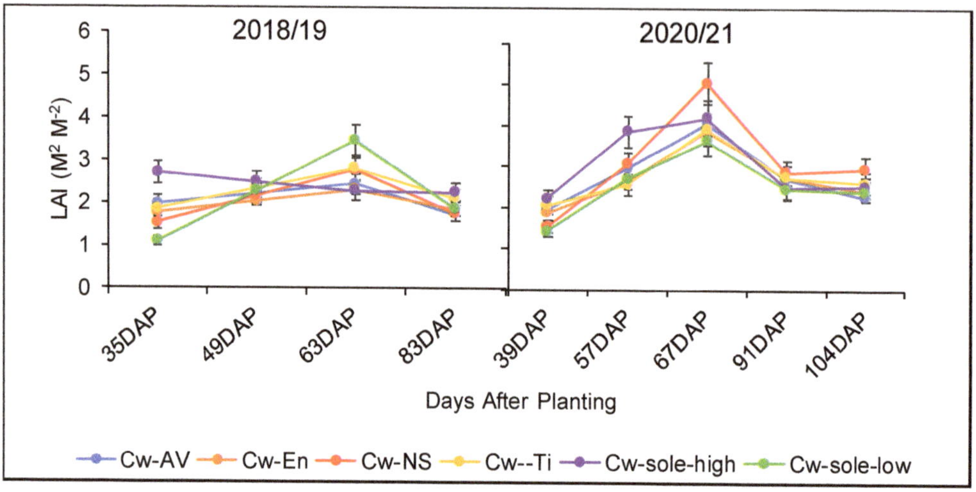

Figure 14. Leaf area index of cowpea treatments in binary and sole cultures at Ofcolaco during.

4. Discussion

The variation in temperatures and rainfall received during the cropping seasons as a result of climate change influenced the agronomic performance of grain sorghum cultivars at the two locations. Ofcolaco was generally warmer than Syferkuil during the 2018/19 and 2020/21 cropping seasons, which may have resulted in variation in crop performance and grain yield. Other studies have reported that the differences in grain yield of sorghum were due to distinct agro-ecological regions which varied across seasons [21,22]. From our study, grain sorghum generally performed better in 2018/19 compared to the 2020/21 cropping seasons, and vice versa for cowpea. The cropping system and the density of the companion crop cowpea did not influence the grain yield of sorghum cultivars at the two test locations across different seasons. The results were contrary to what was observed in another study [23]. The authors reported that sorghum was significantly influenced by the treatment combination in the sorghum-legume intercrop. Grain sorghum cultivars showed a significant variation in terms of grain yield due to the adaptive mechanism of the crop, which varied with cultivar, location, and season. Similar results have been reported elsewhere [24–26].

The density of cowpea and the cropping system significantly influenced grain yield and yield components of cowpea in the two agro-ecological regions and across the cropping seasons. The findings were in line with other studies in which the authors reported that the yield of cowpea was highly influenced by crop density [27]. However, cowpea density did not improve in the pearl millet–cowpea intercrop [28]. In this study, cowpea produced a higher grain yield when grown under high density (74,074 plants per ha), either in binary or sole compared to low density (37,037 plants per ha). Increased density probably allowed more cowpea plants to compete for light and water in binary cultures through improved root density and, ultimately, high yield accumulation. Similar results have been reported by other studies [29–31]. In the sole, cowpea produced more grain yield in the sole compared to the binary culture at Syferkuil and Ofcolaco during the 2018/19 and 2020/21 cropping seasons. This is mainly due to increased canopy size (LAI), which is important for monitoring crop growth and accumulation of grain yield [32].

The results also revealed that cowpea performed better at Syferkuil when intercropped with Titan compared to when intercropped with other grain sorghum cultivars. However, at Ofcolaco, cowpea had a higher yield when intercropped with Avenger, although the results were based on one season of data. High interspecific competition between crops is required for the efficient use of growth resources [33]. However, the efficient use of those resources must be greater than the interspecific competition [34]. In this study, there was high competition for resources such as water, light, etc., between grain sorghum cultivars and cowpea intercrop at Syferkuil, which hindered cowpea yield accumulation under low density when intercropped with Enforcer and Ns5511. However, at Ofcolaco, there was complementarity between cowpea and the two grain sorghum cultivars (Avenger and Titan) in the binary system.

Yield components are important variables used to determine the yield potential of crops in response to different agro-ecological regions [35]. In this study, yield components varied from one location to another and across seasons. For instance, at Syferkuil, Enforcer and NS5511 obtained the highest seed weight per head compared to Avenger and Titan, ultimately resulting in a higher grain yield during the two cropping seasons. Therefore, under the growing conditions of Syferkuil, the seed weight per head can be used to recommend cultivars Enforcer and NS5511 for high-grain-yield production. At Ofcolaco, Enforcer and Titan were superior cultivars in 2018/19, whereas in the 2020/21 cropping season, NS5511 and Avenger obtained higher grain yields. These indicate that the adaptation of grain sorghum cultivars at Ofcolaco is highly dependent on the growing conditions of a particular season. During the two cropping seasons, NS5511 and Avenger had higher seed weight per head compared to Enforcer and Titan. Hence, head weight and seed weight per head can be used by breeders as selection criteria for the recommendation of cultivars to local growers [36]. The higher grain yield of cowpea was explained by the pod weight per plot, which was consistent throughout the cropping seasons at the two test locations.

The leaf area index of a crop canopy is an important parameter that can be used to predict growth and yield [37]. At Syferkuil, the leaf area index of grain sorghum was significantly affected by the cropping system as well as the cultivar. During the two cropping seasons, Enforcer and NS5511, which ultimately accumulated more grain yield, had a higher leaf area index compared to the other cultivars. At Ofcolaco, the leaf area index was significantly influenced by the growing period during 2018/19, whereas in 2020/21, the binary had a higher leaf area index compared to the sole cultures. This further explains the variation in grain yield among grain sorghum cultivars at Ofcolaco. The leaf area index of cowpea was influenced by the cropping system, DAP, as well as cropping seasons. The LAI was higher at 40 and 63 DAP, depending on the cowpea treatment. The capturing of light by canopies at late flowering to mid pod formation stages is important for optimum grain accumulation [32,37].

LER was used in this study to measure the grain sorghum and cowpea intercrop efficiency relative to sole cropping. According to the results, the total LERs were found to vary with the growing seasons and treatments for grain sorghum and cowpea. However, the total LER values calculated were all greater than 1.0 in the test locations and across different seasons, indicating a high yield advantage in the binary cultures and more efficient productivity compared to the sole cultures. Several studies have reported LER values greater than 1.0 in sorghum–cowpea [8,38], sorghum–soybean [39] and maize–cowpea [40]. The results further indicated that the LER was influenced by the density of cowpea as well as the grain sorghum cultivar in intercrop at each experimental site. LER variation due to mixture in various planting patterns has also been reported elsewhere [28,41,42]. The high LER observed in this study was due to the efficient use of resources such as light, water, and nitrogen between grain sorghum cultivars and cowpea [43]. The goal of growers, as well as breeders, is high grain yield, which depends on other yield variables. Hence, the relationship between yield and yield components is important, whether it be positive or negative. According to the results, the strength of the correlation between grain yield and yield components varied with cultivar, intercropping system, and cropping season as well as the agro-ecological region. In conclusion, grain sorghum cultivars were not affected by either cropping system or the density of a companion crop cowpea. Enforcer and NS5511 produced higher grain yield at the two test locations compared to Avenger and Titan. The productivity of cowpea was influenced by the cropping system as well as the crop density. Cowpea performed better in terms of grain yield in sole compared to binary cultures. However, the yield of cowpea improved in binary cultures when the density was 74,074 plants per ha. Head weight of sorghum and pod weight of cowpea can be used as selection criteria for recommendation of cultivars to grow at Syferkuil and Ofcolaco. Based on the results of this study, grain sorghum–cowpea intercrop can be adopted as a climate-smart practice to improve yield compared to mono-cropping. However, the density of cowpea and grain sorghum cultivars should be taken into consideration, as they affect the productivity of the two crops. The research also discovered that in binary cultures, more organic carbon was left in the soil (Table 1), implying that the system could improve soil fertility and benefit subsequent crops. The data generated from this study could be useful in simulating the productivity of intercropping practice as a climate-smart method using crop modelling techniques. It is further suggested that a similar study be carried out to investigate the biological nitrogen fixation of the legume crop cowpea in response to intercropping with different cultivars of grain sorghum. In addition, for better recommendations, the impact of intercropping systems on soil carbon dynamics should be investigated.

Author Contributions: T.E.M. came up with the way to promote intercropping system as an approach to produce sorghum and cowpea sustainably under different agro-ecological conditions. K.K.A., L.M. and Y.G.K. verified the methods and analytical procedure used in conducting the field experiment. All authors have read and agreed to the published version of the manuscript.

Funding: The research was funded by National Research Foundation and Department of Science and Innovation through Risk and Vulnerability Science Centre (RVSC) of the University of Limpopo, South Africa.

Institutional Review Board Statement: This research was approved by the University of Limpopo Department of Plant Production, Soil Science and Agricultural Engineering. Ethical approval was not necessary for this research.

Informed Consent Statement: Not applicable.

Data Availability Statement: All data and materials used in the write up of the manuscript were acquired through existing facilities at RVSC, data generated from the research and climatic data from the Agricultural Research Council, South Africa. The data used in this study are available at RVSC of the University of Limpopo, which can be accessed through the corresponding author.

Acknowledgments: The authors acknowledge the National Research Foundation in partnership with Department of Science and Innovation in South Africa for funding RVSC and the research. The researchers also acknowledge the contribution made by the VLIR-IUC (Belgium) collaborative programme at the University of Limpopo.

Conflicts of Interest: The authors declare that they have no competing financial interest or personal relationships that could have influenced the work reported in this paper.

References

1. Taylor, J.R.N. Overview: Importance of sorghum in Africa. In Proceedings of the AFRIPRO Workshop on the Proteins of Sorghum and Millets: Enhancing Nutritional and Functional Properties for Africa, Pretoria, South Africa, 2–4 April 2003.
2. DAFF (Department of Agriculture, Forestry and Fisheries). *Sorghum: Production Guideline*; DAFF: Pretoria, South Africa, 2011.
3. Mpandeli, N.S. Coping with Climate Variability in Limpopo Province. Unpublished. Ph.D. Thesis, University of Witwatersrand, Johannesburg, South Africa, 2006.
4. Touch, V.; Martin, R.J.; Scott, J.F.; Cowie, A.; Li, D. Climate change adaptation options in rainfed upland cropping systems in the wet tropics: A case study of smallholder farms in North-West Cambodia. *J. Environ. Manag.* **2016**, *182*, 238–246. [CrossRef]
5. Rockstrom, J.; Kaumbutho, P.; Mwalley, J.; Nzabi, A.W.M.; Temesgen, M.L.; Mawenya, L.; Barron, J.; Mutua, J.; Damgaard-Larsen, S. Conservation farming strategies in East and Southern Africa: Yields and rain water productivity from on-farm action research. *Soil. Till. Res.* **2009**, *103*, 23–32. [CrossRef]

6. Stocker, T.F.; Qin, D.; Plattner, G.K.; Tignor, M.; Allen, S.K.; Boschung, J.; Midgley, P.M. *Climate change: The physical science basis. Intergovernmental Panel on Climate Change, Working Group I Contribution to the IPCC Fifth Assessment Report (AR5)*; Cambridge University Press: New York, NY, USA, 2013.
7. Burt, T.; Boardman, J.; Foster, I.; Howden, N. More rain, less soil: Long-term changes in rainfall intensity with climate change. *Earth Surf. Process. Landf.* **2015**, *41*, 563–566. [CrossRef]
8. Oseni, T.O. Evaluation of Sorghum-Cowpea Intercrop Productivity in Savanna Agro-ecology using Competition Indices. *J. Agric. Sci.* **2010**, *2*, 229–234. [CrossRef]
9. Singh, R.; Singh, G.S. Traditional agriculture: A climate-smart approach for sustainable food production. *Energy Ecol. Environ.* **2017**, *2*, 296–316. [CrossRef]
10. Srivastava, P.; Singh, R.; Tripathi, S.; Raghubanshi, A.S. An urgent need for sustainable thinking in agriculture: An Indian scenario. *Ecol. Indic.* **2016**, *67*, 611–622. [CrossRef]
11. Hauggaard-Nielsen, H.; Jørnsgaard, B.; Kinane, J.; Jensen, E.S. Grain legume–cereal intercropping: The practical application of diversity, competition and facilitation in arable and organic cropping systems. *Renew. Agric. Food Sys.* **2008**, *23*, 3–12. [CrossRef]
12. Derpsch, R.; Franzluebbers, A.J.; Duiker, S.W.; Reicosky, D.C.; Koeller, K.; Friedrich, T.; Sturny, W.G.; Sá, J.C.M.; Weiss, K. Why do we need to standardize no-tillage research? *Soil Till. Res.* **2014**, *137*, 16–22. [CrossRef]
13. Ning, C.; Qu, J.; He, L.; Yang, R.; Chen, Q.; Luo, S.; Cai, K. Improvement of yield, pest control and Si nutrition of rice by rice-water spinach intercropping. *Field Crops Res.* **2017**, *208*, 34–43. [CrossRef]
14. Rippke, U.; Ramirez-Villegas, J.; Jarvis, A.; Vermeulen, S.J.; Parker, L.; Mer, F.; Derkkruger, B.; Challinor, A.J.; Howden, M. Timescales of transformational climate change adaptation in sub-Saharan African Agriculture. *Nat. Clim. Change* **2016**, *6*, 605–609. [CrossRef]
15. Kephe, P.N.; Ayisi, K.K.; Petja, B.M. Challenges and opportunities in crop simulation modelling under seasonal and projected climate scenarios for crop production in South Africa. *J. Agric. Food Secur.* **2021**, *10*, 10. [CrossRef]
16. Department of Agricultural Development. *Soil Classification: A Taxonomic System for South Africa*; Soil Classification Working Group, Soil and Irrigation Research Institute, Department of Agricultural Development: Pretoria, South Africa, 1991.
17. Van Reeuwijk, L.P. (Ed.) *Procedure for Soil Analysis*, 6th ed.; International Soil Reference and Information Center (ISRIC)/Food and Agricultural Organization: Wageningen, The Netherlands, 2002; 120p.
18. Prikner, P.; Lachnit, F.; Dvořák, F. A new soil core sampler for determination bulk density in soil profile. *Plant Soil Environ.* **2011**, *50*, 250–256. [CrossRef]
19. Rayment, G.E.; Higginson, F.R. *Australian Laboratory Handbook of Soil and Water Chemical Methods*; Inkata Press: Port Melbourne, Australia, 1992.
20. Anderson, J.M.; Ingram, J.S.I. Microbial biomass. In *Tropical Soil Biology and Fertility: A Handbook of Methods*, 2nd ed.; Anderson, J.M., Ingram, J.S.I., Eds.; CAB International: Wallingford, UK, 1993; pp. 68–70. ISBN 0-85198-821-0.
21. Gasura, E.; Setimela, P.S.; Souta, C.M. Evaluation of the performance of sorghum genotypes using GGE biplot. *Can. J. Plant Sci.* **2015**, *95*, 1205–1214. [CrossRef]
22. Asfaw, A. The Role of Introduced Sorghum and Millets in Ethiopian Agriculture. *J. SAT Agric. Res.* **2007**, *3*.
23. Somu, G.; Meena, N.; Shashikumar, C.; Shivaray, N.; Druvakumar, M.; Kanavi, M.S.P. Performance of sorghum under sorghum legume intercropping system. *J. Pharmacogn. Phytochem.* **2020**, *9*, 2320–2322.
24. Ghani, A.; Saeed, M.; Hussain, D.; Arshad, M.; Shafique, M.M.; Shah, S.A.S. Evaluation of different sorghum (*Sorghum bicolor* L. moench) varieties for grain yield and related characteristics. *Sci. Lett.* **2015**, *3*, 72–74.
25. Nida, H.; Seyoum, A.; Gebreyohannes, A. Evaluation of yield performance of intermediate altitude sorghum (*Sorghum bicolor* (L.) Moench) genotypes using genotype x environment interaction analysis and GGE biplot in Ethiopia. *Int. J. Trend Res. Dev.* **2016**, *3*, 27–35.
26. Hadebe, S.T.; Modi, A.T.; Mabhaudhi, T. Water use characteristics of hybrid, open-pollinated, and landrace sorghum genotypes under rainfed conditions. *Water S. Afr.* **2017**, *43*, 91–103. [CrossRef]
27. Chimonyo, V.G.P.; Modi, A.T.; Mabhaudhi, T. Simulating yield and water use of a sorghum–cowpea inter-crop using APSIM. *Agric. Water Manag.* **2016**, *177*, 317–328. [CrossRef]
28. Nelson, W.C.D.; Hoffmann, M.P.; Vadez, V.; Roetter, R.P.; Whitbread, A.M. Testing pearl millet and cow-pea intercropping systems under high temperatures. *J. Field Crop Res.* **2018**, *217*, 150–166. [CrossRef]
29. Makoi, J.H.J.R.; Chimphango, S.B.M.; Dakora, F.D. Photosynthesis, water-use efficiency and δ 13 C of five cowpea genotypes grown in mixed culture and at different densities with sorghum. *Photosynthetica* **2010**, *48*, 143–155. [CrossRef]
30. Moriri, S.; Owoeye, L.G.; Mariga, I.K. Influence of component crop densities and planting patterns on maize production in dry land maize/cowpea intercropping systems. *African J. Agric. Res.* **2010**, *5*, 1200–1207.
31. Masvaya, E.N.; Nyamangara, J.; Descheemaeker, K.; Giller, K.E. Is maize-cowpea intercropping a viable option for smallholder farms in the risky environments of semi-arid Southern Africa? *Field Crops Res.* **2017**, *209*, 73–87. [CrossRef]
32. Kamara, A.Y.; Tofa, A.; Ademulegun, T.; Solomon, R.; Shehu, H.; Kamai, N.; Omoigui, L. Mize-soybean intercropping for sustainable intensification of cereal-legume cropping systems in Northern Nigeria. *Exp. Agric.* **2017**, *55*, 73–87. [CrossRef]
33. Takim, F.O. Advantages of maize-cowpea intercropping over sole cropping through competition indices. *J. Agric. Biol. Res.* **2012**, *1*, 53–59.

34. Zhang, F.; Li, L. Using competitive and facilitative interactions in intercropping systems enhances crop productivity and nutrient-use efficiency. *Plant Soil* **2003**, *248*, 305–312. [CrossRef]
35. Kozak, M.; Mądry, W. Note on yield component analysis. *Cereal Res. Commun.* **2006**, *34*, 933–940. [CrossRef]
36. El-Aref, K.A.O.; Ahmed, H.A.; Abd-El-Hameed, W.M. Studies on intercropping peanut and cowpea on grain sorghum. *Minia J. Agric. Res. Dev.* **2019**, *39*, 175–189.
37. Yin, X.; Lantinga, E.A.; Schapendonk, A.H.C.M.; Zhong, X. Some quantitative relationships between leaf area index and canopy nitrogen content and distribution. *Ann. Bot.* **2003**, *9*, 893–903. [CrossRef]
38. Yesuf, M. Effect of Planting Arrangement and Population Densities of Haricot Bean on Productivity of Sorghum/Haricot Bean Additive Mixture. Master's Thesis, Alemaya University, Haramaya, Ethiopia, 2003.
39. Yu, Y.; Stomph, T.-J.; Makowski, D.; Van Der Werf, W. Temporal niche differentiation increases the land equivalent ratio of annual intercrops: A meta-analysis. *Field Crop. Res.* **2015**, *184*, 133–144. [CrossRef]
40. Gebremichael, A.; Bekele, B.; Tadesse, B. Evaluation of the effect of sorghum-legume intercropping and its residual effect on yield of sorghum in yeki woreda, sheka zone, Ethiopia. *Int. J. Agric. Res. Innov. Technol.* **2020**, *9*, 62–66. [CrossRef]
41. Musa, M.G.M.; El-Aref, K.A.O.; Bakheit, M.A.; Mahdy, A.Y. Effect of intercropping and plant distribution of sorghum with soybean on growth and yield of Sorghum bicolor. *Arch. Agric. Sci. J.* **2021**, *4*, 228–239. [CrossRef]
42. Dahmardeh, M.; Ghanbari, A.; Syahsar, B.A.; Ramrodi, M. The role of intercropping maize (*Zea mays* L.) and cowpea (*Vigna unguiculata* L.) on yield and soil chemical properties. *Afr. J. Agric. Res.* **2010**, *5*, 631–636.
43. Reddy, R.S. *Principles of Crop Production*; Kalyani Publishers: New Delhi, India, 2008; pp. 45–47.

agriculture

Review

Research Progress on the Theory and Practice of Grassland Eco-Compensation in China

Zhidong Li [1,2], Boru Su [1,2] and Moucheng Liu [2,*]

1 Institute of Geographic Sciences and Natural Resources Research, Chinese Academy of Sciences, Beijing 100101, China; lizd.18s@igsnrr.ac.cn (Z.L.); suboru21@mails.ucas.ac.cn (B.S.)
2 University of Chinese Academy of Sciences, Beijing 100049, China
* Correspondence: liumc@igsnrr.ac.cn

Abstract: In order to curb the phenomenon of grassland degradation caused by human activity, China has begun the exploration of grassland eco-compensation, setting an example for the ecological protection of grasslands and sustainable use of resources around the world. At this stage, China has invested more than 170 billion yuan in grassland eco-compensation, benefiting 12 million farmer and herder households. The related research involves various perspectives, scopes, and methods, but lacks systematic reviewing. This study reviews the relevant theoretical and practical research and explores the connotations and effects of grassland eco-compensation in China. In general, the current grassland eco-compensation in China is a large-scale ecological-economic institutional arrangement with the following five characteristics: (1) the goals are to maintain the grassland ecosystem services and increase the income of herder households; (2) the main bodies are governments and herder households; (3) the main method is financial transfer payments; (4) the compensation standards are based on the opportunity costs of the herder households' responses as the lower limits and the grassland ecosystem service values as the upper limits; and (5) it is a comprehensive compensation system that requires legal, regulatory, technological support and long-term mechanisms. Since 2011, driven by the grassland eco-compensation policy, the income levels of herder households in each pilot area have generally increased, and the overall ecology of grasslands has slightly improved. However, there are still some areas where overload is common. Additionally, there are regional differences in the satisfaction degree of herder households, which is mainly affected by factors such as family income, compensation cognition and family holding grassland scale. Our analysis shows that the shortcomings of current theoretical research are mainly reflected in the low precision of scientific compensation standards, the lack of a basis for differentiated standards, and the single compensation method. The shortcoming of practical research is that most effect evaluations cannot reflect the role of eco-compensation in it. This study suggests that future work should focus on the response mechanism of herder households and the improvement of the compensation measures. At the same time, the scope of research should be expanded, and we should learn from advanced compensation experience in other fields.

Keywords: payment for ecosystem services; grassland eco-compensation; eco-compensation policies in China

Citation: Li, Z.; Su, B.; Liu, M. Research Progress on the Theory and Practice of Grassland Eco-Compensation in China. *Agriculture* **2022**, *12*, 721. https://doi.org/10.3390/agriculture12050721

Academic Editor: Mariusz J. Stolarski

Received: 16 April 2022
Accepted: 17 May 2022
Published: 19 May 2022

1. Introduction

With the deterioration of the natural ecosystem and the development of research on ecosystem services, humans have gradually realized the important economic value of the ecosystem services [1]. However, in the interaction between humans and nature, human activities often lead to external effects on others [2]. For example, planting trees, watershed management and soil remediation always produce positive externalities. Overgrazing, excessive fertilization and untreated sewage discharge always produce negative externalities [3,4]. Without intervention, the protectors often terminate the protective behaviors, because it is difficult to obtain benefits from the positive externalities. Meanwhile, destroyers benefit from not being punished by negative externalities, and thus tend to keep destroying. For a long time, this lack of ecological justice has ultimately led to the overuse of resources, which harms the interests of all [5]. In order to achieve sustainable supplies of ecosystem services and internalize the externalities of the ecosystem services, many countries have begun to explore eco-compensation [4].

In the second half of the 20th century, the concept of sustainable development gradually reached a consensus in the international community. Some developed countries have taken the lead in the exploration and practice of eco-compensation. Internationally, a concept similar to eco-compensation is "payment for ecosystem services (PES)". PES was first widely practiced in forest vegetation restoration, and related research includes a legal framework [6], transaction costs [7], case analysis [8] and so on. While it has produced some eco-economic benefits in forestry systems, PES has gradually been introduced into more and more other ecological conservation fields [9–12].

Eco-compensation research in China started relatively late. However, under the national conditions of promoting the construction of ecological civilization, the Chinese government attaches great importance to improving the eco-compensation mechanisms. At present, China has formed an overall layout of eco-compensation that is dominated by the government, with central financial transfer payments as the main source of funds, and governments at all levels as the main body of implementation [13]. Additionally, it has achieved remarkable results in various ecosystems [14–16], ecological function areas [17], resource extraction areas [18] and agricultural planting areas [19,20].

As the world's largest terrestrial ecosystem, grasslands account for about 37% of the world's non-glacial area [21] and 30–40% of China's land area [22], and have important ecological functions [23]. In recent years, grassland degradation has been widespread in many countries [24–26], posing huge challenges to the sustainable provision of grassland ecosystem services [27–29]. However, compared with forests, watersheds, farmland and other fields, there are relatively few studies on grassland eco-compensation around the world. In 2011, in order to restore the ecological function of grassland and promote the sustainable development of livestock husbandry, China officially established the grassland eco-compensation mechanism [30]. It has set a model for global grassland ecological protection and has attracted the attention of many scholars [31,32]. Facing this new field, this study aims to explore the connotation and effects of grassland eco-compensation in China by reviewing the current research. Then, according to the results, we point out the shortcomings of the current research and provide ideas for the future work.

2. Overview of Grassland Eco-Compensation in China

Grassland eco-compensation in China was first officially proposed at the executive meeting of the State Council on 12 October 2010. The meeting pointed out that due to long-term overgrazing and insufficient investments in grassland ecological protection, China's grasslands are seriously degraded. At the same time, due to the single employment (livestock husbandry), the income growth of herder households is slow. Therefore, since 2011, the central government has paid a large amount of funds every year (over 170 billion yuan by 2021) to implement grassland eco-compensation in China's pilot pastoral areas. The framework of grassland eco-compensation is that the government provides financial support to herder households, encourages them to transform livestock husbandry, and then reduces the grazing intensity of natural grasslands to restore ecological functions. The core measure is to divide the natural grasslands in the pilot area into grazing prohibition (GP) areas and grass–livestock balance (GLB) areas. GP areas are prohibited from grazing or allow very little grazing in some areas, and the government provides subsidies for the grassland contractors. As for GLB areas, the local management department gives reasonable grazing limits according to the current situation of grassland resources. Then, the government provides rewards to contractors who comply with the limits.

Since a large number of the papers are presented in Chinese, this study briefly describes the research overview of eco-compensation through the CNKI (China national knowledge infrastructure) database. We set the topic as "Payments for (Grassland) Ecosystem Service or (Grassland) Eco-Compensation or (Grassland) Ecological Protection Compensation or (Grassland) Ecological Product Value Realization". Journal sources include SCI, EI, CSCD and CSSCI, and the papers sampled included those published up to 2021. In the end, a total of 3902 research papers were retrieved (Figure 1). The results show that eco-compensation research can be traced back to 1998, entered a rapid development stage from 2004 to 2009, and then stabilized. Among these papers, 267 are related to grassland eco-compensation, accounting for 6.84% of the total. They were first seen in 2005, and have remained relatively stable in number since 2009.

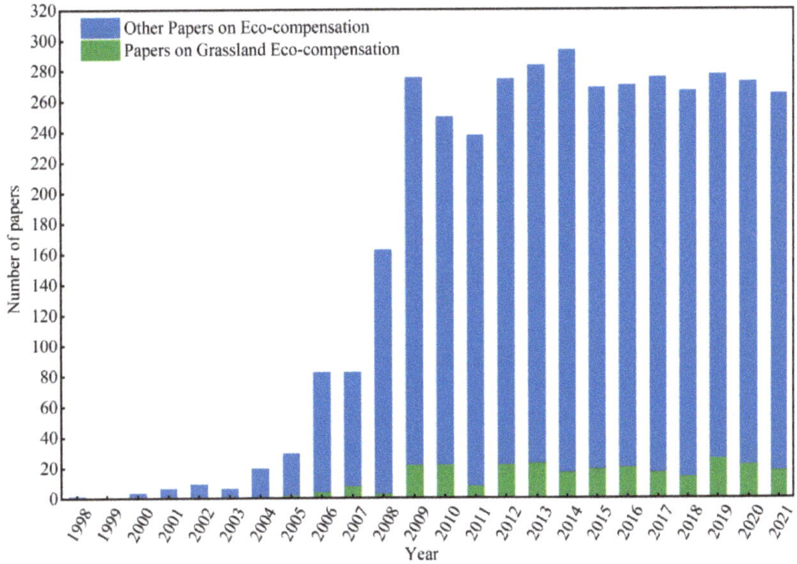

Figure 1. Overview of Eco-Compensation Research.

Therefore, although the Chinese government attaches great importance to grassland eco-compensation, it is still relatively lacking in the eco-compensation research field as a whole, and it is necessary to sort out the existing papers and provide ideas for subsequent research. This study aims to explore the connotations and effects of grassland eco-compensation in China. Section 3 summarizes the research progress of grassland eco-compensation theory in China. Specifically, the concept of eco-compensation is first obtained by comparing with PES, which is used internationally (Section 3.1). Then, combined with the characteristics of grassland eco-compensation (Section 3.2), the connotations of grassland eco-compensation in China are summarized (Section 3.3). Section 4 presents the research progress in grassland eco-compensation practice in China. Specifically, the connotations of eco-compensation are used to interpret the current China grassland eco-compensation policy (Section 4.1), and then, combined with the evaluation research of the policy in four aspects (Section 4.2), the effects of China's grassland eco-compensation are obtained (Section 4.3). Finally, according to the conclusion (Section 5) and the insufficiencies of the current research (Sections 3.3 and 4.3), this study provides three important directions for future work (Section 6).

3. Progress in Theoretical Research on Grassland Eco-Compensation in China

3.1. The Connotation of Ecological Compensation in China

Eco-compensation in China is similar to the concept of PES widely used around the world, but there are still some differences between the two [33,34]. For PES, although scholars have not formed a unified understanding, many studies agree that PES is an effective economic means to ensure the sustainable supply of ecosystem services [35–37]. It links the private interests of landowners with the public benefits of conservation managements [38]. The basic framework of PES is to provide financial incentives for private landowners to implement conservation measures that continue to provide critical ecosystem services (e.g., climate regulation, nutrient cycling, water conservation, etc.) [39].

As for China's eco-compensation, the widely accepted definition holds that "eco-compensation is a public institutional arrangement that uses government and market means to regulate the interests of ecological protection stakeholders. It aims to protect the ecology and promote the harmonious development of humans and nature, and formulate standards according to the value of ecosystem services, ecological protection costs, and development opportunity costs" [13]. From this definition, both eco-compensation and PES are processes that treat ecosystem services as commodities and are traded among stakeholders. Additionally, the goals of both are to achieve the sustainable supply of ecosystem services and protect the interests of suppliers [37]. However, the two are not exactly equivalent. First, eco-compensation has a wider application range than PES. PES is mainly about rewarding conservation behaviors of ecosystem services, but eco-compensation also includes charging for behaviors that damage ecosystem services [40]. Secondly, PES emphasizes voluntariness and belongs to a typical incentive mechanism. However, eco-compensation in China is often a strict public system arrangement. It is a large-scale eco-economic project that is led, managed, and guided by governments [41]. We take the grassland eco-compensation involved in this study as an example. Among the two core measures, GLB is to reward the herder households who reach the reasonable grazing limits, which is similar to PES. However, for GP, the regulation stipulates that grazing is not allowed in the designated area, which has a certain degree of compulsion. This distinction is based on China's current national conditions and the special historical period of ecological civilization construction. A certain degree of compulsion, on the one hand, can help many residents in ecologically fragile areas with relatively low levels of education realize the importance of ecological protection, and on the other hand can support them to increase their income. In conclusion, although there are many similarities, China's eco-compensation has wider application scope and stricter policy measures than PES.

3.2. Research Progress of Grassland Eco-Compensation Theory in China

Referring to relevant definitions and research frameworks, the current theoretical research on eco-compensation mainly includes five aspects: compensation goals, compensation main bodies, compensation standards, compensation methods, and compensation systems [34,42–44]. Therefore, this study will sort the research based on these five aspects.

3.2.1. Research on Compensation Goals

In recent decades, China's grasslands have been severely degraded [45]. Human activities, represented by overgrazing, are believed to be the dominant factor behind this phenomenon [46,47]. Therefore, the restoration of grassland ecosystem services through management has naturally become the basic goal of eco-compensation [48]. However, herder households are one of the core stakeholders in grassland eco-compensation. To solve the problem of negative externalities of grassland ecosystem services, it is necessary to coordinate the relationship between herder households and grasslands. Therefore, some scholars believe that improving the livelihood of herder households should also be one of the goals of grassland eco-compensation [49]. On the other hand, according to the instructions of the central government and related documents, grassland eco-compensation should consider ecological protection and income growth at the same time [30]. Compensation funds should be directed towards poverty-stricken areas and populations [50]. On the whole, the goals of grassland eco-compensation should include the restoration of ecological functions and increasing the income of herder households.

3.2.2. Research on Compensation Main Bodies

The main bodies of eco-compensation include the suppliers and buyers of ecosystem services. Different land use patterns affect the provision of ecosystem services. Therefore, it is generally believed that the potential provider of ecosystem services is the owner/user of the land [51]. For grassland eco-compensation, the contributors to grassland protection, the losers relating to grassland ecosystem destruction, and the builders of grassland ecological industries are mainly herder households [48]. Therefore, herder households are undoubtedly the providers of grassland ecosystem services. As for the buyers of ecosystem services, these can be either a clear ecosystem service user or a third-party (such as the government) ecosystem service user [51]. However, studies have shown that government compensation is more effective than user compensation as the scope of compensation expands [52,53]. Therefore, for grassland eco-compensation in China, having the government as the buyer of ecosystem services is a better choice [54]. In summary, the main bodies of grassland eco-compensation in China are the government and herder households.

3.2.3. Research on Compensation Standards

Compensation standards are one of the core contents of eco-compensation. China is rich in grassland resources, and various types face different degrees of degradation. Scholars have chosen different pilot pastoral areas to study grassland eco-compensation standards (Table 1).

Table 1. Summary table of studies on grassland eco-compensation standards in China.

Management Measure	Scholar	Research Area	Calculation Method	Theoretical Standard (Yuan/ha)
Grazing prohibition	Qi et al. [55]	Xilin Gol, Inner Mongolia	Willingness to be paid	270
	Hu et al. [56]	3 counties including Siziwang Banner, Inner Mongolia	Opportunity cost	123.15
	Wei and Zong [57]	Maqu County, Gansu	minimum data	1751.7
Grass-livestock balance	Yang et al. [58]	Xilin Gol, Inner Mongolia	Willingness to be paid	85.95
	Gong et al. [59]	Inner Mongolia Autonomous Region	Opportunity cost	713.25
	Qi et al. [55]	Xilin Gol, Inner Mongolia	Willingness to be paid	135
	Wei and Qi [60]	Maqu County, Gansu	Opportunity cost	330
	Wei and Qi [61]	Maqu County, Gansu	Willingness to be paid	189.15
	Zhou et al. [62]	5 counties including Shanshan, Xinjiang	Willingness to be paid	130.5

Among the calculation methods, the willingness to be paid method allows the households to personally assess the impacts of the eco-compensation measures on the original production methods, and then gives the expected compensation standards. The opportunity cost method involves calculating the economic losses of the households due to the response to eco-compensation measures through the researchers' surveys, which are mainly the livestock income that should have been generated by livestock reduction. The minimum data method first specifies the target ecosystem service values to be restored, and then calculates the required compensation standards. There are pros and cons to each of these methods, but generally, algorithms that consider the value of ecosystem services will receive higher theoretical compensation standards. The theoretical compensation standards obtained through the willingness to pay method or the opportunity cost method are relatively low. Accordingly, the task force on eco-compensation mechanisms and policies suggests that the basic criterion for determining the eco-compensation standard should be lower than the ecosystem service values and higher than or equal to the opportunity costs and restoration costs [5].

3.2.4. Research on Compensation Methods

Compensation methods determine the efficiency of compensation, which can be divided into financial compensation and industrial compensation [19,63]. Financial compensation is a way to directly provide subsidies for herder households and encourage them to transform traditional livestock husbandry. However, such compensation's effect is always inefficient, because it is difficult for herder households to spontaneously change the current production status without the support of training, supervision, equipment, etc. [64,65]. Industrial compensation is currently in the exploratory stage, and its purpose is to help herder households get rid of their dependence on traditional livestock husbandry. The government can help them to upgrade their industries or obtain alternative incomes by providing policies, technologies, and equipment [44]. This is conducive to fundamentally solving the problem, but it is obviously more difficult than financial compensation.

However, no matter what compensation method is ultimately chosen, financial compensation in the short term is inevitable. There are two ways to allocate funds for grassland eco-compensation, namely the quota based on the grassland scale of the households or the quota based on household population [66]. Both ways have their pros and cons. A quota based on the grassland scale can better reflect the ecosystem service values and opportunity cost provided by herder households, but it is easy to widen the income gap between households [31]. A quota based on the household population will be relatively balanced, but it ignores opportunity costs [66]. At present, a fund allocation method that is both balanced and can reflect the opportunity costs of households is still being explored [67].

3.2.5. Research on Compensation Systems

A complete compensation system is a necessary condition to ensure the progress of grassland eco-compensation [68]. After sorting out previous studies, China's grassland eco-compensation system should include the following important contents. The first is the legal system. With the continuous development of compensation, existing laws and regulations should also keep pace with the times to form a legal system that can cover the entire process of grassland eco-compensation [69]. The second is a strong supervision system. Without strict supervision measures, the implementation efficiency of grassland eco- compensation is likely to be low, and it is difficult to achieve the desired effects [70]. The third is strong scientific and technological support. The transformation of production methods has greatly increased the requirements for production technology. Herder households need more production training to adapt to such transformation [71]. The fourth is a long-term mechanism. Fundamentally curbing overload cannot be achieved in a short period of time, and requires a long-term compensation mechanism to continue to advance [71].

3.3. Summary and Analysis of Theoretical Research

Combined with relevant conclusions, we can conclude that the current grassland eco-compensation in China is a large-scale ecological-economic institutional arrangement with the following five characteristics: (1) the goals are to maintain the grassland ecosystem services and increase the income of herder households; (2) the main bodies are governments and herder households; (3) the main method is financial transfer payments; (4) the compensation standards are based on the opportunity costs of the herder households' responses as the lower limits and the grassland ecosystem service values as the upper limits; and (5) it is a comprehensive compensation system that requires legal, regulatory, technological support and long-term mechanisms.

Although a lot of progress has been made in the research on grassland eco-compensation in China, there are still some aspects to be improved. According to the framework of the compensation, the government provides financial support to herder households, encourages them to transform livestock husbandry, and then reduces the grazing intensity of natural grasslands to restore ecological functions. We can divide such a mechanism into two processes, namely the compensation process of the government (government–household) and the response process of herder households (household–grassland).

The current research on the government compensation process has covered the five basic aspects. Among them, the determination of compensation goals, the identification of the main compensation bodies and the needs of compensation systems have almost reached a consensus between scholars. However, the research on compensation standards and compensation methods is still weak. As one of the core issues of the eco-compensation mechanism, the current research on compensation standards only provides a reasonable compensation range (more than or equal to the opportunity costs and less than the ecosystem service values) through different calculation methods. However, such a large range is not enough to be a scientific basis for guiding practice. The final scientific standards should be precise. In addition, due to the large area of pastoral areas in China, there are many differences between nature and social status. It is necessary for governments at all levels to formulate differentiated compensation standards according to regional characteristics. However, the current research is not enough to meet such a requirement. After the compensation standards are clarified, multiple compensation methods are also essential. If a single government compensation method is maintained for a long time, it will inevitably bring a huge burden to the government's finances. Finding how to give full play to the advantages of the market is an important basis for realizing the long-term mechanisms of grassland eco-compensation, but this cannot be supported by the current research on compensation methods.

As for the response process of herder households, there are very few related studies. Some scholars found that when many herder households in pilot areas received compensation and did not reduce livestock as required, they often attributed the reason to the lack of a supervision system [64,65]. We agree that strong supervision will certainly help herder households to reduce livestock and improve compensation efficiency. However, achieving full supervision in the 255 million ha pilot area would mean significant costs. In addition, current research cannot guarantee whether the long-term strict supervision will bring about other social and economic problems. Therefore, it is necessary to explore a rational method for motivating households to respond to compensation.

4. Research Progress on Grassland Eco-Compensation Practice

4.1. Status of Compensation Policy

China's grassland eco-compensation is promoted by the policy of subsidies and rewards for grassland ecological protection (PSRGEP), officially launched in 2011 (Figure 2). Currently, about 255 million hectares of natural grassland in 13 pilot provinces/autonomous regions are divided into GP areas and GLB areas. The unified standard given by the central government is a GP subsidy of 112.5 yuan/ha and a GLP reward of 37.5 yuan/ha. However, in the process of implementation, the government also encourages each pilot area to make appropriate adjustments from these unified standards according to local circumstances. In addition, in order to promote the transformation and development of livestock husbandry, the government also provides subsidies for planting grass, optimizing species, and updating production materials for households in the pilot areas.

4.2. Resrearch of Compensation Effects

China's PSRGEP has been implemented for more than 11 years. Scholars chose different perspectives to evaluate the compensation effects, mainly focusing on the following four aspects: changes in household income, changes in grassland ecology, effect of households' reduction in livestock, and degree of households' satisfaction. We have reviewed typical effect evaluation studies. In order to visualize the results, the evaluation result grades were set according to the conclusions of the studies (Table 2).

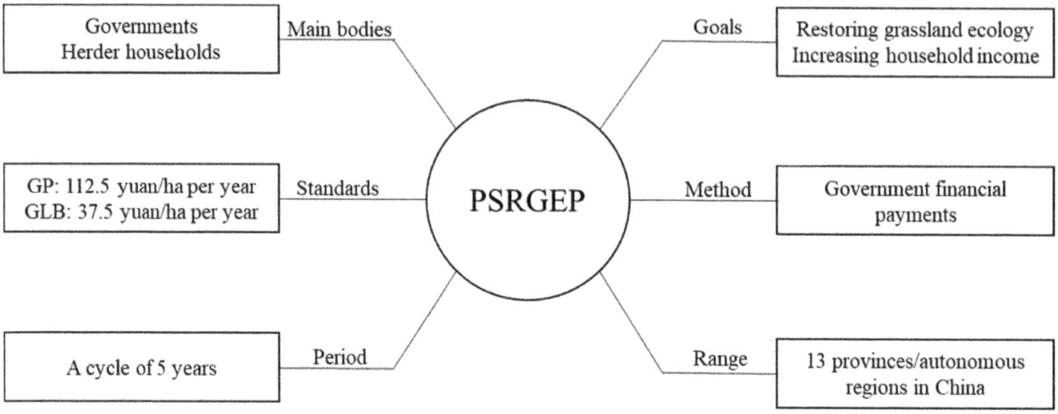

Figure 2. Overview of PSRGEP.

Table 2. Grading criteria for compensation effects.

Indicator	Good	Moderate	Poor
Income change	Average income increases of more than 10%	Average income changes less than 10%	Average income decreases of more than 10%
Ecological improvement	Evaluation results are positive and significant	Evaluation results are not significant	Evaluation results are negatively significant
Livestock reduction	Average reduction in livestock by more than 10%	Average stocking rates vary by less than 10%	Average increase in livestock by more than 10%
Satisfaction level	Satisfied with more than 60% of households	Satisfied households between 40–60%	Satisfied with less than 40% of households

Since 2011, the Chinese government has invested more than 170 billion yuan in the grassland eco-compensation mechanism, benefiting more than 12 million farmer and herder households, and rehabilitating 255 million hectares of grasslands [72]. Overall, China's grassland eco-compensation has achieved remarkable results [73] (Table 3). First, through the method of financial compensation, grassland eco-compensation has directly and effectively improved the income level of herder households [74,75]. Additionally, there is a positive correlation between the extent of the improvement and the amount of compensation funds [65]. As for the ecological changes to grassland, in the past 10 years, scholars have shown the overall improvement of grassland ecology in China through various indicators such as the NDVI index, grassland comprehensive vegetation coverage, theoretical stocking capacity calculated by remote sensing technology and grassland monitoring data [76,77]. However, the effect is relatively slight [78]. Therefore, the grassland eco-compensation mechanism has positively achieved the two main compensation goals overall. Of course, there are also some areas at the micro level that have negative income or ecological effects [77,79].

As for the livestock reduction and satisfaction level of herder households, although they do not directly feedback the goals of compensation, they also affect the efficiency and effects of eco-compensation. Many households in the pilot areas have a long-term dependence on traditional livestock husbandry. Therefore, compared with the change in income, the current effect of reducing livestock is not obvious [64,65,80]. Finally, the satisfaction degree of the households to grassland eco-compensation varies in different regions. The influencing factors mainly include family income, policy cognition, family grassland scale and so on [81,82].

Table 3. Results of PSRGEP effect evaluations.

Indicator	Scholars	Research Area	Grade
Income change	Yin [74]	Urat Back Banner, Inner Mongolia	Good
	Zhang et al. [83]	Xinjiang Autonomous Region	Good
	Liu and Zhang [79]	4 cities including Ordos, Inner Mongolia	Good
Ecological improvement	Liu et al. [76]	54 counties, Inner Mongolia	Good
	Hou et al. [78]	All pilots in China	Good
	Liu [77]	73 counties, Inner Mongolia	Moderate
Livestock reduction	Gao et al. [65]	70 villages, Inner Mongolia	Moderate
	Yin et al. [32]	15 counties including New Barag Left Banner, Inner Mongolia	Poor
	Zhang et al. [84]	8 counties including Siziwang Banner, Inner Mongolia	Good
Satisfaction level	Li et al. [85]	Siziwang Banner, Inner Mongolia	Moderate
	Yang et al. [82]	6 counties including Tianzhu, Gansu	Good
	Hu et al. [86]	3 counties including Siziwang Banner, Inner Mongolia	Moderate

4.3. *Summary and Analysis of Practical Research*

China's PSRGEP is a practice closely integrated with grassland eco-compensation theory. Because of the huge investment, wide coverage and large number of beneficiaries, scholars pay great attention to its effects. Current studies accurately reflect the status of PSRGEP in pilot areas in China. The results show that with the background of PSRGEP, the income of herder households increased significantly and the grassland ecology slightly improved. It is difficult to fundamentally curb the phenomenon of overgrazing, and there are regional differences in the degree of satisfaction of herder households with the policy.

However, the status is not a real effect. Taking grassland ecological improvement as an example, on the one hand, the analysis of the status cannot reflect the efficiency of compensation. The current study suggests that the grassland ecological quality has slightly improved after compensation—for example, the increase in grassland vegetation coverage obtained through the NDVI index measured by remote sensing and the increase in theoretical stocking capacity obtained through grassland biomass monitoring [76–78]. However, no study can draw definite conclusions: is this improvement enough to match the financial investment of more than 170 billion yuan? Is the current level of compensation the most appropriate? Regrettably, current research cannot link compensation measures and compensation effects well. On the other hand, the analysis of status cannot highlight the role of compensation. Scholars always only use eco-compensation as a time boundary for comparison when evaluating grassland ecological improvement. Such results are caused by both natural and human factors. For example, the impact of climate change on ecosystems cannot be ignored. So how do we strip away other factors and focus on the real effect of eco-compensation? There are few relevant studies. Likewise, grassland eco-compensation significantly increases the income of herder households. However, it is worth noting that what the government provides for GP and GLB is compensation, not donations. The income increases of herder households due to grassland eco-compensation should be reflected in sustainable industrial transformation and upgrading. However, the existing studies rarely integrate the change in income with the actual production. If the households just received the compensation funds and did not respond to the compensation, such an increase in income would not be sustainable and cannot reflect the real effects of eco-compensation. Therefore, it is difficult to reflect the role of eco-compensation in the current effect evaluations.

5. Conclusion and Discussion

In summary, this study sorts the relevant theoretical and practical research in recent years, explains the connotations of grassland eco-compensation in China from five aspects, and then evaluates the effects from four perspectives. The results show that the current grassland eco-compensation in China is a large-scale ecological-economic institutional arrangement with the following five characteristics: (1) The goals are to maintain the grassland ecosystem services and increase the income of herder households; (2) the main bodies are governments and herder households; (3) the main method is financial transfer payments; (4) the compensation standards are based on the opportunity costs of the herder households' responses as the lower limits and the grassland ecosystem service values as the upper limits; and (5) it is a comprehensive compensation system that requires legal, regulatory, technological support and long-term mechanisms. Since 2011, driven by the PSRGEP, the income levels of herder households in each pilot area have generally increased, and the overall ecology of grassland has been slightly improved. However, there are still some areas where overload is common. Additionally, there are regional differences in the satisfaction degree of herder households, which is mainly affected by factors such as family income, compensation cognition and family grassland scale. In general, the shortcomings of current theoretical research are mainly reflected in the low precision of scientific compensation standards, the lack of a basis for differentiated standards, and the single compensation method. The shortcoming of practical research is that most effect evaluations cannot reflect the role of eco-compensation in it.

In December 2021, China officially started the third round of its grassland eco-compensation policy. At the important beginning stage of the third round, this study can provide a reference for policymakers to comprehensively review China's grassland eco-compensation mechanism in the first two rounds.

As for the academic contribution of this research, it mainly includes the following two aspects: the first is providing supplements for the field of eco-compensation. Grassland is the largest terrestrial ecosystem, but it is relatively lacking in the field of eco-compensation research. This study sorts the theory and practice of grassland eco-compensation in China, and points out the insufficiency of the current research, so as to provide directions for the improvement of grassland eco-compensation research system. The second is setting out a model for grassland ecological protection. Grassland degradation is occurring in many regions of the world. The exploration of grassland eco-compensation in China can provide a reference for global grassland ecological protection.

6. Future work

Combined with the current research status and the problems in the compensation process, we suggest that future research on grassland eco-compensation theory and practice could focus on the following aspects:

1. Research on the response mechanism of herder households

The response of herder households is the core link of grassland eco-compensation in China, which determines the efficiency and effect of compensation to a large extent. Whether it is the problem of overgrazing being difficult to solve, or the compensation effect being difficult to describe, the key reason is that the response mechanism of herder households is still unclear. We believe that this mechanism can be divided into three steps. First, how do the households respond? Current research is almost exclusively concerned with livestock reduction. However, in fact, herder households have various forms of response compensation, such as the optimization of livestock structure, the optimization of livestock breeds, land transfer, and the development of grass growing industry [30]. These forms are also advocated by PSRGEP, which are beneficial to grassland ecological protection and worth attention from scholars. Second: what factors influence the response of the herder households? The response of herder households to eco-compensation is a complex process that may involve many theories such as livelihood strategies and planned behaviors. Taking livelihood strategies as an example, according to the sustainable livelihood framework, the factors affecting the response strategies (such as livestock reduction) of farmers may not be limited to public policies, but may also include other direct and indirect factors such as the vulnerable environment [87], livelihood assets [88], and other institutional changes [89]. Finding out the influencing factors or processes affecting livestock reduction by herder households may help to improve the supporting policies and achieve a more ideal compensation effect. Finally, the livelihood and ecological effects of the households' response are important. Compensation has changed the production and living conditions of the herder households. To study the resulting livelihood and ecological effects is to evaluate the compensation effect from the perspective of compensation mechanism, which obviously highlights the role of eco-compensation more than the current evaluation results [90]. Macroscopically, pastoral areas not included in the compensation pilots can be used as the reference group for adjacent compensation pilots. Microscopically, the herder households who did not respond to the compensation in the same pilot area can be the reference group for the households who responded to the compensation. Such a series of studies will help us to better improve the compensation theory and examine the effects of compensation. In addition, the existing ecological effect research only focuses on the grassland resources itself. However, other ecological effects brought about by compensation management are also worthy of attention, such as the impacts on the soil environment and the impacts on carbon emissions from livestock husbandry.

2. Improvement of the compensation measures

Compensation standards and compensation methods are the core contents of eco-compensation measures [91,92]. The current calculation methods of compensation standards include the willingness to pay method, the opportunity cost method, and the ecological service accounting method, which correspond to the relevant theories of psychology, economics and ecology. The results obtained by a single theory are very different and have obvious limitations. Therefore, we suggest that future research should try to combine multidisciplinary theories to form a unified comprehensive accounting system. The system should consider the existing mature theories as well as the government's financial ability to pay, the livelihood status of the herder households, the ecological status, and other restrictive factors. Performing this work not only helps to improve the scientific quality of theoretical standards, but also enhances the comparability between regions and provides a basis for differential compensation. For the compensation methods, it is difficult to form a stable long-term mechanism with a single government compensation. Future research could focus on market compensation mechanisms, which can include the following three aspects: first, research on the confirmation and registration of grassland resources, specifically how to establish clear ownership of grassland and improve the property rights system of grassland assets to provide conditions for the establishment of the market mechanism; second, research on market-based financing methods, exploring the feasibility green stocks and insurance products based on grassland ecological functions; and third, research on industries with grassland characteristics, exploring the grassland ecological industry chain financial model and the livestock husbandry franchise management system.

3. Expand the scope of research and learn from successful experiences

At present, there are 13 pilot provinces for grassland eco-compensation in China, but the research area selected by scholars are mainly concentrated in Inner Mongolia, Gansu and Ningxia, relatively few in Xinjiang and Tibet, and almost none in other provinces. Here, we suggest that the scope of research should be expanded. On the one hand, different regions may expose different problems in the compensation process, and exploring more pilots can be an easy way to discover details that have been overlooked in theoretical research. On the other hand, when the research scope is expanded to a certain level, it helps to enhance the comparability between regions with similar background conditions. Researchers can select successful cases to provide a model for guiding compensation practices in other regions. In addition, China's grassland eco-compensation started in 2011 and is still in the stage of exploration and development. However, eco-compensation in other areas can be traced back to the 1990s or even earlier. Future research should try to combine eco-compensation experience in other fields with grassland research. For example, forest ecological compensation has practical experience in several major projects [93–95]. Watershed eco-compensation has cooperation experience between different regions [96–98]. Marine eco-compensation has legislative experience [99,100]. Farmland ecological compensation has good effect evaluation experience [101–103]. Furthermore, there is also international experience in specialized payments for various ecosystem services, such as biodiversity conservation [104,105], water provision [106], carbon dioxide fixation [107]. Absorbing these advanced experiences will accelerate the improvement of grassland eco-compensation in China.

Author Contributions: Conceptualization, M.L. and Z.L.; methodology, M.L.; software, B.S.; validation, B.S.; formal analysis, Z.L.; investigation, Z.L.; resources, Z.L.; data curation, Z.L.; writing—original draft preparation, Z.L.; writing—review and editing, M.L.; visualization, B.S.; supervision, M.L.; project administration, Z.L.; funding acquisition, M.L. All authors have read and agreed to the published version of the manuscript.

Funding: This research and APC were jointly supported by the National Natural Science Foundation of China (grant 42171279) and the Mobility Programme DFG-NSFC (grant M-0342).

Institutional Review Board Statement: Not applicable.

Informed Consent Statement: Not applicable.

Data Availability Statement: Most of the data are available in all tables and figures of the manuscripts. If scholars need more specific data, they can send an email to the corresponding author or the first author.

Conflicts of Interest: The authors declare no conflict of interest.

References

1. Costanza, R.; d'Arge, R.; de Groot, R.; Farber, S.; Grasso, M.; Hannon, B.; Limburg, K.; Naeem, S.; O'Neill, R.V.; Paruelo, J.; et al. The value of the world's ecosystem services and natural capital. *Nature* **1997**, *387*, 253–260. [CrossRef]
2. Buchanan, J.M.; Stubblebine, W.C. Externality. *Economica* **1962**, *29*, 371–384. [CrossRef]
3. Kosoy, N.; Martinez-Tuna, M.; Muradian, R.; Martinez-Alier, J. Payments for environmental services in watersheds: Insights from a comparative study of three cases in Central America. *Ecol. Econ.* **2007**, *61*, 446–455. [CrossRef]
4. Shang, W.; Gong, Y.; Wang, Z.; Stewardson, M.J. Eco-Compensation in China: Theory, practices and suggestions for the future. *J. Environ. Manag.* **2018**, *210*, 162–170. [CrossRef] [PubMed]
5. TFEMP. *Eco-Compensation Mechanisms and Policies in China*; Science Press: Beijing, China, 2008.
6. Ventrubová, K.; Dvořák, P. Legal framework for payments for forest ecosystem services in the Czech Republic. *J. For. Sci.* **2012**, *58*, 131–136. [CrossRef]
7. Phan, T.-H.D.; Brouwer, R.; Hoang, L.P.; Davidson, M.D. A comparative study of transaction costs of payments for forest ecosystem services in Vietnam. *For. Policy Econ.* **2017**, *80*, 141–149. [CrossRef]
8. To, P.; Dressler, W. Rethinking 'Success': The politics of payment for forest ecosystem services in Vietnam. *Land Use Policy* **2019**, *81*, 582–593. [CrossRef]
9. Cuperus, R.; Bakermans, M.M.G.J.; Haes, H.A.U.D.; Canters, K.J. Ecological Compensation in Dutch Highway Planning. *Environ. Manag.* **2001**, *27*, 75–89. [CrossRef]
10. Duelli, P.; Obrist, M.K. Regional biodiversity in an agricultural landscape: The contribution of seminatural habitat islands. *Basic Appl. Ecol.* **2003**, *4*, 129–138. [CrossRef]
11. Knop, E.V.A.; Kleijn, D.; Herzog, F.; Schmid, B. Effectiveness of the Swiss agri-environment scheme in promoting biodiversity. *J. Appl. Ecol.* **2005**, *43*, 120–127. [CrossRef]
12. Johst, K.; Drechsler, M.; Wätzold, F. An ecological-economic modelling procedure to design compensation payments for the efficient spatio-temporal allocation of species protection measures. *Ecol. Econ.* **2002**, *41*, 37–49. [CrossRef]
13. Li, W.; Liu, M. Several strategic thoughts on China's eco-compensation mechanism. *Resour. Sci.* **2010**, *32*, 791–796.
14. Sheng, W.; Zhen, L.; Xie, G.; Xiao, Y. Determining eco-compensation standards based on the ecosystem services value of the mountain ecological forests in Beijing, China. *Ecosyst. Serv.* **2017**, *26*, 422–430. [CrossRef]
15. Xiong, Y.; Wang, K. Eco-Compensation effects of the wetland recovery in Dongting Lake area. *J. Geogr. Sci.* **2010**, *20*, 389–405. [CrossRef]
16. Guan, X.; Liu, W.; Chen, M. Study on the ecological compensation standard for river basin water environment based on total pollutants control. *Ecol. Indic.* **2016**, *69*, 446–452. [CrossRef]
17. Yang, Y.; Yao, C.; Xu, D. Ecological compensation standards of national scenic spots in western China: A case study of Taibai Mountain. *Tour. Manag.* **2020**, *76*, 103950. [CrossRef]
18. Rao, H.; Lin, C.; Kong, H.; Jin, D.; Peng, B. Ecological damage compensation for coastal sea area uses. *Ecol. Indic.* **2014**, *38*, 149–158. [CrossRef]

19. Liu, M.; Liu, W.; Yang, L.; Jiao, W.; He, S.; Min, Q. A dynamic eco-compensation standard for Hani Rice Terraces System in southwest China. *Ecosyst. Serv.* **2019**, *36*, 100897. [CrossRef]
20. Yang, X.; Zhou, X.; Cao, S.; Zhang, A. Preferences in Farmland Eco-Compensation Methods: A Case Study of Wuhan, China. *Land* **2021**, *10*, 1159. [CrossRef]
21. O'Mara, F.P. The role of grasslands in food security and climate change. *Ann. Bot.* **2012**, *110*, 1263–1270. [CrossRef]
22. Ni, J. Forage Yield-Based Carbon Storage in Grasslands of China. *Clim. Change* **2004**, *67*, 237–246. [CrossRef]
23. Ni, J. Carbon storage in grassland of China. *J. Arid Environ.* **2002**, *50*, 205–218. [CrossRef]
24. Bardgett, R.D.; Bullock, J.M.; Lavorel, S.; Manning, P.; Schaffner, U.; Ostle, N.; Chomel, M.; Durigan, G.; Fry, E.L.; Johnson, D.; et al. Combatting global grassland degradation. *Nat. Rev. Earth Environ.* **2021**, *2*, 720–735. [CrossRef]
25. Cao, J.; Adamowski, J.F.; Deo, R.C.; Xu, X.; Gong, Y.; Feng, Q. Grassland Degradation on the Qinghai-Tibetan Plateau: Reevaluation of Causative Factors. *Rangel. Ecol. Manag.* **2019**, *72*, 988–995. [CrossRef]
26. Quinlan, T. Grassland degradation and livestock rearing in Lesotho. *J. S. Afr. Stud.* **2007**, *21*, 491–507. [CrossRef]
27. Abdalla, K.; Mutema, M.; Chivenge, P.; Everson, C.; Chaplot, V. Grassland degradation significantly enhances soil CO_2 emission. *Catena* **2018**, *167*, 284–292. [CrossRef]
28. Xu, X.; Hu, G.; Liu, X.; Lu, S.; Li, S.; Zhao, N. Impacts of nitrogen enrichment on vegetation growth dynamics are regulated by grassland degradation status. *Land Degrad. Dev.* **2021**, *32*, 4056–4066. [CrossRef]
29. Yang, S.; Hao, Q.; Liu, H.; Zhang, X.; Yu, C.; Yang, X.; Xia, S.; Yang, W.; Li, J.; Song, Z. Impact of grassland degradation on the distribution and bioavailability of soil silicon: Implications for the Si cycle in grasslands. *Sci. Total Environ.* **2019**, *657*, 811–818. [CrossRef]
30. The Executive Meeting of the State Council Decided to Establish a Subsidy and Reward Mechanism for Grassland Ecological Protection. Available online: http://www.mof.gov.cn/zhengwuxinxi/caijingshidian/zyzfmhwz/201010/t20101013_342428.htm (accessed on 10 March 2022).
31. Li, Z.; Rao, D.; Liu, M. The Impact of China's Grassland Ecological Compensation Policy on the Income Gap between Herder Households? A Case Study from a Typical Pilot Area. *Land* **2021**, *10*, 1405. [CrossRef]
32. Yin, Y.; Hou, Y.; Langford, C.; Bai, H.; Hou, X. Herder stocking rate and household income under the Grassland Ecological Protection Award Policy in northern China. *Land Use Policy* **2019**, *82*, 120–129. [CrossRef]
33. Chen, Y.; Dou, S.; Xu, D. The effectiveness of eco-compensation in environmental protection—A hybrid of the government and market. *J. Environ. Manag.* **2021**, *280*, 111840. [CrossRef] [PubMed]
34. Liu, D.; Hu, Z.; Jin, L. Review on analytical framework of eco-compensation. *Acta Ecol. Sin.* **2018**, *38*, 380–392. [CrossRef]
35. Garbach, K.; Lubell, M.; DeClerck, F.A.J. Payment for Ecosystem Services: The roles of positive incentives and information sharing in stimulating adoption of silvopastoral conservation practices. *Agric. Ecosyst. Environ.* **2012**, *156*, 27–36. [CrossRef]
36. Jack, B.K.; Kousky, C.; Sims, K.R.E. Designing payments for ecosystem services: Lessons from previous experience with incentive-based mechanisms. *Proc. Natl. Acad. Sci. USA* **2008**, *105*, 9465–9470. [CrossRef]
37. Yu, H.; Xie, W.; Yang, L.; Du, A.; Almeida, C.M.V.B.; Wang, Y. From payments for ecosystem services to eco-compensation: Conceptual change or paradigm shift? *Sci. Total Environ.* **2020**, *700*, 134627. [CrossRef]
38. Ferraro, P.J.; Kiss, A. Direct Payments to Conserve Biodiversity. *Science* **2002**, *298*, 1718–1719. [CrossRef]
39. Millennium Ecosystem Assessment (MEA). *Ecosystems and Human Well-Being*; Island Press: Washington, DC, USA, 2005.
40. Mao, X.; Zhong, Y.; Zhang, S. Conception, theory and mechanism of eco-compensation. *China Popul. Resour. Environ.* **2022**, *12*, 38–41.
41. Teets, J.C.; Gao, M.; Wysocki, M.; Ye, W. The impact of environmental federalism: An analysis of watershed eco-compensation policy design in China. *Environ. Policy Gov.* **2021**, *31*, 580–591. [CrossRef]
42. Cheng, X.; Fang, L.; Mu, L.; Li, J.; Wang, H. Watershed Eco-Compensation Mechanism in China: Policies, Practices and Recommendations. *Water* **2022**, *14*, 777. [CrossRef]
43. Liu, M.; Yang, L.; Min, Q.; Sang, W. Theoretical framework for eco-compensation to national parks in China. *Glob. Ecol. Conserv.* **2020**, *24*, e01296. [CrossRef]
44. Liu, M.; Rao, D.; Yang, L.; Min, Q. Subsidy, training or material supply? The impact path of eco-compensation method on farmers' livelihood assets. *J. Environ. Manag.* **2021**, *287*, 112339. [CrossRef]
45. Ruan, H.; Wu, X.; Wang, S.; Yang, J.; Zhu, H.; Guo, Q.; Wang, L.; Wang, D. The responses of different insect guilds to grassland degradation in northeastern China. *Ecol. Indic.* **2021**, *133*, 108369. [CrossRef]
46. Zhou, H.; Zhao, X.; Tang, Y.; Gu, S.; Zhou, L. Alpine grassland degradation and its control in the source region of the Yangtze and Yellow Rivers, China. *Grassl. Sci.* **2005**, *51*, 191–203. [CrossRef]
47. Li, S.; Verburg, P.H.; Lv, S.; Wu, J.; Li, X. Spatial analysis of the driving factors of grassland degradation under conditions of climate change and intensive use in Inner Mongolia, China. *Reg. Environ. Chang.* **2011**, *12*, 461–474. [CrossRef]
48. Yang, Q.; Nan, Z.; Chen, Q. Research progress of grassland ecological compensation in China. *Acta Ecol. Sin.* **2020**, *40*, 2489–2495. [CrossRef]
49. Jin, L.; Zhentong, H. Grassland ecological compensation policy and chooses of the herdsmen. *Reform* **2014**, *11*, 100–107.
50. The General Office of the State Council issued the "Opinions on Improving the Ecological Protection Compensation Mechanism". Available online: http://www.gov.cn/xinwen/2016-05/13/content_5073164.htm (accessed on 3 March 2022).

51. Engel, S.; Pagiola, S.; Wunder, S. Designing payments for environmental services in theory and practice: An overview of the issues. *Ecol. Econ.* **2008**, *65*, 663–674. [CrossRef]
52. Wunder, S.; Engel, S.; Pagiola, S. Taking stock: A comparative analysis of payments for environmental services programs in developed and developing countries. *Ecol. Econ.* **2008**, *65*, 834–852. [CrossRef]
53. Schomers, S.; Matzdorf, B. Payments for ecosystem services: A review and comparison of developing and industrialized countries. *Ecosyst. Serv.* **2013**, *6*, 16–30. [CrossRef]
54. Pan, J. Subject and relationship of rights and obligations' connotataion for grassland eco-compensation. *J. HIT (Soc. Sci. Ed.)* **2015**, *17*, 37–44. [CrossRef]
55. Qi, X.; Gao, B.; Wang, H.; Zhou, J.; Qiao, G. The study on the compensation and award standards for forage-livestock balance and grazing prohibition based on herder's perspective of grassland ecological protection subsidies and incentives policies—Take Xilin Gol League as an example. *J. Arid Land Resour. Environ.* **2016**, *30*, 30–35. [CrossRef]
56. Hu, Z.; Liu, D.; Kong, D.; Jin, L. Rate calculation of "subsidies for grazing prohibition" in grassland eco-compensation based on opportunity cost method. *J. Arid Land Resour. Environ.* **2017**, *31*, 63–68. [CrossRef]
57. Wei, H.; Zong, X. Ecological compensation standard for graze-prohibited grassland: Application of the minimum data method in Maqu County. *J. Nat. Resour.* **2016**, *31*, 28–38. [CrossRef]
58. Yang, G.; Min, Q.; Li, W.; Liu, L.; Rong, J.; Wu, X. Herdsmen's willingness to accept (WTA) compensation for implement of prohibiting-graze policy in Xinlinguole steppe. *Ecol. Environ.* **2006**, *15*, 747–751. [CrossRef]
59. Gong, F.; Chang, Q.; Wang, F.; Liu, X. Empiricall study on compensation standard for grassland ecology in Inner Mongolia. *J. Arid Land Resour. Environ.* **2011**, *25*, 151–155. [CrossRef]
60. Wei, H.; Qi, Y. Analysis of grassland eco-compensation standard based on the differentiation of the opportunity losses caused by reducing livestock. *J. China Agric. Univ.* **2017**, *22*, 199–207. [CrossRef]
61. Wei, H.; Qi, Y. The analysisi of herddsmen's willingness to accept the reducing—Livestock policy based on the CVM. *J. Arid Land Resour. Environ.* **2017**, *31*, 45–50. [CrossRef]
62. Zhou, J.; Maimaiti, Z.; Pei, Y.; Zou, L. Analysis of herdsmen's willingness to accept the compensation standard of grassland-livestock balance: Based on a survey of 223 herdsmen in Xinjiang. *J. Arid Land Resour. Environ.* **2019**, *33*, 79–84. [CrossRef]
63. Liu, M.; Bai, Y.; Ma, N.; Rao, D.; Yang, L.; Min, Q. Blood transfusion or hematopoiesis? How to select between the subsidy mode and the long-term mode of eco-compensation. *Environ. Res. Lett.* **2020**, *15*, 094059. [CrossRef]
64. Feng, X.; Liu, M.; Qiu, H. Impact of grassland eco-compensation policy on herders' overgazing behavior. *China Popul. Resour. Environ.* **2019**, *29*, 157–165.
65. Gao, L.; Kinnucan, H.W.; Zhang, Y.; Qiao, G. The effects of a subsidy for grassland protection on livestock numbers, grazing intensity, and herders' income in inner Mongolia. *Land Use Policy* **2016**, *54*, 302–312. [CrossRef]
66. Kong, D.; Hu, Z.; Jin, L. Research on the allocation model for grassland eco-compensation fund: Based on the empirical analysis of 34 Gacha in Inner Mongolia. *J. Arid Land Resour. Environ.* **2016**, *30*, 1–6. [CrossRef]
67. Li, Z.; Liu, M. Research progress in the evaluation of policy of subsidy and reward for grassland ecological portection in China. *Acta Agrestia Sin.* **2021**, *29*, 1125–1135. [CrossRef]
68. Jiang, X.; Eaton, S.; Kostka, G. Not at the table but stuck paying the bill: Perceptions of injustice in China's Xin'anjiang eco-compensation program. *J. Environ. Policy Plan.* **2021**, 1–17. [CrossRef]
69. Li, J. The problems and strategy analysis of grassland ecological compensation system—A case of Gansu Province. *Pratacult. Sci.* **2015**, *32*, 1027–1032. Available online: https://kns.cnki.net/kcms/detail/detail.aspx?dbcode=CJFD&dbname=CJFDLAST2015&filename=CYKX201506025&uniplatform=NZKPT&v=0WwFkMcZ2UOSm3EN75siHpJLDW7Ei-ckeyK9yxzjpXQqf1SITYAg4OQmTKCb2RTM (accessed on 16 May 2022).
70. Hu, Z.; Kong, D.; Jin, L. Grasslamd eco-compensation: Game analysis under weak supervision. *Issues Agric. Econ.* **2016**, *37*, 95–102. [CrossRef]
71. Wang, J.; Wang, Z.; Xu, L.; Ding, Y. Problems and countermeasures in the implementation of grassland ecologic grant-premium mechanism based on investigation of household in Xilinhot. *Chin. J. Grassl.* **2016**, *38*, 1–7. [CrossRef]
72. The Office of the State Forestry and Grassland Administration and the General Office of the Ministry of Agriculture and Rural Affairs Jointly Issued a Notice. Implement the Third Round of Grassland Ecological Subsidy and Reward Policies. Available online: http://www.forestry.gov.cn/main/586/20211208/101325708851570.html (accessed on 11 March 2022).
73. The Grassland Ecological Protection Reward Policy has been Implemented for Ten Years, Benefiting More Than 12 Million Farmers and Herdsmen. Available online: http://www.forestry.gov.cn/main/586/20211206/084610261891386.html (accessed on 11 March 2022).
74. Yin, X. Implementation performance and suggestions of grassland eco0compensation policies: Based on Urat Back Banner, Inner Mongolia. *Ecol. Econ.* **2017**, *33*, 39–45.
75. Zhou, S.; Gao, Y.; Zhao, K. The impact of grassland ecological compensation on rural poor farmers and herdsmen's income. *J. Northwest AF Univ. (Soc. Sci. Ed.)* **2020**, *20*, 138–147. [CrossRef]
76. Liu, M.; Dries, L.; Heijman, W.; Huang, J.; Zhu, X.; Hu, Y.; Chen, H. The Impact of Ecological Construction Programs on Grassland Conservation in Inner Mongolia, China. *Land Degrad. Dev.* **2018**, *29*, 326–336. [CrossRef]
77. Liu, A. Inner Mongolia grassland ecological protection subsidy and reward effect and its problem analysis. *Grassl. Pratacult.* **2014**, *26*, 4–8.

78. Hou, L.; Xia, F.; Chen, Q.; Huang, J.; He, Y.; Rose, N.; Rozelle, S. Grassland ecological compensation policy in China improves grassland quality and increases herders' income. *Nat. Commun* **2021**, *12*, 4683. [CrossRef] [PubMed]
79. Liu, Y.; Zhang, X. Effect of grassland ecological protection subsidy policy on households' income. *J. Arid Land Resour. Environ.* **2019**, *33*, 60–67. [CrossRef]
80. Wang, H.; Gao, B.; Qi, X.; Qiao, G. Empirical analysis on the impact of the grassland ecological protection subsidies and incentives policies on herdsmen's reduced-livestock behavior: Based on the 260 herdsmen households in Inner Mongolia. *Issues Agric. Econ.* **2017**, *38*, 73–80. [CrossRef]
81. Ding, W.; Yang, Z.; Ma, C.; Li, X.; Yin, Y.; Hou, X. Satisfaction level, and factors influencing satisfaction of herdsmen with the grassland ecological protection subsidy incentive policy. *Acta Pratacult. Sin.* **2019**, *28*, 12–22. [CrossRef]
82. Yang, Q.; Nan, Z.; Chen, Q.; Tang, Z. Satisfaction and influencing factor to grassland eco-compensation and reward policies for herders: Empirical study in Qinghai-Tibet Plateau and western desert area of Gansu. *Acta Ecol. Sin.* **2020**, *40*, 1436–1444. [CrossRef]
83. Zhang, X.; Lu, J.; Gu, S.; Wang, L. Assessing the implementation effects of grassland eco-compensation in Xinjiang. *J. Arid Land Resour. Environ.* **2017**, *31*, 39–44. [CrossRef]
84. Zhang, J.; Brown, C.; Qiao, G.; Zhang, B. Effect of Eco-compensation Schemes on Household Income Structures and Herder Satisfaction: Lessons from the Grassland Ecosystem Subsidy and Award Scheme in Inner Mongolia. *Ecol. Econ.* **2019**, *159*, 46–53. [CrossRef]
85. Li, Y.; Wei, T.; Jin, L. Herdspeople attitudes towards grassland eco-compensation policies in Siziwang Banner, Inner Mongolia. *Resour. Sci.* **2014**, *36*, 2442–2450.
86. Hu, Z.; Liu, D.; Jin, L. Grassland eco-compensation: Ecological performance, income effect and policy satisfaction. *China Popul. Resour. Environ.* **2016**, *26*, 165–176.
87. Mubaya, C.P.; Mafongoya, P. Local-Level climate change adaptation decision-making and livelihoods in semi-arid areas in Zimbabwe. *Environ. Dev. Sustain.* **2017**, *19*, 2377–2403. [CrossRef]
88. Ellis, F. Household strategies and rural livelihood diversification. *J. Dev. Stud.* **1998**, *35*, 1–38. [CrossRef]
89. Korah, P.I.; Nunbogu, A.M.; Akanbang, B.A.A. Spatio-Temporal dynamics and livelihoods transformation in Wa, Ghana. *Land Use Policy* **2018**, *77*, 174–185. [CrossRef]
90. Li, Z.; Liu, M. Livelihood Diversification Helps Herder Households on the Mongolian Plateau Reduce Emissions: A Case Study of a Typical Pastoral Area. *Agronomy* **2022**, *12*, 267. [CrossRef]
91. Gao, X.; Shen, J.; He, W.; Sun, F.; Zhang, Z.; Zhang, X.; Zhang, C.; Kong, Y.; An, M.; Yuan, L.; et al. Changes in Ecosystem Services Value and Establishment of Watershed Ecological Compensation Standards. *Int. J. Environ. Res. Public Health* **2019**, *16*, 2951. [CrossRef]
92. Gastineau, P.; Mossay, P.; Taugourdeau, E. Ecological compensation: How much and where? *Ecol. Econ.* **2021**, *190*, 107191. [CrossRef]
93. Ouyang, Z.; Zheng, H.; Xiao, Y.; Polasky, S.; Liu, J.; Xu, W.; Wang, Q.; Zhang, L.; Xiao, Y.; Rao, E.; et al. Improvements in ecosystem services from investments in natural capital. *Science* **2016**, *352*, 1455–1459. [CrossRef]
94. Deng, H.; Zheng, P.; Liu, T.; Liu, X. Forest Ecosystem Services and Eco-Compensation Mechanisms in China. *Environ. Manag.* **2011**, *48*, 1079–1085. [CrossRef]
95. Trac, C.J.; Schmidt, A.H.; Harrell, S.; Hinckley, T.M. Environmental Reviews and Case Studies: Is the Returning Farmland to Forest Program a Success? Three Case Studies from Sichuan. *Environ. Pract.* **2017**, *15*, 350–366. [CrossRef]
96. Wang, Q.; Wang, N.; Wang, H.; Xiu, Y. Study on Influencing Factors and Simulation of Watershed Ecological Compensation Based on Evolutionary Game. *Sustainability* **2022**, *14*, 3374. [CrossRef]
97. Shen, J.; Gao, X.; He, W.; Sun, F.; Zhang, Z.; Kong, Y.; Wan, Z.; Zhang, X.; Li, Z.; Wang, J.; et al. Prospect theory in an evolutionary game: Construction of watershed ecological compensation system in Taihu Lake Basin. *J. Clean. Prod.* **2021**, *291*, 125929. [CrossRef]
98. Gao, X.; Shen, J.; He, W.; Sun, F.; Zhang, Z.; Guo, W.; Zhang, X.; Kong, Y. An evolutionary game analysis of governments' decision-making behaviors and factors influencing watershed ecological compensation in China. *J. Environ. Manag.* **2019**, *251*, 109592. [CrossRef] [PubMed]
99. Wang, Y.; Zou, K. Compensation for Marine Ecological Damage: From 'Tasman Sea' to 'Sanchi'. *Sustainability* **2021**, *13*, 13353. [CrossRef]
100. Qu, Q.; Tsai, S.-B.; Tang, M.; Xu, C.; Dong, W. Marine Ecological Environment Management Based on Ecological Compensation Mechanisms. *Sustainability* **2016**, *8*, 1267. [CrossRef]
101. Zheng, H.; Robinson, B.E.; Liang, Y.-C.; Polasky, S.; Ma, D.-C.; Wang, F.-C.; Ruckelshaus, M.; Ouyang, Z.-Y.; Daily, G.C. Benefits, costs, and livelihood implications of a regional payment for ecosystem service program. *Proc. Natl. Acad. Sci. USA* **2013**, *110*, 16681–16686. [CrossRef]
102. Zhu, L.; Zhang, C.; Cai, Y. Varieties of agri-environmental schemes in China: A quantitative assessment. *Land Use Policy* **2018**, *71*, 505–517. [CrossRef]
103. Liu, M.; Chen, C.; Yang, L.; Min, Q.; Xiong, Y. Agricultural eco-compensation may not necessarily reduce chemical inputs. *Sci. Total Environ.* **2020**, *741*, 139847. [CrossRef]
104. Narloch, U.; Drucker, A.G.; Pascual, U. Payments for agrobiodiversity conservation services for sustained on-farm utilization of plant and animal genetic resources. *Ecol. Econ.* **2011**, *70*, 1837–1845. [CrossRef]

105. Wätzold, F.; Drechsler, M.; Johst, K.; Mewes, M.; Sturm, A. A Novel, Spatiotemporally Explicit Ecological-economic Modeling Procedure for the Design of Cost-effective Agri-environment Schemes to Conserve Biodiversity. *Am. J. Agric. Econ.* **2015**, *98*, 489–512. [CrossRef]
106. Prokopy, L.S.; Floress, K.; Klotthor-Weinkauf, D.; Baumgart-Getz, A. Determinants of agricultural best management practice adoption: Evidence from the literature. *J. Soil Water Conserv.* **2008**, *63*, 300–311. [CrossRef]
107. Corbera, E.; Kosoy, N.; Martínez Tuna, M. Equity implications of marketing ecosystem services in protected areas and rural communities: Case studies from Meso-America. *Glob. Environ. Chang.* **2007**, *17*, 365–380. [CrossRef]

www.ingramcontent.com/pod-product-compliance
Lightning Source LLC
Chambersburg PA
CBHW061813210926
51850CB00006B/689

MDPI
St. Alban-Anlage 66
4052 Basel
Switzerland
www.mdpi.com

Agriculture Editorial Office
E-mail: agriculture@mdpi.com
www.mdpi.com/journal/agriculture